Group Theory and General Relativity

Group Theory and General Relativity

Representations of the Lorentz Group and Their Applications to the Gravitational Field

Moshe Carmeli

Professor of Physics
Ben Gurion University
Beer Sheva, Israel

World Scientific
Singapore • New Jersey • London • Hong Kong

Published by

Imperial College Press
57 Shelton Street
Covent Garden
London WC2H 9HE

Distributed by

World Scientific Publishing Co. Pte. Ltd.
P O Box 128, Farrer Road, Singapore 912805
USA office: Suite 1B, 1060 Main Street, River Edge, NJ 07661
UK office: 57 Shelton Street, Covent Garden, London WC2H 9HE

British Library Cataloguing-in-Publication Data
A catalogue record for this book is available from the British Library.

First published in 1977 by McGraw-Hill Inc.

GROUP THEORY AND GENERAL RELATIVITY

ISBN 1-86094-234-2

Printed in Singapore by Uto-Print

To

NATHAN ROSEN

Teacher and Colleague

CONTENTS

INTRODUCTION

The use of group theory in physics became very widespread in the last two decades. The reasons are not hard to find. Firstly, the theory happens to be a natural mathematical language suitable for physical concepts to be expressed in. Secondly, the emergence of very complicated mathematical equations in physics that needed simplification in any possible way, and symmetry in physics is just one such aspect that can be used in simplification.

The first factor can easily be seen in quantum field theory. The description of state functions is made by means of a vector in a Hilbert space. On the other hand, the theory requires Poincaré symmetry, or invariance. But representation theory of the Poincaré group provides us with exactly this kind of mathematical tool that combines special relativity and quantum mechanics. A more sophisticated use of representations of the Poincaré group is subsequently achieved by associating a particle with each representation. This then leads to Wigner's famous classification of particles according to their spins and masses [E. P. Wigner, On Unitary Representations of the Inhomogeneous Lorentz Group, *Ann. Math.* **40,** 149 (1939)]. A further step leads to the classification of all invariant equations according to the representation. This leads to the very important result of finding the equation of motion associated with each representation. In this way one obtains the Dirac equation, the Proca equation, etc. [E. P. Wigner, Invariant Quantum Mechanical Equations of Motion, in *Theoretical Physics*, International Atomic Agency, Vienna, 1963, pp. 59–82].

Wigner's ideas changed our conceptional approach to physics. An example is provided by the use of compact groups, such as SU_2, SU_3, etc., in strong and weak interactions. Once again the theory of groups is mystically fit to describe mathematically the particle's quantum numbers, such as isopin, strangeness, etc.

Although many excellent books on general relativity have been written, the present book is the first book on the subject of group theory and general relativity and deals with the applications of group theory to general relativity. As is well known, the theory of general relativity was founded at a time when group theory was very little used in physics. During the last two decades it has become obvious that general relativity is one of many theories known as gauge theories, and at the same time a lot of work has been done in general relativity using group theory. To be sure we do not have any other gauge theory (except for the obvious case of the electromagnetic theory) that is compared, in its perfection and fitness to experimental results, to general relativity. However, the idea of gauge theories in particle physics is very widespread. An example of this is Weinberg's gauge theory that combines the electromagnetic and weak interactions [S. Weinberg, *Phys. Rev. Letters* **19,** 1264 (1967)]. In this book an extensive discussion on the theory of general relativity from the point of view of gauge fields is given, and an attempt is made to put together in one volume many scattered original works on the use of group theory in general relativity theory. The point of view of looking at gravitation theory as a gauge theory was extensively discussed by the author in the NATO Advanced Study Institute on Mathematical Physics [M. Carmeli, *SL*(2, *C*) Symmetry of the Gravitational Field, in *Group Theory in Non-Linear Problems*, A. O. Barut, Ed., D. Reidel, Dordrecht, Holland; Boston, U.S.A., 1974, pp. 59–110]. However, there is no other volume that encompasses the original articles, now scattered in the professional literature, which fits into the subject category of group theory and general relativity.

This book is based on lectures given by the author in the last four years to advanced undergraduate and graduate students of mathematics and physics at the Ben Gurion University. There are twelve chapters, divided into forty-six sections, five appendices, and an extensive bibliography. Each chapter concludes with a set of problems. The first six chapters are devoted to the theory of representations of the rotation and Lorentz groups. The other six chapters deal with the application of groups, mostly the Lorentz group, to the theory of general relativity. They cover topics that start from the fundamentals of general relativity and end with exact solutions of the gravitational field equations and representations of the Bondi–Metzner–Sachs group. No discussion on cosmology is included. Also, the chapter on the representations of the Bondi–Metzner–Sachs group is just a brief introduction to the subject. A more detailed account of this important group would need the use of the theory of representations of the Poincaré group, in particular Wigner's little group method; that was not the purpose of the present book. [The reader who is interested in more detail about the Bondi–Metzner–Sachs group is referred to R. Penrose' lucid review: Relativistic Symmetry Groups, in *Group Theory in Non-Linear Problems*, A. O. Barut, Ed., D. Reidel, Dordrecht, Holland; Boston, U.S.A., 1974, pp. 1–58, although no discussion is given on the representations themselves.] The whole book is written in a self-contained way in both topics of group theory and general relativity theory. No prior knowledge of either subject by the reader is assumed. The book could be used as a textbook for a two-semester course for students of mathematics and physics at the graduate level, or for research purposes. Parts of the book could also be used as a basis for a one-semester course; for example the first six chapters can be used as a text for a one-semester course on the theory of representations of the rotation and Lorentz groups for advanced undergraduate and graduate students of mathematics and physics. As is well known, the theory of representations of the Lorentz group has traditionally been used as an introduction to the general theory of representations of groups. Another example is that the last six chapters of the book can be used as a text for graduate students of physics and mathematics on the theory of general relativity. The detail of the chapters is as follows:

Chapter 1 is devoted to the theory of representations of the rotation group. It includes such elementary concepts as the pure rotation group, the group SU_2, the very important concept of invariant integral over a group and, of course, the Wigner matrices of irreducible representations of the rotation group. While the rotation group has been widely covered in other texts, the parametrization of these representations here is not done through the traditional Euler angles but by other angles that describe rotations. In chapter 2 the discussion of the Lorentz group begins. This chapter includes an elementary discussion of the problem in general. Chapter 3 includes the important case of the finite-dimensional spinor representations of the Lorentz group. Here the group $SL(2, C)$ is introduced, and its relation to the Lorentz group is outlined. Chapters 4, 5, and 6 are devoted to the infinite-dimensional representations of the group $SL(2, C)$. These representations are the principal series, the complementary series, and the complete series.

The discussion of the infinite-dimensional representations starts in Chap. 4 by outlining the spaces of representations. These include several Hilbert spaces. Here also the theory of Fourier transform on the group SU_2 is introduced. The group operators are subsequently introduced, and the representation of the principal series is realized in these spaces. The complementary series is subsequently introduced in Chap. 5, where an operator formulation is also given. Chapter 6 then concludes the discussion of the infinite-dimensional representations. In this chapter some harmonic analysis of the group $SL(2, C)$ is also given.

The theory of general relativity first appears in Chap. 7 where the standard elements of the theory are given. Applications of the spinor representations to general relativity theory are given in Chap. 8. The Maxwell and Weyl spinors are introduced and classified accordingly. In Chap. 9 the general aspects of the theory of gauge fields are described. This includes the concept of isotopic spin and isotopic gauge transformations. Generalizations are then made to the Lorentz and Poincaré groups, and finally to the group $SL(2, C)$. This leads to our obtaining the gravitational field equations in the familiar form of Newman and Penrose. Thus our approach here is obtaining the Newman–Penrose equations for general relativity from gauge-theory principles. Later chapters of the book are devoted to solving the field equations of general relativity.

In Chap. 10 we analyse the gravitational field variables, proving the Goldberg–Sachs theorem, and dealing with choosing coordinate systems and asymptotic behaviour. In Chap. 11 we give exact solutions to the Newman–Penrose equations of general relativity. These include the Robinson–Trautman solution, the Newman–Tamburino solutions, the NUT–Taub solution, and all type D vacuum solutions, including the familiar Kerr solution. Finally, Chap. 12 concludes the text with the representations of the Bondi–Metzner–Sachs group. The five appendices give reviews of the theory of groups, reviews of the theories of finite and infinite-dimensional representations, whereas the last two appendices summarize the Newman–Penrose equations for the gravitational field of general relativity. The book is then concluded with the bibliography.

Although both the theory of representations of the Lorentz group and the theory of general relativity are presented here, and although the material includes reviews of some of the most recent developments in both topics, the present book does not cover all possible subjects on both topics. Among the following list of excellent books and monographs the remedies for some of these deficiencies can be found:

(1) B. L. van der Waerden, *Modern Algebra*, Fredric Ungar, New York, 1953.
(2) L. Pontrjagin, *Topological Groups*, Princeton University Press, New Jersey, U.S.A., 1946.
(3) E. P. Wigner, *Group Theory and its Application to the Quantum Mechanics of Atomic Spectra*, Academic Press, New York, 1959.
(4) M. A. Naimark, *Linear Representations of the Lorentz Group*, Pergamon Press, New York, 1964.
(5) I. M. Gelfand, M. I. Graev, and N. Ya. Vilenkin, *Generalized Functions*, Vol. 5: *Integral Geometry and Representation Theory*, Academic Press, New York, 1966.
(6) I. M. Gelfand, R. A. Minlos, and Z. Ya. Shapiro, *Representations of the Rotation and Lorentz Groups and their Applications*, Pergamon Press, New York, 1963.
(7) W. Rühl, *The Lorentz Group and Harmonic Analysis*, W. A. Benjamin, New York, 1970.
(8) A. Trautman, F. A. E. Pirani, and H. Bondi, *Lectures on General Relativity* (Brandeis 1964 Summer Institute on Theoretical Physics, Vol. 1), Prentice-Hall, Englewood Cliffs, N.J., U.S.A., 1965.
(9) J. L. Anderson, *Principles of Relativity Physics*, Academic Press, New York, 1967.
(10) W. R. Davis, *Classical Theory of Particles and Fields and the Theory of Relativity*, Gordon and Breach, New York, 1970.
(11) S. Weinberg, *Gravitation and Cosmology: Principles and Applications of the General Theory of Relativity*, John Wiley, New York, 1972.
(12) S. W. Hawking and G. F. R. Ellis, *The Large Scale Structure of Space–Time*, Cambridge University Press, Cambridge, England, 1973.
(13) C. W. Misner, K. S. Thorne, and J. A. Wheeler, *Gravitation*, W. H. Freeman, San Francisco, 1973.
(14) A. Papapetrou, *Lectures on General Relativity*, D. Reidel, Dordrecht, Holland; Boston, U.S.A., 1974.

I would like to conclude this Introduction by citing a very relevant statement on the important role of group theory in physics, even though most researchers now realize and understand this role. However, it is a citation that has its own historical significance and is here very relevant in relation to gravitational theory, just as it was originally. It is a quotation from an introduction of A. Salam in a moment of deep insight, commenting on G. Racah's lectures on Lie groups given at the Institute for Advanced Studies at Princeton [A. Salam, The Formalism of Lie Groups, in *Theoretical Physics*, International Atomic Agency, Vienna, 1963, pp. 173–196]:

> Throughout the history of quantum theory, a battle has raged between the amateurs and professional group theorists. The amateurs have maintained that everything one needs in the theory of groups can be discovered by the light of nature provided one knows how to multiply two matrices. In support of this claim, they of course, justifiably, point to the successes of that prince of amateurs in this field, Dirac, particularly with the spinor representations of the Lorentz group. As an amateur myself, I strongly believe in the truth of the non-professionalist creed. I think perhaps there is not much one has to learn in the way of methodology from the group theorists except caution. But this does not mean one should not be aware of the riches which have been amassed over the course of years particularly in the most highly developed of all mathematical disciplines—the theory of Lie Groups. My lectures then are an amateur's attempt to gather some of the fascinating results for compact simple Lie groups which are likely to be of physical interest. I shall state theorems; and with a physicist's typical unconcern rarely, if ever, shall I prove them. Throughout, the emphasis will be to show the close similarity of these general groups with the most familiar of all groups, the group of rotations in three dimensions. In 1951 I had the good fortune to listen to Prof. Racah lecture on Lie groups at Princeton. After attending these lectures I thought this is really too hard; I cannot learn this; one is hardly ever likely to need all this complicated matter. I was completely wrong. Eleven years later the wheel has gone full cycle and it is my turn to lecture on this subject. I am sure many of you will feel after these lectures that all this is too damned hard and unphysical. The only thing I can say is: I do very much hope and wish you do not have to learn this beautiful theory eleven years too late.

* * *

Many people have helped me to prepare this book, from the first stages of writing to the final stage of reading it, partially or completely. I am in particular indebted to Professor S. Malin, without whose help, continuous encouragement, and reading of the manuscript, the book would probably have never been finished. I am indebted to Professor A. O. Barut whose kind invitation to the NATO Advanced Summer Institute in Istanbul gave me the opportunity to present the content of Chaps. 8 and 9 of the book in my two-week series of lectures there, and for his kind hospitality there. I am indebted to my teacher and colleague Professor Nathan Rosen, both for reading the manuscript and for his comments and encouraging remarks on it which I am sure has led to a better presentation of the material. Many thanks are also due to Professor L. Witten for both encouragement on the idea of the book and for critically reading and commenting on it, and to Professor G. Tauber and Dr. J. Bekenstein for comments and remarks. Last, but not least, I am indebted to my students M. Kaye and C. Charach for systematically reading the manuscript and for their comments and suggestions. Finally, I am grateful to Mrs. Y. Ahuvia for the excellent job she has made of typing the manuscript.

MOSHE CARMELI
Beer Sheva, Israel

CHAPTER

ONE

THE ROTATION GROUP

In the following we find the irreducible representations of the three-dimensional pure rotation group, O_3. This is done by Weyl's method, which makes use of the homomorphism of the special unitary group of order two, SU_2, onto the rotation group. The representations are discussed in terms of two different parameterizations: (1) the angle of rotation in a specified direction and the spherical angles of the direction of rotation; and (2) the traditional Euler angles.

1-1 THE THREE-DIMENSIONAL PURE ROTATION GROUP

A linear transformation g of the variables x_1, x_2, and x_3, which leaves the form $x_1^2 + x_2^2 + x_3^2$ invariant, is called a *three-dimensional rotation.* The aggregate of all such linear transformations g provides a continuous group, which is formed from the set of all *real orthogonal* 3-dimensional matrices[1] and is called the *three-dimensional rotation group.* The determinant of every orthogonal matrix is equal to either $+1$, in which case the transformation describes *pure rotation,* or to -1, in which case it describes a *rotation-reflection.* The aggregate of all pure rotations forms a group, which is a subgroup of the 3-dimensional group, and is called the *pure rotation group. This chapter is concerned with the 3-dimensional pure rotation group.* This group is denoted by O_3.

The Euler Angles

Let g be an element of the group O_3, i.e., a 3-dimensional orthogonal matrix with determinant unity. One then can express each such element in terms of a set of three parameters. An example of such parameters is that of Euler angles, which are

[1] A matrix g is called orthogonal if $g^t g = 1$, where g^t is the transposed of g.

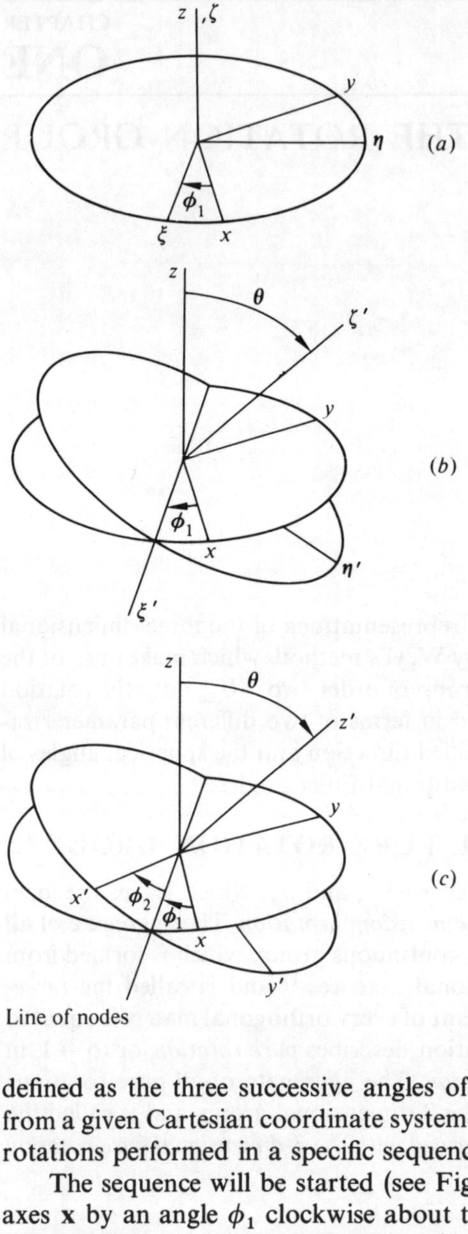

Figure 1.1 The three rotations defining the Euler angles.

defined as the three successive angles of rotation describing the transformation from a given Cartesian coordinate system to another by means of three successive rotations performed in a specific sequence.

The sequence will be started (see Fig. 1.1) by rotating the original system of axes $\mathbf{x}$ by an angle ϕ_1 clockwise about the z axis.[2] The new coordinate system will be labelled $\boldsymbol{\xi}$. One thus has $\boldsymbol{\xi} = g(\phi_1)\mathbf{x}$, where

$$g(\phi_1) = \begin{pmatrix} \cos\phi_1 & -\sin\phi_1 & 0 \\ \sin\phi_1 & \cos\phi_1 & 0 \\ 0 & 0 & 1 \end{pmatrix} \tag{1-1}$$

[2] We use the notation $\mathbf{x} = (x, y, z) = (x_1, x_2, x_3)$, $\boldsymbol{\xi} = (\xi, \eta, \zeta)$, $\boldsymbol{\xi}' = (\xi', \eta', \zeta')$, and $\mathbf{x}' = (x', y', z') = (x'_1, x'_2, x'_3)$.

In the second stage the intermediate axes $\boldsymbol{\xi}$ are rotated about its ξ axis clockwise by an angle θ to another intermediate set which is denoted $\boldsymbol{\xi}'$, and one has $\boldsymbol{\xi}' = g(\theta)\boldsymbol{\xi}$, where

$$g(\theta) = \begin{pmatrix} 1 & 0 & 0 \\ 0 & \cos\theta & -\sin\theta \\ 0 & \sin\theta & \cos\theta \end{pmatrix} \tag{1-2}$$

The ξ' axis is called the *line of nodes*. Finally the $\boldsymbol{\xi}'$ axes are rotated clockwise by an angle ϕ_2 about the ζ' axis to produce the desired $\mathbf{x}'$ system of axes, $\mathbf{x}' = g(\phi_2)\boldsymbol{\xi}'$, where

$$g(\phi_2) = \begin{pmatrix} \cos\phi_2 & -\sin\phi_2 & 0 \\ \sin\phi_2 & \cos\phi_2 & 0 \\ 0 & 0 & 1 \end{pmatrix} \tag{1-3}$$

The matrix of the complete transformation $\mathbf{x}' = g\mathbf{x}$ is given, therefore, by the product of the successive matrices $g = g(\phi_2)g(\theta)g(\phi_1)$ and is given by

$$g = \begin{pmatrix} \cos\phi_2\cos\phi_1 & -\cos\phi_2\sin\phi_1 & \sin\phi_2\sin\theta \\ \quad -\cos\theta\sin\phi_1\sin\phi_2 & \quad -\cos\theta\cos\phi_1\sin\phi_2 & \\ \sin\phi_2\cos\phi_1 & -\sin\phi_2\sin\phi_1 & -\cos\phi_2\sin\theta \\ \quad +\cos\theta\sin\phi_1\cos\phi_2 & \quad +\cos\theta\cos\phi_1\cos\phi_2 & \\ \sin\theta\sin\phi_1 & \sin\theta\cos\phi_1 & \cos\theta \end{pmatrix} \tag{1.4}$$

The angles ϕ_1, θ, ϕ_2 are independent parameters, fully determining the rotation g. They are called the *Euler Angles*. By their definition, one has $0 \leqslant \phi_1 \leqslant 2\pi$, $0 \leqslant \theta \leqslant \pi$, and $0 \leqslant \phi_2 \leqslant 2\pi$.

1-2 THE GROUP SU_2

Rotations can also be specified by *unitary* matrices of order two and determinant unity. The aggregate of all such matrices provides a group which is denoted by SU_2. The relation between the groups O_3 and SU_2 can be established as follows.

The Groups O_3 and SU_2

Let x_l and x'_k, with $k, l = 1, 2, 3$, denote the coordinates of two Cartesian frames related by $x'_k = g_{kl}x_l$, where g_{kl} are elements of the matrix $g \in O_3$, and the summation convention is being employed here and in the sequel. With each coordinate system x_k one associates a 2×2 Hermitian matrix P defined by $P = x_k\sigma^k$, where σ^k are the Pauli spin matrices,

$$\sigma^1 = \begin{pmatrix} 0 & 1 \\ 1 & 0 \end{pmatrix}, \qquad \sigma^2 = \begin{pmatrix} 0 & i \\ -i & 0 \end{pmatrix}, \qquad \sigma^3 = \begin{pmatrix} 1 & 0 \\ 0 & -1 \end{pmatrix} \tag{1-5}$$

In terms of P one requires that the coordinates transform according to $P' = uPu^\dagger$, where u is an element of the group SU_2, $P' = x'_l\sigma^l$, and $u^\dagger$ is the Hermitian conjugate of u. The relation between u of SU_2 and g of O_3 are given by

$$g_{rs} = \tfrac{1}{2}\mathrm{Tr}(\sigma^r u \sigma^s u^\dagger) \tag{1-6}$$

$$u = \mp(1 + \sigma^r\sigma^s g_{rs})/2(1 + \mathrm{Tr}g)^{1/2} \tag{1-7}$$

where Tr stands for trace.

Homomorphism of the Group SU_2 onto the Group O_3

Accordingly, to each rotation g of the group O_3 there corresponds, by Eq. (1-7), two matrices $\mp u$ of the group SU_2 and, conversely, to each unitary matrix u of SU_2 there corresponds, by Eq. (1-6), some rotation g of O_3. It thus follows that the group O_3 is homomorphic (see Appendix A) to the group SU_2. For example, the unitary matrices corresponding to the rotations $g(\phi_1)$, $g(\theta)$, and $g(\phi_2)$ given by Eqs. (1-1)–(1-3) are easily found, using Eq. (1-7). They are

$$u(\phi_1) = \mp\begin{pmatrix} e^{i\phi_1/2} & 0 \\ 0 & e^{-i\phi_1/2}\end{pmatrix} \tag{1-8}$$

$$u(\theta) = \mp\begin{pmatrix} \cos\dfrac{\theta}{2} & i\sin\dfrac{\theta}{2} \\ i\sin\dfrac{\theta}{2} & \cos\dfrac{\theta}{2}\end{pmatrix} \tag{1-9}$$

$$u(\phi_2) = \mp\begin{pmatrix} e^{i\phi_2/2} & 0 \\ 0 & e^{-i\phi_2/2}\end{pmatrix} \tag{1-10}$$

A general rotation g, described by the matrix (1.4), will then correspond to the unitary matrix $u = u(\phi_2)\,u(\theta)\,u(\phi_1)$ and is given by

$$u = \mp\begin{pmatrix} \cos\dfrac{\theta}{2}e^{i(\phi_2+\phi_1)/2} & i\sin\dfrac{\theta}{2}e^{i(\phi_2-\phi_1)/2} \\ i\sin\dfrac{\theta}{2}e^{-i(\phi_2-\phi_1)/2} & \cos\dfrac{\theta}{2}e^{-i(\phi_2+\phi_1)/2}\end{pmatrix} \tag{1-11}$$

1-3 INVARIANT INTEGRALS OVER THE GROUPS O_3 AND SU_2[3]

A function $y = f(g)$ is said to be defined over the group G, if to each element g of G there corresponds a number y. If the group is taken to be the rotation group

[3] The subject of invariant integrals is treated here very briefly. For more details the reader is referred to E. P. Wigner, *Group Theory and its Application to the Quantum Mechanics of Atomic Spectra*, Academic Press, New York 1959; A. Weil, *Actualites Sci. Ind.*, no. 869 (1938); M. A. Naimark, *Linear Representations of the Lorentz Group*, Pergamon Press, New York, 1964; L. S. Pontrjagin, *Topological Groups*, Princeton University Press, New Jersey, U.S.A., 1946.

O_3 and one uses the Euler angles as parameters, then $f(g)$, where $g \in O_3$, becomes simply a function of the angles ϕ_1, θ, ϕ_2, i.e., $f(g) = f(\phi_1, \theta, \phi_2)$. The function f then satisfies

$$\begin{aligned} f(\phi_1 + 2\pi, \theta, \phi_2) &= f(\phi_1, \theta, \phi_2) \\ f(\phi_1, \theta, \phi_2 + 2\pi) &= f(\phi_1, \theta, \phi_2) \end{aligned} \tag{1-12}$$

Invariant Integral over the Group O_3

The integral $\int f(g)\, dg$ is then called the *invariant integral* of the function $f(g)$ over the group O_3, if it satisfies

$$\begin{aligned} \int f(g\, g_0)\, dg &= \int f(g_0 . g)\, dg \\ &= \int f(g)\, dg \end{aligned} \tag{1-13}$$

for any $g_0 \in O_3$, and

$$\int f(g^{-1})\, dg = \int f(g)\, dg \tag{1-14}$$

The expression dg is called a *measure*. When the Euler angles are used to parameterize the elements g of O_3, one can write dg in terms of the angles ϕ_1, θ, ϕ_2 as $dg = (1/8\pi^2) \sin\theta \, d\phi_1 \, d\theta \, d\phi_2$. It satisfies

$$\int dg = 1 \tag{1-15}$$

The integration limits extend over the whole domain of definitions of the variables, i.e., $0 \leqslant \phi_1 \leqslant 2\pi$, $0 \leqslant \theta \leqslant \pi$, and $0 \leqslant \phi_2 \leqslant 2\pi$.

Invariant integral over the Group SU_2

The concepts of functions defined over the group O_3 and invariant integrals defined over the rotation group O_3 can easily be extended to the unitary group SU_2. Again, a function $f(u)$ defined over the group SU_2 can be considered as a function of the angles ϕ_1, θ, ϕ_2, i.e., $f(u) = f(\phi_1, \theta, \phi_2)$. The analogous periodicity conditions to those of Eq. (1-12) for functions defined over O_3 will now be

$$\begin{aligned} f(\phi_1 + 4\pi, \theta, \phi_2) &= f(\phi_1, \theta, \phi_2) \\ f(\phi_1, \theta, \phi_2 + 4\pi) &= f(\phi_1, \theta, \phi_2) \\ f(\phi_1 + 2\pi, \theta, \phi_2 + 2\pi) &= f(\phi_1, \theta, \phi_2) \end{aligned} \tag{1-16}$$

The invariant integral over the group SU_2 then satisfies

$$\int f(uu_0)\,\mathrm{d}u = \int f(u_0 u)\,\mathrm{d}u = \int f(u)\,\mathrm{d}u \tag{1-17}$$

for any $u \in SU_2$, and

$$\int f(u^{-1})\,\mathrm{d}u = \int f(u)\,\mathrm{d}u \tag{1-18}$$

The measure $\mathrm{d}u$ can then be expressed in terms of the Euler angles as $\mathrm{d}u = (1/16\pi^2)\sin\theta\,\mathrm{d}\phi_1\,\mathrm{d}\theta\,\mathrm{d}\phi_2$ and satisfies

$$\int \mathrm{d}u = 1 \tag{1-19}$$

The integration limits here will be: $0 \leqslant \phi_1 \leqslant 4\pi$, $0 \leqslant \theta \leqslant \pi$, and $0 \leqslant \phi_2 \leqslant 2\pi$.

1-4 REPRESENTATIONS OF THE GROUPS O_3 AND SU_2

We have seen that the pure rotation group O_3 is homomorphic to the unimodular unitary group of order two, SU_2, such that to every rotation g of O_3 there correspond two matrices $+u$ and $-u$ of SU_2 and, conversely, to every element u of SU_2 there corresponds some rotation g of O_3.

Single- and Double-Valued Representations

It thus follows that the description of the representations (see Appendix B) of the group O_3 is equivalent to that of the group SU_2; a representation $g \to D(g)$ of the group O_3 is single- or double-valued according to whether or not $D(u)$ is equal to $D(-u)$. The use of the group SU_2 for finding the representations of the group O_3 was originally suggested by H. Weyl and has been widely adopted when the Euler angles are used to parameterize the groups. The advantage of Weyl's method is in giving the *double valued* representations along with the proper representations. The double valued representations are important in physical problems dealing with spin-like properties. In the following we will use Weyl's method as adopted by Carmeli.[4]

We point out that, by using Weyl's method, one can obtain a general invariant result that is a function of $u \in SU_2$, valid for any parameterization one uses to describe the rotation. To find the representations of the group O_3 in terms of a set of parameters, one has merely to express u in terms of these parameters, as in the

[4] M. Carmeli, *J. Math. Phys.* **9**, 1987 (1968). See also H. E. Moses, *Ann. Phys.* (*N.Y.*) **37**, 224, (1966); **42**, 343, (1967); *Nuovo Cimento* **40A**, 1120 (1965).

case when the Euler angles are employed. In addition, by having the results as functions over the group SU_2, certain relations will be obtained which are invariant under change of the parameters. As an example, the orthogonality relations between the matrix elements of the irreducible representation can be written as an invariant integral over SU_2. Hence the relations are valid for any parameterization.

Infinitesimal Generators

An orthogonal matrix describing a rotation with angle ψ about some direction $\mathbf{n} = (\sin\theta\cos\phi, \sin\theta\sin\phi, \cos\theta)$ is given by

$$g_{rs} = \delta_{rs}\cos\psi + n_r n_s(1-\cos\psi) - \varepsilon_{rst} n^t \sin\psi \tag{1-20}$$

where r, s, and t run from 1 to 3. Rotations $g_1(\psi)$, $g_2(\psi)$, and $g_3(\psi)$ around Ox_1, Ox_2, and Ox_3 axes are obtained from (1-20) by putting the proper values for the polar angles θ and ϕ.[5] The *infinitesimal matrices*, g_r, corresponding to rotations about the axis Ox_r are defined by[6]

$$g_r = \left[\frac{dg_r(\psi)}{d\psi}\right]_{\psi=0} \tag{1-21}$$

and satisfy the *commutation relations*

$$[g_r, g_s] = \varepsilon_{rst} g_t \tag{1-22}$$

where $[a, b] = ab - ba$, and ε_{rst} are the structure constants of the group.

Let us denote a representation of the group O_3 in an n-dimensional Euclidean space R by $g \to D(g)$ and for convenience we put[7]

$$A_r(\psi) = D(g_r(\psi)) \tag{1-23}$$

[5] These matrices are given by

$$g_1(\psi) = \begin{pmatrix} 1 & 0 & 0 \\ 0 & \cos\psi & -\sin\psi \\ 0 & \sin\psi & \cos\psi \end{pmatrix}, \quad g_2(\psi) = \begin{pmatrix} \cos\psi & 0 & \sin\psi \\ 0 & 1 & 0 \\ -\sin\psi & 0 & \cos\psi \end{pmatrix}$$

$$g_3(\psi) = \begin{pmatrix} \cos\psi & -\sin\psi & 0 \\ \sin\psi & \cos\psi & 0 \\ 0 & 0 & 1 \end{pmatrix}$$

[6] The g_r are related to $g_r(\psi)$ by $g_r(\psi) = \exp(\psi g_r)$, and are given by

$$g_1 = \begin{pmatrix} 0 & 0 & 0 \\ 0 & 0 & -1 \\ 0 & 1 & 0 \end{pmatrix}, \quad g_2 = \begin{pmatrix} 0 & 0 & 1 \\ 0 & 0 & 0 \\ -1 & 0 & 0 \end{pmatrix}, \quad g_3 = \begin{pmatrix} 0 & -1 & 0 \\ 1 & 0 & 0 \\ 0 & 0 & 0 \end{pmatrix}$$

[7] $A_r(\psi)$ are called the *basic one-parameter groups* of the given representation and define one-parameter groups of operators that satisfy $A_r(\psi_1)A_r(\psi_2) = A_r(\psi_1 + \psi_2)$; they are differentiable functions of ψ and may be expanded as $A_r(\psi) = \exp(\psi A_r)$, where A_r is defined by Eq. (1-24).

The *basic infinitesimal operators* of the representations are then obtained by

$$A_r = \left[\frac{\mathrm{d}A_r(\psi)}{\mathrm{d}\psi}\right]_{\psi=0} \tag{1-24}$$

A representation of the group O_3 is *uniquely determined* by its basic infinitesimal operators A_r. The determination of all the finite-dimensional representations of the group O_3 is based on the fact that the operators A_r satisfy the *same* commutation relations that exists among the infinitesimal matrices g_r:

$$[A_r, A_s] = \varepsilon_{rst} A_t \tag{1-25}$$

The A_r are skew-Hermitian operators,[8] $A_r^\dagger = -A_r$, since, without loss of generality, every finite-dimensional representation of O_3 can be considered to be unitary (see Problems 1.1 and 1.2).

Canonical Basis

Defining the new operators

$$L_\mp = iA_1 \pm A_2, \qquad L_3 = iA_3 \tag{1-26}$$

one finds that

$$\begin{gathered} [L_\mp, L_3] = \pm L_\mp, \qquad [L_+, L_-] = 2L_3 \\ L_+^\dagger = L_-, \; L_3^\dagger = L_3 \end{gathered} \tag{1-27}$$

The problem then reduces to the determination of the operators $L_\mp$, L_3 satisfying conditions (1-27). This is answered by the following: Every finite-dimensional representation of the group O_3 is uniquely determined by a non-negative integer or half-integer j, the *weight* of the representation. The space of the representation corresponding to such a number j has the dimension $2j + 1$; the operators $L_\mp$, L_3 of the representation are given relative to its *canonical basis* $f_{-j}, f_{-j+1}, \ldots, f_j$ by

$$\begin{aligned} L_\pm f_m &= [(j \mp m)(j \pm m + 1)]^{1/2} f_{m\pm 1} \\ L_3 f_m &= m f_m \end{aligned} \tag{1-28}$$

where $m = -j, -j+1, \ldots, j$.[9]

[8] An operator B in a finite-dimensional Euclidean space R is called *adjoint* to the operator A in the same space if $(Ax, y) = (x, By)$ for all x, y of R. The adjoint of an operator A is usually denoted by $A^\dagger$. It can be shown that for any linear operator A there exists one and only one adjoint operator $A^\dagger$, and that the adjoint operator to $A^\dagger$ is A. An operator A is called Hermitian if $A^\dagger = A$. An operator A is called *unitary* if and only if $A^\dagger A = 1$.

[9] It also follows that for each j there corresponds an irreducible representation of O_3. If the operators $L_\pm$ and L_3 of a representation of O_3 in a $(2j+1)$-dimensional space are given relative to some basis $f_{-j}, f_{-j+1}, \ldots, f_j$, then by Eqs. (1-28) that representation is irreducible.

We now find the unitary matrix u corresponding to the rotation g, Eq. (1-20). They are related by Eqs. (1-6) and (1-7). A direct calculation gives

$$u = \mp\begin{pmatrix} \cos\dfrac{\psi}{2} + i\sin\dfrac{\psi}{2}\cos\theta & i\sin\dfrac{\psi}{2}\sin\theta\, e^{i\phi} \\ i\sin\dfrac{\psi}{2}\sin\theta\, e^{-i\phi} & \cos\dfrac{\psi}{2} - i\sin\dfrac{\psi}{2}\cos\theta \end{pmatrix} \tag{1-29}$$

This is the unitary matrix $u \in SU_2$ corresponding to a rotation with angle ψ around the direction **n** specified by θ and ϕ. The corresponding matrix when the Euler angles are employed was given in Eq. (1-11). It will be noted that $u(-\psi, \theta, \phi) = u^{-1}(\psi, \theta, \phi)$.

The unitary matrices $u_1(\psi)$, $u_2(\psi)$, and $u_3(\psi)$ corresponding to the rotations $g_1(\psi)$, $g_2(\psi)$, and $g_3(\psi)$ around the axes of coordinates Ox_1, Ox_2, and Ox_3 (see footnote 5, p. 7), are obtained from (1-29) by putting the appropriate values for θ and ϕ. They are:[10]

$$u_1(\psi) = \mp\begin{pmatrix} \cos\dfrac{\psi}{2} & i\sin\dfrac{\psi}{2} \\ i\sin\dfrac{\psi}{2} & \cos\dfrac{\psi}{2} \end{pmatrix} \tag{1-30a}$$

$$u_2(\psi) = \mp\begin{pmatrix} \cos\dfrac{\psi}{2} & -\sin\dfrac{\psi}{2} \\ \sin\dfrac{\psi}{2} & \cos\dfrac{\psi}{2} \end{pmatrix} \tag{1-30b}$$

$$u_3(\psi) = \mp\begin{pmatrix} e^{i\psi/2} & 0 \\ 0 & e^{-i\psi/2} \end{pmatrix} \tag{1-30c}$$

Using these matrices, the operators $A_r(\psi)$ of the group SU_2 will be determined in the next section.

1-5 MATRIX ELEMENTS OF IRREDUCIBLE REPRESENTATIONS

A matrix u of SU_2 can be considered as that of a linear transformation of the space of all pairs of complex numbers (ξ^1, ξ^2):

$$\xi'^p = \sum_{q=1}^{2} u_{pq}\xi^q \qquad (p = 1,2) \tag{1-31}$$

[10] The infinitesimal matrices u_r corresponding to rotations around Ox_r,

$$u_r = [du_r(\psi)/d\psi]_{\psi=0}$$

are related to the Pauli matrices, Eq. (1-5), by $u_r = \mp(i/2)\sigma^r$.

A representation of the group SU_2 can be obtained if one considers several pairs $(\xi_1^1, \xi_1^2), \ldots, (\xi_k^1, \xi_k^2)$ and forms all products $\xi_1^{p_1} \cdots \xi_k^{p_k}$, letting $p_1, \ldots, p_k$ take the values 1, 2, independently. Under the transformation (1-31), this product transforms like

$$\xi_1'^{p_1} \cdots \xi_k'^{p_k} = \sum_{q_1, \ldots, q_k = 1}^{2} u_{p_1 q_1} \cdots u_{p_k q_k} \xi_1^{q_1} \cdots \xi_k^{q_k} \tag{1-32}$$

The product $\xi_1^{p_1} \cdots \xi_k^{p_k}$ may be considered as a vector in the linear space R_k of all 2^k complex numbers $\xi^{p_1 \cdots p_k}$. The linear transformation $D^{(k)}(u)$ of the space R_k is then given by

$$\xi'^{p_1 \cdots p_k} = \sum_{q_1, \ldots, q_k = 1}^{2} u_{p_1 q_1} \cdots u_{p_k q_k} \xi^{q_1 \cdots q_k} \tag{1-33}$$

Spinor Representation of the Group SU_2

The correspondence $u \to D^{(k)}(u)$ is a representation of the group SU_2, *not* irreducible in general, since the subspace S_k of R_k of all symmetrical vectors ξ is invariant with respect to all the operators $D^{(k)}(u)$. The correspondence $u \to D^{(k)}(u)$ is irreducible, however, in the space S_k. We denote this representation by Z_k. It is called the *spinor representation* of the group SU_2 of weight $k/2$.

An equivalent realization of the representation Z_k is obtained if one identifies the space S_k with the $(k+1)$-dimensional space of homogeneous polynomials $p(z_1, z_2)$ of degree k in the two complex variables z_1 and z_2 and sets up a one-to-one correspondence between ξ of S_k and $p(z_1, z_2)$ in the form

$$p(z_1, z_2) = \sum_{p_1, \ldots, p_k = 1}^{2} \xi^{p_1 \cdots p_k} z_{p_1} \cdots z_{p_k} \tag{1-34}$$

The operator $D^{(k)}(u)$ for this new realization of the space S_k is then given by

$$D^{(k)}(u)p(z_1, z_2) = p(z_1', z_2')$$

$$z_q' = \sum_{p=1}^{2} u_{pq} z_p \qquad (q = 1,2) \tag{1-35}$$

Introducing a new variable $z = z_1/z_2$, the polynomial $p(z_1, z_2)$ can then be written as $z_2^k p(z)$, where $p(z)$ is a polynomial in the variable z of degree not exceeding k. The operators $D^{(k)}(u)$ of the representation Z_k are, accordingly, given by

$$D^{(k)}(u)p(z) = (u_{12}z + u_{22})^k p\left(\frac{u_{11}z + u_{21}}{u_{12}z + u_{22}}\right) \tag{1-36}$$

This equation gives, in particular, the operators $A_r(\psi) = D(u_r(\psi))$ when the matrices $u_r(\psi)$, Eqs. (1-30), are used[11] (see Problem 1.3).

[11] For the determination of the operators $A(\psi)$, one needs $u_r(\psi)$ only for small values of ψ. The signs in Eqs. (1-30) are determined by the conditions $\lim u_r(\psi) = 1$ when $\psi \to 0$. Hence the + sign must be used.

Matrix Elements of the Operator $D(u)$

It follows that every irreducible finite-dimensional representation of the group SU_2 is uniquely determined by some non-negative integer or half-integer $j = k/2$, the *weight* of the representation.[12] The functions

$$f_m(z) = \frac{(-z)^{j-m}}{[(j-m)!(j+m)!]^{1/2}} \tag{1-37}$$

where $m = -j, -j+1, \ldots, j$, form a canonical basis (see Problem 1.4) for the representation Z_k in the space S_k. Using Eq. (1-36), one finds

$$D^{(k)}(u) f_n(z) = \sum_{m=-j}^{j} D^j_{mn}(u) f_m(z) \tag{1-38}$$

where $D^j_{mn}(u)$ are the matrix elements of the operator $D(u)$ of the irreducible representation of weight j relative to the canonical basis, which corresponds to an arbitrary rotation g. Its explicit expression is[13]

$$D^j_{mn}(u) = (-1)^{2j-m-n}\left[\frac{(j-m)!(j+m)!}{(j-n)!(j+n)!}\right]^{1/2}$$
$$\times \sum \binom{j-n}{a}\binom{j+n}{j-m-a} u_{11}^{a} u_{12}^{j-m-a} u_{21}^{j-n-a} u_{22}^{m+n+a} \tag{1-39}$$

where the summation runs from $a = \max(0, -m-n)$ to $\min(j-m, j-n)$, and

$$\binom{m}{n} = \frac{m!}{(m-n)!n!}$$

In Eq. (1-39) the indices m and n take the values $-j, -j+1, \ldots, j$ and $j = 0, \frac{1}{2}, 1, \frac{3}{2}, 2, \ldots$.

To find the matrix elements (1-39) in terms of the variables ψ, θ, and ϕ we simply substitute for u_{pq} their expressions as functions of these variables as given by Eq. (1-29).[14] One obtains for $D^j_{mn}(u) = D^j_{mn}(\psi, \theta, \phi)$:

$$D^j_{mn}(\psi, \theta, \phi) = (-1)^{2j-m-n}\left[\frac{(j-m)!(j+m)!}{(j-n)!(j+n)!}\right]^{1/2}\left(i \sin\frac{\psi}{2}\sin\theta e^{-i\phi}\right)^{m-n}$$
$$\times \left(\cos\frac{\psi}{2} - i\sin\frac{\psi}{2}\cos\theta\right)^{m+n} S(j, m, n; x) \tag{1-40}$$

[12] Conversely, for a non-negative integer or half-integer j, there exists an irreducible representation of the group SU_2 of weight j. A representation of weight j can be realized as the spinor representation Z_k, where $k = 2j$; and every finite-dimensional irreducible representation of the group SU_2 is equivalent to one of the representations Z_k.

[13] It will be noted that $D^j_{mn}(-u) = (-1)^{2j} D^j_{mn}(u)$. Thus the representation is single-valued for integer j and double-valued for half-integer j. In the sequel the matrix u of Eq. (1-29) will be taken with the + sign.

[14] See Problem 1.5 for the expression of D^j in terms of Euler's angles.

Here[15]

$$S(j, m, n; x) = 2^{m-j}(j-n)!(j+n)! \times \sum \frac{(x+1)^a (x-1)^{j-m-a}}{a!(j-n-a)!(j-m-a)!(a+m+n)!} \tag{1-41}$$

and $x = 1 - 2\sin^2(\psi/2)\sin^2\theta$.

Properties of the Matrices $D^j(u)$

The matrices $D^j(u)$ are unitary, and the correspondence $u \to D^j(u)$ is a representation of the group SU_2. Accordingly one has

$$D^j(u_1 u_2) = D^j(u_1) D^j(u_2)$$

or, in terms of matrix elements, one obtains

$$D^j_{mn}(u_1 u_2) = \sum_{q=-j}^{j} D^j_{mq}(u_1) D^j_{qn}(u_2)$$

Furthermore, one observes that

$$D^j(u^{-1}) = [D^j(u)]^{-1} = [D^j(u)]^\dagger$$

or, explicitly,

$$D^j_{mn}(u^{-1}) = \bar{D}^j_{nm}(u)$$

Let us now introduce the unitary matrix

$$\gamma = \begin{pmatrix} e^{-it/2} & 0 \\ 0 & e^{it/2} \end{pmatrix}$$

where t is a real number. The matrix elements $D^j_{mn}(\gamma)$ can then be found using Eqs. (1-36) and (1-37). One obtains

$$D(\gamma) f_n(z) = (-1)^{j-n} e^{ijt} \frac{(e^{-it} z)^{j-n}}{[(j-n)!(j+n)!]^{1/2}} = e^{int} f_n(z)$$

Consequently, one finds that the matrix $D^j(\gamma)$ is diagonal, and that the diagonal elements are given by

$$D^j_{mm}(\gamma) = e^{imt}$$

Furthermore, one obtains

$$D^j_{mn}(\gamma u) = e^{imt} D^j_{mn}(u), \qquad D^j_{mn}(u\gamma) = e^{int} D^j_{mn}(u)$$

[15] It will be noted that the function $S(j, m, n; x)$ is equal to the Jacobi polynomial $P_s^{\alpha\beta}(x)$ when $s = j - \frac{1}{2}(|m+n| + |m-n|)$, $\alpha = |m-n|$, and $\beta = |m+n|$.

Orthogonality Relation

We conclude this section by giving the orthogonality relation that the matrices D^j satisfy:[16]

$$\int D^{j_1}_{m_1 n_1}(u)\bar{D}^{j_2}_{m_2 n_2}(u)\, du = (2j_1+1)^{-1}\,\delta_{j_1 j_2}\;\delta_{m_1 m_2}\;\delta_{n_1 n_2} \tag{1-42}$$

1-6 DIFFERENTIAL OPERATORS OF INFINITESIMAL ROTATIONS

We are now in a position to find the differential operators corresponding to infinitesimal rotations about the coordinate axis, namely, the operators A_1, A_2, and A_3 and, consequently, the operators $L_{\mp}$ and L_3. These operators are well known in the literature when the Euler angles are employed. We here derive these operators in terms of the variables ψ, θ, and ϕ.

Representations of O_3 in Space of Functions

Let $g \to D(g)$ be an irreducible representation of weight j of the group O_3 and let $D_{mn} = D^j_{mn}$ be its matrix elements. We consider these elements as functions of the rotation g, $D_{mn} = D_{mn}(g)$. Since $g \to D(g)$ is a representation, one has $D(gg') = D(g)D(g')$. In terms of matrix elements, the last relation is

$$D_{mn}(gg') = \sum_{q=-j}^{j} D_{mq}(g)D_{qn}(g') \tag{1-43}$$

where $D_{mn}(gg')$ are the matrix elements of the operators $D(gg')$. Define now a transformation U such that

$$U(g')D_{mn}(g) = D_{mn}(gg') \tag{1-44}$$

Comparing Eqs. (1-43) and (1-44) we obtain

$$U(g')D_{mn}(g) = \sum_{q=-j}^{j} D_{qn}(g')D_{mq}(g) \tag{1-45}$$

Furthermore, one can show that

$$U(g')U(g'') = U(g'g'') \tag{1-46}$$

[16] Relations similar to (1-42) are valid for any compact group. See, for example, L. S. Pontrjagin, *Topological Groups*, Princeton University Press, New Jersey, U.S.A., 1946.

It thus follows that the transformation $U(g')$ realizes a representation of the group O_3 in the space of $2j+1$ functions of the mth row of the matrix $D(g)$ [compare Eq. (1-38)], and that the matrix elements of $U(g')$ are $D_{qn}(g')$.[17]

The Basic Infinitesimal Operators

To find the operators A_r we take g' as the rotation through some angle α around the axis Ox_r and expand the relation (1-44) in powers of α. Expansion of $D_{mn}(gg')$, which we denote by $D_{mn}(\tilde{\psi}, \tilde{\theta}, \tilde{\phi})$, gives

$$D_{mn}(\tilde{\psi}, \tilde{\theta}, \tilde{\phi}) = D_{mn}(\psi, \theta, \phi) + \alpha\left[\frac{\partial D_{mn}}{\partial \psi}\frac{d\tilde{\psi}}{d\alpha} + \frac{\partial D_{mn}}{\partial \theta}\frac{d\tilde{\theta}}{d\alpha} + \frac{\partial D_{mn}}{\partial \phi}\frac{d\tilde{\phi}}{d\alpha}\right]_{\alpha=0} + \cdots \tag{1-47}$$

To determine A_r we have to determine

$$\left.\frac{d\tilde{\psi}}{d\alpha}\right|_{\alpha=0}, \quad \left.\frac{d\tilde{\theta}}{d\alpha}\right|_{\alpha=0}, \quad \text{and} \quad \left.\frac{d\tilde{\phi}}{d\alpha}\right|_{\alpha=0} \tag{1-48}$$

for each rotation.

Now the matrix of the rotation g is a function of the angles ψ, θ, and ϕ which, by Eq. (1-20), has the form

$$g(\psi, \theta, \phi) = \begin{pmatrix} \cos\psi + \sin^2\theta\cos^2\phi(1-\cos\psi) & \sin^2\theta\cos\phi\sin\phi(1-\cos\psi) - \cos\theta\sin\psi & \sin\theta\cos\theta\cos\phi(1-\cos\psi) + \sin\theta\sin\phi\sin\psi \\ \sin^2\theta\sin\phi\cos\phi(1-\cos\psi) + \cos\theta\sin\psi & \cos\psi + \sin^2\theta\sin^2\phi(1-\cos\psi) & \sin\theta\cos\theta\sin\phi(1-\cos\psi) - \sin\theta\cos\phi\sin\psi \\ \sin\theta\cos\theta\cos\phi(1-\cos\psi) - \sin\theta\sin\phi\sin\psi & \sin\theta\cos\theta\sin\phi(1-\cos\psi) + \sin\theta\cos\phi\sin\psi & \cos\psi + \cos^2\theta(1-\cos\psi) \end{pmatrix} \tag{1-49}$$

The matrix of rotation gg' is given by some angles $\tilde{\psi}$, $\tilde{\theta}$, and $\tilde{\phi}$ which depend on the rotation angle α and which are equal to ψ, θ, and ϕ when $\alpha = 0$. Expansion of the matrix gg' in a power series in α gives

$$gg' = g(\psi, \theta, \phi) + \alpha\left\{\left.\frac{\partial g}{\partial \psi}\frac{d\tilde{\psi}}{d\alpha}\right|_{\alpha=0} + \left.\frac{\partial g}{\partial \theta}\frac{d\tilde{\theta}}{d\alpha}\right|_{\alpha=0} + \left.\frac{\partial g}{\partial \phi}\frac{d\tilde{\phi}}{d\alpha}\right|_{\alpha=0}\right\} + \cdots \tag{1-50}$$

---

[17] The representation $g' \to U(g')$ in the space of functions $D_{mq}(g)$, $q = -j, -j+1, \ldots, j$, is irreducible, and the $D_{mq}(g)$ form a canonical basis in this space. Hence the operators $L_\pm$ and L_3 of this representation satisfy the relation (1-28), i.e.,

$$L_\pm D^j_{mn}(g) = [(j \pm n + 1)(j \mp n)]^{1/2} D^j_{m,n\pm1}(g)$$
$$L_3 D^j_{mn}(g) = nD^j_{mn}(g)$$

To find the operator A_1 we identify g' with the rotation with angle α around Ox_1 given by

$$g_1(\alpha) = \begin{pmatrix} 1 & 0 & 0 \\ 0 & \cos\alpha & -\sin\alpha \\ 0 & \sin\alpha & \cos\alpha \end{pmatrix}$$

$$= \begin{pmatrix} 1 & 0 & 0 \\ 0 & 1 & 0 \\ 0 & 0 & 1 \end{pmatrix} + \alpha \begin{pmatrix} 0 & 0 & 0 \\ 0 & 0 & -1 \\ 0 & 1 & 0 \end{pmatrix} + \cdots \tag{1-51}$$

Therefore

$$gg_1 = g(\psi, \theta, \phi) + \alpha \begin{pmatrix} 0 & g_{13} & -g_{12} \\ 0 & g_{23} & -g_{22} \\ 0 & g_{33} & -g_{32} \end{pmatrix} + \cdots \tag{1-52}$$

On the other hand, gg_1 is given by Eq. (1-50) when $g_1 = g'$. Comparing these two expressions for gg_1, we obtain equations from which the three expressions given in (1-48) can be determined for the case of rotation about Ox_1. We obtain[18]

$$2\sin\theta\cos\phi\sin\frac{\psi}{2}\left(-\sin\theta\sin\phi\left.\frac{d\tilde{\phi}}{d\alpha}\right|_{\alpha=0} + \cos\theta\cos\phi\left.\frac{d\tilde{\theta}}{d\alpha}\right|_{\alpha=0}\right)$$
$$- \cos\frac{\psi}{2}(1 - \sin^2\theta\cos^2\phi)\left.\frac{d\tilde{\psi}}{d\alpha}\right|_{\alpha=0} = 0 \tag{1-53a}$$

$$2\sin\theta\sin\phi\sin\frac{\psi}{2}\left(\sin\theta\cos\phi\left.\frac{d\tilde{\phi}}{d\alpha}\right|_{\alpha=0} + \cos\theta\sin\phi\left.\frac{d\tilde{\theta}}{d\alpha}\right|_{\alpha=0}\right)$$
$$- \cos\frac{\psi}{2}(1 - \sin^2\theta\sin^2\phi)\left.\frac{d\tilde{\psi}}{d\alpha}\right|_{\alpha=0} = \sin\theta\left(\cos\theta\sin\phi\sin\frac{\psi}{2} - \cos\phi\cos\frac{\psi}{2}\right) \tag{1-53b}$$

$$2\cos\theta\sin\frac{\psi}{2}\left.\frac{d\tilde{\theta}}{d\alpha}\right|_{\alpha=0} + \cos\frac{\psi}{2}\sin\theta\left.\frac{d\tilde{\psi}}{d\alpha}\right|_{\alpha=0} = \cos\theta\sin\phi\sin\frac{\psi}{2} + \cos\phi\cos\frac{\psi}{2} \tag{1-53c}$$

The solution of Eqs. (1-53) is

$$\left.\frac{d\tilde{\psi}}{d\alpha}\right|_{\alpha=0} = \cos\phi\sin\theta$$
$$\left.\frac{d\tilde{\theta}}{d\alpha}\right|_{\alpha=0} = \frac{1}{2}\left(\sin\phi + \cot\frac{\psi}{2}\cos\theta\cos\phi\right) \tag{1-54}$$
$$\left.\frac{d\tilde{\phi}}{d\alpha}\right|_{\alpha=0} = \frac{1}{2}\operatorname{cosec}\theta\left(\cos\theta\cos\phi - \cot\frac{\psi}{2}\sin\phi\right)$$

[18] One obtains nine equations; only three of them are independent. Equations (1-53) are obtained by equating the diagonal elements of the matrices (1-50) and (1-52).

Using Eqs. (1-54) in Eq. (1-47), we find the operator A_1 corresponding to the rotation around Ox_1,

$$A_1 = \cos\phi \sin\theta \frac{\partial}{\partial\psi} + \frac{1}{2}\left(\sin\phi + \cot\frac{\psi}{2}\cos\theta\cos\phi\right)\frac{\partial}{\partial\theta} + \frac{1}{2}\operatorname{cosec}\theta\left(\cos\theta\cos\phi - \cot\frac{\psi}{2}\sin\phi\right)\frac{\partial}{\partial\phi} \tag{1-55a}$$

The operators A_2 and A_3 are found in a similar way:

$$A_2 = \sin\phi\sin\theta\frac{\partial}{\partial\psi} - \frac{1}{2}\left(\cos\phi - \cot\frac{\psi}{2}\cos\theta\sin\phi\right)\frac{\partial}{\partial\theta} + \frac{1}{2}\operatorname{cosec}\theta\left(\cos\theta\sin\phi + \cot\frac{\psi}{2}\cos\phi\right)\frac{\partial}{\partial\phi} \tag{1-55b}$$

$$A_3 = \cos\theta\frac{\partial}{\partial\psi} - \frac{1}{2}\cot\frac{\psi}{2}\sin\theta\frac{\partial}{\partial\theta} - \frac{\partial}{\partial\phi} \tag{1-55c}$$

Angular Momentum Operators

Using the last three equations in Eq. (1-26) one obtains for the angular momentum operators L_+, L_-, and L_3:

$$L_\pm = i\,e^{\pm i\phi}\left\{\sin\theta\frac{\partial}{\partial\psi} + \frac{1}{2}\left(\mp i + \cot\frac{\psi}{2}\cos\theta\right)\frac{\partial}{\partial\theta} + \frac{1}{2}\operatorname{cosec}\theta\left(\cos\theta \pm i\cot\frac{\psi}{2}\right)\frac{\partial}{\partial\phi}\right\} \tag{1-56a}$$

$$L_3 = i\left(\cos\theta\frac{\partial}{\partial\psi} - \frac{1}{2}\cot\frac{\psi}{2}\sin\theta\frac{\partial}{\partial\theta} - \frac{1}{2}\frac{\partial}{\partial\phi}\right) \tag{1-56b}$$

The angular momentum operators derived above were expressed in terms of the angle of rotation ψ and the spherical angles of direction of rotation θ and ϕ. One can use, however, the Euler angles and obtain the standard expressions of angular momentum operators given in other treatises (see Problem 1.6).

PROBLEMS

1.1 Show that every finite-dimensional representation $g \to D(g)$ of the group O_3 can be made to be unitary. Show that a scalar product can be defined in the space of representation in such a way that all the operators $D(g)$ become unitary operators.

1.2 Show that if the representation $g \to D(g)$ is unitary then $A_r^\dagger = -A_r$.

1.3 Find the three operators $A_r(\psi)$ for the representation (1-36). From them derive the basic infinitesimal operators A_r. Finally, show that

$$L_+ p = -\partial p/\partial z$$
$$L_- p = z^2(\partial p/\partial z) - kzp$$
$$L_3 p = -z(\partial p/\partial z) + (k/2)p$$

1.4 Show that the functions (1-37) form a canonical basis for the representation Z_k, by checking that $L_\pm f_m$ and $H_3 f_m$ are indeed given by Eqs. (1-28) when f_m is put for p in the corresponding equations of Problem 1.3.

1.5 Write the matrix elements D^j of Eq. (1-39) in terms of the Euler angles by use of Eq. (1-11). Show that $D^j_{mn}(\phi_1, \theta, \phi_2)$ is given by

$$D^j_{mn}(\phi_1, \theta, \phi_2) = (-i)^{2j-m-n}\left[\frac{(j-m)!(j+m)!}{(j-n)!(j+n)!}\right]^{1/2} \sum i^{-2a}\binom{j-n}{a}\binom{j+n}{j-m-a}$$
$$\times \left(\cos\frac{\theta}{2}\right)^{m+n+2a}\left(\sin\frac{\theta}{2}\right)^{2j-m-n-2a} e^{-i(m\phi_2+n\phi_1)}$$

1.6 Find the angular momentum operators $L_\pm$, L_3 when expressed in terms of Euler's angles. Use the matrix (1-4) instead of (1-49) and show that the angular momentum operators are now given by:

$$L_\pm = e^{\mp i\phi_1}\left(\pm\cot\theta\frac{\partial}{\partial\phi_1} + i\frac{\partial}{\partial\theta} \mp \operatorname{cosec}\theta\frac{\partial}{\partial\phi_2}\right)$$
$$L_3 = i\frac{\partial}{\partial\phi_1}$$

1.7 Verify that the angular momentum operators L_+, L_-, and L_3 of Problem 1.6 satisfy the following equations:

$$L_\pm D^j_{mn} = [(j \pm n + 1)(j \mp n)]^{1/2} D^j_{m, n\pm 1}$$
$$L_3 D^j_{mn} = nD^j_{mn}$$

where D^j_{mn} are given in terms of Euler's angles (see Problem 1.5), and m, n take the values $-j, -j+1, \ldots, j$. Show that, by changing the variable ϕ_1 into ϕ_2 and vice versa, one obtains the following relations:

$$\tilde{L}_\pm D^j_{mn} = [(j \pm m + 1)(j \mp m)]^{1/2} D^j_{m\pm 1, n}$$
$$\tilde{L}_3 D^j_{mn} = mD^j_{mn}$$

where the angular momentum operators $\tilde{L}_\pm$, $\tilde{L}_3$ are given by

$$\tilde{L}_\pm = e^{\mp i\phi_2}\left(\pm\cot\theta\frac{\partial}{\partial\phi_2} + i\frac{\partial}{\partial\theta} \mp \operatorname{cosec}\theta\frac{\partial}{\partial\phi_1}\right)$$
$$\tilde{L}_3 = i\frac{\partial}{\partial\phi_2}$$

1.8 Show that every function $f(u)$ in the group SU_2 whose modulus square is integrable with respect to the measure du can be expanded in terms of the functions $D^j_{mn}(u)$.

1.9 Write the matrix $u \in SU_2$ in the form

$$u = \begin{pmatrix} \alpha & \beta \\ -\bar{\beta} & \bar{\alpha} \end{pmatrix}$$

where $|\alpha|^2 + |\beta|^2 = 1$. From this show that the matrix u is uniquely determined by the parameters

$$t = |\alpha|^2, \qquad \phi = \arg \alpha, \qquad \psi = \arg \beta$$

Show that in terms of the new parameters one has

$$\int f(u)\, du = \frac{1}{4\pi^2} \int_0^1 dt \int_0^{2\pi} d\phi \int_0^{2\pi} f(t, \phi, \psi)\, d\psi.$$

1.10 Introduce the four coordinates x_1, x_2, x_3, x_4, related to the parameters of Problem 1.9 by $\alpha = x_1 + ix_2$, $\beta = x_3 + ix_4$. Then the relation $|\alpha|^2 + |\beta|^2 = 1$ implies $x_1^2 + x_2^2 + x_3^2 + x_4^2 = 1$, i.e., the surface of the unit sphere S_4 in a four-dimensional space, and hence each element u of SU_2 uniquely corresponds to a point on S_4. Show that the expression $\frac{1}{2}\, dt\, d\phi\, d\psi$ has the simple geometrical meaning of a surface element of the sphere S_4.

1.11 Show that

$$D^l_{m'm}(u) D^{\bar{l}}_{n'n}(u) = \sum_{L=|l-\bar{l}|}^{l+\bar{l}} \bar{s}_{Lm'n'} D^L_{m'+n',\, m+n}(u) s_{Lmn}$$

Show that the coefficients s_{Lmn} can be chosen to be real and find their form. [See E. P. Wigner, *Group Theory and its Application to the Quantum Mechanics of Atomic Spectra*, Academic Press, New York, 1959.]

CHAPTER

TWO

THE LORENTZ GROUP

After discussing representations of the rotation group in the previous chapter, we now present the Lorentz group which is discussed in Chapters 2 to 6. The present chapter is an introduction to the theory of the Lorentz group. We first define the Galilean and the Poincaré groups. Infinitesimal Lorentz matrices and their commutation relations are consequently found. The one-parameter groups of operators are consequently introduced and the role of the representations of the rotation group (which is a subgroup of the Lorentz group) in representing the Lorentz group is discussed. Representations of the Lorentz group are then briefly discussed, and the canonical basis is introduced. Finally, conditions under which a representation of the Lorentz group becomes unitary, are given.

2-1 INFINITESIMAL LORENTZ MATRICES

Galilean Group

In classical mechanics one assumes that the laws of motion do not depend on the choice of a particular fixed coordinate system with respect to which motion is considered to take place and, furthermore, that the same laws do not change by going over from one coordinate system into another which has uniform, rectilinear, translational motion relative to the first. These requirements mean that the laws of classical mechanics are invariant under orthogonal transformations of coordinates and hence one has rotational invariance and, furthermore, one has what is called a Galilean invariance. If one chooses both coordinate systems so that the corresponding axes are parallel, and if $\mathbf{v}$ is the velocity of one coordinate system with respect to the other, the Galilean transformation can be expressed as follows:

$$x' = x + v_x t, \qquad y' = y + v_y t, \qquad z' = z + v_z t \tag{2-1}$$

In these transformations the time coordinate t appears as a parameter. The aggregate of all transformations of the form (2-1) forms a group, called the *Galilean group*.

In the theory of relativity, space and time coordinates appear on the same footing. One therefore deals with linear four-dimensional space of vectors with components x^0, x^1, x^2, and x^3.[1] Such a four-dimensional space, with the scalar product

$$x^0y^0 - x^1y^1 - x^2y^2 - x^3y^3 = \eta_{\mu\nu}x^\mu y^\nu \tag{2-2}$$

is called a *Minkowskian space*. In Eq. (2-2) the symbol $\eta_{\mu\nu}$ (and later on $\eta^{\mu\nu}$) is the flat space metric, given by the matrix

$$\eta = \begin{pmatrix} 1 & 0 & 0 & 0 \\ 0 & -1 & 0 & 0 \\ 0 & 0 & -1 & 0 \\ 0 & 0 & 0 & -1 \end{pmatrix} \tag{2-3}$$

We call $x^\mu x_\mu = \eta_{\mu\nu}x^\mu x^\nu$ the square of the length of the vector x^μ. Indices are raised and lowered by means of the $\eta^{\mu\nu}$ and $\eta_{\mu\nu}$. A vector V^μ is called *timelike* if its square length is positive, *spacelike* if it is negative, and *null* if the length vanishes. A timelike vector V^μ is called *positive* or *negative* according to whether V_0 is positive or negative, respectively. The manifold of all null vectors forms the *light cone*.

Poincaré Group

Linear transformations in Minkowski space defined by

$$x'^\mu = \Lambda^\mu{}_\nu x^\nu + y^\mu \tag{2-4}$$

which satisfy the relation

$$\eta_{\mu\nu}\Lambda^\mu{}_\alpha\Lambda^\nu{}_\beta = \eta_{\alpha\beta} \tag{2-5a}$$

or, using matrix notation,

$$\Lambda^t\eta\Lambda = \eta \tag{2-5b}$$

where Λ^t is the transposed matrix to the matrix Λ, are called inhomogeneous Lorentz transformations. These transformations are usually denoted by the symbol (Λ, y). The homogeneous Lorentz transformations $(\Lambda, 0)$ leave the scalar product invariant:

$$x'_\mu y'^\mu = \eta_{\mu\nu}x'^\mu y'^\nu = \eta_{\mu\nu}\Lambda^\mu{}_\alpha\Lambda^\nu{}_\beta x^\alpha y^\beta = x_\alpha y^\alpha$$

[1] The coordinate $x^0 = ct$, where c is the speed of light, and t is the time coordinate.

The inhomogeneous Lorentz transformations form a group. It is called the *inhomogeneous Lorentz group* or the *Poincaré group*. The multiplication law of the group can be written as

$$(\Lambda_1, y_1)(\Lambda_2, y_2) = (\Lambda_1 \Lambda_2, y_1 + \Lambda_1 y_2)$$

and the inverse of the group element (Λ, y) is $(\Lambda^{-1}, -\Lambda^{-1}y)$. The homogeneous Lorentz transformations form a subgroup of the Poincaré group. It is called the *Lorentz group*. The translational part of the inhomogeneous Lorentz transformations (2-4), denoted by $(1, y)$, provides another subgroup of the Poincaré group; it is the *group of translations*. The group of translations is an *invariant subgroup* of the Poincaré group (see Problem 2.1). All of these groups are Lie groups.

Proper, Orthochronous, Lorentz Group

The homogeneous Lorentz group possesses four disconnected pieces that arise as follows.

Equation (2-5b) shows that $(\det \Lambda)^2 = 1$, and accordingly, the determinant of every Lorentz transformation is equal to either $+1$, in which case the transformation is called *proper*, or to -1, in which case the transformation is called *improper*. From Eq. (2-5a), when one chooses $\alpha = \beta = 0$, one obtains

$$(\Lambda^0{}_0)^2 - (\Lambda^1{}_0)^2 - (\Lambda^2{}_0)^2 - (\Lambda^3{}_0)^2 = 1$$

Hence $(\Lambda^0{}_0)^2 \geqslant 1$, and consequently we have either

$$\Lambda^0{}_0 \geqslant 1 \qquad \text{or} \qquad \Lambda^0{}_0 \leqslant -1$$

A Lorentz transformation whose element $\Lambda^0{}_0 \geqslant 1$ is called *orthochronous*. The aggregate of all orthochronous Lorentz transformations provides a subgroup of the homogeneous Lorentz group. A Lorentz transformation satisfies the condition $\Lambda^0{}_0 \geqslant 1$ if and only if it transforms every positive timelike vector into another positive timelike vector (see Problem 2.2).

The four parts of the homogeneous Lorentz group are then given by:[2]

(1) $L_+^\uparrow$: $\det \Lambda = +1, \Lambda^0{}_0 \geqslant 1$. This part contains the identity element of the group. The aggregate of all proper, orthochronous, Lorentz transformations provides a group which is a subgroup of the Lorentz group. It is called the *proper, orthochronous Lorentz group*.[3]

(2) $L_-^\uparrow$: $\det \Lambda = -1, \Lambda^0{}_0 \geqslant +1$. This manifold contains a *space inversion* element S which describes a reflection relative to the three space axes:

$$x'^0 = x^0, \qquad x'^1 = -x^1, \qquad x'^2 = -x^2, \qquad x'^3 = -x^3$$

[2] For details see R. F. Streater and A. S. Wightman, *PCT, Spin and Statistics, and All That*, Benjamin, New York, 1964.

[3] The proper, orthochronous, Lorentz group is identical with the group $SO(3, 1)$.

(3) $L_-^\downarrow$: $\det \Lambda = -1, \Lambda^0{}_0 \leqslant -1$. This manifold contains a *time reversal* element T which describes a reflection relative to the time axes:

$$x'^0 = -x^0, \qquad x'^1 = x^1, \qquad x'^2 = x^2, \qquad x'^3 = x^3$$

(4) $L_+^\downarrow$: $\det \Lambda = +1, \Lambda^0{}_0 \leqslant -1$. This manifold contains the element ST.

Finally, one notices that from the above four parts of the homogeneous Lorentz group one obtains the subgroup $L^\uparrow = L_+^\uparrow \cup L_-^\uparrow$, the *orthochronous Lorentz group*, and the subgroup $L_+ = L_+^\uparrow \cup L_+^\downarrow$, the proper Lorentz group. One, furthermore, notices that every improper Lorentz transformation has the form $\Lambda = S\Lambda_1$, where Λ_1 is a proper Lorentz transformation.

In the following we will be concerned with the proper, orthochronous, Lorentz group. This group will be denoted by L. Obviously the rotation group O_3, discussed in Chapter 1, is a subgroup of the group L.

Infinitesimal Lorentz Matrices

Rotations $a_1(\psi)$, $a_2(\psi)$, and $a_3(\psi)$ and Lorentz transformations $b_1(\psi)$, $b_2(\psi)$, and $b_3(\psi)$, around and along the axes Ox^1, Ox^2, and Ox^3 can be written explicitly. One easily finds the rotation matrices, using the results of Chapter 1, to be given by

$$a_1(\psi) = \begin{pmatrix} 1 & 0 & 0 & 0 \\ 0 & 1 & 0 & 0 \\ 0 & 0 & \cos\psi & -\sin\psi \\ 0 & 0 & \sin\psi & \cos\psi \end{pmatrix}, \qquad a_2(\psi) = \begin{pmatrix} 1 & 0 & 0 & 0 \\ 0 & \cos\psi & 0 & \sin\psi \\ 0 & 0 & 1 & 0 \\ 0 & -\sin\psi & 0 & \cos\psi \end{pmatrix}$$

$$a_3(\psi) = \begin{pmatrix} 1 & 0 & 0 & 0 \\ 0 & \cos\psi & -\sin\psi & 0 \\ 0 & \sin\psi & \cos\psi & 0 \\ 0 & 0 & 0 & 1 \end{pmatrix} \tag{2-6a}$$

The Lorentz transformations can also be found. They are given by (see Problem 2.3):

$$b_1(\psi) = \begin{pmatrix} \cosh\psi & \sinh\psi & 0 & 0 \\ \sinh\psi & \cosh\psi & 0 & 0 \\ 0 & 0 & 1 & 0 \\ 0 & 0 & 0 & 1 \end{pmatrix}, \qquad b_2(\psi) = \begin{pmatrix} \cosh\psi & 0 & \sinh\psi & 0 \\ 0 & 1 & 0 & 0 \\ \sinh\psi & 0 & \cosh\psi & 0 \\ 0 & 0 & 0 & 1 \end{pmatrix}$$

$$b_3(\psi) = \begin{pmatrix} \cosh\psi & 0 & 0 & \sinh\psi \\ 0 & 1 & 0 & 0 \\ 0 & 0 & 1 & 0 \\ \sinh\psi & 0 & 0 & \cosh\psi \end{pmatrix} \tag{2-6b}$$

They satisfy the relations

$$\begin{aligned} a_k(\psi_1)a_k(\psi_2) &= a_k(\psi_1 + \psi_2) \\ b_k(\psi_1)b_k(\psi_2) &= b_k(\psi_1 + \psi_2) \end{aligned} \tag{2-7}$$

where $k = 1, 2, 3$.

The *infinitesimal matrices* a_1, a_2, a_3 and b_1, b_2, b_3 of the group L are defined by[4]

$$a_k = \left.\frac{\mathrm{d}a_k(\psi)}{\mathrm{d}\psi}\right|_{\psi=0}, \qquad b_k = \left.\frac{\mathrm{d}b_k(\psi)}{\mathrm{d}\psi}\right|_{\psi=0} \tag{2-8}$$

They are related to the rotations and Lorentz transformations $a_k(\psi)$ and $b_k(\psi)$ by

$$a_k(\psi) = \exp(\psi a_k), \qquad b_k(\psi) = \exp(\psi b_k)$$

with $k = 1, 2, 3$.

Commutation Relations

The infinitesimal Lorentz matrices satisfy the following commutation relations:

$$\begin{aligned}[a_i, a_j] &= \varepsilon_{ijk} a_k \\ [b_i, b_j] &= -\varepsilon_{ijk} a_k \\ [a_i, b_j] &= \varepsilon_{ijk} b_k \,. \end{aligned} \tag{2-9}$$

Here $[a, b] = ab - ba$. Equations (2-9) can easily be verified using Eqs. (2-8).

2-2 INFINITESIMAL OPERATORS

One-parameter Group of Operators

We denote an arbitrary linear representation of the proper, orthochronous, Lorentz group L in a Banach space B (see Appendix C) by $\Lambda \to D(\Lambda)$, and for convenience we denote

$$A_k(\psi) = D(a_k(\psi)), \qquad B_k(\psi) = D(b_k(\psi)) \tag{2-10}$$

Using Eqs. (2-7) one obtains

$$\begin{aligned} A_k(\psi_1)A_k(\psi_2) &= A_k(\psi_1 + \psi_2) \\ B_k(\psi_1)B_k(\psi_2) &= B_k(\psi_1 + \psi_2) \end{aligned} \tag{2-11}$$

[4] Explicitly, they are given by

$$a_1 = \begin{pmatrix} 0 & 0 & 0 & 0 \\ 0 & 0 & 0 & 0 \\ 0 & 0 & 0 & -1 \\ 0 & 0 & 1 & 0 \end{pmatrix}, \quad a_2 = \begin{pmatrix} 0 & 0 & 0 & 0 \\ 0 & 0 & 0 & 1 \\ 0 & 0 & 0 & 0 \\ 0 & -1 & 0 & 0 \end{pmatrix}, \quad a_3 = \begin{pmatrix} 0 & 0 & 0 & 0 \\ 0 & 0 & -1 & 0 \\ 0 & 1 & 0 & 0 \\ 0 & 0 & 0 & 0 \end{pmatrix}$$

and

$$b_1 = \begin{pmatrix} 0 & 1 & 0 & 0 \\ 1 & 0 & 0 & 0 \\ 0 & 0 & 0 & 0 \\ 0 & 0 & 0 & 0 \end{pmatrix}, \quad b_2 = \begin{pmatrix} 0 & 0 & 1 & 0 \\ 0 & 0 & 0 & 0 \\ 1 & 0 & 0 & 0 \\ 0 & 0 & 0 & 0 \end{pmatrix}, \quad b_3 = \begin{pmatrix} 0 & 0 & 0 & 1 \\ 0 & 0 & 0 & 0 \\ 0 & 0 & 0 & 0 \\ 1 & 0 & 0 & 0 \end{pmatrix}$$

where $i = 1, 2, 3$. Furthermore, $A_k(0) = 1$, $B_k(0) = 1$. $A_k(\psi)$ and $B_k(\psi)$ are continuous functions of ψ and are called *one-parameter groups of operators*. If the representation $\Lambda \to D(\Lambda)$ is finite-dimensional, then the operators $A_k(\psi)$ and $B_k(\psi)$ are differentiable functions of ψ. If the representation is infinite-dimensional, however, these operators might be non-differentiable, but there may still exist a vector x in the space B for which $A_k(\psi)x$ and $B_k(\psi)x$ are differentiable vector-functions.[5]

The *infinitesimal operators* of the one-parameter groups $A_k(\psi)$ and $B_k(\psi)$ are defined by[6]

$$A_k = \left.\frac{dA_k(\psi)}{d\psi}\right|_{\psi=0}, \qquad B_k = \left.\frac{dB_k(\psi)}{d\psi}\right|_{\psi=0} \tag{2-12}$$

In the following, the representations $\Lambda \to D(\Lambda)$ of the group L are assumed to satisfy the following properties:

In the space of representation B there exists a dense subspace X which is invariant under all representations operators $D(\Lambda)$. All infinitesimal operators, and their products, of the representations, apply then to the vectors x of X. For any vector x of X, the vector $D(\Lambda)x$ is an infinitely differentiable function on the group L. Furthermore, for any vector x of X and any one-parameter group $a(\psi)$ of the group L, the vector $A(\psi)x = D(a(\psi))x$ satisfies the condition

$$\frac{dA(\psi)x}{d\psi} = AA(\psi)x = A(\psi)Ax \tag{2-13}$$

where the operator A is related to $A(\psi)$ by

$$A(\psi)x = x + \psi Ax + \frac{\psi^2}{2!}A^2x + \frac{\psi^3}{3!}\tilde{A}(\psi)A^3x \tag{2-14a}$$

$$A(\psi)x = x + \psi Ax + \frac{\psi^2}{2!}A^2x + \cdots + \frac{\psi^n}{n!}A^nx + \cdots \tag{2-14b}$$

Here

$$\tilde{A}(\psi) = \frac{3}{\psi^2}\int_0^\psi (1-t)A^2(t)\,dt$$

[5] In general, let $A(\psi)$ be a continuous one-parameter group of operators in a Banach space B, and denote by $X(A)$ the set of all vectors x of B for which the limit of $(A(\psi)x - x)/\psi$, when $\psi \to 0$ exists in the sense of the norm in B. Obviously the set $X(A)$ contains the vector $x = 0$. Define now the operator A for all x of $X(A)$ by $Ax = \lim\{(A(\psi)x - x)/\psi\}$ at the limit $\psi \to 0$. The domain of definition, $X(A)$, of the operator A is a subspace of B, and A is linear, namely $A(\lambda_1 x_1 + \lambda_2 x_2) = \lambda_1 Ax_1 + \lambda_2 Ax_2$ for x_1 and x_2 of $X(A)$. Such an operator A is called the infinitesimal operator of the one-parameter group $A(\psi)$. If $A(\psi) = D(a(\psi))$ is the group of operators of the representation $\Lambda \to D(\Lambda)$, corresponding to a one-parameter subgroup $a(\psi)$ of the group L, the corresponding operator A is then called the *infinitesimal operator* of the representation $\Lambda \to D(\Lambda)$.

[6] $A_k(\psi)$ and $B_k(\psi)$ might then be expanded in terms of A_k and B_k as $A_k(\psi) = \exp(\psi A_k)$ and $B_k(\psi) = \exp(\psi B_k)$.

is bounded in norm in any finite interval. The series on the right-hand side of Eq. (2-14b) converges in norm, absolutely. If $P(\psi)$ is a finite product of the operators $A_k(\psi)$ and A_j, and if x is a vector in X, then $|P(\psi)x|$ is a bounded function of ψ in any finite interval. From Eq. (2-14b) it follows that, on the vectors x of X, the operator $A(\psi)$ is uniquely determined by the infinitesimal operator A. Since these vectors form a set, dense in the space B, and $A(\psi)$ is bounded, it follows by continuity that $A(\psi)$ is uniquely determined on the whole space B. Hence one has also that the one-parameter group $A(\psi)$ is uniquely determined on the whole space B by the infinitesimal operator A.

Now since the group O_3 is a subgroup of the group L, obviously the representation $\Lambda \to D(\Lambda)$ of the group L is also a representation of the group SU_2. Accordingly, let us consider the representation only for the group SU_2.

Decomposition of a Representation of the Group SU_2 into Irreducible Representations[7]

Let $u \to D(u)$ be an arbitrary continuous representation of the group SU_2 in a reflexive Banach space B, and define the operator

$$E^j_{mn} = (2j+1) \int \bar{D}^j_{mn}(u) D(u)\, du \tag{2-15}$$

Since $\bar{D}^j_{mn}(u)D(u)$ is a continuous operator function on the group SU_2, the integral on the right-hand side of Eq. (2-15) exists, and E^j_{mn} is a bounded operator in the space B.

Lemma 2.1 *The operators E^j_{mn} satisfy the relations*

$$D(u)E^j_{mn} = \sum_{m'=-j}^{j} D^j_{m'm}(u) E^j_{m'n} \tag{2-16a}$$

$$E^j_{mn} D(u) = \sum_{m'=-j}^{j} D^j_{nm'}(u) E^j_{mm'} \tag{2-16b}$$

$$E^j_{mn} E^{j'}_{m'n'} = E^j_{mn'}\, \delta^{jj'}\, \delta_{nm'} \tag{2-17}$$

To prove Eq. (2-16a) one calculates $D(u)E^j_{mn}$. Using the properties of $D^j_{mn}(u)$ (see Chapter 1) one obtains

$$D(u)E^j_{mn} = D(u)\{(2j+1) \int \bar{D}^j_{mn}(u') D(u')\, du'\}$$

$$= (2j+1) \int \bar{D}^j_{mn}(u') D(uu')\, du'$$

[7] See, for example, M. A. Naimark, *Linear Representations of the Lorentz Group*, Pergamon Press, New York, 1964.

Changing variables $uu' \to u'$ and using the properties of $D^j_{mn}(u)$ one obtains

$$D(u)E^j_{mn} = (2j+1)\int \bar{D}^j_{mn}(u^{-1}u')D(u')\,du'$$

$$= (2j+1)\sum_{k=-j}^{j}\int \bar{D}^j_{mk}(u^{-1})\bar{D}^j_{kn}(u')D(u')\,du'$$

$$= \sum_{k=-j}^{j} \bar{D}^j_{mk}(u^{-1})E^j_{kn}$$

$$= \sum_{k=-j}^{j} D^j_{km}(u)E^j_{kn}$$

The proof of Eq. (2-16b) is similar. Finally Eq. (2-17) can be shown by calculating the product $E^j_{mn}E^{j'}_{m'n'}$ and using Eq. (2-16a). One obtains

$$E^j_{mn}E^{j'}_{m'n'} = \left\{(2j+1)\int \bar{D}^j_{mn}(u)D(u)\,du\right\}E^{j'}_{m'n'}$$

$$= (2j+1)\int \bar{D}^j_{mn}(u)D(u)E^{j'}_{m'n'}\,du$$

$$= (2j+1)\int \bar{D}^j_{mn}(u)\sum_{k=-j}^{j} D^{j'}_{km'}(u)E^{j'}_{kn'}\,du$$

Using the orthagonality relation Eq. (1-42) between the matrices D^j_{mn}, one obtains Eq. (2-17).

As a particular case of Eqs. (2-16) one easily obtains

$$D(\gamma)E^j_{mn} = e^{im\omega}E^j_{mn} \tag{2-18a}$$

$$E^j_{mn}D(\gamma) = e^{in\omega}E^j_{mn} \tag{2-18b}$$

for a matrix γ given by

$$\gamma = \begin{pmatrix} e^{-i\omega/2} & 0 \\ 0 & e^{i\omega/2} \end{pmatrix}$$

Lemma 2.2 *The operators E^j, defined by*

$$E^j = \sum_{m=-j}^{j} E^j_{mm} \tag{2-19}$$

satisfy the relations

$$E^jE^{j'} = E^j\,\delta^{jj'} \tag{2-20}$$

$$[D(u), E^j] = 0, \qquad u \in SU_2 \tag{2-21}$$

Equation (2-19) follows from Eq. (2-17). Using Eqs. (2-16) we obtain

$$D(u)E^j = \sum_{m,\,n=-j}^{j} D^j_{mn}(u)E^j_{mn}$$

$$E^jD(u) = \sum_{m,\,n=-j}^{j} D^j_{nm}(u)D^j_{nm}$$

Hence the commutator $[D(u), E^j] = 0$ and the proof is complete.

Definition We denote by M^j_m the aggregate of all vectors x of the space B satisfying the condition $E^j_{mm}x = x$, and denote by M^j the aggregate of all vectors x of B satisfying the condition $E^jx = x$.

Lemma 2.3 *The spaces M^j_m and M^j are closed subspaces of the space B.*

Suppose $E^j_{mm}x_1 = x_1$ and $E^j_{mm}x_2 = x_2$, then $E^j_{mm}(\alpha_1 x_1 + \alpha_2 x_2) = \alpha_1 x_1 + \alpha_2 x_2$ for any complex numbers α_1 and α_2. Hence M^j_m is a subspace of the space B. Furthermore, let $E^j_{mm}x_n = x_n$ and $|x_n - x| \to 0$ as $n \to \infty$, then, proceeding to the limit and making use of the continuity of the operators E^j_{mm}, we obtain $E^j_{mm}x = x$. Hence the subspace M^j_m is closed. The subspace M^j is also a closed subspace of the space B as can easily be seen.

Lemma 2.4 *The subspaces M^j_m, for all possible j and m, and the subspaces M^j, for all possible j, are linearly independent.*[8]

Suppose x belongs to the subspace $M^{j'}_{m'}$. Then, using Eq. (2-17), one obtains

$$E^j_{mm}x = E^j_{mm}E^{j'}_{m'm'}x = E^{j'}_{m'm'}\,\delta^{jj'}\,\delta_{mm'}\,x = \delta^{jj'}\,\delta_{mm'}\,x \tag{2-22}$$

In the same way one shows that if x belongs to $M^{j'}$ then

$$E^jx = \delta^{jj'}x \tag{2-23}$$

Suppose now that $x_1 + x_2 + \cdots + x_n = 0$, where x_1 belongs to $M^{j_1}_{m_1}$, x_2 belongs to $M^{j_2}_{m_2}$, etc., and $(j_1, m_1) \neq (j_2, m_2) \neq \cdots$. Applying now $E^{j_1}_{m_1m_1}$ to $(x_1 + \cdots + x_n) = 0$, and using Eq. (2-22), one obtains $x_1 = 0$. In this way one shows that $x_1 = x_2 = \cdots = x_n = 0$, and consequently the subspaces M^j_m are linearly independent. In a similar way one shows that the subspaces M^j are also linearly independent.

[8] Subspaces M_1, M_2, ... are called linearly independent if an equation of the form $x_1 + x_2 + \cdots + x_n = 0$, where x_j is a vector in M_j, is valid if and only if one has $x_1 = x_2 = \cdots = x_n = 0$.

Lemma 2.5 *The space M^j is the direct sum of the subspaces*[9] *M^j_m for $m = -j, -j+1, \ldots, j$:*

$$M^j = M^j_{-j} \dot{+} M^j_{-j+1} \dot{+} \cdots \dot{+} M^j_j \tag{2-24}$$

Lemma 2.6 *The space B is the closed direct sum of all the subspaces M^j_m, and hence of all subspaces M^j.*

The proofs of Lemmas 5 and 6 are left for the reader.

Definition A representation $u \to D(u)$ of the group SU_2 in a Banach space B is called *multiple of the irreducible representation $u \to D^{(j)}(u)$ of weight j*, if the space B is the direct sum of closed subspaces B_m, with $m = -j, -j+1, \ldots, j$, such that for any B_k and x of B_k there exist vectors x_m of B_m, $m = -j, -j+1, \ldots, j$, such that $x_k = x$ and

$$D(u)x_m = \sum_{n=-j}^{j} D^j_{nm}(u)x_n \tag{2-25}$$

Lemma 2.7 *Every subspace M^j is invariant with respect to the representation $u \to D(u)$. If the space M^j is not empty then the representation $u \to D(u)$, considered only on the space M^j, is a multiple of the irreducible representation $u \to D^{(j)}(u)$ of weight j.*

In fact, by Lemma 2.5, the space M^j is the direct sum of the subspaces M^j_m. Now let x belong to the space M^j_n for some n and substitute $x_m = E^j_{mn}x$ for $m = -j, -j+1, \ldots, j$. Using Eq. (2-17) one has $E^j_{mm}x_m = E^j_{mm}E^j_{mn}x = E^j_{mn}x = x_m$. Hence x_m belongs to the space M^j_m. In particular $x_n = E^j_{nn}x = x$ since x belongs to the space M^j_n. Applying now Eq. (2-16a) we obtain

$$D(u)x_m = D(u)E^j_{mn}x = \sum_{m'=-j}^{j} D^j_{m'm}(u)E^j_{m'n}x = \sum_{m'=-j}^{j} D^j_{m'm}x_{m'}$$

Hence the space M^j is invariant with respect to the operator $D(u)$ and the operator $D(u)$ satisfies Eq. (2-25). Hence the representation $u \to D(u)$, when considered as a representation in the space M^j, is a multiple of the representation $u \to D^{(j)}(u)$.

Definition A bounded linear operator A in a Banach space B is called the direct sum of the bounded linear operators A_m in the closed subspaces B_m, denoted by $A = A_1 \dot{+} A_2 \dot{+} \cdots$, if $B = B_1 \dot{+} B_2 \dot{+} \cdots$ and $A(x_1 + x_2 + \cdots + x_n) = A_1x_1 + A_2x_2 + \cdots + A_nx_n$ for any finite number of vectors x_n of B_n.

[9] A subspace M is called the direct sum of linearly independent subspaces $M_1, M_2, \ldots, M_n$ denoted $M = M_1 + \cdots + M_n$, if M is the aggregate of all sums of the form $x_1 + \cdots + x_n$, where x_k belongs to M_k. A closed subspace M will be called the closed direct sum of the finite or infinite (possibly nondenumerable) number of closed subspaces $M_1, M_2, \ldots$, denoted by $M = M_1 \dot{+} M_2 \dot{+} \cdots$, if M is the closure of the aggregate of all finite sums of the form $x_1 + x_2 + \cdots + x_n$, where x_k belongs to M_k, and $k = 1, 2, 3, \ldots$, if the sequence $x^{(m)} = x^{(m)}_1 + x^{(m)}_2 + \cdots + x^{(m)}_{n_m}$, with $x^{(m)}_k$ belongs to M_k, converges in norm to zero, while for a fixed k the sequence $x^{(m)}_k$, $m = 1, 2, 3, \ldots$, converges in norm, then the sequence $x^{(m)}_k$ converges in norm to zero.

Definition A representation $g \to D(g)$ of a group SU_2 in a Banach space B is called the direct sum of the representations $g \to D^{(m)}(g)$ in the subspaces B_m if $D(g) = D^{(1)}(g) \dotplus D^{(2)}(g) \dotplus D^{(3)}(g) \dotplus \cdots$ for all elements g of the group G.

Lemma 2.8 *Every representation $u \to D(u)$ of the group SU_2 is the direct sum of representations that are multiples of irreducible representations $u \to D^{(j)}(u)$.*

In fact this is a direct result of Lemmas 2.6 and 2.7.

Further Assumptions

We now come back to the representations $\Lambda \to D(\Lambda)$ of the group L in the Banach space B. Let us consider this representation as a representation $g \to D(g)$ of the group SU_2. By Lemma 2.8, the space B is the closed direct sum of the subspaces M^j, invariant relative to the operators $D(g)$, where g is an element of the group SU_2. Here the representation $g \to D(g)$ of the subgroup SU_2, when considered only on the space M^j, is a multiple of the irreducible representation of weight j of the subgroup SU_2.

Definition A given representation $\Lambda \to D(\Lambda)$ of the group L is said *not* to contain irreducible representation of weight j of the group SU_2 if the space M^j is empty, and is said to contain it k times if the representation $\Lambda \to D(\Lambda)$, for Λ of SU_2, considered only on M^j, is a k-th multiple of the irreducible representation of weight j of the group SU_2.

The further assumptions on the representations of the group L are that each irreducible representation of the group SU_2 is contained at most once in the representation of the group L, and that each subspace M^j is completely contained in the set X.[10]

Lemma 2.9 *If the infinitesimal operators A_i, B_i and A'_i, B'_i, $i = 1, 2, 3$, of the two representations $\Lambda \to D(\Lambda)$ and $\Lambda \to D'(\Lambda)$ of the group L in a space B coincide on a subspace X' of X which is dense in the space B and invariant relative to the operators A_i, B_i, A'_i, B'_i, then the two representations $\Lambda \to D(\Lambda)$ and $\Lambda \to D'(\Lambda)$ coincide on the whole space B.*

In fact from our previous assumptions it follows that the one-parameter groups of operators obtained from A_i, B_i, A'_i, B'_i will satisfy:

$$A_i(\psi) = A'_i(\psi), \qquad B_i(\psi) = B'_i(\psi); \qquad (i = 1, 2, 3)$$

on the whole space B. But every proper orthochronous Lorentz transformation Λ is the product of matrices $a_i(\psi)$ and $b_i(\psi)$ (see Problem 2.4). Hence the operators $D(\Lambda)$ and $D'(\Lambda)$ are the products of operators $A_i(\psi)$, $B_i(\psi)$ and $A'_i(\psi)$, $B'_i(\psi)$,

[10] In Chapter 6 we shall see that the formulae obtained are also valid without these assumptions.

respectively. Consequently, the operators A_i, B_i and A_i', B_i' coincide on the whole space B.

As a result of Lemma 2.9 it follows that a representation $\Lambda \to D(\Lambda)$ of the group L is completely determined by its infinitesimal operators A_i, B_i, $i = 1, 2, 3$.

Lemma 2.10 *If a nonzero subspace X' of X is invariant relative to all the infinitesimal operators A_i, B_i, $i = 1, 2, 3$, of an irreducible representation $\Lambda \to D(\Lambda)$ of the group L, then X' is dense in the space B.*

The proof of this lemma is left to the reader (Problem 2.7).

Commutation Relations

The determination of the irreducible representations of the group L is based on the fact that the basic infinitesimal operators of a representation satisfy the same commutation relations that exist among the infinitesimal matrices a_i and b_i given by Eqs. (2-9). Hence one has:

$$[A_i , A_j] = \varepsilon_{ijk} A_k$$
$$[B_i , B_j] = -\varepsilon_{ijk} A_k \tag{2-26}$$
$$[A_i , B_j] = \varepsilon_{ijk} B_k$$

Defining now the operators

$$L_{\mp} = iA_1 \pm A_2 , \qquad L_3 = iA_3$$
$$K_{\mp} = iB_1 \pm B_2 , \qquad K_3 = iB_3 \tag{2-27}$$

one finds for the commutation relations (2-26) the following:

$$[L_{\mp} , L_3] = \pm L_{\mp} , \qquad [L_+ , L_-] = 2L_3$$
$$[K_{\mp} , K_3] = \mp L_{\mp} , \qquad [K_+ , K_-] = -2L_3$$
$$[L_{\pm} , K_{\pm}] = [L_3 , K_3] = 0, \qquad [L_{\pm} , K_{\mp}] = \pm 2K_3 \tag{2-28}$$
$$[L_{\pm} , K_3] = \mp K_{\pm} , \qquad [K_{\pm} , L_3] = \mp K_{\pm}$$

Hence the problem reduces to the determination of the operators L and K satisfying the conditions stated in Eq. (2-28).

2-3 REPRESENTATIONS OF THE GROUP L

Canonical Basis

A given representation $\Lambda \to D(\Lambda)$ of the proper, orthochronous, homogeneous, Lorentz group L is obviously also a representation of the rotation group O_3. Clearly, if a given representation of the group L is irreducible it need not be irreducible when considered as a representation of the group O_3. In fact, any infinite-dimensional representation of the group L, when regarded as a representation of the group O_3, is highly reducible; it is equivalent to a direct sum of an infinite number of irreducible representations. The space B of any irreducible representation of the group L is, therefore, a closed direct sum of subspaces M^j, where M^j is the $(2j+1)$-dimensional space in which the irreducible representation of weight j of the group O_3 is realized.

Following the standard convention, one chooses the $2j + 1$ normalized eigenvectors of the operator L_3 as the canonical basis for the subspace M^j. Let these base vectors be denoted as f^j_m, where $m = -j, -j+1, \ldots, j$, where the superscript j indicates the subspace to which the vector f^j_m belongs,[11] and the subscript m is the eigenvalue of the operators L_3. A detailed investigation of the commutation relations (2-28) in terms of the canonical basis f^j_m then leads to the following conclusions:[12]

(1) Each irreducible representation of the group L is characterized by a pair of numbers (j_0, c), where j_0 is a non-negative integral or half-integral number, and c is a complex number;

(2) The space $B(j_0, c)$ of any given irreducible infinite-dimensional representation of the group L is characterized by the non-negative integral or half-integral j_0 such that $B(j_0, c) = M^{j_0} \dotplus M^{j_0+1} \dotplus M^{j_0+2} \dotplus \cdots$. The whole space $B(j_0, c)$ is spanned, therefore, by the set of base-vectors f^j_m, where $j = j_0, j_0+1, j_0+2, \ldots$, and $m = -j, -j+1, -j+2, \ldots, j$. If the given irreducible representation is finite-dimensional, then the direct sum of the subspaces M's terminates after a finite number of terms.

(3) A given representation is finite-dimensional if and only if it satisfies the condition $c^2 = (j_0 + n)^2$ for some natural number n.

(4) The irreducible representation corresponding to a given pair (j_0, c) is, with a suitable choice of a basis f^j_m in the space of representation, given by the formulae:[13]

$$\begin{aligned}
L_\pm f^j_m &= [(j \pm m + 1)(j \mp m)]^{1/2} f^j_{m\pm1} \\
L_3 f^j_m &= m f^j_m \\
K_\pm f^j_m &= \pm[(j \mp m)(j \mp m - 1)]^{1/2} C_j f^{j-1}_{m\pm1} \\
&\quad - [(j \mp m)(j \pm m + 1)]^{1/2} A_j f^j_{m\pm1} \\
&\quad \pm [(j \pm m + 1)(j \pm m + 2)]^{1/2} C_{j+1} f^{j+1}_{m\pm1} \\
K_3 f^j_m &= [(j-m)(j+m)]^{1/2} C_j f^{j-1}_m - m A_j f^j_m \\
&\quad - [(j+m+1)(j-m+1)]^{1/2} C_{j+1} f^{j+1}_m
\end{aligned} \tag{2-29}$$

[11] The superscript in f^j_m specifies the subspace uniquely since each irreducible representation of the group O_3 is contained at most once in any given irreducible representation of the group L.

[12] See, for example, M. Carmeli and S. Malin, *Fortschr. Physik.* **21**, (1973).

[13] Equations (2-29), for the unitary representations case, and under certain assumptions, were first obtained by Gelfand (See M. A. Naimark, *Linear Representations of the Lorentz Group*, Pergamon Press, New York, 1964, p. 117); they were later on rederived by Harish-Chandra, *Proc. Roy. Soc.* (A) **189**, 372, (1947) and *Phys. Rev.* **71**, 793 (1947), and by I. M. Gelfand and A. M. Yaglom, *Zh. Eksp. Teor. Fiz.* **18**, 703 (1948); English translation in: Technical Translation TT-345, National Research Council of Canada, Ottawa, 1953.

In Eqs. (2-29)

$$A_j = icj_0\left(\frac{1}{j} - \frac{1}{j+1}\right)$$
$$C_j = \frac{i}{j}\left[\frac{(j^2 - j_0^2)(j^2 - c^2)}{4j^2 - 1}\right]^{1/2} \tag{2-30}$$

(5) To each pair of numbers (j_0, c), where j_0 is a nonnegative integral or half-integral number and c is a complex number, there corresponds a representation $\Lambda \to D(\Lambda)$ of the group L whose infinitesimal operators are given by Eqs. (2-29) and (2-30). Since only the products of the numbers j_0 and c and the squares of these numbers appear in Eqs. (2-30) it is clear that the pairs (j_0, c) and $(-j_0, -c)$ characterize the same representation.

Unitarity Conditions

If the representation $\Lambda \to D(\Lambda)$ of the group L is unitary[14] then Eqs. (2-29) satisfy certain conditions that are summarized below.

Let A be an infinitesimal operator of a unitary representation $\Lambda \to D(\Lambda)$ of the group L. Then $A(\psi) = D(a(\psi))$ is a unitary operator and therefore its adjoint[15] operator $[(A(\psi)]^\dagger = [A(\psi)]^{-1} = A(-\psi)$. Consequently, we have $(A(\psi)f, g) = (f, A(-\psi)g)$. Differentiating both sides of this equation with respect to the parameter ψ we obtain, for $\psi = 0$,

$$(Af, g) = -(f, Ag) \tag{2-31}$$

Using Eq. (2-31) one then finds that

$$\begin{aligned}(L_+ f, g) &= (f, L_- g)\\ (L_3 f, g) &= (f, L_3 g)\\ (K_+ f, g) &= (f, K_- g)\\ (K_3 f, g) &= (f, K_3 g)\end{aligned} \tag{2-32}$$

A systematic use of Eqs. (2-32) in Eqs. (2-29) then leads to the following:

If the irreducible representation $\Lambda \to D(\Lambda)$ of the group L is unitary, then the pair of numbers (j_0, c) characterizing the representation satisfies either: (a) c is purely imaginary and j_0 is an arbitrary non-negative integral or half-integral number, or (b) c is a real number in the interval $0 < |c| \leqslant 1$ and $j_0 = 0$. Representations characterized by $(0, c)$ and $(0, -c)$ are equivalent.

[14] A representation $\Lambda \to D(\Lambda)$ of a group G in a space R is called unitary if and only if R is a Hilbert space and $D(\Lambda)$ is a unitary operator for all Λ of G. This implies that $(D(\Lambda)x, D(\Lambda)y) = (x, y)$ for all Λ of G and all x and y of R, where (x, y) denotes the scalar product in the space R. For further details see Appendix C.

[15] An operator B is called an adjoint to the operator A if it satisfies $(Ax, y) = (x, By)$ for all x and y of the space R.

The representations corresponding to the case (a) are called the *principal series of representations*, and those corresponding to the case (b) are called the *complementary series of representations*. Both series of these infinite-dimensional representations will be discussed in Chapters 4 and 5. In the next chapter, however, we will discover the finite-dimensional representations of the group L.

PROBLEMS

2.1 Show that the group of translations is an invariant subgroup of the Poincaré group. Show also that if T denotes the group of translations, P the Poincaré group, and L the homogeneous Lorentz group, then P is the semidirect product of L and T, namely $P = L \times T$.

2.2 Show that a Lorentz transformation satisfies the conditions $\Lambda^0{}_0 \geqslant 1$ if and only if it transforms every positive timelike vector into another positive timelike vector.

2.3 Prove Equations (2-6b).

2.4 Show that every proper, orthochronous, Lorentz matrix Λ can be represented in the form $\Lambda = gb_1(\psi)k$, where g and k are three-dimensional rotations, and $b_1(\psi)$ is given by Eq. (2-6b). As a result, show that Λ can be represented in the form of a product of transformations of the form $a_1(\psi)$, $a_3(\psi)$, and $b_1(\psi)$.

2.5 Prove Lemma 2.5

2.6 Prove Lemma 2.6

2.7 Prove Lemma 2.10

2.8 Prove Equations (2-26). [See M. A. Naimark, *Linear Representations of the Lorentz Group*, Pergamon Press, New York, 1964.]

2.9 Prove Equations (2-29) and (2-30).

2.10 Prove Equations (2-32).

CHAPTER

THREE

SPINOR REPRESENTATION OF THE LORENTZ GROUP

After giving the general aspects of representations of the proper, orthochronous, homogeneous Lorentz group in Chapter 2, we now give the simplest realization of these representations; this is the spinor representation of the Lorentz group. The spinor representation includes all finite-dimensional representations and is non-unitary. The discussion is started by introducing the group of all 2×2 complex matrices with determinant unity, the group $SL(2, C)$, and establishing a homomorphism between the groups $SL(2, C)$ and the proper, orthochronous, Lorentz group. Subgroups of the group $SL(2, C)$ are then discussed, and the connection with the Lobachevskian motion is pointed out. Subsequently, the spinor representation of the group $SL(2, C)$ is introduced. A realization of the representation is carried out in spaces of polynomials, and the important concept of two-component spinor is introduced. The spinor representation is then realized by means of the group SU_2. Matrix elements of the spinor representations operators are found and their relation to those of irreducible representations of the group SU_2 is pointed out. Finally, the infinitesimal operators of the spinor representation are found, and irreducibility and other properties of these representations are established.

3-1 THE GROUP $SL(2, C)$ AND THE LORENTZ GROUP

The Group $SL(2, C)$

In what follows we establish the fact that elements of the proper, orthochronous, homogeneous Lorentz group L discussed in the last chapter can be described by means of elements of $SL(2, C)$, the group of all 2×2 complex matrices

$$g = \begin{pmatrix} a & b \\ c & d \end{pmatrix} \tag{3-1}$$

with

$$\det g = ad - bc = 1 \tag{3-2}$$

In the natural topology of matrices the group $SL(2, C)$ is simply connected. The relation between the two groups can be established as follows.

One associates with each four vector x^μ a Hermitian matrix

$$Q = \begin{pmatrix} x_0 + x_3 & x_1 + ix_2 \\ x_1 - ix_2 & x_0 - x_3 \end{pmatrix} \tag{3-3}$$

In this way one defines a one-to-one linear correspondence between all four-vectors and all 2×2 Hermitian matrices. Equation (3-3) can also be written as

$$Q = x_\alpha \sigma^\alpha \tag{3-4}$$

where σ^k, $k = 1, 2, 3$, are the three Pauli matrices given by Eqs. (1-5) and σ^0 is the 2×2 unit matrix:

$$\sigma^0 = \begin{pmatrix} 1 & 0 \\ 0 & 1 \end{pmatrix}, \quad \sigma^1 = \begin{pmatrix} 0 & 1 \\ 1 & 0 \end{pmatrix}, \quad \sigma^2 = \begin{pmatrix} 0 & i \\ -i & 0 \end{pmatrix}, \quad \sigma^3 = \begin{pmatrix} 1 & 0 \\ 0 & -1 \end{pmatrix} \tag{3-5}$$

It is often also very convenient to parametrize the elements g of the group $SL(2, C)$ by

$$g = g_\mu \sigma^\mu$$

where g_0, g_k, $k = 1, 2, 3$, are complex numbers.

Corresponding to every element g of the group $SL(2, C)$ consider the following transformation in the space of the Hermitian matrices Q:

$$Q' = gQg^\dagger \tag{3-6}$$

where $g^\dagger$ is the Hermitian conjugate of g, and $Q' = x'_\alpha \sigma^\alpha$. The corresponding operation in the Minkowskian space of four-vectors is a linear transformation

$$x'^\alpha = \Lambda^\alpha{}_\beta(g) x^\beta \tag{3-7a}$$

or, in matrix notation,

$$x' = \Lambda(g)x \tag{3-7b}$$

where the transformation matrix Λ can be expressed in terms of the matrix g of the group $SL(2, C)$. The transformation (3-7) preserves the scalar product since

$$(x')^2 = \det Q' = \det Q = (x)^2 \tag{3-8}$$

Homomorphism of the Group $SL(2, C)$ on the Group L

The matrix elements $\Lambda^{\alpha}{}_{\beta}$ can be expressed in terms of the corresponding matrix g of the group SL(2, C). Using the properties of the Pauli spin matrices, and using Eq. (3-6) and Eqs. (3-7), one has

$$\begin{aligned} x'^{\alpha} &= \delta^{\alpha}_{\beta}\, x'^{\beta} \\ &= \frac{1}{2}\,\mathrm{Tr}(\sigma^{\alpha}\sigma^{\beta})x'_{\beta} \\ &= \frac{1}{2}\,\mathrm{Tr}(\sigma^{\alpha}Q') \\ &= \frac{1}{2}\,\mathrm{Tr}(\sigma^{\alpha}gQg^{\dagger}) \\ &= \frac{1}{2}\,\mathrm{Tr}(\sigma^{\alpha}g\sigma^{\beta}g^{\dagger})x_{\beta} \end{aligned}$$

Comparing this result with Eq. (3-7a) one obtains

$$\Lambda^{\alpha}{}_{\beta} = \frac{1}{2}\,\mathrm{Tr}(\sigma^{\alpha}g\sigma^{\beta}g^{\dagger}) \tag{3-9a}$$

where $g^{\dagger}$ is the Hermitian conjugate of the matrix g, and Tr stands for trace.

The explicit expression of the transformation $\Lambda(g)$ in terms of the parameters g_0 and g_k of the matrix g is as follows (Problem 3.1):

$$\begin{aligned} \Lambda^{0}{}_{0} &= |g_0|^2 + \sum_{k=1}^{3} |g_k|^2 \\ \Lambda^{k}{}_{0} &= g_0\bar{g}_k + \bar{g}_0 g_k - i\varepsilon^{klm} g_l \bar{g}_m \\ \Lambda^{0}{}_{k} &= g_0\bar{g}_k + \bar{g}_0 g_k + i\varepsilon^{klm} g_l \bar{g}_m \\ \Lambda^{l}{}_{k} &= \delta^{l}_{k}\left(|g_0|^2 - \sum_{s=1}^{3}|g_s|^2\right) + g_k\bar{g}_l + \bar{g}_k g_l - i\varepsilon^{klm}(\bar{g}_0 g_m - g_0\bar{g}_m) \end{aligned} \tag{3-9b}$$

where the ε symbols are fixed by $\varepsilon^{123} = \varepsilon^{0123} = +1$. In particular, one notices the useful relation

$$\mathrm{Tr}\,\Lambda(g) = |\mathrm{Tr}\, g|^2 = 4|g_0|^2$$

One also notices that because the group $SL(2, C)$ is connected, and the mapping into the homogeneous Lorentz group is a continuous homomorphism, the image of the group $SL(2, C)$ must be a subgroup of the proper orthochronous, Lorentz group L.

Equations (3-9) show that to an arbitrary matrix g of $SL(2, C)$ there corresponds a 4×4 matrix Λ. We now show that the matrix Λ belongs to the proper, orthochronous, Lorentz group L.

First, from Eq. (3-8), one sees the quadratic form $x_0^2 - x_1^2 - x_2^2 - x_3^2$ is invariant under the transformation Λ, and therefore the matrix Λ is an element of the homogeneous Lorentz group. As a consequence, $\det \Lambda = \pm 1$. But for the special case for which g is the 2×2 unit matrix, the corresponding Λ is the identity transformation, and hence $\det \Lambda = 1$. Since $\det \Lambda$ is a continuous function of the four variables a, b, c, d of the matrix g of the group $SL(2, C)$, and since the domain of variation of these four variables is simply connected, a discontinuous jump from $\det \Lambda = +1$ to $\det \Lambda = -1$ is excluded. Consequently, $\det \Lambda = +1$ for all values of a, b, c, d, subject to the restriction (3-2). Hence Λ belongs to the proper Lorentz group. Finally, from Eqs. (3-9) one sees that $\Lambda^0{}_0$ cannot be negative. Accordingly, Λ is orthochronous. Consequently, Λ is an element of the proper, orthochronous, Lorentz group L.

Suppose now that an element Λ of the group L is given. Let us try to invert the relations (3-9). If $\mathrm{Tr}\,\Lambda \neq 0$ we obtain[1] (Problem 3.2):

$$\begin{aligned} g &= g_0 \sigma^0 + \sum_{k=1}^{3} g_k \sigma^k \\ &= D^{-1}\left\{ \mathrm{Tr}\,\Lambda \sigma^0 + \sum_{k=1}^{3} (\Lambda^k{}_0 + \Lambda^0{}_k - i\varepsilon_\sigma{}^{0k\rho} \Lambda^\sigma{}_\rho)\sigma^k \right\} \end{aligned} \tag{3-10}$$

where

$$D^2 = 4 - \mathrm{Tr}\,\Lambda^2 + (\mathrm{Tr}\,\Lambda)^2 - i\varepsilon^{\mu\lambda}_{\rho\sigma} \Lambda^\sigma{}_\lambda \Lambda^\rho{}_\mu \tag{3-11}$$

The sign of the denominator D is undetermined. Since the smallest subgroup of the group L that contains all elements with $\mathrm{Tr}\,\Lambda \neq 0$ is L itself, the image of the group $SL(2, C)$ is the whole of the group L.

It is possible to find the elements g of $SL(2, C)$ which go into L in the case $\mathrm{Tr}\,\Lambda = 0$ also. If

$$\sum_{k=1}^{3} (\Lambda^k{}_0)^2 \neq 0$$

the matrix Λ then describes a rotation with an angle π, and one has

$$g_0 = 0, \qquad g = \sum_{k=1}^{3} g_k \sigma^k, \qquad g^2 = -e$$

$$\Lambda(g)\Lambda(g) = \Lambda(g^2) = \Lambda(-e) = I$$

where e is the 2×2 unit matrix. The 3×3 matrix $M^l{}_k = \delta^l{}_k \Lambda^0{}_0 + \Lambda^l{}_k$ is symmetric and possesses the three eigenvalues 0, $\Lambda^0{}_0 - 1$, and $\Lambda^0{}_0 + 1$. The eigenvalue 0 belongs to the normalized eigenvector $\mathbf{v}_0$ defined by

$$(v_0)^k = \left[\sum_{l=1}^{3} (\Lambda^l{}_0)^2 \right]^{-1/2} \Lambda^k{}_0$$

[1] W. Rühl, *The Lorentz Group and Harmonic Analysis*, Benjamin, New York, 1970.

If we denote the normalized real eigenvector, corresponding to the eigenvalue $\Lambda^0{}_0 + 1$, by $\mathbf{v}_1$, we can then express the components g_k in terms of the vectors $\mathbf{v}_1$ and $\mathbf{v}_1 \times \mathbf{v}_0$ as follows:

$$g_k = \pm\{[\tfrac{1}{2}(\Lambda^0{}_0 - 1)]^{1/2}(\mathbf{v}_1 \times \mathbf{v}_0)^k + i[\tfrac{1}{2}(\Lambda^0{}_0 + 1)]^{1/2}v_1^k\} \tag{3-12}$$

Again the matrix g is determined only up to a sign. The remaining case, for which $\mathrm{Tr}\,\Lambda = \Lambda^k{}_0 = 0$, $k = 1, 2, 3$, is contained in Eq. (3-12) as the limit $\Lambda^0{}_0 = 1$. One obtains (Problem 3.3):

$$g_k = \pm i v_1^k \tag{3-13}$$

In this fashion one reaches the conclusion that there exists a two-to-one mapping between all the elements of the group $SL(2, C)$ and all the elements Λ of the proper, orthochronous, Lorentz group L such that to each element Λ of the group L there correspond two elements $\pm g$ of the group $SL(2, C)$, and to each element g of the group $SL(2, C)$ there corresponds an element Λ of the group L. The mapping conserves the group multiplication and constitutes, therefore, a homomorphism of the group $SL(2, C)$ on the group L. As a result of this, the description of the representations of the group L is equivalent to that of the group $SL(2, C)$; a representation $g \to D(g)$ of L is single- or double-valued according to whether or not $D(g)$ is equal to $D(-g)$ or not.

Kernel of Homomorphism

The sign ambiguity of $g = g(\Lambda)$ means, in particular, that the unit matrix I of the group L is the image of both central[2] elements $e_\pm$ of the group $SL(2, C)$, where $e_\pm = \pm e$, and e is the 2×2 unit matrix. Hence we have established an isomorphism between the proper, orthochronous, Lorentz group L and the group $SL(2, C)/Z_2$, where Z_2 denotes the centre of the group $SL(2, C)$ consisting of the elements $e_\pm$.

Subgroups of the Group $SL(2, C)$

The group $SL(2, C)$ possesses some important subgroups, some of which play crucial roles in further investigations. These subgroups correspond to subgroups of the proper, orthochronous, homogeneous Lorentz group L as well. Since the group $SL(2, C)$ is more natural to handle than the group L, one prefers to deal with the group $SL(2, C) \times T_4$, where T_4 is the translational group. The group $SL(2, C) \times T_4$ is sometimes called the *inhomogeneous $SL(2, C)$ group*.

[2] Group elements are called central if they commute with all group elements. Central elements form the center of the group. See Appendix A for further details.

The group SU_2 has already been mentioned as a subgroup of $SL(2, C)$. It consists, of course, of those elements u satisfying $u^\dagger = u^{-1}$. A possible parametrization of the group SU_2 is as follows:

$$u = u_\alpha \sigma^\alpha \tag{3-14}$$

with the condition

$$u_0^2 + \sum_{k=1}^{3} u_k^2 = 1 \tag{3-15}$$

Here u_0 and u_k, $k = 1, 2, 3$, are real numbers.

Another subgroup of $SL(2, C)$ is the group $SU(1, 1)$. It consists of those elements v of $SL(2, C)$ satisfying the condition $v^\dagger \sigma^3 v = \sigma^3$. A possible parametrization is as follows:

$$v = v_0 \sigma^0 + v_1 \sigma^1 + v_2 \sigma^2 + i v_3 \sigma^3 \tag{3-16}$$

with the condition

$$v_0^2 - v_1^2 - v_2^2 + v_3^2 = 1 \tag{3-17}$$

Here the numbers v_0 and v_k, $k = 1, 2, 3$, are real.

A third subgroup of $SL(2, C)$ is the group $SL(2, R)$. It consists of elements a of the group $SL(2, C)$ satisfying $a^\dagger \sigma^2 a = \sigma^2$. They can be presented as

$$a = a_0 \sigma^0 + a_1 \sigma^1 + i a_2 \sigma^2 + a_3 \sigma^3 \tag{3-18}$$

Here the numbers a_0 and a_k, $k = 1, 2, 3$, are real and satisfy the condition

$$a_0^2 - a_1^2 + a_2^2 - a_3^2 = 1 \tag{3-19}$$

The matrix a is a real 2×2 matrix.

By a rotation $\exp(i\pi\sigma^1/4) = 2^{-1/2}(\sigma^0 + i\sigma^1)$ in the $x_2 - x_3$ plane, we can map the group $SU(1, 1)$ on the group $SL(2, R)$:

$$a = \exp(-i\pi\sigma^1/4) v \exp(i\pi\sigma^1/4) \tag{3-20}$$

from which one infers that $a_0 = v_0$, $a_1 = v_1$, $a_2 = v_3$, and $a_3 = -v_2$. This one-to-one mapping of the two groups $SU(1, 1)$ and $SL(2, R)$ onto each other is sometimes called the *standard isomorphism.*

Finally, the group of triangular matrices

$$(\phi, \mu) = \begin{pmatrix} e^{-i\phi/2} & 0 \\ \mu e^{-i\phi/2} & e^{i\phi/2} \end{pmatrix} \tag{3-21}$$

where μ is complex and $0 \leqslant \phi \leqslant 4\pi$, with the group multiplication law $(\phi_1, \mu_1) \times (\phi_2, \mu_2) = (\phi_1 + \phi_2(\pm 4\pi), \mu_1 + e^{i\phi_1}\mu_2)$, is isomorphic to the group of Euclidean motions on the Riemannian plane of functions $z^{1/2}$. The corresponding subgroup of L is isomorphic to the group of motions in the complex z-plane itself. The notation of this subgroup is $U(1) \times T_2$ if one means the subgroup of L, and $U(1)' \times T_2$ if one means the subgroup of $SL(2, C)$.

Connection with Lobachevskian Motion

We have seen that each complex, unimodular, two-dimensional matrix g induces a Lorentz transformation in the Minkowskian space according to $Q' = gQg^\dagger$, where Q is given by Eq. (3-3). These Lorentz transformations map the surfaces

$$x_0^2 - x_1^2 - x_2^2 - x_3^2 = c \tag{3-22}$$

into themselves, for they preserve the corresponding quadratic form. There are three types of such surfaces. These are either sheet of a two-sheeted hyperboloid when $c > 0$, a single-sheeted hyperboloid when $c < 0$, and either the positive or negative cone when $c = 0$.[3]

The Lorentz transformations induce transformations that are called *motions* of these surfaces. In this way to each complex unimodular two-dimensional matrix g there corresponds a motion on each of the surfaces above. One can show that a given motion corresponds to two matrices g_1 and g_2 if and only if $g_1 = \pm g_2$.

The upper sheet of a two-sheeted hyperboloid together with the motions defined in this way is one model of *Lobachevskian space*. This means that the *group of complex two-dimensional unimodular matrices is locally isomorphic to the group of Lobachevskian motions.* In addition to Lobachevskian space, there exist two related spaces with groups of motions locally isomorphic to the same group of matrices. Models of these spaces are the single-sheeted hyperboloid and the positive cone.

We conclude this brief discussion on the Lobachevskian space by pointing out that the group of motions on each of these surfaces is *transitive*, that is every point of the space can be transformed by some motion to any other point. Let us prove this assertion for the upper sheet of the two-sheeted hyperboloid $x_0^2 - x_1^2 - x_2^2 - x_3^2 = 1$, as the proof for the other surfaces is similar. Using Eq. (3-3), then the points on our surface correspond to positive definite *unimodular* Hermitian matrices. Since every such matrix can be written in the form $Q = gg^\dagger = geg^\dagger$, where g is a complex unimodular matrix and e is the 2×2 unit matrix. This proves that there exists a motion transforming the fixed unit matrix into Q.

3-2 SPINOR REPRESENTATION OF THE GROUP $SL(2, C)$

Spinor Representation in Space of Polynomials

We now construct the spinor representation of the group $SL(2, C)$. This representation contains all the *irreducible, finite-dimensional, nonunitary representations* of the group $SL(2, C)$.[4]

[3] If, instead of considering points x in Minkowskian space, we deal with Hermitian matrices Q, the surfaces would be the following three types of manifolds in the space of Hermitian matrices: all positive definite (or negative definite) Hermitian matrices with fixed determinant $c > 0$; all Hermitian matrices with fixed determinant $c < 0$; and all Hermitian matrices $Q \geqslant 0$ (or $Q \leqslant 0$), that is matrices Q whose corresponding Hermitian form takes on nonnegative (or nonpositive) values, with determinant zero. See, for example, I. M. Gelfand, M. I. Graev, and N. Ya. Vilenkin, *Integral Geometry and Representation Theory*, Academic Press, New York, 1966.

[4] M. Carmeli and S. Malin, *Fortschr. der Phys.* **21**, 397 (1973).

We denote by P_{mn} the aggregate of all polynomials $p(z, \bar{z})$ in the variables z and its complex conjugate $\bar{z}$, of degree not exceeding m in the variable z and n in $\bar{z}$, where m and n are fixed non-negative integers determining the representation. The space P_{mn} is a linear vector space, where the operation of addition, and multiplication by a number, are defined in the usual way for polynomials.

An element of the group $SL(2, C)$ will be denoted, according to Eq. (3-1), by

$$g = \begin{pmatrix} a & b \\ c & d \end{pmatrix}$$

where a, b, c, and d are complex numbers satisfying the condition $ad - bc = 1$. We define now the operator $D(g)$ in the space P_{mn} by

$$D(g)p(z, \bar{z}) = (bz + d)^m(\bar{b}\bar{z} + \bar{d})^n p\left(\frac{az + c}{bz + d}, \frac{\bar{a}\bar{z} + \bar{c}}{\bar{b}\bar{z} + \bar{d}}\right) \tag{3-23}$$

The correspondence $g \to D(g)$ is a linear representation of the group $SL(2, C)$ as can be verified. This is the *spinor representation* of the group $SL(2, C)$ of dimension $(m + 1)(n + 1)$.

Two-Component Spinors

In order to relate the spinor representation introduced above to the *two-component spinors,*[5] one realizes the representation in a somewhat different way as described below.

One considers all systems of numbers $\phi_{A_1 \cdots A_m \dot{X}_1 \cdots \dot{X}_n}$ symmetrical in both the indices $A_1 \cdots A_m$ and in $\dot{X}_1 \cdots \dot{X}_n$ taking the values 0, 1 and $\dot{0}$, $\dot{1}$. The set of all such systems of numbers provides a linear space, denoted by S_{mn}, of dimension $(m + 1)(n + 1)$.

A one-to-one linear mapping between the spaces P_{mn} and S_{mn} can easily be established. To each system of numbers $\phi_{A_1 \cdots A_m \dot{X}_1 \cdots \dot{X}_n}$ of the space S_{mn} there corresponds the polynomial

$$p(z, \bar{z}) = \sum_{\substack{A_1, \ldots, A_m \\ \dot{X}_1, \ldots, \dot{X}_n}} \phi_{A_1 \cdots A_m \dot{X}_1 \cdots \dot{X}_n} z^{A_1 + \cdots + A_m} \bar{z}^{\dot{X}_1 + \cdots + \dot{X}_n} \tag{3-24}$$

of degree not exceeding m in the variable z and n in $\bar{z}$. Therefore $p(z, \bar{z})$ belongs to the space P_{mn}. On the other hand every polynomial

$$p(z, \bar{z}) = \sum_{r, s} p_{rs} z^r \bar{z}^s \tag{3-25}$$

in the space P_{mn} can be written in the form (3-24) if one relates the ϕ and p by means of

$$\binom{m}{r}\binom{n}{s}\phi_{A_1 \ldots A_m \dot{X}_1 \ldots \dot{X}_n} = p_{rs} \tag{3-26}$$

with $A_1 + \cdots + A_m = r$ and $\dot{X}_1 + \cdots + \dot{X}_n = s$.

[5] Throughout this chapter the term spinor is used to mean symmetrical spinor.

A second form of the spinor representation is then obtained if one applies the polynomials (3-24) in Eq. (3-23). One obtains

$$D(g)p(z, \bar{z}) = \sum_{\substack{A_1, \ldots, A_m \\ \dot{X}_1, \ldots, \dot{X}_n}} \phi'_{A_1 \cdots A_m \dot{X}_1 \cdots \dot{X}_n} z^{A_1 + \cdots + A_m} \bar{z}^{\dot{X}_1 + \cdots + \dot{X}_n} \tag{3-27}$$

where we have used the notation

$$\phi'_{A_1 \cdots A_m \dot{X}_1 \cdots \dot{X}_n} = \sum_{\substack{B_1, \ldots, B_m \\ \dot{Y}_1, \ldots, \dot{Y}_n}} g_{A_1 B_1} \cdots g_{A_m B_m} \bar{g}_{\dot{X}_1 \dot{Y}_1} \cdots \bar{g}_{\dot{X}_n \dot{Y}_n} \phi_{B_1 \cdots B_m \dot{Y}_1 \cdots \dot{Y}_n} \tag{3-28}$$

and where use has been made of the notation $g_{11} = a$, $g_{10} = b$, $g_{01} = c$, and $g_{00} = d$.

The quantity $\phi_{A_1 \cdots A_m \dot{X}_1 \cdots \dot{X}_n}$ is called a two-component spinor, symmetric in its m undotted indices and in its n dotted indices. Equation (3-28) expresses the law of transformation of symmetrical spinors under the transformation g of the group $SL(2, C)$. With a change in notation, one obtains the same transformation law (3-28) but with $g_{00} = a$, $g_{01} = b$, $g_{10} = c$, and $g_{11} = d$ since the matrix

$$\begin{pmatrix} d & c \\ b & a \end{pmatrix}$$

occuring in the transformation (3-28) is also an element of the group $SL(2, C)$. Hence we will use the same transformation law (3-28) but with $g_{00} = a$, $g_{01} = b$, $g_{10} = c$, and $g_{11} = d$, where a, b, c, and d are the elements of the matrix g of the group $SL(2, C)$.

Example 3.1 The simplest example of a spinor is a one index spinor. ϕ_A, with $A = 0, 1$. Its law of transformation is, according to Eq. (3-28), given by

$$\phi'_A = \sum_{B=0}^{1} g_{AB} \phi_B \tag{3-29}$$

with $A = 0, 1$.

Example 3.2 Taking the complex conjugate of Eq. (3-29), one obtains

$$\overline{\phi'_A} = \sum_{B=0}^{1} \bar{g}_{AB} \overline{\phi_B}$$

The complex conjugate of a spinor ϕ_A is sometimes denoted by $\bar{\phi}_{\dot{A}}$. Hence the above equation will usually be written as

$$\bar{\phi}'_{\dot{A}} = \sum_{\dot{B}=\dot{0}}^{\dot{1}} \bar{g}_{\dot{A}\dot{B}} \bar{\phi}_{\dot{B}} \tag{3-30}$$

Example 3.3 A mixed spinor $\phi_{A\dot{B}}$ will transform as

$$\phi'_{A\dot{B}} = \sum g_{AC}\bar{g}_{\dot{B}\dot{D}}\phi_{C\dot{D}} \tag{3-31}$$

If one denotes by $D^{(1/2,\,1/2)}$ the representation corresponding to the transformation (3-31) of the spinor components $(\phi_{0\dot{0}}, \phi_{0\dot{1}}, \phi_{1\dot{0}}, \phi_{1\dot{1}})$, then the transformation matrix is given by

$$D^{(1/2,\,1/2)}(g) = \begin{pmatrix} a\bar{a} & a\bar{b} & b\bar{a} & b\bar{b} \\ a\bar{c} & a\bar{d} & b\bar{c} & b\bar{d} \\ c\bar{a} & c\bar{b} & d\bar{a} & d\bar{b} \\ c\bar{c} & c\bar{d} & d\bar{c} & d\bar{d} \end{pmatrix} \tag{3-32}$$

Example 3.4 If we use the space of polynomials P_{mn} for $m = n = 1$, then

$$p(z, \bar{z}) = p_{00} + p_{01}\bar{z} + p_{10}z + p_{11}z\bar{z} \tag{3-33}$$

The law of transformation of the coefficients p_{rs} is again given by means of the matrix $D^{(1/2,\,1/2)}(g)$ of Eq. (3-32).

Spinor Representation by means of the Group SU_2

We now establish a third form for the spinor representation using the group SU_2. Starting from Eq. (3-25) we denote the polynomial $p(z, \bar{z})$ by $p(z)$, and define the function

$$\alpha(g) = g_{22}^{m}\bar{g}_{22}^{n} \tag{3-34}$$

where the matrix g of the group $SL(2, C)$ is denoted as

$$g = \begin{pmatrix} g_{11} & g_{12} \\ g_{21} & g_{22} \end{pmatrix}$$

The spinor representation formula (3-23) can then be written in the form

$$D(g)p(z) = \alpha(zg)p(z(g)) \tag{3-35}$$

Here z denotes a complex variable as well as the matrix

$$z = \begin{pmatrix} 1 & 0 \\ z & 1 \end{pmatrix} \tag{3-36}$$

and the variable $z(g)$ in Eq. (3-35) is given by $z(g) = g'_{21}/g'_{22}$, where the matrix g' of $SL(2, C)$ is given by

$$g' = zg = \begin{pmatrix} g_{11} & g_{12} \\ g_{11}z + g_{21} & g_{12}z + g_{22} \end{pmatrix} \tag{3-37}$$

Therefore the new variable $z(g)$ in Eq. (3-35), using Eq. (3-37), is given by

$$z(g) = \frac{g'_{21}}{g'_{22}} = \frac{g_{11}z + g_{21}}{g_{12}z + g_{22}} \tag{3-38}$$

The new form of the spinor representation is obtained by expressing Eq. (3-35) in terms of matrix elements of the group SU_2. Let

$$u = \begin{pmatrix} u_{11} & u_{12} \\ u_{21} & u_{22} \end{pmatrix} \tag{3-39}$$

denote an element of the group SU_2. Let $\tilde{P}_{mn}$ denote the space of all polynomials $q(u)$ that are homogeneous in the variables u_{21} and u_{22} of degree m, and in $\bar{u}_{21}$ and $\bar{u}_{22}$ of degree n, satisfying the condition

$$q(\gamma u) = e^{i(m-n)\psi/2} q(u) \tag{3-40}$$

where the matrix $\gamma \in SU_2$ is given by

$$\gamma = \begin{pmatrix} e^{-i\psi/2} & 0 \\ 0 & e^{i\psi/2} \end{pmatrix} \tag{3-41}$$

In the following we will establish a mapping between the space $\tilde{P}_{mn}$ of polynomials $q(u)$ and the space P_{mn} of polynomials $p(z)$. To this end we proceed as follows.

We first introduce the set of all matrices k of the form

$$k = \begin{pmatrix} \lambda^{-1} & \mu \\ 0 & \lambda \end{pmatrix} \tag{3-42}$$

where λ and μ are complex numbers with $\lambda \neq 0$. It is easily seen that the aggregate of all the matrices k forms a subgroup of the group $SL(2, C)$. This subgroup is denoted by K. Let us, furthermore, notice that the set of all matrices z of the form given by Eq. (3-36) also provides a subgroup of the group $SL(2, C)$. This subgroup is denoted by Z.

Lemma 3.1 *Any element g of the group $SL(2, C)$, satisfying the condition $g_{22} \neq 0$, can be uniquely decomposed into the form*

$$g = kz, \qquad k \in K, \qquad z \in Z \tag{3-43}$$

In fact, suppose a matrix g of the group $SL(2, C)$ is given. Then Eq. (3-43) means

$$\begin{pmatrix} g_{11} & g_{12} \\ g_{21} & g_{22} \end{pmatrix} = \begin{pmatrix} \lambda^{-1} & \mu \\ 0 & \lambda \end{pmatrix} \begin{pmatrix} 1 & 0 \\ z & 1 \end{pmatrix}$$

A straightforward calculation then shows that we have the unique solution

$$\lambda = g_{22}, \qquad \mu = g_{12} \tag{3-44a}$$

$$z = g_{21}/g_{22} \tag{3-44b}$$

Consider now the set of all right cosets $\tilde{z}$ (see Appendix A) of the group $SL(2, C)$ with respect to the subgroup K. It follows from Lemma 3.1 that there exists a one-to-one correspondence between all these right cosets and all the complex numbers z of the matrices z; Eq. (3-44b) shows that all the elements of the group $SL(2, C)$ for which the ratio $g_{21}/g_{22} = z$ is the same, belong to the same right coset. We will prove, furthermore, that every right coset $\tilde{z}$ of the group $SL(2, C)$ with respect to the subgroup K contains elements of the group SU_2.

Lemma 3.2 *Every element g of the group $SL(2, C)$ can be decomposed into the form*

$$g = ku, \qquad k \in K, \qquad u \in SU_2 \tag{3-45}$$

In fact, since the general form of the element u of SU_2 is given by

$$u = \begin{pmatrix} \alpha & \beta \\ -\bar{\beta} & \bar{\alpha} \end{pmatrix}; \qquad \alpha\bar{\alpha} + \beta\bar{\beta} = 1 \tag{3-46}$$

Equation (3-45) has the form

$$\begin{pmatrix} g_{11} & g_{12} \\ g_{21} & g_{22} \end{pmatrix} = \begin{pmatrix} \lambda^{-1} & \mu \\ 0 & \lambda \end{pmatrix} \begin{pmatrix} \alpha & \beta \\ -\bar{\beta} & \bar{\alpha} \end{pmatrix} \tag{3-47}$$

Solving Eq. (3-47) for the variables λ, μ, α, and β one obtains, using the condition (3-46),

$$|\lambda|^2 = |g_{21}|^2 + |g_{22}|^2 \tag{3-48}$$

$$\alpha = \frac{\bar{g}_{22}}{\lambda} \tag{3-49}$$

$$\beta = -\frac{\bar{g}_{21}}{\lambda} \tag{3-50}$$

$$\mu = \begin{cases} \dfrac{g_{12} - \beta/\lambda}{\bar{\alpha}}; & \alpha \neq 0 \\[2ex] -\dfrac{g_{11}}{\bar{\beta}}; & \alpha = 0 \end{cases} \tag{3-51}$$

The decomposition of the matrix $g = ku$ of Eq. (3-45) is, therefore, not unique since the phase ω of λ is left undetermined by Eqs. (3-48) to (3-51). In fact, if $\gamma \in SU_2$ is an arbitrary matrix of the form

$$\gamma = \begin{pmatrix} e^{-i\omega} & 0 \\ 0 & e^{i\omega} \end{pmatrix} \tag{3-52}$$

with a real phase ω, and k belongs to the subgroup K, then $k\gamma$ also belongs to K. Hence

$$g = ku = (k\gamma)(\gamma^{-1}u) = k'u' \tag{3-53}$$

with $k' = k\gamma$ belongs to K, $u' = \gamma^{-1}u$ belongs to the group SU_2, and γ is an arbitrary matrix of the form (3-52). One easily finds that the transition from k and u to $k\gamma$ and $\gamma^{-1}u$ is equivalent to the transition from λ, α, and β to $e^{i\omega}\lambda$, $e^{i\omega}\alpha$, and $e^{i\omega}\beta$, respectively.

If now g is an element that belongs to the right coset $\tilde{z}$, then Eq. (3-45) shows that u too belongs to $\tilde{z}$. But if also u_1 belongs to $\tilde{z}$ then $u_1 = k'u$, where k' belongs to the subgroup K. Hence $g = ku = kk'^{-1}u_1$ and, using Eq. (3-53), one finds $u_1 = \gamma^{-1}u$. Consequently, every right coset $\tilde{z}$ contains unitary matrices of the group SU_2, and unitary matrices from the same right coset differ by a left factor γ. Finally, if z and u belong to the same left coset $\tilde{z}$, then for some k we have $z = ku$. Applying Eqs. (3-48) to (3-51) to the matrix

$$z = \begin{pmatrix} 1 & 0 \\ z & 1 \end{pmatrix}$$

one finds that

$$|\lambda|^2 = 1 + |z|^2, \qquad \alpha = \frac{1}{\bar{\lambda}}, \qquad \beta = -\frac{\bar{z}}{\bar{\lambda}} \tag{3-54}$$

The above considerations show that the set of all matrices u' that belong to the right coset $\tilde{z}$ forms a right coset Γu of the group SU_2 with respect to its subgroup Γ, where Γ is the aggregate of all matrices of the form (3-52). We can denote this right coset by $\tilde{u}$ and make the correspondence between every right coset $\tilde{z}$ and every right coset $\tilde{u}$, the one containing the matrix u. The transformation $\tilde{z} \to \tilde{z}(g)$ can then be considered as a transformation $\tilde{u} \to \tilde{u}(g)$ of the corresponding right cosets $\tilde{u}$.

The mapping between the space $\tilde{P}_{mn}$ of polynomials $q(u)$ and the space P_{mn} of polynomials $p(z)$ can now be established by putting

$$q(u) = \pi^{1/2}\alpha(u)p(z) \tag{3-55a}$$

where $\alpha(u) = u_{22}^m \bar{u}_{22}^n$ and the matrices u and

$$z = \begin{pmatrix} 1 & 0 \\ z & 1 \end{pmatrix}$$

belong to the same right coset $\tilde{z}$. Since by Eqs. (3-54) $z = -\bar{\beta}/\bar{\alpha} = u_{21}/u_{22}$, the mapping between the two spaces $\tilde{P}_{mn}$ and P_{mn} is easily established, and one obtains

$$q(u) = \pi^{1/2} \sum_{r,s} p_{rs} u_{21}^r u_{22}^{m-r} \bar{u}_{21}^s \bar{u}_{22}^{n-s} \tag{3-55b}$$

The representation formula (3-35) can now be considered as a representation in the space $\tilde{P}_{mn}$ by using the mapping (3-55). One obtains

$$D(g)q(u) = \frac{\alpha(ug)}{\alpha(u(g))} q(u(g)) \tag{3-56}$$

where $q(u)$ belongs to $\tilde{P}_{mn}$ and $u(g)$ is a matrix of SU_2 which belongs to the right coset $\tilde{z}(g) = \tilde{z}'$ where z' is given by Eq. (3-38). The matrix $u(g)$ can be explicitly obtained in terms of the matrices u of SU_2 and g of $SL(2, C)$. To see this we proceed as follows:[6]

Let the matrix $u(g)$ be denoted by u',

$$u' = \begin{pmatrix} \alpha' & \beta' \\ -\bar{\beta}' & \bar{\alpha}' \end{pmatrix}; \qquad |\alpha'|^2 + |\beta'|^2 = 1 \tag{3-57}$$

According to Eq. (3-45) ug can be written in the form $ug = ku(g) = ku'$, where k is a matrix having the form given by Eq. (3-42). If one denotes now ug by g', then one has $g' = ku'$, or explicitly,

$$\begin{pmatrix} g'_{11} & g'_{12} \\ g'_{21} & g'_{22} \end{pmatrix} = \begin{pmatrix} \lambda^{-1} & \mu \\ 0 & \lambda \end{pmatrix} \begin{pmatrix} \alpha' & \beta' \\ -\bar{\beta}' & \bar{\alpha}' \end{pmatrix} \tag{3-58}$$

This gives

$$g'_{21} = -\lambda\bar{\beta}', \qquad g'_{22} = \lambda\bar{\alpha}'$$

from which one obtains

$$\alpha' = \frac{\bar{g}'_{22}}{\bar{\lambda}}, \qquad \beta' = -\frac{\bar{g}'_{21}}{\bar{\lambda}} \tag{3-59}$$

Furthermore, using the condition (3-57) one obtains

$$|\lambda|^2 = |g'_{21}|^2 + |g'_{22}|^2 \tag{3-60}$$

But $g' = ug$. Let us denote the matrix u by

$$u = \begin{pmatrix} \alpha & \beta \\ -\bar{\beta} & \bar{\alpha} \end{pmatrix}; \qquad |\alpha|^2 + |\beta|^2 = 1$$

and the matrix g by

$$g = \begin{pmatrix} g_{11} & g_{12} \\ g_{21} & g_{22} \end{pmatrix}; \qquad g_{11}g_{22} - g_{12}g_{21} = 1$$

Then one has from $g' = ug$ the following:

$$\begin{pmatrix} g'_{11} & g'_{12} \\ g'_{21} & g'_{22} \end{pmatrix} = \begin{pmatrix} \alpha g_{11} + \beta g_{21} & \alpha g_{12} + \beta g_{22} \\ -\bar{\beta} g_{11} + \bar{\alpha} g_{21} & -\bar{\beta} g_{12} + \bar{\alpha} g_{22} \end{pmatrix} \tag{3-61}$$

[6] M. Carmeli and S. Malin, *J. Math. Phys.* **12**, 225 (1971).

If we now write $\lambda = |\lambda|\, e^{i\Lambda}$, where Λ is some real number (phase), then one finally obtains for Eq. (3-59) and (3-60)

$$\begin{aligned} \alpha' &= (-\beta\bar{g}_{12} + \alpha\bar{g}_{22})|\lambda|^{-1}\, e^{i\Lambda} \\ \beta' &= (\beta\bar{g}_{11} - \alpha\bar{g}_{21})|\lambda|^{-1}\, e^{i\Lambda} \end{aligned} \tag{3-62}$$

and

$$|\lambda|^2 = |\beta\bar{g}_{11} - \alpha\bar{g}_{21}|^2 + |-\beta\bar{g}_{12} + \alpha\bar{g}_{22}|^2 \tag{3-63}$$

Hence $u(g)$ is determined by means of u and g up to an arbitrary phase factor. It is readily verified that the right-hand side of Eq. (3-56) is independent of the arbitrary phase factor since it cancels out. The spinor representation is, therefore, well defined on the space $\tilde{P}_{mn}$ by Eq. (3-56).

Example 3.5 Let us suppose that g is a matrix u_0 that belongs to the group SU_2:

$$g = u_0 = \begin{pmatrix} \alpha_0 & \beta_0 \\ -\bar{\beta}_0 & \bar{\alpha}_0 \end{pmatrix}; \qquad |\alpha_0|^2 + |\beta_0|^2 = 1 \tag{3-64}$$

Then one obtains for the matrix $u' = u(g) = u(u_0)$, using Eqs. (3-62) and (3-63),

$$\begin{aligned} \alpha' &= (-\beta\bar{\beta}_0 + \alpha\alpha_0)\, e^{i\Lambda} \\ \beta' &= (\beta\bar{\alpha}_0 + \alpha\beta_0)\, e^{i\Lambda} \\ |\lambda| &= 1 \end{aligned} \tag{3-65}$$

Consequently, one has

$$\frac{\alpha(uu_0)}{\alpha(u(u_0))} = e^{i(m-n)\Lambda}. \tag{3-66}$$

Example 3.6 Suppose now that the matrix g is of the form

$$g = \varepsilon = \begin{pmatrix} \varepsilon_{22}^{-1} & 0 \\ 0 & \varepsilon_{22} \end{pmatrix} \tag{3-67}$$

where ε_{22} is real. One now obtains

$$\begin{aligned} \alpha' &= \alpha\varepsilon_{22}|\lambda|^{-1}\, e^{i\Lambda} \\ \beta' &= \beta\varepsilon_{22}^{-1}|\lambda|^{-1}\, e^{i\Lambda} \\ |\lambda|^2 &= |\beta|^2\varepsilon_{22}^{-2} + |\alpha|^2\varepsilon_{22}^2 \end{aligned} \tag{3-68}$$

whereas

$$\frac{\alpha(u\varepsilon)}{\alpha(u(\varepsilon))} = |\lambda|^{m+n}\, e^{i(m-n)\Lambda} \tag{3-69}$$

Matrix Elements of the Spinor Operator $D(g)$

To conclude this section we find the matrices of the spinor operators $D(g)$.[7] Consider the complex two-vectors $\xi = (\xi^1, \xi^2)$, which transform as

$$\xi' = \xi g \tag{3-70}$$

under application of a matrix g of the group $SL(2, C)$. Let us construct a linear vector space of complex polynomials in the two variables ξ^1 and ξ^2 and denote these polynomials by $p(\xi) = p(\xi^1, \xi^2)$ which are homogeneous in ξ^1 and ξ^2 of degree $2J$, where $2J$ is an arbitrary nonnegative integer. In this space, having the dimension $(2J + 1)$, we define the transformation $D(g)$ for any g of $SL(2, C)$ by

$$D(g)p(\xi) = p(\xi') = p(\xi g) \tag{3-71}$$

These transformations provide a $(2J + 1)$-dimensional representation of the group $SL(2, C)$.

In order to relate the transformation (3-71) to a more familiar notation we expand the polynomial p into powers of ξ^1 and ξ^2:

$$p(\xi) = \sum_{M=-J}^{J} \chi_M^J N_M^J (\xi^1)^{J+M} (\xi^2)^{J-M} \tag{3-72}$$

Here N_M^J are normalized constants defined by

$$N_M^J = \{(2J)!/(J + M)!((J - M)!\}^{1/2}$$

and χ_M^J are expansion coefficients. In terms of χ_M^J the transformation $D(g)$ can be expressed as

$$(D(g)\chi^J)_M = \sum_{M'=-J}^{J} D_{MM'}^J(g)\chi_{M'}^J \tag{3-73}$$

From Eqs. (3-71), (3-72), and (3-73) one finds

$$D_{MM'}^J(g) = \left[\frac{(J + M)!\,(J - M)!}{(J + M')!\,(J - M')!}\right]^{1/2} \sum_n \binom{J + M'}{n}\binom{J - M'}{J + M - n} \times g_{11}^n g_{12}^{J+M-n} g_{21}^{J+M'-n} g_{22}^{n-M-M'} \tag{3-74}$$

where the sum over n extends over all integers for which neither of the binomial coefficients vanish.

If we restrict the group $SL(2, C)$ to the subgroup SU_2, we obtain the matrix $D_{MM'}^J(u)$ defining the unitary irreducible representation of the group SU_2 (see Chapter 1). Each element $D_{MM'}^J(u)$ is a homogeneous polynomial of degree $2J$ in the matrix elements of u, and the coefficients of the polynomial are real. A substitution $u_{ij} \to g_{ij}$, with $i, j = 1, 2$, leads us back to the matrix elements $D_{MM'}^J(g)$. If we

[7] See, for example, W. Rühl, *The Lorentz Group and Harmonic Analysis*, Benjamin, New York, 1970.

substitute $u_{11} = \alpha + i\beta$, $u_{12} = \gamma + i\delta$, $u_{21} = -\gamma + i\delta$, and $u_{22} = \alpha - i\beta$, with real α, β, γ, δ, such that $\alpha^2 + \beta^2 + \gamma^2 + \delta^2 = 1$, the substitution $u_{ij} \to g_{ij}$ can be interpreted as an extension of the real parameters α, β, γ, δ into the complex domain. This is called a *complexification* of the group SU_2.

3-3 INFINITESIMAL OPERATORS OF THE SPINOR REPRESENTATION

We now find the infinitesimal operators $L_{\mp}$, L_3, $K_{\mp}$, K_3 of the spinor representation for the group $SL(2, C)$ discussed in the last section.

One-parameter Subgroups

First we find the one-parameter subgroups of the group $SL(2, C)$ corresponding to the one-parameter subgroups $a_k(\psi)$ and $b_k(\psi)$ of the proper, orthochronous, homogeneous Lorentz group L. These can easily be derived using Eqs. (3-10). One finds for these one-parameter groups:

$$a_1(\psi) = \begin{pmatrix} \cos\dfrac{\psi}{2} & i\sin\dfrac{\psi}{2} \\ i\sin\dfrac{\psi}{2} & \cos\dfrac{\psi}{2} \end{pmatrix}, \quad a_2(\psi) = \begin{pmatrix} \cos\dfrac{\psi}{2} & -\sin\dfrac{\psi}{2} \\ \sin\dfrac{\psi}{2} & \cos\dfrac{\psi}{2} \end{pmatrix}$$

$$a_3(\psi) = \begin{pmatrix} e^{i\psi/2} & 0 \\ 0 & e^{-i\psi/2} \end{pmatrix} \tag{3-75a}$$

$$b_1(\psi) = \begin{pmatrix} \cosh\dfrac{\psi}{2} & \sinh\dfrac{\psi}{2} \\ \sinh\dfrac{\psi}{2} & \cosh\dfrac{\psi}{2} \end{pmatrix}, \quad b_2(\psi) = \begin{pmatrix} \cosh\dfrac{\psi}{2} & i\sinh\dfrac{\psi}{2} \\ -i\sinh\dfrac{\psi}{2} & \cosh\dfrac{\psi}{2} \end{pmatrix}$$

$$b_3(\psi) = \begin{pmatrix} e^{\psi/2} & 0 \\ 0 & e^{-\psi/2} \end{pmatrix} \tag{3-75b}$$

In terms of the infinitesimal matrices a_k and b_k, where $k = 1, 2, 3$, of the group $SL(2, C)$, they can be written as

$$a_k(\psi) = e^{\psi a_k}, \quad b_k(\psi) = e^{\psi b_k},$$

where a_k and b_k are given in terms of the three Pauli matrices (3-5) by $a_k = i\sigma^k/2$ and $b_k = \sigma^k/2$.

Infinitesimal Operators

Using now Eqs. (3-23) one can find the operators $A_k(\psi)$ and $B_k(\psi)$, where $k = 1, 2, 3$. For example

$$A_1(\psi)p(z, \bar{z}) = \left(i \sin \frac{\psi}{2} z + \cos \frac{\psi}{2}\right)^m \left(-i \sin \frac{\psi}{2} \bar{z} + \cos \frac{\psi}{2}\right)^n \times p\left(\frac{\cos \frac{\psi}{2} z + i \sin \frac{\psi}{2}}{i \sin \frac{\psi}{2} z + \cos \frac{\psi}{2}}, \frac{\cos \frac{\psi}{2} \bar{z} - i \sin \frac{\psi}{2}}{-i \sin \frac{\psi}{2} \bar{z} + \cos \frac{\psi}{2}}\right)$$

$$A_2(\psi)p(z, \bar{z}) = \left(-\sin \frac{\psi}{2} z + \cos \frac{\psi}{2}\right)^m \left(-\sin \frac{\psi}{2} \bar{z} + \cos \frac{\psi}{2}\right)^n \times p\left(\frac{\cos \frac{\psi}{2} z + \sin \frac{\psi}{2}}{-\sin \frac{\psi}{2} z + \cos \frac{\psi}{2}}, \frac{\cos \frac{\psi}{2} \bar{z} + \sin \frac{\psi}{2}}{-\sin \frac{\psi}{2} \bar{z} + \cos \frac{\psi}{2}}\right) \tag{3-76}$$

$$A_3(\psi)p(z, \bar{z}) = e^{-im\psi/2}\, e^{in\psi/2} p(e^{i\psi} z, e^{-i\psi}\bar{z}).$$

In the same way one finds the operators $B_k(\psi)$, $k = 1, 2, 3$. Differentiating both sides of these equations with respect to the parameter ψ, and putting $\psi = 0$, one obtains:

$$A_1 p = \left\{\frac{i}{2}(1 - z^2)\frac{\partial}{\partial z} - \frac{i}{2}(1 - \bar{z}^2)\frac{\partial}{\partial \bar{z}} + \frac{i}{2}(mz - n\bar{z})\right\}p$$

$$A_2 p = \left\{\frac{1}{2}(1 + z^2)\frac{\partial}{\partial z} + \frac{1}{2}(1 + \bar{z}^2)\frac{\partial}{\partial \bar{z}} - \frac{1}{2}(mz + n\bar{z})\right\}p \tag{3-77a}$$

$$A_3 p = \left\{iz\frac{\partial}{\partial z} - i\bar{z}\frac{\partial}{\partial \bar{z}} - \frac{i}{2}(m - n)\right\}p$$

$$B_1 p = \left\{\frac{1}{2}(1 - z^2)\frac{\partial}{\partial z} + \frac{1}{2}(1 - \bar{z}^2)\frac{\partial}{\partial \bar{z}} + \frac{1}{2}(mz + n\bar{z})\right\}p$$

$$B_2 p = \left\{-\frac{i}{2}(1 + z^2)\frac{\partial}{\partial z} + \frac{i}{2}(1 + \bar{z}^2)\frac{\partial}{\partial \bar{z}} + \frac{i}{2}(mz - n\bar{z})\right\}p \tag{3-77b}$$

$$B_3 p = \left\{z\frac{\partial}{\partial z} + \bar{z}\frac{\partial}{\partial \bar{z}} - \frac{1}{2}(m + n)\right\}p$$

The operators $L_\pm$, L_3, $K_\pm$, and K_3 can now be found, using Eqs. (2-77):

$$\begin{aligned}
L_+ p &= \left\{ -\frac{\partial}{\partial z} - \bar{z}^2 \frac{\partial}{\partial \bar{z}} + n\bar{z} \right\} p \\
L_- p &= \left\{ z^2 \frac{\partial}{\partial z} + \frac{\partial}{\partial \bar{z}} - mz \right\} p \\
L_3 p &= \left\{ -z \frac{\partial}{\partial z} + \bar{z} \frac{\partial}{\partial \bar{z}} + \frac{1}{2}(m - n) \right\} p \\
K_+ p &= \left\{ i \frac{\partial}{\partial z} - i\bar{z}^2 \frac{\partial}{\partial \bar{z}} + in\bar{z} \right\} p \\
K_- p &= \left\{ -iz^2 \frac{\partial}{\partial z} + i \frac{\partial}{\partial \bar{z}} + imz \right\} p \\
K_3 p &= \left\{ iz \frac{\partial}{\partial z} + i\bar{z} \frac{\partial}{\partial \bar{z}} - \frac{i}{2}(m + n) \right\} p
\end{aligned} \tag{3-78}$$

Further Properties of Spinor Representations

We conclude this section by the following lemmas.

Lemma 3.3 *The spinor representations of the group SL(2, C) are irreducible.*

Lemma 3.4 *Every finite-dimensional irreducible representation of the group SL(2, C) is equivalent to some spinor representation. The pair of numbers j_0 and c of the representation formula (2-30) is then related to the pair of numbers m and n of the spinor representation by*

$$j_0 = \tfrac{1}{2}|m - n| \tag{3-79a}$$

$$c = \begin{cases} [\text{sign}\,(m - n)][\tfrac{1}{2}(m + n) + 1]; & m \neq n \\ \pm[\tfrac{1}{2}(m + n) + 1]; & m = n \end{cases} \tag{3-79b}$$

Lemma 3.5 *The spinor representations are all nonunitary.*

The proofs of Lemmas 3.3, 3.4, and 3.5 are left for the reader (Problems 3.12, 3.13, and 3.14).

PROBLEMS

3.1 Use Eq. (3-9a) to prove Eqs. (3-9b).

3.2 Prove Eqs. (3-10). [See W. Rühl, *The Lorentz Group and Harmonic Analysis*, Benjamin, New York, 1970.]

3.3 Prove Eq. (3-13). [See W. Rühl, *op. cit.*]

3.4 Prove that the kernel of the homomorphism of the group $SL(2, C)$ onto the group L coincides with the centre of the group $SL(2, C)$. [Notice that by definition the element g is in the kernel of the homomorphism if for all Hermitian matrices Q one has $Q = gQg^\dagger$.]

3.5 Use Eq. (3-9a) to show that if a, b, c, and d are the four elements of the matrix g of the group $SL(2, C)$, with $ad - bc = 1$, then the corresponding matrix Λ of the proper, orthochronous, homogeneous, Lorentz group is given by

$$\Lambda = \begin{pmatrix} \frac{1}{2}(a\bar{a} + b\bar{b} + c\bar{c} + d\bar{d}) & \mathrm{Re}\,(a\bar{b} + c\bar{d}) & \mathrm{Im}\,(\bar{a}b - c\bar{d}) & \frac{1}{2}(a\bar{a} - b\bar{b} + c\bar{c} - d\bar{d}) \\ \mathrm{Re}\,(a\bar{c} + b\bar{d}) & \mathrm{Re}\,(a\bar{d} + b\bar{c}) & \mathrm{Im}\,(\bar{a}d + b\bar{c}) & \mathrm{Re}\,(a\bar{c} - b\bar{d}) \\ \mathrm{Im}\,(a\bar{c} + b\bar{d}) & \mathrm{Im}\,(a\bar{d} + b\bar{c}) & \mathrm{Re}\,(\bar{a}d - b\bar{c}) & \mathrm{Im}\,(a\bar{c} - b\bar{d}) \\ \frac{1}{2}(a\bar{a} + b\bar{b} - c\bar{c} - d\bar{d}) & \mathrm{Re}\,(a\bar{b} - c\bar{d}) & \mathrm{Im}\,(\bar{a}b + c\bar{d}) & \frac{1}{2}(a\bar{a} - b\bar{b} - c\bar{c} + d\bar{d}) \end{pmatrix}$$

Show that the same matrix can also be obtained directly from either Eq. (3-6) or Eq. (3-9b).

3.6 Show that the formula (3-23) provides a representation of the group $SL(2, C)$.

3.7 Prove Eqs. (3-27) and (3-28).

3.8 Show that the transformation (3-71) is another form of the spinor representation.

3.9 Show that the matrix elements $D^J_{MM'}(g)$ of Eq. (3-74) satisfy the properties:

$$D^J_{MM'}(\bar{g}) = \bar{D}^J_{MM'}(g)$$

$$D^J_{MM'}(g^t) = D^J_{M'M}(g)$$

$$D^J_{MM'}((g^{-1})^t) = (-1)^{M-M'} D^J_{-M,\,-M'}(g)$$

3.10 Show that the parameter ψ of Eqs. (3-75b) is related to the relative speed v of the Lorentz transformation by

$$\psi = \cosh^{-1}\left[(1 - v^2/c^2)^{-1/2}\right]$$

3.11 Find the operators $B_k(\psi)$ for the spinor representation of the group $SL(2, C)$ and from them prove Eqs. (3-77b).

3.12 Prove Lemma 3.3.

3.13 Prove Lemma 3.4.

3.14 Prove Lemma 3.5. [Thus the group $SL(2, C)$ does not contain finite-dimensional unitary representations.]

CHAPTER

FOUR

PRINCIPAL SERIES OF REPRESENTATIONS OF $SL(2, C)$

In Chap. 3 we found all finite-dimensional representations of the Lorentz group. In this chapter and the next two chapters we give all infinite-dimensional representations of that group. In this chapter we discuss the important representation known as the representation of the principal series, which is unitary and irreducible. We start our discussion by introducing some infinite-dimensional spaces in which the representations of the principal series are later realized. This is followed by a discussion of the concept of Fourier transform on the group SU_2. We then realize the representation of the principal series, which can be considered as a generalization of the finite-dimensional spinor representation. A comparison with the infinitesimal approach, discussed in Chap. 2, is then made. Finally, we discuss some preliminary aspects of the theory of harmonic analysis on the group $SL(2, C)$.

4-1 LINEAR SPACES OF REPRESENTATIONS

We start our discussion of infinite-dimensional representations of the group $SL(2, C)$ by defining certain infinite-dimensional linear spaces in which the representations are realized.[1]

[1] See, for example, M. A. Naimark, *Linear Representations of the Lorentz Group*, Pergamon Press, New York, 1964.

The Hilbert Space $L_2(Z)$

We denote by $L_2(Z)$ the Hilbert space of all measurable functions $f(z)$ defined on the group Z, satisfying the condition

$$\int |f(z)|^2 \, dz < \infty \tag{4-1}$$

The scalar product in the space $L_2(Z)$ is given by

$$(f_1, f_2) = \int f_1(z)\bar{f}_2(z) \, dz \tag{4-2}$$

Here Z denotes the aggregate of all matrices z of the form (see Chapter 3):

$$z = \begin{pmatrix} 1 & 0 \\ z & 1 \end{pmatrix} \tag{4-3}$$

Obviously Z provides a subgroup of the group $SL(2, C)$. The integrals in Eqs. (4-1) and (4-2) are invariant integrals on the group Z, with the measure

$$dz = dx \, dy; \qquad z = x + iy \tag{4-4}$$

The measure dz is invariant under the group displacement $z \to zz_0$,

$$\int f(zz_0) \, dz = \int f(z) \, dz \tag{4-5}$$

It is also possible to consider the function $f(z)$ as a function of the variables x and y and denoting $f(z)$ by $f(x, y)$, where $z = x + iy$. By Eq. (4-4) one then has

$$\int f(z) \, dz = \int_{-\infty}^{\infty} \int_{-\infty}^{\infty} f(x, y) \, dx \, dy \tag{4-6}$$

But the displacement $z \to zz_0$ carries the parameters x and y into $x + x_0$ and $y + y_0$ since

$$zz_0 = \begin{pmatrix} 1 & 0 \\ z & 1 \end{pmatrix} \begin{pmatrix} 1 & 0 \\ z_0 & 1 \end{pmatrix} = \begin{pmatrix} 1 & 0 \\ z + z_0 & 1 \end{pmatrix}$$

Hence we have

$$\begin{aligned} \int f(zz_0) \, dz &= \int_{-\infty}^{\infty} \int_{-\infty}^{\infty} f(x + x_0, y + y_0) \, dx \, dy \\ &= \int_{-\infty}^{\infty} \int_{-\infty}^{\infty} f(x, y) \, dx \, dy = \int f(z) \, dz \end{aligned}$$

Furthermore, one can easily show (see Problem 4.1) that under the transformation $z \to z' = z(g)$,

$$z(g) = \frac{g_{11}z + g_{21}}{g_{12}z + g_{22}} \tag{4-7}$$

from Eq. (3-38), g being an element of the group $SL(2, C)$, the measure dz is transformed into the new measure

$$dz' = |g_{12}z + g_{22}|^{-4}\, dz \tag{4-8a}$$

Equation (4-8a) can also be given a group theoretical meaning. This is so since if we define the function

$$\beta(g) = |g_{22}|^4 \tag{4-9}$$

and if k is an element of the group K, i.e. a matrix of the form

$$k = \begin{pmatrix} \lambda^{-1} & \mu \\ 0 & \lambda \end{pmatrix} \tag{4-10}$$

(see Chapter 3), then

$$\beta(kg) = |\lambda g_{22}|^4 = |\lambda|^4 |g_{22}|^4 = \beta(k)\beta(g)$$

Furthermore, $\beta(z) = 1$, where z is a matrix of the form (4-3). Now let $g' = zg$, and apply Eq. (3-37), we see that $\beta(zg) = |g_{12}z + g_{22}|^4$. Hence Eq. (4-8a) can be written as

$$dz' = \beta^{-1}(zg)\, dz \tag{4-8b}$$

The Hilbert Space $L^2(SU_2)$

The Hilbert space $L^2(SU_2)$ is defined as the set of all functions $f(u)$, where u is an element of the group SU_2, which are measurable and satisfy the conditions $\int |f(u)|^2\, du < \infty$. The scalar product in the space $L^2(SU_2)$ is defined by

$$(f_1, f_2) = \int f_1(u)\bar{f}_2(u)\, du$$

The proof of the completeness of the space $L^2(SU_2)$ is left for the reader (compare Appendix C).

The Hilbert Space $L_2^{2s}(SU_2)$

We denote by $L_2^{2s}(SU_2)$ the set of all functions $\phi(u)$ on the group SU_2, which are measurable, and satisfy the conditions

$$\phi(\gamma u) = e^{is\psi}\phi(u) \tag{4-11}$$

where γ of SU_2 is given by

$$\gamma = \begin{pmatrix} e^{-i\psi/2} & 0 \\ 0 & e^{i\psi/2} \end{pmatrix} \tag{4-12}$$

and $\int |\phi(u)|^2 \, du < \infty$. We introduce in the space $L_2^{2s}(SU_2)$ a scalar product defined by

$$(\phi_1, \phi_2) = \int \phi_1(u)\bar{\phi}_2(u) \, du$$

$L_2^{2s}(SU_2)$ then becomes a Hilbert space; it is a closed subspace of the Hilbert space $L^2(SU_2)$ discussed above, and is complete. In fact $L^2(SU_2)$ is the direct sum of all the $L_2^{2s}(SU_2)$. Here s should be an integer or half-integer in order to guarantee singlevaluedness of the function $\phi(u)$.

Fourier Transform on the Group SU_2

We now generalize the concept of Fourier transform of functions to functions defined over groups.

Let $f(x)$ be a measurable function with integrable square modulus on the real line $-\infty < x < \infty$, and consider its Fourier transform

$$f(k) = \frac{1}{\sqrt{(2\pi)}} \int_{-\infty}^{\infty} f(x) \, e^{-ikx} \, dx \tag{4-13}$$

The theorem of Plancherel then states that the integral in Eq. (4-13) converges in the mean and the functions f satisfy

$$\int_{-\infty}^{\infty} |f(k)|^2 \, dk = \int_{-\infty}^{\infty} |f(x)|^2 \, dx \tag{4-14}$$

Equation (4-14) is known as Plancherel's formula. In the following we use the above formalism for analogous generalizations to Lie groups. In the particular case of compact Lie groups, such as the group SU_2, the Plancherel theorem is included in the theorem of Peter and Weyl.[2]

To this end we first give a group theoretical meaning to Plancherel's theorem. One notices that the set of all real numbers x can be considered as a group, the *additive group of real numbers*, R, in which the group operation is defined as the *addition* of numbers. Consider now, for a fixed real number α, the function $D(x; k) = e^{ikx}$. Obviously, $D(x_1 + x_2;\ k) = D(x_1;\ k)D(x_2; k)$ and $D(0; k) = 1$. Consequently, the operation of multiplication by $D(x; k)$ in a one-dimensional space realizes a one-dimensional representation of the group R.

Consider next the measurable functions $f(x)$ which are integrable square on the real line $-\infty < x < \infty$. They provide a Hilbert space, denoted by $L^2(-\infty, \infty)$. A unitary representation of the group R, known as the *regular representation of the group R*, can be realized in the space $L^2(-\infty, \infty)$ by the representation formula $D(y)f(x) = f(x + y)$. The function $f(k)$ of Eq. (4-13), whose values belong to the space C of all complex numbers, and satisfy $\int_{-\infty}^{\infty} |f(k)|^2 \, dk < \infty$, can be con-

[2] See, for example, L. S. Pontrjagin, *Topological Groups*, Princeton University Press, Princeton, New Jersey, 1946.

sidered as the continued direct sum of continuously many images of the space C. Plancherel's formula (4-14) shows that the space $L^2(-\infty, \infty)$ of functions $f(x)$ is isometric (see Appendix C) to the space of functions $f(k)$ through the isometry transformation (4-13). The operator $D(y)$ of the regular representation, under the isometric mapping (4-13), goes into the operator $\tilde{D}(y)$ defined in the space of functions $f(k)$. To find the operator $\tilde{D}(y)$ we notice that an application to the function $f(x)$ of the operator $D(y)$ of the regular representation takes the function $f(k)$ into

$$\frac{1}{\sqrt{(2\pi)}}\int_{-\infty}^{\infty} f(x+y)\,\mathrm{e}^{-ikx}\,\mathrm{d}x = \frac{1}{\sqrt{(2\pi)}}\int_{-\infty}^{\infty} f(x)\,\mathrm{e}^{-ik(x-y)}\,\mathrm{d}x = \mathrm{e}^{iky}f(k)$$

Hence the application of the operator $\tilde{D}(y)$ to the function $f(k)$ yields a multiplication of the k-th image $f(k)$ by $D(y; k) = \mathrm{e}^{iky}$, i.e. a transformation of each separate one-dimensional image of C according to a one-dimensional representation of the group R. In analogy with the definition of the usual (discrete) orthogonal direct sum of representations, the representation $x \to \tilde{D}(x)$ is called the continued direct sum of the one-dimensional representations $x \to D(x; k)$. The Fourier transform (4-13) realizes an isometric mapping of the regular representation of the additive group R of real numbers into the continued direct sum of its irreducible, one-dimensional, representations, thus realizing a decomposition of the regular representation of the group R into its irreducible representations. These considerations point out the group theoretical significance of the Fourier transform and Plancherel's formula.[3]

Consider now the Hilbert space $L^2(SU_2)$. Let $D^j_{mn}(u)$ be the matrix elements of the irreducible representations of the group SU_2 in their canonical basis (see Chapter 1), and define the function

$$\phi^j_{mn}(u) = \sqrt{(2j+1)}D^j_{mn}(u) \tag{4-15}$$

As has been shown in Chapter 1, these functions provide a complete orthogonal system in the space $L^2(SU_2)$. Hence one has

$$\int |f(u)|^2\,\mathrm{d}u = (f, f) = \sum_j \sum_{m,n} |\alpha^j_{mn}|^2 \tag{4-16}$$

where

$$\alpha^j_{mn} = (f, \phi^j_{mn}) = \int f(u)\bar{\phi}^j_{mn}(u)\,\mathrm{d}u \tag{4-17}$$

Here the summation runs over $j = 0, \frac{1}{2}, 1, \frac{3}{2}, \ldots$, and $m, n = -j, -j+1, -j+2, \ldots, j$. Equations (4-16) and (4-17) can also be written as

$$F^j = \int f(u)[D^j(u)]^\dagger\,\mathrm{d}u \tag{4-18}$$

[3] These results are valid to arbitrary locally compact commutative groups. See, for example, M. A. Naimark, *Normed Rings*, Noordhoff, Groningen, The Netherlands, 1959.

where $D^j(u)$ is the matrix whose elements are $D^j_{mn}(u)$. Hence F^j is a matrix whose elements are

$$F^j_{mn} = \int f(u)\bar{D}^j_{nm}(u)\, du \tag{4-19}$$

Equation (4-18) can now be considered as the generalization of the Fourier transform (4-13); the comparable part of the numerical function $f(x)$, labelling the irreducible representations of the additive group of real numbers R, is being taken over in the present case by the matrix F^j of discrete variables j, labelling the weight of irreducible representation of the group SU_2. From Eqs. (4-15), (4-17), and (4-19) one finds that $\alpha^j_{mn} = \sqrt{(2j+1)}F^j_{nm}$, whereas Eq. (4-16) gives

$$\int |f(u)|^2\, du = \sum_j (2j+1) \sum_{m,\, n} |F^j_{mn}|^2 \tag{4-20}$$

Equation (4-20) is Plancherel's formula for the group SU_2.[4]

We define the *regular* (more accurately, *right regular*) representation $u \to D(u)$ of the group SU_2 in the Hilbert space $L^2(SU_2)$ by associating to every element u_0 of SU_2 the operator $D(u_0)$, called the *right translation* operator on the space $L^2(SU_2)$, which transforms all the $f(u)$ according to

$$D(u_0)f(u) = f(uu_0)$$

Obviously $D(u_1 u_2) = D(u_1)D(u_2)$, so that $D(u)$ is a representation of the group SU_2. Because the measure du is invariant under right translations, the regular representation is *unitary*. One may then ask for the decomposition of this representation into irreducible representations.

The matrix elements F^j_{mn} are transformed in the transition from $f(u)$ to $f(uu_0)$, using Eq. (4-19), into

$$F'^j_{mn} = \sum_{m'=-j}^{j} D^j_{mm'}(u_0)F^j_{m'n} \tag{4-21}$$

Let us denote by R^j the $(2j+1)$-dimensional space of all possible systems $x = \{x_{-j}, x_{-j+1}, \ldots, x_j\}$ of numbers $x_{-j}, x_{-j+1}, \ldots, x_j$ with the scalar product defined by $(x, y) = (2j+1) \sum x_m \bar{y}_m$, where m takes the values from $-j$ to j. For fixed values of j and m, the numbers F^j_{mn}, with $n = -j, -j+1, \ldots, j$, are elements $x^j_m = (F^j_{m,-j}, F^j_{m,-j+1}, \ldots, F^j_{mj})$ of the space R^j. Hence the function F^j can be considered as a vector-valued function $x = \{x^j_m\}$ with values in the space $R^j_m = R^j$. One then has $\int |f(u)|^2\, du = \sum |x^j_m|^2$. Hence the transition from f to F^j by Eq. (4-19) is an isometric mapping of the space $L^2(SU_2)$ onto the direct sum of spaces R^j in which each space R^j occurs $2j+1$ times. If $\tilde{D}(u)$ is the operator to which the operator $D(u)$ corresponds under this mapping, Eq. (4-21) shows that each of the subspaces $R^j_m = R^j$ is invariant with respect to the operator $\tilde{D}(u)$ and the matrix of the operator $\tilde{D}(u)$ in this subspace is the matrix $D^j_{mn}(u)$. Consequently, the generalized Fourier transform (4-18) realizes an isometric mapping of the

[4] Equation (4-20) can be generalized for any compact topological group.

regular representation $u \to D(u)$ onto the direct sum of irreducible representations $u \to D^j(u)$, where each representation $u \to D^j(u)$ is included in this direct sum $2j + 1$ times.

The Hilbert Space l_2^{2s}

We now use the generalized Fourier transform, introduced above, in order to transform each Hilbert space $L_2^{2s}(SU_2)$ into the Hilbert space l_2^{2s} which is defined as follows.[5]

Consider all possible systems of numbers ϕ_m^j, where $m = -j, -j+1, \ldots, j$ and $j = |s|, |s|+1, |s|+2, \ldots$, with the condition

$$\sum_{j=|s|}^{\infty} (2j+1) \sum_{m=-j}^{j} |\phi_m^j| < \infty \tag{4-22}$$

The aggregate of all such systems ϕ_m^j of numbers forms a Hilbert space, denoted by l_2^{2s}, in which the scalar product is defined by

$$\sum_{j=|s|}^{\infty} (2j+1) \sum_{m=-j}^{j} \phi_m^j \bar{\psi}_m^j \tag{4-23}$$

for any two vectors ϕ_m^j and ψ_m^j of l_2^{2s}. For any integer or half-integer value of s, the two Hilbert spaces $L_2^{2s}(SU_2)$ and l_2^{2s} are isometric. Let us find the transformation between the two spaces.

Let $D_{sm}^j(u)$ be the matrix elements of the irreducible representation of the group SU_2. As has been shown in Chapter 1, the functions $D_{sm}^j(u)$ satisfy

$$D_{sm}^j(\gamma u) = e^{is\psi} D_{sm}^j(u) \tag{4-24}$$

where the matrix γ of SU_2 is given by Eq. (4-12). For a fixed value of s, the functions $D_{sm}^j(u)$ provide a complete orthogonal set for the Hilbert space $L_2^{2s}(SU_2)$, where $m = -j, -j+1, \ldots, j$ and $j = |s|, |s|+1, |s|+2, \ldots$ The functions $D_{sm}^j(u)$ also satisfy the orthogonality relation

$$\int D_{sm}^j(u) \bar{D}_{s'm'}^{j'}(u)\, du = \frac{1}{2j+1} \delta^{jj'} \delta_{ss'} \delta_{mm'} \tag{4-25}$$

Consequently, any function $\phi(u)$ of the space $L_2^{2s}(SU_2)$ can be uniquely expanded in the form

$$\phi(u) = \sum_{j=|s|}^{\infty} (2j+1) \sum_{m=-j}^{j} \phi_m^j D_{sm}^j(u) \tag{4-26}$$

where

$$\phi_m^j = \int \phi(u) \bar{D}_{sm}^j(u)\, du \tag{4-27}$$

[5] M. Carmeli, *J. Math. Phys.* **11**, 1917 (1970).

The system of numbers ϕ_m^j, consequently, satisfy the condition (4-22) if and only if the corresponding function $\phi(u)$ satisfies the condition $\int |\phi(u)|^2 \, du < \infty$ of the Hilbert space $L_2^{2s}(SU_2)$. The Hilbert spaces $L_2^{2s}(SU_2)$ and l_2^{2s} are, therefore, isometric and the mapping between them is given by the generalized Fourier transforms (4-26) and (4-27).

Linear Spaces of Homogeneous Functions

Infinite-dimensional representations of the group $SL(2, C)$ can also be realized on spaces of homogeneous functions.[6] The homogeneous functions can be considered as generalizations of the spaces of polynomials which are homogeneous, with integral degrees, in pairs of variables. Such spaces are denoted by $D(\chi)$, and the homogeneous functions of two complex variables are denoted by $F(z_1, z_2, \bar{z}_1, \bar{z}_2)$. For convenience, these functions will be denoted by $F(z_1, z_2)$. Hence the function $F(z_1, z_2)$ is considered as homogeneous of degree (λ, μ), where λ and μ are complex numbers differing by an integer. This means, for any complex number $a \neq 0$, we have

$$F(az_1, az_2) = a^{\lambda}\bar{a}^{\mu}F(z_1, z_2) \tag{4-28}$$

In order to make the function f single-valued, the homogeneity condition $F(e^{i\theta}z_1, e^{i\theta}z_2) = e^{i(\lambda-\mu)\theta}F(z_1, z_2)$ must reduce to an identity for $\theta = 2n\pi$, where n is an integer. Hence the condition $\lambda - \mu = m$, with m as an integer, should be satisfied. We shall make this assumption throughout.

Let n_1 and n_2 be a pair of complex numbers whose difference is an integer. Instead of the degrees λ and μ we shall characterize homogeneous functions by the pair of numbers n_1 and n_2, and for simplicity we put $\chi = (n_1, n_2)$, where $n_1 = \lambda + 1$ and $n_2 = \mu + 1$. In addition, we use the notation $-\chi = (-n_1, -n_2)$ if $\chi = (n_1, n_2)$. The linear spaces $D(\chi)$ of homogeneous functions $F(z_1, z_2)$ are then defined by the following postulates:

(1) $F(z_1, z_2)$ is homogeneous of degree $(n_1 - 1, n_2 - 1)$;

(2) $F(z_1, z_2)$ is infinitely differentiable in the variables z_1, z_2 and their complex conjugates throughout the entire region of variation of z_1 and z_2, except at the point (0, 0).

We also introduce a topology into the space $D(\chi)$. A sequence $\{F_m(z_1, z_2)\}$ of functions in $D(\chi)$ will be said to converge to zero if on every closed bounded set that does not include the point (0, 0) these functions converge uniformly to zero together with all their derivatives. One can show that the space $D(\chi)$ is complete with respect to this topology, and hence $D(\chi)$ is a topological vector space.

[6] See, for example, I. M. Gelfand, M. I. Graev, and N. Ya. Vilenkin, *Generalized Functions*, Vol. 5. *Integral Geometry and Representation Theory*, Academic Press, New York, 1966.

Other Realizations of the Space $D(\chi)$

Instead of using the spaces of homogeneous functions, it is sometimes useful to consider other equivalent realizations of the spaces $D(\chi)$.

Consider the complex line $z_2 = 1$ in a space of two complex dimensions. This line intersects each line passing through the origin (except $z_2 = 0$) at one and only one point. Therefore, each $F(z_1, z_2)$ of $D(\chi)$ is determined uniquely by its values on this line. In this way with each function $F(z_1, z_2)$ of $D(\chi)$ we associate the function

$$f(z) = F(z, 1) \tag{4-29}$$

Due to the homogeneity of $F(z_1, z_2)$, knowledge of the function $f(z)$ suffices to reconstruct $F(z_1, z_2)$,

$$F(z_1, z_2) = z_2^{n_1-1}\bar{z}_2^{n_2-1} f(z_1/z_2) \tag{4-30}$$

The function $f(z)$ depends on only one complex variable z. It is infinitely differentiable in z and $\bar{z}$, the infinite far point included. At this infinite far point $f(z)$ has an asymptotic series expansion of the form

$$f(z) \sim z^{n_1-1}\bar{z}^{n_2-1} \sum_{j,\,k=0}^{\infty} a_{jk} z^{-j}\bar{z}^{-k} \tag{4-31}$$

The topology of the space $D(\chi)$ of functions $F(z_1, z_2)$ can be carried over to the new space of functions $f(z)$. In view of Eq. (4-30) we may call this space also $D(\chi)$.

Another realization of the space $D(\chi)$ is available if one considers the manifold Ω of all points (z_1, z_2) such that $|z_1|^2 + |z_2|^2 = 1$. Every line passing through the origin and intersecting the manifold Ω at (ω_1, ω_2) contains also all points of the form $(e^{i\psi}\omega_1, e^{i\psi}\omega_2)$, where ψ is a real number such that $0 \leqslant \psi < 2\pi$. The homogeneity condition then leads to

$$F(e^{i\psi}\omega_1, e^{i\psi}\omega_2) = e^{i(n_1-n_2)\psi} F(\omega_1, \omega_2) \tag{4-32}$$

Hence, with every function $F(z_1, z_2)$ of the space $D(\chi)$ we can associate an infinitely differentiable function $F(\omega_1, \omega_2)$ on the sphere Ω whose equation is $|\omega_1|^2 + |\omega_2|^2 = 1$ and which satisfies the homogeneity condition of Eq. (4-32). $F(z_1, z_2)$ can be obtained from $F(\omega_1, \omega_2)$ by

$$F(z_1, z_2) = r^{n_1+n_2-2} F(z_1/r, z_2/r) \tag{4-33}$$

where $r = (|z_1|^2 + |z_2|^2)^{1/2}$. Conversely, if $F(\omega_1, \omega_2)$ is any infinitely differentiable function on Ω and satisfies Eq. (4-32), the function $F(z_1, z_2)$ of Eq. (4-33) belongs to $D(\chi)$.[7]

[7] The realization of the space $D(\chi)$ in terms of the functions $F(\omega_1, \omega_2)$ is interesting since Ω is a compact manifold.

4-2 THE GROUP OPERATORS

Representation of $SL(2, C)$ on $D(\chi)$

We now determine the representation of the group $SL(2, C)$, the group of all complex unimodular 2×2 matrices

$$g = \begin{pmatrix} a & b \\ c & d \end{pmatrix}; \qquad ad - bc = 1,$$

on the space $D(\chi)$. Each element g of $SL(2, C)$ induces the linear transformation

$$z_1' = az_1 + cz_2$$
$$z_2' = bz_1 + dz_2$$

on the space of the two complex variables z_1 and z_2. This transformation, in turn, may be associated with the transformation

$$F(z_1, z_2) \to F(az_1 + cz_2, bz_1 + dz_2)$$

on the space $D(\chi)$. Thus the operator $D(g, \chi)$ that defines the above transformation,

$$D(g; \chi)F(z_1, z_2) = F(az_1 + cz_2, bz_1 + dz_2) \tag{4-34}$$

is defined exactly as in the case of spinor representation. Since the origin $z_1 = z_2 = 0$ goes into itself, namely $z_1' = z_2' = 0$, it is clear that $D(g; \chi)F(z_1, z_2)$ is in $D(\chi)$ if $F(z_1, z_2)$ itself is in $D(\chi)$, since the former function is also infinitely differentiable and homogeneous of degree $(n_1 - 1, n_2 - 1)$.

The operator $D(g; \chi)$ is continuous in the space $D(\chi)$. This means if $\{F_m(z_1, z_2)\}$ in $D(\chi)$ converges to zero, then the sequence $\{D(g; \chi)F_m(z_1, z_2)\}$ also converges to zero. Furthermore, if $\lim g_m = g$ when $m \to \infty$, then $\lim D(g_m; \chi)F(z_1, z_2) = D(g; \chi)F(z_1, z_2)$ when $m \to \infty$, for any F of $D(\chi)$. Hence the operator $D(g; \chi)$ is a continuous operator of the element g of $SL(2, C)$. One can easily show that

$$\begin{aligned} D(g_1 g_2; \chi) &= D(g_1; \chi)D(g_2; \chi) \\ D(e; \chi) &= 1 \end{aligned} \tag{4-35}$$

where e is the unit 2×2 matrix and 1 is the unit operator, for any two elements g_1 and g_2 of the group $SL(2, C)$. Consequently, the set of transformations $D(g; \chi)$ on the space $D(\chi)$ forms a representation of the group $SL(2, C)$. Hence to any pair of complex numbers $\chi = (n_1, n_2)$, whose difference is an integer, there corresponds a representation $D(g; \chi)$ of the group $SL(2, C)$ realized on the space $D(\chi)$ of infinitely differentiable functions $F(z_1, z_2)$, homogeneous of degree $(n_1 - 1, n_2 - 1)$ in accordance with Eq. (4-34). The pair χ is called the *weight* of the representation.

Other Realizations for $D(g; \chi)$

Since the space $D(\chi)$ can also be realized as the space of all infinitely differentiable functions $f(z) = F(z, 1)$ of z and $\bar{z}$, one can express the operators $D(g; \chi)$ on these functions. Using the homogeneity property of the function $F(z_1, z_2)$ one obtains:

$$D(g; \chi)F(z_1, z_2) = F(az_1 + cz_2, bz_1 + dz_2)$$

$$= (bz_1 + dz_2)^{n_1 - 1}(\bar{b}\bar{z}_1 + \bar{d}\bar{z}_1)^{n_2 - 1} f\left(\frac{az_1 + cz_2}{bz_1 + dz_2}\right)$$

But $D(g; \chi)f(z) = D(g; \chi)F(z, 1)$ is the function of one variable corresponding to $D(g; \chi)F(z_1, z_2)$. Hence we obtain for the operator $D(g; \chi)$, operating on $f(z)$:

$$D(g; \chi)f(z) = (bz + d)^{n_1 - 1}(\bar{b}\bar{z} + \bar{d})^{n_2 - 1} f\left(\frac{az + c}{bz + d}\right) \tag{4-36}$$

Conjugate Representations

Let $D'(\chi)$ be the space conjugate to $D(\chi)$, i.e. the space of linear functionals on $D(\chi)$ (see Appendix C). One can then define a *bilinear* functional (Φ, f), where Φ belongs to $D'(\chi)$ and f belongs to $D(\chi)$, in a natural way for these two spaces. (Φ, f) is the value of the functional Φ on the function f.

A representation of the group $SL(2, C)$ can then be defined on $D'(\chi)$ by defining its operators by

$$(\tilde{D}(g; \chi)\Phi, f) = (\Phi, D(g^{-1}; \chi)f) \tag{4-37}$$

One can easily show that $\tilde{D}(g_1 g_2; \chi) = \tilde{D}(g_1; \chi)\tilde{D}(g_2; \chi)$. The representation (4-37) is called conjugate to the representation whose operator is $D(g; \chi)$. If one replaces f by $D(g; \chi)f$ in Eq. (4-37) one obtains $(\tilde{D}(g; \chi)\Phi, D(g; \chi)f) = (\Phi, f)$ and therefore the bilinear functional (Φ, f) is invariant under the operators $D(g; \chi)$ and its conjugate $\tilde{D}(g; \chi)$.

Consider now two spaces $D(\chi_1)$ and $D(\chi_2)$ and construct the invariant bilinear functional $B(\phi, \psi)$, where ϕ belongs to $D(\chi_1)$ and ψ to $D(\chi_2)$. Since the bilinear functional is invariant, one has $B(D(g; \chi_1)\phi, D(g; \chi_2)\psi) = B(\phi, \psi)$ for any element g of the group $SL(2, C)$. Assume, furthermore, that $B(\phi, \psi)$ is nondegenerate on $D(\chi)$.[8] Then $D(\chi_1)$ can be mapped in a one-to-one way into the space $D'(\chi_2)$ so that $D(g; \chi_1)$ is mapped into $\tilde{D}(g; \chi_2)$. In this way $\tilde{D}(g; \chi_2)$, the representation conjugate to $D(g; \chi_2)$, can be considered as an extension of $D(g; \chi_1)$.

To see this one notices that any function ϕ of $D(\chi)$ defines a linear functional $B(\phi, \psi)$ on $D(\chi_2)$. In this way $D(\chi_1)$ is mapped into $D'(\chi_2)$. Since by assumption $B(\phi, \psi)$ is nondegenerate on $D(\chi_1)$, only $\phi = 0$ is mapped into zero under this mapping. Hence $D(\chi_1) \to D'(\chi_2)$ is a one-to-one mapping. We next show that $\tilde{D}(g; \chi_2)$ is an extension of $D(g; \chi_1)$, i.e. on $D(\chi_1)$ it coincides with $D(g; \chi_1)$. For

[8] The functional $B(\phi, \psi)$ is nondegenerate on $D(\chi)$ if $B(\phi, \psi) = 0$ for fixed ϕ of $D(\chi_1)$ and any ψ of $D(\chi_2)$ yields $\phi = 0$.

consider a function ϕ to be an element of $D'(\chi_2)$. Then $(\tilde{D}(g; \chi_2)\phi, \psi) = (\phi, D(g^{-1}; \chi_2)\phi)$. On the other hand, from the definition of the mapping $D(\chi_1) \to D'(\chi_2)$ one has $(\phi, \psi) = B(\phi, \psi)$. Therefore,

$$(\phi, D(g^{-1}; \chi_2)\psi) = B(\phi, D(g^{-1}; \chi_2)\psi) = B(D(g; \chi_1)\phi, \psi) = (D(g; \chi_1)\phi, \psi)$$

One therefore finds that $(\tilde{D}(g; \chi_2)\phi, \psi) = (D(g; \chi_1)\phi, \psi)$, or $D(g; \chi_1)\phi = \tilde{D}(g; \chi_2)\phi$.

We shall see later on that for every pair of complex numbers $\chi = (n_1, n_2)$, the representations $D(g; \chi)$ and $D(g; -\chi)$, where $-\chi = (-n_1, -n_2)$, have a non-degenerate invariant bilinear functional, and hence the conjugate representation $\tilde{D}(g; \chi)$ is an extension of $D(g; -\chi)$.

Realization of the Representation of the Principal Series

We now realize the representation of the principal series in some of the spaces introduced in Section 4-1. First this is done in the Hilbert space $L_2(Z)$.[9]

Again our starting point is the spinor representation discussed in Chapter 3. This representation was written in the form $D(g)p(z) = \alpha(zg)p(z(g))$, where $p(z) = p(z, \bar{z})$ was a polynomial of the space P_{mn}, $\alpha(zg) = (bz + d)^m(b\bar{z} + \bar{d})^n$, and $z(g)$ is given by Eq. (3-38). Let us try, therefore, a representation $g \to D(g)$ of the group $SL(2, C)$ in the space $L_2(Z)$ of the form $D(g)f(z) = \alpha(z, g)f(z(g))$, where $f(z)$ now is an element of the Hilbert space $L_2(Z)$ and $\alpha(z, g)$ is some unknown function of z and g of $SL(2, C)$. But in order that $g \to D(g)$ be a representation, we must have $D(g_1 g_2)f(z) = D(g_1)D(g_2)f(z)$. This imposes a condition on the function $\alpha(z, g)$ of the form $\alpha(z, g_1 g_2) = \alpha(z, g_1)\alpha(z(g_1), g_2)$. Denoting now $\alpha(e, g)$ by $\alpha(g)$, where e is the unit matrix. The latter equation gives, when one puts $z = e$ and $g_2 = g$, $\alpha(e, g_1 g) = \alpha(e, g_1)\alpha(e(g_1), g)$, or $\alpha(g_1 g) = \alpha(g_1)\alpha(e(g_1), g)$. If one choses g_1 to be an element k of the form given by Eq. (4-9), i.e. an element of the group K, then one has $\alpha(kg) = \alpha(k)\alpha(e(k), g)$. But one can easily see that $e(k) = e$. Hence one obtains $\alpha(kg) = \alpha(k)\alpha(g)$. On the other hand if one takes g_1 to be equal to z, one obtains $\alpha(zg) = \alpha(z)\alpha(e(z), g)$. But $e(z) = z$.[10] Hence we obtain $\alpha(zg) = \alpha(z)\alpha(z, g)$. The assumed form for the principal series of representation is, therefore, given by $D(g)f(z) = \{\alpha(zg)/\alpha(z)\}f(z(g))$. Finally, if we take $k = g = e$ we obtain $\alpha(e) = \alpha(e)\alpha(e)$ which yields either $\alpha(e) = 1$ or $\alpha(e) = 0$. We shall see below that $\alpha(e) = 1$ and the other case is impossible.

Let us now assume that the operator $D(g)$ is isometric (see Appendix C), thus $(D(g)f_1, D(g)f_2) = (f_1, f_2)$. Taking $f_1 = f_2 = f$ then gives

$$\int |\alpha(z, g)|^2 |f(z(g))|^2 \, dz = \int |f(z)|^2 \, dz$$

[9] I. M. Gelfand and M. A. Naimark, *Izv. Akad. Nauk SSSR, Ser. Mat.* **11,** 411, (1947); M. A. Naimark, *Linear Representations of the Lorentz Group*, Pergamon Press, New York, 1964.

[10] We recall that the variable $z(g) = g'_{21}/g'_{22}$, where the matrix g' of $SL(2, C)$ is given by $g' = zg$. Hence to calculate $e(k)$ one has $g' = ek = k$ and therefore $z(g) = g'_{21}/g'_{22} = k_{21}/k_{22} = 0$ and the matrix $z(g) = e$. In the case of calculating $e(z)$ the matrix $g' = ez = z$ and therefore the variable $z(g) = g'_{21}/g'_{22} = z_{21}/z_{22} = z$ and the matrix $z(g) = z$.

Changing variables $z \to z' = z(g)$ in the integral on the right-hand side of the latter equation, and using Eq. (4-8b), gives

$$\int |\alpha(z, g)|^2 |f(z(g))|^2 \,dz = \int \beta^{-1}(zg) |f(z(g))|^2 \,dz$$

where $\beta(g)$ is defined by Eq. (4-9). Since the function $f(z)$ of $L_2(Z)$ is arbitrary, we conclude that

$$|\alpha(z, g)| = \beta^{-1/2}(zg) \tag{4-38}$$

In particular, using Eq. (4-9), we obtain $|\alpha(e)| = |\alpha(e, e)| = \beta^{-1/2}(e) = 1$ and therefore $\alpha(e) \neq 0$. Conversely, if Eq. (4-38) is satisfied, then the operator $D(g)$ is isometric. Taking g as z^{-1} in Eq. (4-38), and using $\alpha(zg) = \alpha(z)\alpha(z, g)$, we obtain $|\alpha(z)| = 1$. One can show that it is always possible to take $\alpha(z) = 1$ (Problem 4.6). Consequently, the representation formula for the principal series can be written as $D(g)f(z) = \alpha(zg)f(z(g))$, with the function $\alpha(zg)$ determined by $|\alpha(zg)| = \beta^{-1/2}(zg)$, and $\beta(g) = |g_{22}|^4$. It remains to find out the function $\alpha(zg)$.

We have seen that $\alpha(kg) = \alpha(k)\alpha(g)$. It follows, since every g of $SL(2, C)$ can be written as $g = kz$ (see Lemma 3.1, Chapter 3), that $\alpha(g) = \alpha(kz) = \alpha(k)\alpha(z) = \alpha(k)$. Hence it is enough to find out the function $\alpha(k)$. We take now $k = k_1$, and $g = k_2$ in $\alpha(kg) = \alpha(k)\alpha(g)$, and obtain $\alpha(k_1 k_2) = \alpha(k_1)\alpha(k_2)$. Hence $\alpha(k_1^{-1}kk_1) = \alpha(k_1^{-1})\alpha(k)\alpha(k_1) = \alpha(k)$ since $\alpha(k_1^{-1})\alpha(k_1) = \alpha(e) = 1$. Writing now the matrix k explicitly (see Eq. (3-42)),

$$k = \begin{pmatrix} \lambda^{-1} & \mu \\ 0 & \lambda \end{pmatrix}$$

one sees that the function $\alpha(k)$ is a function of the two variables λ, μ, $\alpha(k) = \alpha(\lambda, \mu)$. We show below that α does not depend on the variable μ. For suppose

$$k_1 = \begin{pmatrix} \lambda_1^{-1} & 0 \\ 0 & \lambda_1 \end{pmatrix}$$

then

$$k_1^{-1}kk_1 = \begin{pmatrix} \lambda^{-1} & \lambda_1^2 \mu \\ 0 & \lambda \end{pmatrix}$$

and the relation $\alpha(k_1^{-1}kk_1) = \alpha(k)$ means $\alpha(\lambda, \lambda_1^2 \mu) = \alpha(\lambda, \mu)$. But λ_1 is an arbitrary complex variable. Hence α does not depend on μ, i.e. $\alpha(k) = \alpha(\lambda)$. The relation $\alpha(k_1 k_2) = \alpha(k_1)\alpha(k_2)$ then means $\alpha(\lambda_1 \lambda_2) = \alpha(\lambda_1)\alpha(\lambda_2)$, whereas the relation $|\alpha(g)| = \beta^{-1/2}(g) = |\mathring{g}_{22}|^{-2}$ gives $|\alpha(\lambda)| = |\lambda|^{-2}$.

Consequently, let us now define the function

$$\chi(\lambda) = |\lambda|^2 \alpha(\lambda)$$

satisfying $\chi(\lambda_1 \lambda_2) = \chi(\lambda_1)\chi(\lambda_2)$ and $|\chi(\lambda)| = 1$. If we denote $\lambda = \mathrm{e}^t \mathrm{e}^{i\phi}$, with $-\infty < t < \infty$ and $0 \leqslant \phi \leqslant 2\pi$, and $\lambda_1 = \mathrm{e}^t$, $\lambda_2 = \mathrm{e}^{i\phi}$, we obtain $\chi(\lambda) = \chi(\mathrm{e}^t)\chi(\mathrm{e}^{i\phi})$, and therefore it is enough to find the functions $\chi_1(t) = \chi(\mathrm{e}^t)$ and $\chi_2(\phi) = \chi(\mathrm{e}^{i\phi})$.

They satisfy $\chi_1(t_1+t_2)=\chi_1(t_1)\chi_1(t_2)$, $\chi_2(\phi_1+\phi_2)=\chi_2(\phi_1)\chi_2(\phi_2)$, with $|\chi_1(t)| = |\chi_2(\phi)| = 1$. Because the function $\chi(\lambda)$ is supposed to be single-valued, it must also satisfy $\chi_2(\phi + 2\pi) = \chi_2(\phi)$. The only solutions for χ_1 and χ_2, satisfying these conditions, are

$$\chi_1(t) = e^{i\rho t},\ \chi_2(\phi) = e^{im\phi}$$

where m must be an integer and ρ is a real number. Accordingly, the function $\chi(\lambda)$ will have the form

$$\chi(\lambda) = |\lambda|^{i\rho-m}\lambda^m$$

and the function $\alpha(k) = \alpha(\lambda)$ is given by

$$\alpha(k) = |\lambda|^{i\rho-m-2}\lambda^m$$

The principal series of representation is therefore given by

$$D(g)f(z) = \alpha(zg)f(z(g)) \tag{4-39a}$$

with $\alpha(g) = \alpha(k)$ is given by

$$\alpha(g) = |g_{22}|^{i\rho-m-2}g_{22}^m \tag{4-39b}$$

The variable $z(g)$ is given according to Eq. (3-38) by $(g_{11}z + g_{21})/(g_{12}z + g_{22})$. Conversely, with arbitrary real ρ and arbitrary integral m, Eqs. (4-39) determine an isometric operator $D(g)$ in the Hilbert space $L_2(Z)$. The problem of showing that Eqs. (4-39) indeed provide a representation is left for the reader to verify (Problem 4.7).

4-3 SU_2 DESCRIPTION OF THE PRINCIPAL SERIES

Properties of the Principal Series

In the last section we introduced the representation of the principal series, realized in the Hilbert space $L_2(Z)$. We now discuss some of its important properties and realize it in other spaces.

First one notices that for any functions $f_1(z)$ and $f_2(z)$ of $L_2(Z)$, the scalar expression $(D(g)f_1, f_2)$ is a continuous function on the group $SL(2, C)$. This follows from the definition of the representation formula (4-39).

The relation $D(g)D(g^{-1}) = D(g^{-1})D(g) = D(e) = 1$ shows that the operator $D(g)$ has an inverse, $D(g^{-1})$, which is also isometric in the Hilbert space $L_2(Z)$. Thus the operator $D(g)$ is unitary. Consequently one has: To every pair of numbers m and ρ, where m is an integer and ρ is a real number, there corresponds a unitary representation $g \to D(g)$ in the space $L_2(Z)$ defined by Eqs. (4-39). Furthermore, the detailed expression of the representation formula (4-39) shows a great similarity to the nonunitary spinor representation (see Problem 4.8).

Finally, we point out that the representation of the principal series is irreducible (Problem 4.9).

Realization of the Principal Series by Means of the Group SU_2

We now realize the representation of the principal series in the Hilbert space $L_2^{2s}(SU_2)$. We recall that the space $L_2^{2s}(SU_2)$ was the collection of functions $\phi(u)$ satisfying $\phi(\gamma u) = e^{is\psi}\phi(u)$, where γ is given by Eq. (4-12) (see Section 4.1). The scalar product was defined by $(\phi_1, \phi_2) = \int \phi_1(u)\bar{\phi}_2(u)\, du$. Using Eq. (4-39b) one sees that $\alpha(\gamma) = e^{im\omega}$, where $\omega = \psi/2$ is the phase in the matrix γ of Eq. (4-12). Hence the functions $\phi(u)$ of the space $L_2^{2s}(SU_2)$ satisfy the condition

$$\phi(\gamma u) = \alpha(\gamma)\phi(u) \tag{4-40}$$

where use has been made of $m = 2s$.

We now assign, similar to the correspondence (3-55) for the spinor case, to every function $f(z)$ of the space $L_2(Z)$ the function

$$\phi(u) = \pi^{1/2}\alpha(u)f(z) \tag{4-41}$$

where u of SU_2 and z of Z belong to the right coset $\tilde{z}$ (see Section 3.2). The mapping between the two spaces $L_2^{2s}(SU_2)$ and $L_2(Z)$ is isometric, since u and γu belong to the same coset $\tilde{z}$ and $\phi(\gamma u) = \pi^{1/2}\alpha(\gamma u)f(z) = \pi^{1/2}\alpha(\gamma)\alpha(u)f(z) = \alpha(\gamma)\phi(u)$, thus the condition (4-40) is satisfied. Furthermore,

$$\int |\phi(u)|^2\, du = \int |f(z)|^2\, dz, \tag{4-42}$$

so that the function defined by Eq. (4-41) belongs to the space $L_2^{2s}(SU_2)$ and the mapping from $f(z)$ to $\phi(u)$ preserves the norm. The proof of Eq. (4-42) is left for the reader (Problem 4.10). The inverse of Eq. (4-41) means, given an arbitrary function $\phi(u)$ of the space $L_2^{2s}(SU_2)$, one defines

$$f(z) = \pi^{-1/2}\alpha^{-1}(u)\phi(u) \tag{4-43}$$

again with u and z belonging to the right coset $\tilde{z}$, where $f(z)$ belongs to the space $L_2(Z)$.

We now find the representation of the principal series in the space $L_2^{2s}(SU_2)$. One has

$$D(g)\phi(u) = \pi^{1/2}\alpha(u)D(g)f(z) = \pi^{1/2}\alpha(u)\alpha(zg)f(z(g)) = \alpha(u)\{\alpha(zg)/\alpha(u(g))\}\phi(u(g)),$$

where $u(g)$ is an arbitrary element of the right coset $\tilde{u}(g)$ (see Section 3.2). But z and u belong to $\tilde{z}$ and $z(g)$ and $u(g)$ belong to $\tilde{z}(g)$, hence one can put $z = k_0 u$, $ug = ku(g)$ and therefore $\alpha(zg) = \alpha(k_0 ug) = \alpha(k_0)\alpha(ug) = \alpha(ug)/\alpha(u)$, thus $\alpha(u) \times \{\alpha(zg)/\alpha(u(g))\} = \alpha(ug)/\alpha(u(g))$. Hence the representation of the principal series in the Hilbert space $L_2^{2s}(SU_2)$ is given by

$$D(g)\phi(u) = \frac{\alpha(ug)}{\alpha(u(g))}\phi(u(g)) \tag{4-44a}$$

where $u(g)$ is an arbitrary matrix of the right coset $\tilde{u}(g)$ and is given by Eq. (3-57) and (3-62), and $\alpha(g)$ is given by

$$\alpha(g) = |g_{22}|^{i\rho - 2s - 2} g_{22}^{2s} \tag{4-44b}$$

Here $2s = m$ is an integer.

Realization of the Principal Series in the Space l_2^{2s}

Using the generalized Fourier transform, introduced in Section 4.1, we now express the representation of the principal series as infinite-dimensional matrices, the elements of which will be explicitly given as integrals over the group SU_2.[11]

To this end one first notices that the matrix elements $D^j_{sm}(u)$ of the irreducible representations of the group SU_2, for a fixed s, are elements of the Hilbert space $L_2^{2s}(SU_2)$ since they also satisfy $D^j_{sm}(\gamma u) = e^{is\psi} D^j_{sm}(u)$ (see Chapter 1) just as the functions $\phi(u)$ of the space $L_2^{2s}(SU_2)$ do. Consequently, the representation formula (4-44) can be applied to the functions $D^j_{sm}(u)$ to yield

$$D(g)D^j_{sm}(u) = \frac{\alpha(ug)}{\alpha(u(g))} D^j_{sm}(u(g)) \tag{4-45}$$

Using now the expansion (4-26) in the representation formula (4-44a), and using Eq. (4-45), one obtains

$$D(g)\phi(u) = \sum_j (2j + 1) \sum_m \phi^j_m \frac{\alpha(ug)}{\alpha(u(g))} D^j_{sm}(u(g)) \tag{4-46}$$

Since Eq. (4-45) is a representation of the group $SL(2, C)$ in the space $L_2^{2s}(SU_2)$, hence the function $\{\alpha(ug)/\alpha(u(g))\}D^j_{sm}(u(g))$ belongs to the space $L_2^{2s}(SU_2)$. Accordingly, it can be expanded in the form of an infinite series of the form (4-26). Such an expansion can be written as

$$\frac{\alpha(ug)}{\alpha(u(g))} D^j_{sm}(u(g)) = \sum_{j'} (2j' + 1) \sum_{m'} D^{jj'}_{mm'}(g; s, \rho) D^{j'}_{sm'}(u) \tag{4-47}$$

where, because of Eq. (4-27), one has

$$D^{jj'}_{mm'}(g; s, \rho) = \int \frac{\alpha(ug)}{\alpha(u(g))} D^j_{sm}(u(g)) \bar{D}^{j'}_{sm'}(u) \, du \tag{4-48}$$

Combining now Eqs. (4-46) and (4-47) one finally obtains

$$D(g)\phi(u) = \sum_j (2j + 1) \sum_m \phi'^j_m D^j_{sm}(u) \tag{4-49}$$

where the quantities ϕ'^j_m are defined by

$$\phi'^{j'}_{m'} \sum_{j=|s|}^{\infty} (2j + 1) \sum_{m=-j}^{j} D^{jj'}_{mm'}(g; s, \rho) \phi^j_m \tag{4-50}$$

[11] M. Carmeli, *J. Math. Phys.* **11**, 1917 (1970).

Accordingly, the operator $D(g)$ of the representation of the principal series of the group $SL(2, C)$ in the Hilbert space l_2^{2s} is described by the linear transformation given in Eq. (4-50) describing the law of transformation of the quantities ϕ_m^j, where $j = |s|, |s| + 1, |s| + 2, \ldots$, and $m = -j, -j + 1, -j + 2, \ldots, j$. The coefficients $D_{mm'}^{jj'}(g; s, \rho)$, given by Eq. (4-49), are functions on the group $SL(2, C)$ as well as of ρ and s, where ρ is real and $2s$ is an integer. These functions are the matrix elements of an infinite-dimensional matrix whose rows are labelled by (j, m) and columns by (j', m'). They are given by Eq. (4-48) as integrals over the group SU_2.

It will be noted that the quantities ϕ_m^j, whose transformation law is given by Eq. (4-50), were obtained from the representation formula (4-44) of the principal series in analogy with the way two-component spinors, transforming according to Eq. (3-28), were obtained from the representation formula (3-56) [or, in fact, (3-23)] of the spinor representation. Both ϕ_m^j and the two-component spinors are coefficients appearing in the space of representations. Of course, the spinor representation was finite and nonunitary, whereas the representation of the principal series was infinite-dimensional and unitary.

The Principal Series as a Representation for the Group SU_2

Since the group SU_2 is a subgroup of the group $SL(2, C)$, the representation of the principal series is also a representation for the group SU_2. Equation (4-44a) will then have the form $D(u_0)\phi(u) = \{\alpha(uu_0)/\alpha(u(u_0))\}\phi(u(u_0))$, where u_0 is an arbitrary element of the group u_0. A straightforward calculation, using Eqs. (3-65), shows that $u(u_0) = \gamma uu_0$, where γ is the matrix

$$\gamma = \begin{pmatrix} e^{i\Lambda} & 0 \\ 0 & e^{-i\Lambda} \end{pmatrix}$$

and Λ is an arbitrary real number (phase). If one chooses $\Lambda = 0$, one thus obtains for the representation of the principal series as a representation for SU_2,

$$D(u_0)\phi(u) = \phi(uu_0) \tag{4-51}$$

realized in the Hilbert space $L_2^{2s}(SU_2)$.

Let us now assume that in the space $L_2^{2s}(SU_2)$ there is a subspace R in which the irreducible representation of weight j of the group SU_2 is realized, and let $D_{mn}^j(u)$, with $m, n = -j, -j + 1, \ldots, j$, be the matrix elements of this representation in its canonical basis. Let $\phi_m(u)$, with $m = -j, -j + 1, \ldots, j$, be the canonical basis in the subspace R, thus

$$D(u_0)\phi_n(u) = \sum_{m=-j}^{j} D_{mn}^j(u_0)\phi_m(u) \tag{4-52}$$

Hence, comparing Eqs. (4-51) and (4-52), one obtains

$$\phi_n(uu_0) = \sum D_{mn}^j(u_0)\phi_m(u) \tag{4-53}$$

It is left for the reader (Problem 4.13) to show that

$$\phi_m(u) = \alpha_s D_{sm}^j(u) \tag{4-54}$$

where α_s is a constant. Accordingly, the subspace R is spanned by the functions $D^j_{sm}(u)$, where $m = -j, -j+1, \ldots, j$. Hence the irreducible representation of SU_2 of weight j occurs only once in the representation of the principal series in the Hilbert space $L_2^{2s}(SU_2)$, where s is one of the numbers $-j, -j+1, \ldots, j$. Conversely, if s is one of the numbers $-j, -j+1, \ldots, j$, then in the subspace R spanned by the function $D^j_{sm}(u)$, where $m = -j, -j+1, \ldots, j$, the irreducible representation of SU_2 is included in the principal series.

Accordingly, when considered as a representation of the group SU_2, the representation of the principal series is *reducible*, and decomposes into the sum of irreducible representations.

Functions on the Group $SL(2, C)$

The theory of functions defined on a group was introduced in Chapter 1 and has been applied to the group SU_2. To conclude this section we now apply the theory to functions defined on the group $SL(2, C)$.

A function $f(g)$ on the group $SL(2, C)$,

$$g = \begin{pmatrix} a & b \\ c & d \end{pmatrix}; \qquad ad - bc = 1, \tag{4-55}$$

is called *rapidly decreasing* if it satisfies the condition

$$|f(g)| < C|g|^{-n} \tag{4-56}$$

where $n > 0$, and

$$|g| = \{|a|^2 + |b|^2 + |c|^2 + |d|^2\}^{1/2}. \tag{4-57}$$

The function $|g|$ is called the *norm* of the matrix g.

Integration on the group $SL(2, C)$ can be defined by introducing the measure dg, which is invariant under left and right translations on the group. Hence if g is replaced by $g_0 g$ or gg_0, one has $dg = d(g_0 g) = d(gg_0)$. The integral is, of course, invariant under left and right translation of the function $f(g)$,

$$\int f(g_0 g)\, dg = \int f(gg_0)\, dg = \int f(g)\, dg \tag{4-58}$$

To find out the explicit expression of the measure dg, we proceed as follows. Let us first consider the set of all complex matrices

$$g = \begin{pmatrix} \alpha & \beta \\ \gamma & \delta \end{pmatrix} \tag{4-59}$$

in two dimensions, and associate with each matrix g of the form (4-59) a point (α, β, γ, δ) in a space of four complex dimensions. The unimodular matrices then form a quadratic surface of the form $\alpha\delta - \beta\gamma = 1$ in this four-dimensional space. With this surface we associate the differential form $\sigma(g)$ defined by[12]

$$d\alpha\, d\beta\, d\gamma\, d\delta = d(\alpha\delta - \beta\gamma) \cdot \sigma(g) \tag{4-60}$$

[12] Here $d(\alpha\delta - \beta\gamma)$. $\sigma(g)$ is the exterior product of the differential forms. See, for example, I. M. Gelfand and G. E. Shilov, *Generalized Functions, Vol. 1. Properties and Operations*, Academic Press, New York and London, 1964.

The differential form will then have the form[13]

$$\sigma(g) = \delta^{-1}\, d\beta\, d\gamma\, d\delta = \gamma^{-1}\, d\alpha\, d\gamma\, d\delta = -\beta^{-1}\, d\alpha\, d\beta\, d\delta$$
$$= -\alpha^{-1}\, d\alpha\, d\beta\, d\gamma \tag{4-61}$$

Since α, β, γ, and δ cannot all simultaneously vanish, $\sigma(g)$ has no singularities on the group $SL(2, C)$. Moreover, $\sigma(g)$ is invariant under translations since then α, β, γ, and δ undergo a linear transformation whose determinant is unity, and therefore $d\alpha\, d\beta\, d\gamma\, d\delta$ is invariant under such transformations. On the other hand, the determinant $\alpha\delta - \beta\gamma$ of the matrix g is invariant under such transformations. Therefore, $d(\alpha\delta - \beta\gamma)$ is also invariant, and accordingly $\sigma(g)$ is invariant under translations. The measure on $SL(2, C)$ is then defined by

$$dg = \left(\frac{i}{2}\right)^3 \sigma\bar{\sigma} \tag{4-62}$$

leading to the explicit form

$$dg = \left(\frac{i}{2}\right)^3 \frac{d\beta\, d\bar{\beta}\, d\gamma\, d\bar{\gamma}\, d\delta\, d\bar{\delta}}{|\delta|^2} = \left(\frac{i}{2}\right)^3 \frac{d\alpha\, d\bar{\alpha}\, d\gamma\, d\bar{\gamma}\, d\delta\, d\bar{\delta}}{|\gamma|^2}$$
$$= \left(\frac{i}{2}\right)^3 \frac{d\alpha\, d\bar{\alpha}\, d\beta\, d\bar{\beta}\, d\delta\, d\bar{\delta}}{|\beta|^2} = \left(\frac{i}{2}\right)^3 \frac{d\alpha\, d\bar{\alpha}\, d\beta\, d\bar{\beta}\, d\gamma\, d\bar{\gamma}}{|\alpha|^2} \tag{4-63}$$

It will be noted that dg is invariant also under inversions, $d(g^{-1}) = dg$. This is so since if the matrix g has determinant unity, then

$$g^{-1} = \begin{pmatrix} \delta & -\beta \\ -\gamma & \alpha \end{pmatrix}$$

and $d(g^{-1})$ is given by Eq. (4-63) when α, β, γ, δ are replaced by δ, $-\beta$, $-\gamma$, α, respectively, a replacement that leaves Eq. (4-63) invariant. Hence one has

$$\int f(g^{-1})\, dg = \int f(g)\, dg \tag{4-64}$$

Finally, a function $f(g)$ is called *summable* on $SL(2, C)$ if $\int |f(g)|\, dg$ converges. One can show that every rapidly decreasing function is summable. Furthermore, with each element g of $SL(2, C)$ of the form (4-59) we can associate a point on the surface $\alpha\delta - \beta\gamma = 1$ in the space of four complex dimensions. Any three of the parameters α, β, γ, δ can be chosen to be local coordinates in the neighborhood of any such point. The function $f(g)$ is then called *infinitely differentiable* in a neighbourhood of g_0 if, considered as a function of the three coordinates chosen in this neighbourhood, it has derivatives of all orders in this neighbourhood. Such a definition is independent of the choice of coordinate system.

[13] See, for example, I. M. Gelfand, M. I. Graev, and N. Ya. Vilenkin, *op. cit.*

4-4 COMPARISON WITH THE INFINITESIMAL APPROACH

Comparison of the Parameters (s, ρ) and (j_0, c)

We have seen in the last section that all irreducible representations of the group $SL(2, C)$ of the principal series are characterized by a pair of numbers (s, ρ), where s is an integer or half-integer and ρ is real. When the representation is given in its global form, the space of the representation depends on the value of s (see Section 4.3) and the operators depend on both s and ρ [compare Eq. (4-48)].

The representation of the principal series was already defined in terms of the infinitesimal operators in Section 2-3. It was found to depend on a pair of parameters (j_0, c), where j_0 is an arbitrary non-negative integral or half-integral number and c is pure imaginary. The values of these parameters occurred in the formulas for the infinitesimal operators (2-29) and (2-30).

By applying the global form of a given representation to infinitesimal elements of the group $SL(2, C)$, one can calculate the infinitesimal operators of the representation. Comparing the infinitesimal operator thus obtained with the results of Section 2-3 one establishes the relationship between the pairs of parameters (s, ρ) and (j_0, c). The result is as follows (Problem 4.17):

$$j_0 = |s|, \quad c = -i(\text{sign } s)\frac{\rho}{2}; \qquad s \neq 0 \tag{4-65a}$$

$$j_0 = 0, \quad c = \pm i\frac{\rho}{2}; \qquad s = 0 \tag{4-65b}$$

As a result of Eqs. (4-65), it follows that two representations of the principal series, characterized by the pairs (s, ρ) and $(-s, -\rho)$, are unitarily equivalent. The proof of this statement is left as an exercise for the reader (Problem 4.18).

The rest of this section will be devoted to discussing some harmonic analysis on the group $SL(2, C)$.

Tangent Space to the Group $SL(2, C)$[14]

Consider the element $h(z)$ of the group $SL(2, C)$,

$$h(z) = \begin{pmatrix} a(z) & b(z) \\ c(z) & d(z) \end{pmatrix}; \qquad a(z)\, d(z) - b(z)c(z) = 1 \tag{4-66}$$

where the matrix elements of $h(z)$ are analytic functions of the complex variable z for small values of $|z|$, and assume that $h(0) = e$, where e is the 2×2 unit matrix. In other words we assume $a(0) = d(0) = 1$ and $b(0) = c(0) = 0$. The matrix $h(z)$ can be considered as a *curve*, whereas the matrix $h(0) = e$ as a *point*, in the space of the group $SL(2, C)$. The matrix

$$h = h'(0) = \begin{pmatrix} a'(0) & b'(0) \\ c'(0) & d'(0) \end{pmatrix} \tag{4-67}$$

[14] See for example, I. M. Gelfand, M. I. Graev, and N. Y. Vilenkin, *op. cit.*

is consequently called the *tangent vector* of the curve $h(z)$ at e. The aggregate of all vectors, tangent to curves passing through the point e, forms a three-dimensional complex linear space since if h_1 and h_2 are tangent vectors at e to the curves $h_1(z)$ and $h_2(z)$, then $\lambda_1 h_1 + \lambda_2 h_2$ is the tangent vector to $h(z) = h_1(\lambda_1 z)h_2(\lambda_2 z)$. This linear space is known as the *tangent space* to the group at e, and is denoted by Λ. A matrix belongs to the space Λ if and only if its trace vanishes.

Lie Operators

The concepts of Lie operators play important roles in the theory of representations. They are defined as follows.

Let $h(z)$ be a curve passing through the point e with a tangent vector h, and define the operator $A(h)f(g) = \{\partial f(h(z)g)/\partial z\}_{z=0}$. If one parametrizes the elements

$$g = \begin{pmatrix} \alpha & \beta \\ \gamma & \delta \end{pmatrix}$$

of the group $SL(2, C)$ by the elements α, β, and δ, then $f(h(z)g) = f(a(z)\alpha + b(z)\gamma, a(z)\beta + b(z)\delta, c(z)\beta + d(z)\delta)$ and therefore

$$\begin{aligned} \left.\frac{\partial f}{\partial z}\right|_{z=0} &= \frac{\partial f}{\partial \alpha}\{\alpha a'(0) + \gamma b'(0)\} + \frac{\partial f}{\partial \beta}\{\beta a'(0) + \delta b'(0)\} + \frac{\partial f}{\partial \delta}\{\beta c'(0) + \delta d'(0)\} \\ &= \left\{a'(0)\left(\alpha\frac{\partial}{\partial \alpha} + \beta\frac{\partial}{\partial \beta} - \delta\frac{\partial}{\partial \delta}\right) + b'(0)\left(\gamma\frac{\partial}{\partial \alpha} + \delta\frac{\partial}{\partial \beta}\right) + c'(0)\beta\frac{\partial}{\partial \delta}\right\} f \end{aligned} \tag{4-68}$$

where use has been made of the fact that $a(0) = d(0) = 1$, $b(0) = c(0) = 0$, and $d'(0) = -a'(0)$. The operator $A(h)$ is called the *left Lie operator*. The aggregate of all the left Lie operators forms a linear space since $A(\lambda_1 h_1 + \lambda_2 h_2) = \lambda_1 A(h_1) + \lambda_2 A(h_2)$. One can choose the one-parameter subgroups (see Chapter 1) as

$$h_1(z) = \begin{pmatrix} 1 & 0 \\ z & 1 \end{pmatrix}, \qquad h_2(z) = \begin{pmatrix} e^z & 0 \\ 0 & e^{-z} \end{pmatrix}, \qquad h_3(z) = \begin{pmatrix} 1 & z \\ 0 & 1 \end{pmatrix} \tag{4-69}$$

The tangent vectors are then given by

$$h_1 = \begin{pmatrix} 0 & 0 \\ 1 & 0 \end{pmatrix}, \qquad h_2 = \begin{pmatrix} 1 & 0 \\ 0 & -1 \end{pmatrix}, \qquad h_3 = \begin{pmatrix} 0 & 1 \\ 0 & 0 \end{pmatrix} \tag{4-70}$$

The three matrices h_1, h_2, and h_3 can then be taken as a basis in the linear space Λ. Using Eq. (4-68) then leads to the left Lie operators A_1, A_2, and A_3 corresponding to h_1, h_2, and h_3, with

$$\begin{aligned} A_1 &= \beta\frac{\partial}{\partial \delta} \\ A_2 &= \alpha\frac{\partial}{\partial \alpha} + \beta\frac{\partial}{\partial \beta} - \delta\frac{\partial}{\partial \delta} \\ A_3 &= \gamma\frac{\partial}{\partial \alpha} + \delta\frac{\partial}{\partial \beta} \end{aligned} \tag{4-71}$$

In the same fashion one can define the *right Lie operators*, defined by $B(h)f(g) = \{\partial f(g(h(z))/\partial z\}_{z=0}$, where $h(z)$ is any curve for which $h(0) = e$ and $h'(0) = h$. One then obtains for these operators

$$B_1 = \beta \frac{\partial}{\partial \alpha}$$

$$B_2 = \alpha \frac{\partial}{\partial \alpha} - \beta \frac{\partial}{\partial \beta} - \delta \frac{\partial}{\partial \delta} \tag{4-72}$$

$$B_3 = \alpha \frac{\partial}{\partial \beta} + \gamma \frac{\partial}{\partial \delta}$$

where $B_1 = B(h_1)$, $B_2 = B(h_2)$, and $B_3 = B(h_3)$. One can then show that the commutation relations between the left Lie operators are given by (Problem 4.19):

$$[A_1, A_2] = -2A_1$$

$$[A_2, A_3] = -2A_3 \tag{4-73a}$$

$$[A_3, A_1] = -A_2$$

whereas the right Lie operators give

$$[B_1, B_2] = 2B_1$$

$$[B_2, B_3] = 2B_3 \tag{4-73b}$$

$$[B_3, B_1] = B_2$$

Moreover, the left Lie operators commute with the right Lie operators. The above commutation relations can then be used to write any polynomial of Lie operators with constant coefficients as a linear combination of ordered products.

Laplacian Operators

Let us now find out the operator that commutes with all the Lie operators. It is given by

$$\Delta = A_1 A_3 + A_3 A_1 + \tfrac{1}{2}A_2^2 \tag{4-74}$$

The operator Δ is called the *Laplacian operator* for the group $SL(2, C)$, and can also be written as

$$\Delta = \text{Tr } A^2 = \text{Tr } B^2 \tag{4-75}$$

where A and B are defined by

$$A = \begin{pmatrix} \frac{1}{2}A_2 & A_3 \\ A_1 & -\frac{1}{2}A_2 \end{pmatrix}, \qquad B = \begin{pmatrix} \frac{1}{2}B_2 & B_3 \\ B_1 & -\frac{1}{2}B_2 \end{pmatrix} \tag{4-76}$$

Equation (4-75) is a result of the fact that $A = (g^t)^{-1}Bg^t$ (see Problem 4.20) and therefore $\text{Tr } A^2 = \text{Tr } B^2$. Since Δ can be written as a polynomial of the $A(h)$ with constant coefficients, it commutes with all the $B(h)$. On the other hand, it can also be written as a polynomial of the $B(h)$ with constant coefficients, thus commutes with all the $A(h)$. Hence Δ commutes with all the Lie operators.

PROBLEMS

4.1 The transformation $z \to z' = z(g)$ of Eq. (4-7) corresponds to a certain transformation of the variables x and y into x' and y', where $z = x + iy$ and $z' = x' + iy'$. Use the Cauchy–Riemann equations to show that the Jacobian $|\partial(x', y')/\partial(x, y)|$ of this transformation is equal to the square of the modulus of the derivative dz'/dz. Use Eq. (4-7) to show that the Jacobian is equal to $|g_{12}z + g_{22}|^{-4}$.

4.2 Prove the completeness of the space $L^2(SU_2)$.

4.3 Show that, using Eq. (4-20),

$$\int |f(u)|^2\, du = \sum_j (2j+1)\, \mathrm{Tr}\,(F^{j\dagger}F^j).$$

4.4 Prove Eq. (4-21).

4.5 Prove Eq. (4-31).

4.6 Show that it is always possible to choose the function $\alpha(z)$, whose modulus $|\alpha(z)| = 1$, to be equal to 1. Prove this by rewriting the representation formula for the principal series, using the new function $\tilde{f}(z) = \alpha(z)f(z)$ which also belongs to the Hilbert space $L_2(Z)$. Show that the operator of the representation can then be written as

$$\tilde{D}(g)\tilde{f}(z) = \tilde{\alpha}(zg)\tilde{f}(z(g))$$

where now $\tilde{\alpha}(z) = 1$. [See M. A. Naimark, *Linear Representations of the Lorentz Group*, Pergamon Press, New York, 1964, p. 145.]

4.7 Prove that the correspondence $g \to D(g)$ defined by Eqs. (4-39) is a representation of the group $SL(2, C)$, i.e., show that $D(e) = 1$ and $D(g_1)D(g_2) = D(g_1 g_2)$.

4.8 Write the representation of the principal series, given by Eq. (4-39), explicitly. Show that it has the form

$$D(g)f(z) = (g_{12}z + g_{22})^{m/2+i\rho/2-1}(\bar{g}_{12}\bar{z} + \bar{g}_{22})^{-m/2+i\rho/2-1} f\left(\frac{g_{11}z + g_{21}}{g_{12}z + g_{22}}\right).$$

Hence it has the same form as the spinor representation given by Eq. (3-23), where the polynomial $p(z, \bar{z})$ is now replaced by $f(z)$ and the powers m and n are replaced now by $m/2 + i\rho/2 - 1$ and $-m/2 + i\rho/2 - 1$, respectively.

4.9 Show that the representation of the principal series is irreducible. [See M. A. Naimark, *Linear Representations of the Lorentz Group*, Pergamon Press, New York, 1964, p. 151.]

4.10 Prove Eq. (4-42). Use the three parameters t, θ, θ' as indicated in Problem 1.9 to parametrize u of SU_2. To relate these variables to z use $z = r\,e^{ik}$, $\arg \lambda = \omega$, and Eqs. (3-54). Show that one obtains the following relations between the two sets of variables (t, θ, θ') and (r, ω, k): $t = 1/(1 + r^2)$, $\theta = -\omega$, $\theta' = \pi + \omega - k$ and therefore $|\alpha(u)|^2\, dt\, d\theta\, d\theta' = 2r\, dr\, dk\, d\omega = 2\, d\omega\, dz$, or $|\beta(u)|^{-1}\, dt\, d\theta\, d\theta' = 2\, d\omega\, dz$.

4.11 Show that as a result of the fact that the representation of the principal series is unitary one obtains:

$$du' = \frac{\beta(u(g))}{\beta(ug)}\, du$$

if $u' = u(g)$.

4.12 Calculate the expression $\alpha(ug)/\alpha(u(g))$, appearing in the representation (4-44) of the principal series, for the case for which $g = u_0$ (unitary matrix of SU_2) and

$$\varepsilon = g = \begin{pmatrix} \varepsilon_{22}^{-1} & 0 \\ 0 & \varepsilon_{22} \end{pmatrix}$$

where ε_{22} is real. Show that for the first case one has

$$\alpha(uu_0)/u(u(u_0)) = e^{2is\Lambda}$$

whereas for the second case one obtains

$$\alpha(u\varepsilon)/\alpha(u(\varepsilon)) = |\lambda|^{i\rho-2}\, e^{2is\Lambda}$$

where $|\lambda|^2 = |\beta|^2\varepsilon_{22}^{-2} + |\alpha|^2\varepsilon_{22}^2$. [Use Eqs. (3-62) and (3-63)].

4.13 Prove Eq. (4-54). Show that $\alpha_s = (2j+1)^{1/2}\alpha_{s\rho}^j$, with $|\alpha_{s\rho}^j| = 1$. Show that

$$\alpha_{s\rho}^j = \prod_{n=j_0}^{j} \frac{-2n + i\rho}{(4n^2 + \rho^2)^{1/2}}$$

4.14 Show that the norm $|g|$ of a matrix g, defined by Eq. (4-57), satisfies the following properties:

$$|g_1 + g_2| \leqslant |g_1| + |g_2|$$

$$|g_1 g_2| \leqslant |g_1|\,|g_2|$$

$$|\alpha g| = |\alpha|\,|g|$$

where α is an arbitrary number.

4.15 Show the invariance of the measure dg of the group $SL(2, C)$, i.e., $dg = d(g_0 g) = d(g g_0)$, by calculating the Jacobian of the transformation from g to $g_0 g$ and $g g_0$.

4.16 Show that the measure dg on the group $SL(2, C)$ can also be written as

$$dg = \frac{d\beta_1\, d\beta_2\, d\gamma_1\, d\gamma_2\, d\delta_1\, d\delta_2}{|\delta|^2} = \frac{d\alpha_1\, d\alpha_2\, d\beta_1\, d\beta_2\, d\gamma_1\, d\gamma_2}{|\gamma|^2}$$

where $\alpha = \alpha_1 + i\alpha_2$, $\beta = \beta_1 + i\beta_2$, $\gamma = \gamma_1 + i\gamma_2$, $\delta = \delta_1 + i\delta_2$, and α, β, γ, δ are the matrix elements of $g \in SL(2, C)$ with $\alpha\delta - \beta\gamma = 1$.

4.17 Prove Eqs. (4-65).

4.18 Show that two representations of the principal series characterized by the pairs (s, ρ) and $(-s, -\rho)$ are unitarily equivalent.

4.19 Prove Eqs. (4-73). Show also that the left Lie operators commute with the right Lie operators.

4.20 Define the two matrices

$$A = \begin{pmatrix} \frac{1}{2}A_2 & A_3 \\ A_1 & -\frac{1}{2}A_2 \end{pmatrix}, \qquad B = \begin{pmatrix} \frac{1}{2}B_2 & B_3 \\ B_1 & -\frac{1}{2}B_2 \end{pmatrix}$$

whose elements are Lie operators. Show that there is a relationship between the left and right Lie operators of the form $g^t A = B g^t$, where Bg denotes symbolic matrix multiplication, i.e., the elements of the matrix g^t are not differentiated.

CHAPTER

FIVE

COMPLEMENTARY SERIES OF REPRESENTATIONS OF $SL(2, C)$

In Chapter 4 the representation of the principal series of the group $SL(2, C)$ was realized in varieties of forms in different infinite-dimensional spaces. The representation of the principal series, however, do not realize all irreducible unitary representations of the group $SL(2, C)$. Rather, every irreducible unitary representation of the group $SL(2, C)$ is equivalent to a representation of either the principal series or the complementary series of representations. In this chapter we give the representation of the complementary series. We first define the space of representation of the complementary series and then realize the representation in this space. Subsequently, a SU_2 description of the representation of the complementary series is given. This is followed by a comparison with the infinitesimal approach. A brief review of an operator formulation of the representation theory of the group $SL(2, C)$ then concludes the chapter.

5-1 REALIZATION OF THE COMPLEMENTARY SERIES

In Chapter 4 the representation of the principal series of the group $SL(2, C)$, which is unitary and irreducible, was realized in varieties of forms in different spaces. One of its forms was given in the Hilbert space $L_2(Z)$ by the representation formula (4-39).

Complementary Series

The representations of the principal series, however, do not realize all irreducible unitary representations of the group $SL(2, C)$. Rather, every irreducible unitary representation of the group $SL(2, C)$ is unitarily equivalent to a representation of either the principal series or the *complementary series* of representations.

Formally, the formula of the representation of the complementary series can be obtained from the formula of the representation of the principal series, Eq. (4-39), if one takes ρ in the form $\rho = i\sigma$ and $s = 0$, where now σ is real and has the values $0 < \sigma < 2$. Unfortunately, the operators thus obtained follow not to be unitary in the scalar product defined by Eq. (4-2) for the Hilbert space $L_2(Z)$. In fact, Eq. (4-39) defines a unitary operator if and only if the function $\alpha(g)$ is given by Eq. (4-39b) with real ρ.

A realization of the representation of the complementary series in terms of unitary operators is, however, possible when defined on a Hilbert space to be defined as follows, and in which the variable ρ is taken to be imaginary.

Space of Representation[1]

To this end let us define a linear space of functions $f(z)$, denoted by R, in which the scalar product is defined by[2]

$$\langle f_1, f_2 \rangle = \iint K(z_1, z_2) f_1(z_1) \bar{f}_2(z_2)\, dz_1\, dz_2 \tag{5-1}$$

for any two functions $f_1(z)$ and $f_2(z)$ of R, and where $K(z_1, z_2)$ is some function of z_1 and z_2, called the *kernel*. Furthermore, we assume that the space R and the kernel K satisfy the following conditions:

(1) The space R contains all bounded measurable functions $f(z)$ which vanish in the exterior of some bounded set in the z-plane;

(2) The integral for the scalar product in the space R, Eq. (5-1), converges absolutely for all $f_1(z)$ and $f_2(z)$ of the space R; and

(3) The space R is invariant with respect to all operators $D(g)$ of the representation of the complementary series.

Now since we require the representation to be unitary, the scalar product given by Eq. (5-1) must be invariant under replacing the functions f_1 and f_2 by

[1] I. M. Gelfand and M. A. Naimark, *Izv. Akad. Nauk SSSR, Ser. Mat.* **11,** 411 (1974); M. A. Naimark, *Linear Representations of the Lorentz Group*, Pergamon Press, New York, 1964.

[2] The scalar product given by Eq. (5-1) is a generalization to our previous scalar product of Eq. (4-2) just as the scalar product $(x, y) = \sum a_{ij} x_i \bar{y}_j$, which is a positive-definite Hermitian quadratic form, of a finite-dimensional space is a generalization of the scalar product $(x, y) = \sum x_i \bar{y}_i$.

$D(g)f_1$ and $D(g)f_2$, respectively. Thus we require that $\langle D(g)f_1, D(g)f_2\rangle = \langle f_1, f_2\rangle$ or, using Eq. (4-39a).

$$\iint K(z_1, z_2)\alpha(z_1 g)\bar{\alpha}(z_2 g) f_1(z_1(g))\overline{f_2}(z_2(g))\, dz_1\, dz_2$$
$$= \iint K(z_1', z_2') f_1(z_1')\overline{f_2}(z_2')\, dz_1'\, dz_2' \qquad (5\text{-}2)$$

If we now make the substitution $z_1' = z_1(g)$ and $z_2' = z_2(g)$, and using Eq. (4-8b), then the right-hand side of Eq. (5-2) will have the form

$$\iint K(z_1(g), z_2(g)) f_1(z_1(g))\overline{f_2}(z_2(g))\beta^{-1}(z_1 g)\beta^{-1}(z_2 g)\, dz_1\, dz_2$$

Equating this expression with the left-hand side of Eq. (5-2), and using the condition (1) for the space R mentioned above, we obtain

$$K(z_1, z_2)\alpha(z_1 g)\bar{\alpha}(z_2 g) = K(z_1(g), z_2(g))\beta^{-1}(z_1 g)\beta^{-1}(z_2 g) \qquad (5\text{-}3)$$

Hence if we use Eq. (4-39b), Eq. (3-38), and Eq. (4-9) to write down explicitly the functions $\alpha(g)$, $z(g)$ and $\beta(g)$, respectively in Eq. (5-3) we obtain

$$K\left(\frac{g_{11}z_1 + g_{21}}{g_{12}z_1 + g_{22}}, \frac{g_{11}g_1 + g_{21}}{g_{12}g_2 + g_{22}}\right) = K(z_1, z_2)\,|g_{12}z_1 + g_{22}|^{-m+i\rho+2}$$
$$\times (g_{12}z_1 + g_{22})^m\,|g_{12}z_2 + g_{22}|^{-m+i\bar{\rho}+2}$$
$$\times (\bar{g}_{12}\bar{z}_2 + \bar{g}_{22})^m \qquad (5\text{-}4)$$

Accordingly, we may choose the matrix g to be given by z_0, i.e., we choose

$$g = \begin{pmatrix} 1 & 0 \\ z_0 & 1 \end{pmatrix}$$

and as a result Eq. (5-4) gives a condition on the function $K(z_1, z_2)$,

$$K(z_1 + z_0, z_2 + z_0) = K(z_1, z_2) \qquad (5\text{-}5)$$

If, in particular, one now takes $z_0 = -z_2$, then one obtains

$$K(z_1 - z_2, 0) = K(z_1, z_2) \qquad (5\text{-}6)$$

We will denote the function $K(z, 0)$ by $K_1(z)$, thus we obtain from Eq. (5-6)

$$K_1(z_1 - z_2) = K(z_1, z_2) \qquad (5\text{-}7)$$

Using this relation in Eq. (5-4), we then obtain

$$K_1\left(\frac{z_1 - z_2}{(g_{12}z_1 + g_{22})(g_{12}z_2 + g_{22})}\right) = K_1(z_1 - z_2)\,|g_{12}z_1 + g_{22}|^{-m+i\rho+2}$$
$$\times (g_{12}z_1 + g_{22})^m\,|g_{12}z_2 + g_{22}|^{-m-i\bar{\rho}+2}$$
$$\times (\bar{g}_{12}\bar{z}_2 + \bar{g}_{22})^m \qquad (5\text{-}8)$$

Equation (5-8) then gives (see Problem 5.1) $m = 0$ and $\bar{\rho} = -\rho$, namely ρ is a pure imaginary number, $\rho = i\sigma$, where σ is real. Equation (5-8) also gives, when one takes $z_2 = 0$, choosing g_{12} so that $g_{12} z_1 + g_{22} = 1$, and putting $g_{22} = z_1 = z$,

$$K_1(z) = K_1(1) |z|^{\sigma - 2} \tag{5-9}$$

Hence if we take $z_1 - z_2 = z$ we obtain

$$K(z_1, z_2) = C |z_1 - z_2|^{\sigma - 2} \tag{5-10}$$

where we have used the notation $K_1(1) = C$, with C as an arbitrary constant.

Accordingly, the formula for the representation of the complementary series will have the form

$$D(g)f(z) = |g_{12} z + g_{22}|^{-2-\sigma} f\left(\frac{g_{11} z + g_{21}}{g_{12} z + g_{22}}\right) \tag{5-11}$$

whereas the scalar product (5-1) will have the form

$$\langle f_1, f_2 \rangle = C \iint |z_1 - z_2|^{\sigma - 2} f_1(z_1) \bar{f_2}(z_2)\, dz_1\, dz_2 \tag{5-12}$$

thus imposing the condition $\sigma > 0$ so as the integral above converges. The set of all measurable functions $f(z)$ satisfying $|f(z)| \leqslant (1 + |z|^2)^{-1-\sigma}$ for almost all complex z then satisfies the three conditions (1)–(3) that the space R has to satisfy.

Value of the Parameter σ

Equation (5-12) defines a scalar product only if $\langle f, f \rangle \geqslant 0$ (see Appendix C). Hence one has the condition

$$\iint |z_1 - z_2|^{\sigma - 2} f(z_1) \bar{f}(z_2)\, dz_1\, dz_2 \geqslant 0 \tag{5-13}$$

If follows that this condition of positive definiteness is satisfied for $0 < \sigma < 2$, along with the possibility of choosing $C = 1$ (Problem 5.2).

In the case for which $\sigma = 2$ and $C = 1$, the scalar product (5-12) will have the form

$$\langle f_1, f_2 \rangle = \int f_1(z_1)\, dz_1 \int \bar{f_2}(z_2)\, dz_2 \tag{5-14}$$

a particular form of which is

$$\langle f, f \rangle = \left| \int f(z)\, dz \right|^2 \geqslant 0 \tag{5-15}$$

thus $\langle f, f \rangle = 0$ only if $\int f(z)\, dz = 0$. In this case the aggregate of all such functions $f(z)$ provides a one-dimensional Hilbert space, where the scalar product is given by Eq. (5-14). In this case $D(g)f = f$, or $D(g) = 1$. Hence the one-dimensional representation $g \to 1$ is the unit representation. In other words the unit representation is the limit of the representation of the complementary series when the parameter $\sigma = 2$ (Problem 5.3).

Finally, the case for which $\sigma > 2$, Eqs. (5-11) and (5-12) do not yield a unitary representation (Problem 5.4).

Realization of the Complementary Series

We denote by $\tilde{R}$ the set of all functions $f(z)$ which vanish outside some circle and are differentiable with respect to x and y, where $z = x + iy$, as many times as desired. Clearly $\tilde{R}$ is a subspace of R. For $0 < \sigma < 2$ we introduce in $\tilde{R}$ the scalar product

$$\langle f_1, f_2 \rangle = \iint |z_1 - z_2|^{\sigma - 2} f_1(z_1) \overline{f_2}(z_2)\, dz_1\, dz_2 \tag{5-16}$$

Hence $\tilde{R}$ becomes a Euclidean space, whose completion[3] we denote by $R(\sigma)$. The space $R(\sigma)$ is a Hilbert space, which is also the completion of the space R, and consequently R is a subspace of $R(\sigma)$.

The unitary representation of the complementary series is then defined in the space $R(\sigma)$ by Eq. (5-11), where $0 < \sigma < 2$. One can show that the representation so obtained is irreducible (Problem 5.5).

In the next section we will use the group SU_2 techniques to describe the representation of the complementary series.

5-2 SU_2 DESCRIPTION OF THE COMPLEMENTARY SERIES

In the last section we described the representation of the complementary series, an infinite-dimensional, unitary, irreducible representation, which was realized in the Hilbert space $R(\sigma)$. In this section we realize the representation of the complementary series in other spaces defined on the group SU_2. This is similar to what was done in Section 4-3 for the representation of the principal series.

The Euclidean Space H

To this end we set for each function $f(z)$ of the space R (see Section 5-1) the function

$$\phi(u) = \pi^{1/2} \alpha(u) f(z) \tag{5-17}$$

defined on the group SU_2. Here u and z belong to the right coset $\tilde{z}$ (see Chapter 3), and the function $\alpha(u)$ is now defined by

$$\alpha(g) = |g_{22}|^{-2-\sigma} \tag{5-18}$$

[3] If $\tilde{R}$ is an incomplete Euclidean space then one can complete it into a Hilbert space. See M. A. Naimark, *Normed Rings*, Noordhoff, Groningen, The Netherlands, 1959.

The function $\phi(u)$ defined by Eq. (5-17) is constructed in a similar way to the function $\phi(u)$ defined by Eq. (4-41) for representations of the principal series. Furthermore, it satisfies the condition

$$\phi(\gamma u) = \phi(u) \tag{5-19}$$

since $\alpha(\gamma u) = \alpha(u)$, where $\gamma \in SU_2$ is given by

$$\gamma = \begin{pmatrix} e^{-i\psi/2} & 0 \\ 0 & e^{i\psi/2} \end{pmatrix} \tag{5-20}$$

We now construct a Hilbert space out of the functions $\phi(u)$. First we define the scalar product of the functions $\phi(u)$. To this end we first write the scalar product

$$\langle f_1, f_2 \rangle = \iint |z_1 - z_2|^{\sigma-2} f_1(z_1)\overline{f_2}(z_2)\, dz_1\, dz_2 \tag{5-21}$$

in the space R in terms of the functions $\phi(u)$ defined by Eq. (5-17). A straightforward calculation then shows that (Problem 5.6).

$$\langle f_1, f_2 \rangle = \pi \int K(u_1 u_2^{-1})\phi_1(u_1)\overline{\phi_2}(u_2)\, du_1\, du_2 \tag{5-22}$$

Hence if we denote by H the aggregate of all bounded, measurable, functions $\phi(u)$, satisfying the condition $\phi(\gamma u) = \phi(u)$, then the integral on the right-hand side of Eq. (5-22) will converge absolutely if $\phi_1(u)$ and $\phi_2(u)$ belong to H. In the space H we can therefore define the scalar product by

$$\langle \phi_1, \phi_2 \rangle = \pi \iint K(u_1 u_2^{-1})\phi_1(u_1)\overline{\phi_2}(u_2)\, du_1\, du_2 \tag{5-23}$$

where the kernel function $K(u_1 u_2^{-1})$ is defined by

$$K(u) = |u_{21}|^{\sigma-2} \tag{5-24}$$

The Hilbert Space $H(\sigma)$

The space H then becomes a Euclidean space. We denote its completion by $H(\sigma)$, where $H(\sigma)$ is a Hilbert space. Because of Eqs. (5-22) and (5-23), Eq. (5-17) provides an isometric mapping of the space R on the space H, which is extended in a unique way by continuity to an isometric mapping of the space $R(\sigma)$ on the space $H(\sigma)$.

Complementary Series in the Space $H(\sigma)$

In order to realize the representation of the complementary series in the Hilbert space $H(\sigma)$, we notice that the isometric mapping (5-17) transforms the operators $D(g)$ of a representation of the complementary series defined on the Hilbert space $R(\sigma)$ over into operators in the Hilbert space $H(\sigma)$. Let us again denote these operators by $D(g)$. One then finds that the operators $D(g)$ of a representation of the complementary series on the Hilbert space $H(\sigma)$ are given by

$$D(g)\phi(u) = \frac{\alpha(ug)}{\alpha(u(g))}\,\phi(u(g)) \tag{5-25a}$$

where the function $\alpha(g)$ is given by

$$\alpha(g) = |g_{22}|^{-2-\sigma} \tag{5-25b}$$

for any g of the group $SL(2, C)$, and $0 < \sigma < 2$. Here the function $u(g)$ is given by Eq. (3-57).

Canonical Basis in the Space H

We now define, following Carmeli and Malin,[4] a set of functions that provides an orthogonal basis in the space H. It is given by

$$e^j_m(u) = N_j D^j_{0m}(u) \tag{5-26}$$

for $j = 0, 1, 2, 3, \ldots$ and $m = -j, -j+1, \ldots, j$, where $D^j_{0m}(u)$ are the matrix elements of the irreducible representations of the group SU_2 (see Chapter 1). The constant N_j is a real normalization factor whose value is given by

$$N_j = \left\{\pi \int K(u) D^j_{00}(u)\, du\right\}^{-1/2} \tag{5-27}$$

and whose explicit expression will be given in the sequel.

To show that the functions e^j_m indeed provide an orthogonal basis in the space H, we calculate the scalar product

$$\begin{aligned}\langle e^{j_1}_{m_1}, e^{j_2}_{m_2}\rangle &= \pi \iint K(u'u''^{-1}) e^{j_1}_{m_1}(u')\bar{e}^{j_2}_{m_2}(u'')\, du'\, du'' \\ &= \pi N_{j_1} N_{j_2} \iint K(u'u''^{-1}) D^{j_1}_{0m_1}(u')\, du' \bar{D}^{j_2}_{0m_2}(u'')\, du''\end{aligned}$$

By making the transition $u' \to u'u''$ in the above integral, one obtains

$$\langle e^{j_1}_{m_1}, e^{j_2}_{m_2}\rangle = \pi N_{j_1} N_{j_2} \iint K(u') D^{j_1}_{0m_1}(u'u'')\, du' \bar{D}^{j_2}_{0m_2}(u'')\, du''$$

Using the relation

$$D^{j_1}_{0m_1}(u'u'') = \sum_{m=-j_1}^{j_1} D^{j_1}_{0m}(u') D^{j_1}_{mm_1}(u'')$$

in the last integral, we obtain

$$\begin{aligned}\langle e^{j_1}_{m_1}, e^{j_2}_{m_2}\rangle = \pi N_{j_1} N_{j_2} \sum_{m=-j_1}^{j_1} &\int K(u') D^{j_1}_{0m}(u')\, du' \\ \times &\int D^{j_1}_{mm_1}(u'') \bar{D}^{j_2}_{0m_2}(u'')\, du''\end{aligned}$$

[4] M. Carmeli and S. Malin, *J. Math. Phys.* **12**, 255 (1971).

Using now the orthogonality relation (1-42) that the matrices D^j satisfy, we obtain

$$\langle e^{j_1}_{m_1}, e^{j_2}_{m_2}\rangle = \pi N_{j_1} N_{j_2}\left\{\int K(u')D^{j_1}_{00}(u')\, du'\right\}\frac{\delta^{j_1j_2}\,\delta^{m_1m_2}}{2j_1+1}$$

which, by virtue of Eq. (5-27), gives

$$\langle e^{j_1}_{m_1}, e^{j_2}_{m_2}\rangle = \frac{\delta^{j_1j_2}\,\delta^{m_1m_2}}{2j_1+1} \tag{5-28}$$

The Normalization Factor N_j

The normalization factor N_j was defined above by

$$N_j^{-2} = \pi \int K(u)D^j_{00}(u)\, du \tag{5-29}$$

where $K(u)$ is a function of $u \in SU_2$ given by $K(u) = |u_{21}|^{\sigma-2}$ and $0 < \sigma < 2$.

By calculating the scalar product of $D^j_{00}(u)$ with itself, we find

$$\langle D^j_{00}, D^j_{00}\rangle = \frac{\pi}{2j+1}\int K(u)D^j_{00}(u)\, du$$

Hence we have

$$N_j^{-2} = (2j+1)\langle D^j_{00}, D^j_{00}\rangle \tag{5-30}$$

Now, the right-hand side of Eq. (5-30) is positive. Hence $N_j^{-2} > 0$, and therefore N_j is real.

The evaluation of the integral in Eq. (5-29) is straightforward. Using the expression (1-39) for $D^j_{mn}(u)$ we find

$$D^j_{00}(u) = (-1)^{2j}\sum_{m=0}^{j}\binom{j}{m}^2 (u_{11}u_{22})^m (u_{12}u_{21})^{j-m}$$

Writing now the unitary matrix u in the form

$$u = \begin{pmatrix} a & b \\ -\bar{b} & \bar{a}\end{pmatrix}; \qquad |a|^2 + |b|^2 = 1$$

we then obtain for the integral in Eq. (5-29) the following:

$$\sum_{m=0}^{j}(-1)^{3j-m}\binom{j}{m}^2\int |a|^{2m}|b|^{2(j-m-1)+\sigma}\, du \tag{5-31}$$

We now express du in terms of some three real parameters. We take these parameters as $t = |a|^2$, $\theta = \arg a$, $\lambda = \arg b$. Then $du = (1/2\pi)^2\, dt\, d\theta\, d\lambda$, and the integral in Eq. (5-31) becomes

$$\left(\frac{1}{2\pi}\right)^2 \int_0^1 dt \int_0^{2\pi} d\theta \int_0^{2\pi} t^m(1-t)^{j-m-1+\sigma/2}\, d\lambda$$

We therefore obtain for N_j^{-2} the expression

$$N_j^{-2} = \pi \sum_{m=0}^{j} (-1)^{3j-m} \binom{j}{m}^2 \int_0^1 t^m(1-t)^{j-m-1+\sigma/2}\, dt \tag{5-32}$$

This last integral can easily be evaluated by parts. One obtains

$$\begin{aligned}\int_0^1 t^m(1-t)^{j-m-1+\sigma/2}\, dt &= \frac{m!}{\left(j+\frac{\sigma}{2}-m\right)\cdots\left(j+\frac{\sigma}{2}-1\right)} \int_0^1 (1-t)^{j+\sigma/2-1}\, dt \\ &= \frac{m!}{\left(j+\frac{\sigma}{2}-m\right)\cdots\left(j+\frac{\sigma}{2}\right)} \\ &= \frac{\Gamma(m+1)\Gamma\left(j+\frac{\sigma}{2}-m\right)}{\Gamma\left(j+\frac{\sigma}{2}+1\right)}\end{aligned} \tag{5-33}$$

Using Eq. (5-33) in Eq. (5-32), we finally obtain

$$N_j^{-2} = \sum_{m=0}^{j} (-1)^{3j-m} \binom{j}{m}^2 B\left(m+1, j+\frac{\sigma}{2}-m\right) \tag{5-34}$$

where $B(x, y)$ is the familiar Euler B-function,

$$B(x, y) = \frac{\Gamma(x)\Gamma(y)}{\Gamma(x+y)} \tag{5-35}$$

The Spaces h and $h(\sigma)$

In analogy with the generalized Fourier transform, introduced in Section 4-1 between the spaces $L_2^{2s}(US_2)$ and l_2^{2s}, there exists for the complementary series an isometric mapping from the Euclidean space of functions H (and its completion, the Hilbert space $H(\sigma)$) to a Euclidean space of systems of numbers h (and its completion, a Hilbert space $h(\sigma)$) which is defined as follows.

The Euclidean space h is defined as the aggregate of all systems of numbers ψ_m^j, where $m = -j, -j+1, \ldots j$ and $j = 0, 1, 2, 3, \ldots$, satisfying the condition

$$\sum_j (2j+1) N_j^{-2} \sum_{m=-j}^{j} |\psi_m^j|^2 < \infty \tag{5-36}$$

The scalar product in the space h is defined by

$$\sum_j (2j+1)N_j^{-2} \sum_{m=-j}^{j} \phi^j_m \overline{\psi^j_m} \tag{5-37}$$

for any two vectors ϕ^j_m and ψ^j_m of the space h. The coefficients N_j are defined by Eq. (5-27). With each vector ϕ^j_m of the space h we associate the function

$$\phi(u) = \sum_j (2j+1)N_j^{-1} \sum_m \phi^j_m e^j_m(u) \tag{5-38}$$

where $e^j_m(u)$ are the basis functions given by Eq. (5-26). Since $e^j_m(\gamma u) = e^j_m(u)$, where γ is given by Eq. (5-20), it follows that the function $\phi(u)$ defined by Eq. (5-38) belongs to the space H. On the other hand, every function in the space H can be written in the form (5-38) since, as we have seen, $e^j_m(u)$ provides a complete[5] orthogonal set in the space H. In fact, the two spaces H and h are isometric, where the transition from one space to the other is made by means of

$$\phi^j_m = N_j \langle \phi, e^j_m \rangle \tag{5-39}$$

A simple calculation also shows that[6]

$$\langle \phi, \psi \rangle = \sum_j (2j+1)N_j^{-2} \sum_m \phi^j_m \overline{\psi^j_m} \tag{5-40}$$

[5] The orthogonal set of functions $e^j_m(u)$ is also complete in H. This can be seen by considering the irreducible unitary representation (5-25a) of the group $SL(2, C)$ in the Hilbert space $H(\sigma)$ as an infinite-dimensional unitary representation for the subgroup SU_2 and by decomposing it into its orthogonal sum of the finite-dimensional irreducible representations. If v is an element of SU_2, then Eq. (5-25a) gives for the representation of SU_2

$$D(v)\phi(u) = \phi(uv)$$

since one can put $u(v) = uv$ in this case (see Chapter 3). Applying the last formula to $e^j_m(u)$, we obtain

$$\begin{aligned} D(v)e^j_m(u) &= e^j_m(uv) = N_j D^j_{0m}(uv) \\ &= N_j \sum_{m'=-j}^{j} D^j_{0m'}(u) D^j_{m'm}(v) \\ &= \sum_{m'=-j}^{j} D^j_{m'm}(v) e^j_{m'}(u) \end{aligned}$$

Hence the operator $D(v)$ realizes a representation of SU_2 in the space R_j of the $(2j+1)$ functions e^j_m with $-j \leqslant m \leqslant j$, where the matrix elements of $D(v)$ are $D^j_{m'm}(v)$. The representation $v \to D(v)$ in the space of functions $e^j_m(u)$, $m = -j, -j+1, \ldots, j$, is irreducible and the e^j_m form a canonical basis in this space. Accordingly, the infinite-dimensional representation of SU_2 in the space $H(\sigma)$ is decomposed into irreducible parts defined in the subspaces R_j of e^j_m, where $m = -j, -j+1, \ldots j$ and $j = 0, 1, 2, 3, \ldots$. In other words, every function of the space $H(\sigma)$, and hence of H, can be decomposed in the form given by Eq. (5-38).

[6] If ψ is taken to be equal to ϕ, then Eq. (5-40) gives Plancherel's formula

$$\langle \phi, \phi \rangle = \sum_j (2j+1)N_j^{-2} \sum_{m=-j}^{j} |\phi^j_m|^2$$

If we denote by $h(\sigma)$ the completion of the Euclidean space h, then the isometric mapping (5-39) of H on h can be extended in a unique way by continuity to an isometric mapping of the space $H(\sigma)$ on the space $h(\sigma)$. The operator $D(g)$ of a representation of the complementary series in the space $H(\sigma)$ passes over into operators in the space $h(\sigma)$, which are also denoted by $D(g)$, and whose explicit expression we find below. This expression will also define a different form for the complementary series of representations.

Realization of the Complementary Series in the Space $h(\sigma)$

Applying the operator $D(g)$ to the function $\phi(u)$ written in the form (5-38) we obtain

$$D(g)\phi(u) = \sum_j (2j+1)N_j^{-1} \sum_m \phi_m^j \frac{\alpha(ug)}{\alpha(u(g))} e_m^j(u(g))$$

or, expanding the expression on the right-hand side into infinite series,

$$D(g)\phi(u) = \sum_j (2j+1) \sum_m \phi_m^j \sum_{j'} (2j'+1)N_{j'}^{-1} \times \sum_{m'} D_{mm'}^{jj'}(g;\sigma)e_{m'}^{j'}(u) \tag{5-41}$$

where the infinite-dimentional matrix $D_{mm'}^{jj'}(g;\sigma)$ is given by

$$D_{mm'}^{jj'}(g;\sigma) = \pi \frac{N_{j'}}{N_j} \iint K(u'u''^{-1}) \frac{\alpha(u'g)}{\alpha(u'(g))} \times e_m^j(u'(g))\bar{e}_{m'}^{j'}(u'')\, du'\, du'' \tag{5-42}$$

Accordingly, Eq. (5-41) has the form

$$D(g)\phi(u) = \sum_j (2j+1)N_j^{-1} \sum_m \phi_m'^j e_m^j(u) \tag{5-43}$$

where

$$\phi_{m'}'^{j'} = \sum_{j=0}^{\infty} (2j+1) \sum_{m=-j}^{j} D_{mm'}^{jj'}(g;\sigma)\phi_m^j \tag{5-44}$$

Consequently, we see that the operator $D(g)$ of the representation of the complementary series of the group $SL(2, C)$ in the Hilbert space $h(\sigma)$ is the linear transformation determined by Eq. (5-44), describing the law of transformation of the quantities ϕ_m^j, where $m = -j, -j+1, \ldots, j$ and $j = 0, 1, 2, 3, \ldots$. The matrix elements $D_{mm'}^{jj'}(g;\sigma)$ are functions of $g \in SL(2, C)$ and of the real parameter σ, where $0 < \sigma < 2$. They can formally be obtained from the matrix elements $D_{mm'}^{jj'}(g; s, \rho)$ of the representation of the principal series given by Eq. (4-48) (see Problem 5.8).

Comparison with the Infinitesimal Approach

The representation of the complementary series, as defined in this chapter, is characterized by a parameter σ whose range of variation is $0 < \sigma < 2$. The value of σ determines the scalar product, as defined by Eq. (5-22), in the Hilbert space of representations and also the operators of the representations, as given by Eqs. (5-25).

The representation of the complementary series was defined, however, at the end of Chapter 2 through the infinitesimal approach. It was shown that all irreducible representations of the group $SL(2, C)$ are characterized by a pair of numbers (j_0, c), where j_0 takes the values $0, \frac{1}{2}, 1, \frac{3}{2}, 2, \ldots$ and c is complex. In particular the representations of the complementary series were found to be characterized by the conditions that c is a real number in the interval $0 < |c| \leqslant 1$ and $j_0 = 0$.

To establish the relationship between the parameters σ and c one applies the global form of a given representation to infinitesimal elements of the group $SL(2, C)$ and compares the infinitesimal elements so obtained with the results of Chapter 2. The result is

$$j_0 = 0, \qquad c = \pm \frac{\sigma}{2} \tag{5-45}$$

The case for which $j_0 = 0$ and $|c| = \sigma/2 = 1$ corresponds to the identity representation.

The discussions in Chapters 4 and 5 show that the representation of the principal and complementary series (and the identity representation) realize all irreducible unitary representations described by Eqs. (2-29). In other words, the representations of the principal and complementary series realize all irreducible unitary representations of the group $SL(2, C)$ to within unitary equivalence, or every irreducible unitary representation of the group $SL(2, C)$ is unitarily equivalent to a representation of the principal or complementary series.[7]

5-3 OPERATOR FORMULATION

So far the infinite-dimensional, unitary, representations of the homogeneous Lorentz group have been discussed via the global approach, in which the representations are realized as operators defined over an abstract space of functions. This method is due to Gelfand and Naimark,[8] and it complements the infinitesimal approach due to Bargmann.[9] In the latter method one finds the matrices corresponding to infinitesimal generators in a given representation, and expresses matrices corresponding to finite group elements as exponential functions of the generators.

[7] These statements were proved under the assumptions made in Section 2-2. A proof of these statements, without the assumptions of Section 2-2, was first given by I. M. Gelfand and M. A. Naimark, *op. cit.*

[8] See I. M. Gelfand and M. A. Naimark, *op. cit.*

[9] V. Bargmann, *Ann. Math.* **48**, 568 (1947).

While the use of the finite-dimensional, nonunitary, spinor representation of the group $SL(2, C)$ is widespread in connection with the theory of general relativity (see Chapters 7 to 12) and covariant fields, the infinite-dimensional representations have not been used so extensively. However, the latter are becoming more and more involved in physical theories in connection with particles theory and are, therefore, most likely to be used in gravitation theory. An example of that is the harmonic analysis of scattering amplitudes with respect to the group $SL(2, C)$ done by Toller.[10] Another example, by Nambu,[11] is the use of infinite component fields to represent an infinite family of particles suggested by the Regge classification of elementary particles. There exist other examples[12] in which representations of the group $SL(2, C)$ play an important role.

In view of these considerations, Bars and Gürsey[13] developed an operator formulation, in accordance with the quantum mechanical treatment of groups of physical relevance to reformulate the theory of representations of $SL(2, C)$ presented so far. Such a formulation might have an important role in future developments of quantum theory of gravitation. The theory may also find applications in the representation theory of other noncompact groups relevant to physics, such as the de Sitter group or the conformal group. In the following we give a brief review of this formulation.

Casimir Operators

We define the rotation operators $J_k = iA_k$ and the boost operators $K_n = iB_n$, where A_k and B_n, with $k, n = 1, 2, 3$, are the infinitesimal generators defined by Eq. (2-12). The commutation relations of Eqs. (2-26) then gives

$$[J_k, J_l] = i\varepsilon_{klm} J_m$$

$$[K_k, K_l] = -i\varepsilon_{klm} J_m$$

$$[J_k, K_l] = i\varepsilon_{klm} K_m$$

These commutation relations can then be simplified into one equation,

$$[J_{\kappa\lambda}, J_{\mu\nu}] = i(\delta_{\kappa\mu} J_{\lambda\nu} + \delta_{\lambda\nu} J_{\kappa\mu} - \delta_{\kappa\nu} J_{\lambda\mu} - \delta_{\lambda\mu} J_{\kappa\nu}) \tag{5-46}$$

where Greek indices run from 0 to 3. The operator $J_{\mu\nu}$, with $J_{\mu\nu} = -J_{\nu\mu}$, is related to J_k and K_n by $J_k = \frac{1}{2}\varepsilon_{klm} J_{lm}$ and $K_n = iJ_{0n}$. We shall use the self-dual and anti-self-dual combination $M_{\mu\nu} = \frac{1}{2}(J_{\mu\nu} + \tilde{J}_{\mu\nu})$, $N_{\mu\nu} = \frac{1}{2}(J_{\mu\nu} - \tilde{J}_{\mu\nu})$, where the operator $\tilde{J}_{\mu\nu} = \frac{1}{2}\varepsilon_{\mu\nu\alpha\beta} J_{\alpha\beta}$. We also define the left- and right-handed vector operators $X_k^L = M_{k0} = \frac{1}{2}\varepsilon_{klm} M_{lm}$, $X_k^R = -N_{k0} = \frac{1}{2}\varepsilon_{klm} N_{lm}$. In vector notation one has $\mathbf{X}_L = \frac{1}{2}(\mathbf{J} + i\mathbf{K})$, $\mathbf{X}_R = \frac{1}{2}(\mathbf{J} - i\mathbf{K})$, along with the commutation relations

$$\begin{aligned} [X_m^L, X_n^R] &= 0 \\ [X_k^L, X_l^L] &= i\varepsilon_{klm} X_m^L \\ [X_k^R, X_l^R] &= i\varepsilon_{klm} X_m^R \end{aligned} \tag{5-47}$$

[10] M. Toller, *Nuovo Cim.* **53**, 671 (1968).
[11] Y. Nambu, *Phys. Rev.* **160**, 1171 (1967).
[12] G. Domokos, S. Kovesi-Domokos, and E. Schonberg, *Phys. Rev.* **D2**, 1026 (1970).
[13] I. Bars and F. Gürsey, *J. Math. Phys.* **13**, 131 (1972).

The group $SL(2, C)$ admits two *Casimir operators* that commute with $J_{\mu\nu}$. These are

$$F_1 = \tfrac{1}{4}J_{\mu\nu}J_{\mu\nu} = \tfrac{1}{2}(\mathbf{J}\cdot\mathbf{J} - \mathbf{K}\cdot\mathbf{K}) = \mathbf{X}_L\cdot\mathbf{X}_L + \mathbf{X}_R\cdot\mathbf{X}_R$$
$$F_2 = \tfrac{1}{4}J_{\mu\nu}\tilde{J}_{\mu\nu} = i\mathbf{J}\cdot\mathbf{K} = \mathbf{X}_L\cdot\mathbf{X}_L - \mathbf{X}_R\cdot\mathbf{X}_R \tag{5-48}$$

Under the parity transformation S (see Section 2-1)

$$\mathbf{J}\to\mathbf{J}, \qquad \mathbf{K}\to-\mathbf{K} \tag{5-49}$$

the operator F_1 behaves like scalar whereas the operator F_2 as pseudoscalar:

$$F_1\to F_1, \qquad F_2\to -F_2 \tag{5-50}$$

It will be also convenient to use the alternative Casimir operators

$$C_1 = \tfrac{1}{2}(F_1 + F_2) = \mathbf{X}_L\cdot\mathbf{X}_L$$
$$C_2 = \tfrac{1}{2}(F_1 - F_2) = \mathbf{X}_R\cdot\mathbf{X}_R \tag{5-51}$$

which are transformed into each other under parity. The irreducible representation can be labelled by the eigenvalues of C_1 and C_2 which define the (complex) numbers j_1 and j_2 through the equations

$$C_1 = j_1(j_1+1), \qquad C_2 = j_2(j_2+1) \tag{5-52}$$

The z-Basis of the Group $SL(2, C)$

We now introduce the z basis as the simultaneous eigenstates of C_1, C_2 and Z, where $Z = Z(C_1, C_2, \mathbf{X}^L)$ is a certain non-Hermitian operator that is a rational function of $J_{\mu\nu}$. This operator should satisfy the following condition:

(1) In order to have four commuting operators C_1, C_2, the Hermitian and anti-Hermitian parts of Z, we must have $[Z + Z^\dagger, Z - Z^\dagger] = 0$, or $[Z, Z^\dagger] = 0$.

(2) Let z be the eigenvalue of Z. In order for z to transform as $z' = gz = (az+b)/(cz+d)$ under an $SL(2, C)$ transformation, with elements

$$g = e^{i(\boldsymbol{\sigma}\cdot\boldsymbol{\alpha})/2} = e^{i[\boldsymbol{\sigma}\cdot(\boldsymbol{\omega}-i\mathbf{v})]/2} = \begin{pmatrix} a & b \\ c & d \end{pmatrix} \tag{5-53}$$

with $ad - bc = 1$, we must have Z transforming as the ratio of two operators Ψ_1 and Ψ_2 that, under the group $SL(2, C)$, transform like the components of a spinor.

One may define Ψ_1 and Ψ_2 as functions of $J_{\mu\nu}$ if Ψ, with components Ψ_1 and Ψ_2, satisfies the equation

$$(\boldsymbol{\sigma}\cdot\mathbf{X}_L)\Psi = \Omega\Psi; \qquad \Psi = \begin{pmatrix}\Psi_1 \\ \Psi_2\end{pmatrix} \tag{5-54}$$

where $\boldsymbol{\sigma}$ are the three 2×2 Pauli matrices, given by Eqs. (3-5). In other words, Ψ is defined as one of the eigenspinors of the 2×2 matrix of generators, and Ω as the eigenvalue operator associated with Ψ. The invariant Ω is a function of C_1 (see below). Using now the fact that $\mathbf{X}_L \times \mathbf{X}_L = i\mathbf{X}_L$, which is equivalent to the second of Eqs. (5-47), one obtains $(\boldsymbol{\sigma}\cdot\mathbf{X}_L)^2\Psi = (\mathbf{X}_L\cdot\mathbf{X}_L - \boldsymbol{\sigma}\cdot\mathbf{X}_L)\Psi$, or, with the help of Eqs. (5-51), $C_1\psi = \boldsymbol{\sigma}\cdot\mathbf{X}_L(\boldsymbol{\sigma}\cdot\mathbf{X}_L + 1)\Psi = \Omega(\Omega+1)\Psi$. Therefore, one finds for Ω two operator solutions

$$\Omega = \Omega_1 = -\tfrac{1}{2} + (C_1 + \tfrac{1}{4})^{1/2} \tag{5-55a}$$

or

$$\Omega = \Omega_1' = -\tfrac{1}{2} - (C_1 + \tfrac{1}{4})^{1/2} \tag{5-55b}$$

These solutions satisfy the equations

$$\Omega_1 + \Omega_1' = -1 \tag{5-56a}$$

$$-\Omega_1\Omega_1' = C_1 = \Omega_1(\Omega_1+1) = \Omega_1'(\Omega_1'+1) \tag{5-56b}$$

In the following Ψ will be called the *spinor operator* associated with Ω_1, and it satisfies the homogeneous operator-matrix equation

$$\begin{pmatrix} X_3^L - \Omega_1 & X_1^L + iX_2^L \\ X_1^L - iX_2^L & -X_3^L - \Omega_1 \end{pmatrix}\begin{pmatrix}\Psi_1\\ \Psi_2\end{pmatrix} = 0 \tag{5-57}$$

The compatibility of the two linear equations is assured by Eqs. (5-51) and (5-56b). The operator Z is now defined by

$$\begin{aligned} Z = \Psi_1\Psi_2^{-1} &= (\Omega_1 - X_3^L)^{-1}(X_1^L + iX_2^L) \\ &= (X_1^L - iX_2^L)^{-1}(\Omega_1 + X_3^L) \end{aligned} \tag{5-58}$$

Consider now a finite Lorentz transformation with *rotation parameters* ω and *boost parameters* v. Let

$$D = D(g) = e^{i\mathbf{J}\cdot\boldsymbol{\omega} + i\mathbf{K}\cdot\mathbf{v}} = e^{i\boldsymbol{\alpha}\cdot\mathbf{K}_L + i\bar{\boldsymbol{\alpha}}\cdot\mathbf{X}_R} \tag{5-59a}$$

where

$$\boldsymbol{\alpha} = \boldsymbol{\omega} - i\mathbf{v}, \qquad \bar{\boldsymbol{\alpha}} = \boldsymbol{\omega} + i\mathbf{v} \tag{5-59b}$$

One finds that

$$D^{-1}\boldsymbol{\sigma}\cdot\mathbf{X}_L D = e^{(i\sigma/2)\cdot\boldsymbol{\alpha}}\boldsymbol{\sigma}\cdot\mathbf{X}_L\, e^{(-i/2)\boldsymbol{\sigma}\cdot\boldsymbol{\alpha}} = g\boldsymbol{\sigma}\cdot\mathbf{X}_L g^{-1} \tag{5-60}$$

where g is an element of the group $SL(2, C)$ and is defined by Eq. (5-53). Similarly, one finds

$$D^{-1}\boldsymbol{\sigma}\cdot\mathbf{X}_R D = e^{(i/2)\boldsymbol{\sigma}\cdot\bar{\boldsymbol{\alpha}}}\boldsymbol{\sigma}\cdot\mathbf{X}_R\, e^{(-i/2)\boldsymbol{\sigma}\cdot\bar{\boldsymbol{\alpha}}} = \tilde{g}\boldsymbol{\sigma}\cdot\mathbf{X}_R\tilde{g}^{-1} \tag{5-61}$$

where the matrix $\tilde{g}$ is given by

$$\tilde{g} = \sigma_2\bar{g}\sigma_2 = \begin{pmatrix} \bar{d} & -\bar{c} \\ -\bar{b} & \bar{a} \end{pmatrix}$$

with g given by Eq. (5-53). The 2×2 matrix operator that transforms with $\bar{g}$ is then $-\bar{\boldsymbol{\sigma}} \cdot \mathbf{X}_R$, since one has

$$D^{-1}(-\bar{\boldsymbol{\sigma}} \cdot \mathbf{X}_R)D = D^{-1}\sigma_2 \boldsymbol{\sigma} \cdot \mathbf{X}_R \sigma_2 D = \sigma_2 \tilde{g} \boldsymbol{\sigma} \cdot \mathbf{X}_R \tilde{g}^{-1}\sigma_2$$
$$= \bar{g}(-\bar{\boldsymbol{\sigma}} \cdot \mathbf{X}_R)\bar{g}^{-1} \tag{5-62}$$

where $\bar{g}$ is the complex conjugate to the matrix g, and g is given by Eq. (5-53).

If one now defines Φ by

$$(-\bar{\boldsymbol{\sigma}} \cdot \mathbf{X}_R)\Phi = \Omega_2 \Phi \tag{5-63}$$

in analogous to Eq. (5-54) for Ψ, one can find the right-handed analog of Eq. (5-57) (see Problem 5.10). The non-Hermitian operator $\bar{Z}$ is then defined by

$$\bar{Z} = \Phi_1 \Phi_2^{-1} = (-\Omega_2 - X_3^R)^{-1}(X_1^R - iX_2^R)$$
$$= (X_1^R + iX_2^R)^{-1}(-\Omega_2 + X_3^R) \tag{5-64}$$

The z basis can now be defined by the simultaneous eigenstates of C_1, C_2, Z and $\bar{Z}$. If z represents the eigenvalue of Z on these states, then the state $|z\rangle$ will transform into

$$D\,|z\rangle = \lambda(g; z)\left|\frac{az + b}{cz + d}\right\rangle \tag{5-65}$$

where $\lambda(g; z)$ is a certain multiplier which can be determined later. If ζ is the eigenvalue of $\bar{Z}$, then in order for ζ to transform as $\zeta \to (\bar{a}\zeta + \bar{b})/(\bar{c}\zeta + \bar{d})$, one must have $\zeta = \bar{z}$. Thus the states $|j_1, j_2; z, \bar{z}\rangle$ are defined by

$$\Omega_1\,|j_1, j_2; z, \bar{z}\rangle = j_1\,|j_1, j_2; z, \bar{z}\rangle$$
$$\Omega_2\,|j_1, j_2; z, \bar{z}\rangle = j_2\,|j_1, j_2; z, \bar{z}\rangle$$
$$Z\,|j_1, j_2; z, \bar{z}\rangle = z\,|j_1, j_2; z, \bar{z}\rangle$$
$$\bar{Z}\,|j_1, j_2; z, \bar{z}\rangle = \bar{z}\,|j_1, j_2; z, \bar{z}\rangle$$

Within a given representation (j_1, j_2), using the commutation relations, one obtains the expressions

$$Z = (j_1 - X_3^L)^{-1}(X_1^L + iX_2^L)$$
$$= (X_1^L + iX_2^L)(j_1 + 1 - X_3^L)^{-1}$$
$$= (X_1^L - iX_2^L)^{-1}(j_1 + X_3^L)$$
$$= (j_1 + 1 + X_3^L)(X_1^L - iX_2^L)^{-1} \tag{5-66a}$$
$$\bar{Z} = (-j_2 - X_3^R)^{-1}(X_1^R - iX_2^R)$$
$$= (X_1^R - iX_2^R)(-j_2 - 1 - X_3^R)^{-1}$$
$$= (X_1^R + iX_2^R)^{-1}(-j_2 + X_3^R)$$
$$= (-j_2 - 1 + X_3^R)(X_1^R + iX_2^R)^{-1} \tag{5-66b}$$

Unitary Representations

We now restrict ourselves to unitary representations, so that $D(g)$ of Eq. (5-59a) is unitary. This means the operators **J** and **K** are Hermitian, $\mathbf{J} = \mathbf{J}^\dagger$ and $\mathbf{K} = \mathbf{K}^\dagger$, and we have $(\mathbf{X}_L)^\dagger = \mathbf{X}_R$ and $(C_1)^\dagger = C_2$. Consequently, for the eigenvalues of C_1 and C_2 we have the relation

$$\bar{j}_1(\bar{j}_1 + 1) = j_2(j_2 + 1) \tag{5-67}$$

which yields

$$j_2 = -\tfrac{1}{2} \pm (\bar{j}_1 + \tfrac{1}{2}) \tag{5-68}$$

Define now the operator **X** by $\mathbf{X}_L = \mathbf{X}$, thus $\mathbf{X}_R = \mathbf{X}^\dagger$, and introduce j_0 and k by $2j + 1 = j_0 + k$ and $2j_2 + 1 = -j_0 + k$. Consequently, the Casimir operators F_1 and F_2 of Eqs. (5-48) have the eigenvalues $F_1 = \frac{1}{2}(\mathbf{J}\cdot\mathbf{J} - \mathbf{K}\cdot\mathbf{K}) = \frac{1}{2}(j_0^2 + k^2 - 1)$ and $F_2 = i\mathbf{J}\cdot\mathbf{K} = (j_1 + j_2 + 1)(j_1 - j_2) = j_0 k$.

Since F_1 and $-iF_2$ are Hermitian for unitary representations and, as we shall see below, j_0 is a real integer or half-integer (see Problem 5.12), we obtain the two cases for which representations of the group $SL(2, C)$ are unitary. These can be characterized by defining $k = \rho + i\sigma$ as follows:

(a) The principal series, for which

$$\rho = j_1 + \bar{j}_2 + 1 = \bar{j}_1 + j_2 + 1 = 0,\ k = -\bar{k} \tag{5-69}$$

(b) The complementary series, for which

$$j_0 + i\sigma = j_1 - \bar{j}_2 = 0,\ k = \bar{k} \tag{5-70}$$

One notices that the last equation, along with j_0 = real, imply

$$j_1 = \bar{j}_1 = j_2 = \bar{j}_2 \tag{5-71}$$

Both of these two cases come from the condition $\rho(j_0 + i\sigma) = 0$, which is equivalent to Eq. (5-67).

From the expressions (5-66) for Z and $\bar{Z}$ we obtain

$$Z = (j_1 - X_3)^{-1}X_- = (j_1 + 1 + X_3)X_+^{-1} \tag{5-72a}$$

$$\bar{Z} = (-j_2 - X_3^\dagger)^{-1}(X_-)^\dagger = (-j_2 - 1 + X_3^\dagger)(X_+^{-1})^\dagger \tag{5-72b}$$

where we have used the notation $X_\pm = X_1 \mp iX_2$ and $(X_\pm)^\dagger = X_1^\dagger \pm iX_2^\dagger$.

We now derive a relation between $\bar{Z}$ and $Z^\dagger$. We have, from Eq. (5-66a) that $Z^\dagger = (\bar{j}_1 + 1 - X_3^\dagger)^{-1}(X_-)^\dagger = (\bar{j}_1 + X_3^\dagger)(X_+^{-1})^\dagger$. Hence $Z^\dagger$ satisfies the commutation relation $[Z, Z^\dagger] = 0$, which allows its Hermitian and anti-Hermitian parts to be diagonalized simultaneously. In particular, we notice the important commutation relations

$$[Z, \Pi] = 1,\ [Z^\dagger, \Pi^\dagger] = -1 \tag{5-73}$$

where Π is defined by $\Pi = X_+$. One can then show that (Problem 5.11)

$$[(\Pi^n)^\dagger, Z^\dagger] = [(X_1^\dagger + iX_2^\dagger)^n, Z^\dagger] = n(X_1^\dagger + iX_2^\dagger)^{n-1} \tag{5-74}$$

which leads to

$$f(\Pi^\dagger)Z = Z^\dagger f(\Pi^\dagger) + f'(\Pi^\dagger) \tag{5-75}$$

where f' is the derivative of the function f with respect to its argument. If one chooses the function f by $f(\Pi^\dagger) = (\Pi^\dagger)^{-\rho} = (X_1^\dagger + iX_2^\dagger)^{-\bar{j}_1 - j_2 - 1}$, where ρ is defined by Eq. (5-69), we obtain from Eq. (5-75) the relation $Z = (\Pi^\dagger)^{-\rho} Z^\dagger (\Pi^\dagger)^\rho$. On the other hand, because of the commutation relation $[X_l, X_m^\dagger] = 0$, one also obtains $[\Pi^\rho, Z^\dagger] = 0$, thus we can write

$$\bar{Z} = GZ^\dagger G^{-1} = Z^\dagger - \rho(X_1^\dagger + iX_2^\dagger)^{-1} \tag{5-76}$$

where G is defined by

$$G = G^\dagger = (\Pi^\dagger)^{-\rho} = (\Pi^\dagger \Pi)^{-\rho} = (X_1^\dagger + iX_2^\dagger)^{-\rho}(X_1 - iX_2)^{-\rho}.$$

From the last equation it follows that $\bar{Z}$ satisfies the commutation relation

$$[\bar{Z}, \Pi^\dagger] = -1 \tag{5-77}$$

Finally, one notices the relations

$$\begin{aligned} \bar{Z}^\dagger &= G^{-1} Z G = Z - \rho(X_1 - iX_2)^{-1} \\ [\bar{Z}^\dagger, \Pi] &= 1 \\ [\bar{Z}^\dagger, Z^\dagger] &= [\Pi, \Pi^\dagger] = 0 \end{aligned} \tag{5-78}$$

which show that both Z and $\bar{Z}^\dagger$ are canconical conjugate to Π, whereas $\bar{Z}$ and $Z^\dagger$ are canonical conjugate to $\Pi^\dagger$.

The operator G reduces into the unit operator in the case of the principal series, in which case we obtain $\bar{Z} = Z^\dagger$ when $\rho = 0$. In general, however, the operator G plays the role of a metric operator in a Hilbert space.

The generators $\mathbf{X}$ and $\mathbf{X}^\dagger$ can now be expressed by means of Π, $\Pi^\dagger$, and their canonical conjugates. The result is

$$\begin{aligned} X_1 - iX_2 &= \Pi \\ X_3 &= -j_1 + \Pi Z \\ X_1 + iX_2 &= 2j_1 Z - \Pi Z^2 \end{aligned} \tag{5-79}$$

The first of Eqs. (5-79) is the definition of Π, the second follows from $Z = (X_1 - iX_2)^{-1}(j_1 + X_3)$, which is one of the forms of Z given by Eq. (5-66a), whereas the third of Eqs. (5-79) follows from Eq. (5-72a) after replacing X_3 by its value as given by the second of Eqs. (5-79).

Similarly, from the expressions of $\bar{Z}$ one finds

$$\begin{aligned} X_1^\dagger + iX_2^\dagger &= \Pi^\dagger \\ X_3^\dagger &= j_2 + \Pi^\dagger \bar{Z} \\ X_1^\dagger - iX_2^\dagger &= -2j_2 \bar{Z} - \Pi^\dagger \bar{Z}^2 \end{aligned} \tag{5-80}$$

Finally, the generators $\mathbf{X}$ and $\mathbf{X}^\dagger$ can also be expressed by means of $\bar{Z}^\dagger$ and $Z^\dagger$. Equation (5-76) yields $\bar{Z} = Z^\dagger - \rho(\Pi^\dagger)^{-1}$, and $Z = \bar{Z}^\dagger + \rho\Pi^{-1}$, whereas Eqs. (5-78) yield $\Pi\bar{Z}^\dagger\Pi^{-1} = \bar{Z}^\dagger - \Pi^{-1}$. Using these results and using Eq. (5-67), one finds

$$\begin{aligned} X_3 &= \bar{J}_2 + 1 + \Pi\bar{Z}^\dagger \\ X_1 + iX_2 &= -2(\bar{J}_2 + 1)\bar{Z}^\dagger - \Pi(\bar{Z}^\dagger)^2 \\ X_3^\dagger &= -(\bar{J}_1 + 1) + \Pi^\dagger Z^\dagger \\ X_1^\dagger - iX_2^\dagger &= 2(\bar{J}_1 + 1)Z^\dagger - \Pi^\dagger(Z^\dagger)^2 \end{aligned} \tag{5-81}$$

Consequently, using the commutation relations (5-73), (5-77), and Eq. (5-24), one can derive the Lie algebra (5-47) and the formulas (5-52) from either Eqs. (5-79) and (5-80) or Eqs. (5-81) together with the first of Eqs. (5-79) and (5-80). Hence one obtains a unitary representation of the Lorentz group in terms of (Π, Z) and $(\Pi^\dagger, \bar{Z})$ or $(\Pi, \bar{Z}^\dagger)$ and $(\Pi^\dagger, Z^\dagger)$.[14]

PROBLEMS

5.1 Use Eq. (5-8) by once taking $z_2 = 0$ and g_{12} so that $g_{12}z_1 + g_{22} = 1$, and secondly taking $z_1 = 0$ and g_{12} so that $g_{12}z_2 + g_{22} = 1$. Equate the two obtained formulae, taking into account the arbitrary nature of the variables z_1 and z_2, in order to show that

$$|g_{22}|^{-i\rho}\bar{g}_{22}^{m} = |g_{22}|^{i\rho}g_{22}^{m}$$

Show that this equation, by choosing $g_{22} = e^{i\theta}$ where θ is real, leads to the conditions $m = 0$ and $\bar{\rho} = -\rho$.

5.2 Prove that

$$\iint |z_1 - z_2|^{\sigma-2} f(z_1)\bar{f}(z_2)\, dz_1\, dz_2 = 2^\sigma\pi\{\Gamma(\sigma/2)/\Gamma(1-\sigma/2)\} \int |\phi(w)|^2 |w|^{-\sigma}\, dw$$

the right-hand side of which is obviously positive for $0 < \sigma < 2$. Here $\phi(w)$ is the Fourier transform

$$\phi(w) = \frac{1}{2\pi}\int f(z)\, e^{-iRe(\bar{z}w)}\, dz$$

[See M. A. Naimark, *Linear Representations of the Lorentz Group*, Pergamon Press, 1964, p. 174.]

5.3 Show that the unit representation can be considered as a limit of the representation of the complementary series at $\sigma = 2$.

5.4 Show that Eqs. (5-11) and (5-12) do not determine a unitary representation for $\sigma > 2$.

5.5 Show that the representation of the complementary series is irreducible.

5.6 Calculate the scalar product (5-21) in terms of the function $\phi(u)$ by using Eq. (5-17). Change the integration variables z_1 and z_2 into the variables u_1 and u_2 by use of $u_1 = k_1 z_1$ and $u_2 = k_2 z_2$. Show that one then obtains for the scalar product the following:

$$\langle f_1, f_2\rangle = \pi \iint |z_1 - z_2|^{\sigma-2} |(u_1)_{22}|^{\sigma-2} |(u_2)_{22}|^{\sigma-2} \times \phi_1(u_1)\bar{\phi}_2(u_2)\, du_1\, du_2$$

[14] For more details on the operator treatment of representations of the Lorentz group, see I. Bars and F. Gürsey, *op. cit.*

Use now $z_1 = (u_1)_{21}/(u_1)_{22}$ and $z_2 = (u_2)_{21}/(u_2)_{22}$ to obtain

$$\langle f_1, f_2 \rangle = \pi \iint K(u_1 u_2^{-1})\phi_1(u_1)\bar{\phi}_2(u_2)\, du_1\, du_2$$

where the kernel function $K(u_1 u_2^{-1})$ is defined by $K(u) = |u_{21}|^{\sigma-2}$.

5.7 Show that the irreducible representations of the group SU_2 of weight j occurs only once in the representation of the complementary series if and only if j is an integer. (Compare Section 4-3).

5.8 Show that the matrix elements $D^{jj'}_{mm'}(g; \sigma)$ of the representation of the complementary series, given by Eq. (5-42), can formally be obtained from the matrix elements $D^{jj'}_{mm'}(g; s, \rho)$ of the representation of the principal series, given by Eq. (4-48). Show that the relation between them is given by $D^{jj'}_{mm'}(g; \sigma) = \{D^{jj'}_{mm'}(g; s, \rho)\}_{s=0,\, \rho=i\sigma}$. [See M. Carmeli and S. Malin, *J. Math. Phys.* **12,** 225, (1971).]

5.9 Prove Eqs. (5-45).

5.10 Let $(-\boldsymbol{\sigma} \cdot \mathbf{X}_R)\Phi = \Omega_2 \Phi$. Show that $C_2 = \mathbf{X}_R \cdot \mathbf{X}_R = \Omega_2(\Omega_2 + 1) = \Omega_2'(\Omega_2' + 1)$, where Ω_2 and Ω_2' are defined as the roots $\Omega_2 = -\frac{1}{2} + (C_2 + \frac{1}{2})^{1/2}$ and $\Omega_2' = -\Omega_2 - 1 = -\frac{1}{2} - (C_2 + \frac{1}{4})^{1/2}$. Show also that Φ satisfies the homogeneous equation

$$\begin{pmatrix} -X_3^R - \Omega_2 & -X_1^R + iX_2^R \\ -X_1^R - iX_2^R & X_3^R - \Omega_2 \end{pmatrix}\begin{pmatrix} \Phi_1 \\ \Phi_2 \end{pmatrix} = 0,$$

which is the right-handed analog of Eq. (5-57).

5.11 Prove Eq. (5-74).

5.12 Show that from Eqs. (5-79) and (5-80) one obtains

$$J_3 = X_3 + X_3^\dagger = -(j_1 - j_2) + \Pi Z + \Pi^\dagger \bar{Z}$$

Show that by introducing the operators $A = (1/\sqrt{2})(Z + \Pi^\dagger)$, $B = (1/\sqrt{2})(\bar{Z} - \Pi)$, $\bar{A} = (1/\sqrt{2})(\bar{Z} + \Pi)$, $\bar{B} = (1/\sqrt{2})(Z - \Pi^\dagger)$, which obey the commutation relation $[A, \bar{B}] = [\bar{A}, B] = [A, B] = [\bar{A}, \bar{B}] = 0$ and $[A, \bar{A}] = [B, \bar{B}] = 1$, typical of harmonic oscillator creation and annihilation operators. The relation given above can then be recast into the form

$$J_3 = -(j_1 - j_2) + \bar{A}A - \bar{B}B$$

Since $\bar{A}A$ and $\bar{B}B$ must have integer values and J_3 has integer or half-integer eigenvalues m, it follows that $j_0 = j_1 - j_2$ is real and allowed to take integer or half-integer values.

CHAPTER

SIX

COMPLETE SERIES OF REPRESENTATIONS OF $SL(2, C)$

This chapter concludes our discussion of the theory of representations of the Lorentz group. We first introduce the representations of the complete series, following Naimark and Gelfand, who showed that these representations include all the completely irreducible, though not necessarily unitary, representations of the Lorentz group. In the first section the complete series is realized in the Hilbert spaces $L_2^{2s}(SU_2)$ and l_2^{2s}. We then find out the relation of the representations of the complete series to the spinor representation. This is followed by a discussion of the theory of invariant bilinear functionals and intertwining operators. The case of unitary representations is accordingly discussed. The last section of the chapter is devoted to a very brief review of the theory of harmonic analysis on the Lorentz group.

6-1 REALIZATION OF THE COMPLETE SERIES[1]

Realization of the Complete Series in the Space $L_2^{2s}(SU_2)$

As has been pointed out in Section 5-1, all the unitary representations of the group $SL(2, C)$ are included in either the principal or the complementary series.[2] Gel-

[1] M. Carmeli and S. Malin, *Fortschr. Physik* **21**, 397, (1973); *Int. J. Theoret. Phys.* **9**, 145 (1974).

[2] It is interesting to note that the definition of the principal and complementary series of representations can be generalized from the group $SL(2, C)$ to $SL(N, C)$ for arbitrary $N > 2$. However, for $N > 2$ there seems to be unitary representations not included in the two series. See E. M. Stein, *Ann. Math.* **86**, 461 (1967).

fand and Naimark,[3] and Naimark,[4] have shown that all the completely irreducible[5] representations of the group $SL(2, C)$, not necessarily unitary, are included, up to equivalence, in a series of representations called the *complete series*.[6]

In this section the complete series is realized in the Hilbert spaces $L_2^{2s}(SU_2)$ and l_2^{2s} (see Section 4-1).

All the representations of the complete series can be characterized by a pair of numbers (s, ρ), where s is an integer or half-integer and ρ satisfies the condition $\rho^2 \neq -4(|s| + k)^2$, for $k = 1, 2, 3, \ldots$, and is otherwise an arbitrary complex number. The pairs (s, ρ) and $(-s, -\rho)$ define the same representation. The representations can be realized in the Hilbert spaces $L_2^{2s}(SU_2)$. Hence the space of realization depends on the parameter s alone and is independent of ρ. A given representation corresponding to a pair (s, ρ) is realized in the Hilbert space $L_2^{2s}(SU_2)$ by a set of operators $D(g)$, where g is an element of the group $SL(2, C)$, defined by

$$D(g)\phi(u) = \frac{\alpha(ug)}{\alpha(u(g))}\phi(u(g)) \tag{6-1a}$$

where $\phi(u)$ is an element of the Hilbert space $L_2^{2s}(SU_2)$, and where $\alpha(g)$ is given by

$$\alpha(g) = g_{22}^{2s}|g_{22}|^{i\rho - 2s - 2} \tag{6-1b}$$

Here the function $u(g)$ was defined in Chapter 3.

The representation formulae (6-1) for the complete series are the same as those of Eqs. (4-44) for the principal series; the difference is that ρ now can take complex values, while in Eqs. (4-44) ρ is real. It can be shown that the operators $D(g)$ defined by Eqs. (6-1) are unitary if and only if ρ is real.

The Complete Series in the Space l_2^{2s}

In complete analogy with Section 4-3 the generalized Fourier transform, introduced in Section 4-1, can now be utilized to obtain a realization of the complete series in the space l_2^{2s}. The result is

$$\phi'^{j'}_{m'} = \sum_{j=|s|}^{\infty} (2j+1) \sum_{m=-j}^{j} D^{jj'}_{mm'}(g; s, \rho)\phi^j_m \tag{6-2}$$

[3] I. M. Gelfand and M. A. Naimark, *Izv. Akad. Nauk SSSR, Ser. Mat.* **11**, 411 (1947).

[4] M. A. Naimark, *Dokl. Adad. Nauk SSSR* **97**, 969 (1954).

[5] The definition of complete irreducibility is as follows. Given a representation $D(g)$ of the group $SL(2, C)$ on a Banaçh space B one first defines a bounded linear operator C as admissible if it has the form $C(\xi) = \sum_{i=1}^{n} f_i(\xi)e_i$, where $f_1, \ldots, f_n \in \Omega'$ and $e_1, \ldots, e_n \in \Omega$. The definitions of the sets Ω and Ω' will be given in the sequel. One then defines the representations as completely irreducible if for every admissible operator C in the Banach space B there exists a sequence x_n of X such that $(D(x_n)\xi, \eta) \to (C\xi, \eta)$ as $n \to \infty$ for all ξ of Ω and η of Ω'. Here X is the group ring, to be defined in the sequel, and the operators $D(x_n)$ are defined in footnote 8 below. It can be shown that every unitary, irreducible, representation in a separable Hilbert space is completely irreducible.

[6] The definition of equivalence of representations in the sense of the present chapter will be given in the sequel.

where the infinite-dimensional matrix $D^{jj'}_{mm'}$ is given by

$$D^{jj'}_{mm'}(g; s, \rho) = \int \frac{\alpha(ug)}{\alpha(u(g))} D^{j}_{sm}(u(g))\bar{D}^{j'}_{sm'}(u)\, du \tag{6-3}$$

These equations are again the same as Eqs. (4-50) and (4-48), respectively, for the representations of the principal series except insofar as the definition of the function $\alpha(g)$, given by Eq. (6-1b) here, is extended to include complex values of the variable ρ.

In the following we find the relation of the complete series to the complementary series.

Equivalence of Representations

As has been pointed out before, the complete series describes all the infinite-dimensional completely irreducible representations, to within equivalence, of the group $SL(2, C)$. The meaning of equivalence here is such that the spaces of two equivalent representations need *not* be isometric, but it is the formulae that are essential for the representations and not the norm of the space. In the present subsection we define equivalence of representations and show that the representations belonging to the complementary series are, from this point of view, equivalent to representations contained in the complete series.

The definition of equivalence of representations realized in Banach spaces requires some preliminary mathematical definitions:

(i) The group ring X. Let X denotes the set of all infinitely differentiable functions $x(g)$, where g is an element of the group $SL(2, C)$, which vanish for all the matrices g satisfying the condition $|g_{11}|^2 + |g_{12}|^2 + |g_{21}|^2 + |g_{22}|^2 > C$, for a big enough number C which may depend on the function $x(g)$. This set forms a *ring*[7] if addition and multiplication by complex numbers are defined in the usual way and multiplication of ring elements is defined by[8]

$$x_1 \cdot x_2(g) = \int x_1(g')x_2(g'^{-1}g)\, dg' \tag{6-4}$$

(ii) Conjugate representations. Given a Banach space B, whose elements are denoted by ξ, its conjugate space B' is defined as the space of all bounded linear functionals $f(\xi)$ in B (see Appendix C). If D is a given operator defined in the Banach space B, then its conjugate operator D' is defined in the conjugate space B' by $D'f(\xi) = f(D\xi)$. Now, given a representation in terms of the operator $D(g)$ on a Banach space B, one defines the operator $\tilde{D}(g) = D'(g^{-1})$ as the conjugate representation in the Banach space B'.

[7] See, for example, B. L. van der Waerden, *Modern Algebra*, New York, 1953.

[8] This definition comes about as follows: given a representation of the group $SL(2, C)$ as a set of operators $D(g)$ one defines as operator $D(x)$ corresponding to every function $x(g)$ of X as $D(x) = \int x(g)D(g)\, dg$. By a straightforward calculation one finds that $D(x_1)D(x_2) = D(x_1 \cdot x_2)$, where $x_1 \cdot x_2$ is defined by Eq. (6-4).

(iii) The set Ω corresponding to a given representation in a Banach space B is defined as the aggregate of all finite linear combinations of the vectors $D(x)\xi$ (the operator $D(x)$ is defined in footnote 8), where ξ belongs to the space M^j (see Section 2-3) for any value of j and x of X. The set corresponding to the conjugate representation is denoted by Ω'.

We now define two representations $D^{(1)}(g)$ and $D^{(2)}(g)$ on the Banach space $B^{(1)}$ and $B^{(2)}$, respectively, as *equivalent* if there exist linear operators $A^{(1)}$ and $A^{(2)}$ that transform $B^{(1)}$ into $B^{(2)}$ and $B^{(2)}$ into $B^{(1)}$ respectively, whose domains of definition are $\Omega^{(1)}$ and $\Omega'^{(2)}$ and domains of variation $\Omega^{(2)}$ and $\Omega'^{(1)}$, respectively, satisfying for all ξ of $\Omega^{(1)}$ and f of $\Omega'^{(2)}$,

$$(A^{(1)}\xi, f) = (\xi, A^{(1)}f) \tag{6-5}$$

Moreover, if $A^{(1)}\xi = 0$ and $A^{(2)}f = 0$ then $\xi = 0$ and $f = 0$; and finally, one has

$$\begin{aligned} A^{(1)}D^{(1)}(x)\xi &= D^{(2)}(x)A^{(1)}\xi \\ A^{(2)}\tilde{D}^{(2)}(x)f &= \tilde{D}^{(1)}(x)A^{(2)}F \end{aligned} \tag{6-6}$$

It is noteworthy that for the representations to be equivalent, the Banach spaces need not be isometric.

Relation to the Complementary Series

The complete series of representations discussed above was characterized by a pair of parameters (s, ρ), where s is an integer or half-integer and ρ is a complex number. We now show that the complete series of representations that is characterized by the particular case $s = 0$ and $0 < -i\rho < 2$ are equivalent to the representation of the complementary series.

The space of representations of the complete series, corresponding to $s = 0$, was defined above as the Hilbert space $L_2^0(SU_2)$. The space of representations of the complementary series was defined as the Hilbert space $H(\sigma)$. These spaces correspond to the spaces $B^{(1)}$ and $B^{(2)}$, respectively, in the definition of equivalence of representations discussed above. Now, one faces the following situation: if $D^{(1)}(g)$ is a representation of the complementary series corresponding to a value σ of the parameter, and $D^{(2)}(g)$ is representation of the complete series corresponding to the value $s = 0$, $\rho = i\sigma$, then the representations $D^{(1)}(g)$ and $D^{(2)}(g)$ are given in the two Banach spaces by the same formulas (6-1) and (5-25), respectively. It follows, therefore, that the sets $\Omega^{(1)}$ and $\Omega^{(2)}$, corresponding to a given representation in the Banach spaces $L_2^0(SU_2)$ and $H(\sigma)$, are the same because both consist of all the finite linear combinations of the vectors $D(x)\xi$, where ξ is any of the functions $D^j_{\text{on}}(u)$. The operators $A^{(1)}$ and $A^{(2)}$, appearing in the definition of equivalence, are trivially defined now as the identity operators in $\Omega^{(1)} = \Omega^{(2)}$ and $\Omega'^{(1)} = \Omega'^{(2)}$, respectively. One can check that they satisfy Eqs. (6-5) and (6-6). Therefore, any complementary series of representations, corresponding to a value σ, is equivalent to the complete series of representations characterized by the pair of parameters $s = 0$ and $\rho = i\sigma$.

Condition of Reducibility

In introducing the representations of the complete series, we restricted the values of its parameters (s, ρ) by excluding the representations for which

$$\rho^2 = -4(|s| + k)^2; \qquad k = 1, 2, 3, \ldots \tag{6-7}$$

We now consider the representations for which the condition (6-7) is valid. We show that the representations realized by the general formula for the complete series, Eqs. (6-1), are not irreducible if Eq. (6-7) is satisfied.

To see that indeed when $\rho = -4(|s| + k)^2$, where $k = 1, 2, 3, \ldots$, the representation (6-1) is not irreducible we proceed as follows.

Suppose that $\rho = -2i(|s| + k)$ and denote by P_{MN} the set of all homogeneous polynomials in the variables u_{21}, $\bar{u}_{21}$, u_{22} and $\bar{u}_{22}$:

$$p(u) = \sum_{\alpha, \beta, \gamma, \delta} a_{\alpha\beta\gamma\delta} u_{21}^{\alpha} \bar{u}_{21}^{\beta} u_{22}^{\gamma} \bar{u}_{22}^{\delta} \tag{6-8}$$

with the conditions

$$\alpha - \beta + \gamma - \delta = 2s \tag{6-9}$$

$$\alpha + \beta + \gamma + \delta = 2|s| + 2k - 2 \tag{6-10}$$

where $k = 1, 2, 3, \ldots$. One can easily see, using Eq. (6-9), that $p(\gamma u) = e^{is\psi} p(u)$, where γ is given by Eq. (5-20). Therefore, P_{MN} is a subspace of the Hilbert space $L_2^{2s}(SU_2)$ (see Section 4-1). We show that the space P_{MN} is invariant with respect to the operator $D(g)$ of equation (6-1).

To this end one factorizes the matrix g of $SL(2, C)$ into the product $g = u_1 \varepsilon u_2$, where u_1 and u_2 are elements of the group SU_2 and ε is the matrix given by

$$\varepsilon = \begin{pmatrix} \varepsilon_{22}^{-1} & 0 \\ 0 & \varepsilon_{22} \end{pmatrix}$$

with ε_{22} a real number. Such a factorization is possible for all matrices g of $SL(2, C)$ (see Problem 6.2). Since $g \to D(g)$ is a representation, one has $D(g) = D(u_1)D(\varepsilon)D(u_2)$. Consequently, it is sufficient to show that the space P_{MN} is invariant under each of the operators $D(u_1)$, $D(\varepsilon)$, and $D(u_2)$. Now $D(u_1)p(u) = \{\alpha(uu_1)/\alpha(u(u_1))\}p(u(u_1))$. It is shown in Problem 4.12 that the ratio $\alpha(uu_1)/\alpha(u(u_1))$ is equal to $\exp(2is\Lambda)$, where Λ is an arbitrary real number. Also, a direct calculation, using Eqs. (3-65), shows that

$$p(u(u_1)) = \sum_{\substack{\alpha, \beta, \\ \gamma, \delta,}} \exp[i\Lambda(-\alpha + \beta - \gamma + \delta)] a_{\alpha\beta\gamma\delta} \times (uu_1)_{21}^{\alpha} (\overline{uu}_1)_{21}^{\beta} (uu_1)_{22}^{\gamma} (\overline{uu}_1)_{22}^{\delta}$$

Hence, using the condition (6-9), one obtains $D(u_1)p(u) = p(uu_1)$, which shows that the space P_{MN} of homogeneous polynomials is invariant with respect to the operator $D(u_1)$. The same, of course, holds for the operator $D(u_2)$.

Similarily, P_{MN} is invariant with respect to the operator $D(\varepsilon)$, where $D(\varepsilon)p(u) = \{\alpha(u\varepsilon)/\alpha(u(\varepsilon))\}p(u(\varepsilon))$. It is shown in Problem 4.12 that $\alpha(u\varepsilon)/\alpha(u(\varepsilon))$ is equal to $\exp(2is\Lambda)|\lambda|^{i\rho-2}$, where $|\lambda|$ is given by $|\lambda|^2 = |\beta|^2\varepsilon_{22}^{-2} + |\alpha|^2\varepsilon_{22}^2$. Furthermore, one can easily verify that

$$p(u(\varepsilon)) = \sum_{\substack{\alpha,\beta,\\ \gamma,\delta}} \exp[i\Lambda(-\alpha+\beta-\gamma+\delta)]|\lambda|^{-(\alpha+\beta+\gamma+\delta)} \times \varepsilon_{22}^{-\alpha-\beta+\gamma+\delta} a_{\alpha\beta\gamma\delta} u_{21}^{\alpha}\bar{u}_{21}^{\beta}u_{22}^{\gamma}\bar{u}_{22}^{\delta}$$

Using the conditions (6-9) and (6-10), and the fact that $\rho = -2i(|s| + k)$, one finds

$$D(\varepsilon)p(u) = \sum_{\substack{\alpha,\beta,\\ \gamma,\delta}} \varepsilon_{22}^{-\alpha-\beta+\gamma+\delta} a_{\alpha\beta\gamma\delta} u_{21}^{\alpha}\bar{u}_{21}^{\beta}u_{22}^{\gamma}\bar{u}_{22}^{\delta}$$

This shows that $D(\varepsilon)p(u)$ is a polynomial in the space P_{MN}.

Consequently, the space P_{MN} of homogeneous polynomials is invariant with respect to the operator $D(g)$, and therefore the representation (6-1) is not irreducible when $\rho = -2i(|s| + k)$, with $k = 1, 2, 3, \ldots$. The representation (6-1) is not irreducible also when $\rho = 2i(|s| + k)$, with $k = 1, 2, 3, \ldots$, since the pairs (s, ρ) and $(-s, -\rho)$ define the same completely irreducible representation.

Hence we conclude that the representation (6-1) is not irreducible when $\rho^2 = -4(|s| + k)^2$, with $k = 1, 2, 3, \ldots$.

In the next section we show that the general formulae (6-1) of the complete series, under the condition (6-7), realize the usual finite-dimensional spinor representation. This fact will enable us to establish a direct relation between certain quantities appearing in the generalized Fourier transform of the homogeneous polynomials and the 2-component spinors. We show also that this relation is a single linear transformation.

6-2 COMPLETE SERIES AND SPINORS

In the last section we considered the representation formula (6-1) of the complete series, but under the condition (6-7) that $\rho^2 = -4(|s| + k)^2$, where $k = 1, 2, 3, \ldots$. We showed that the representation obtained is not irreducible in that case.

Relation to Spinors

We now show that the general formula (6-1), under the condition (6-7), realizes the spinor representation when applied to the finite-dimensional linear space of polynomials over the group SU_2 instead of the infinite-dimensional Hilbert space $L_2^{2s}(SU_2)$. This fact enables us to establish a direct relation between certain quantities appearing here and the 2-component spinors. We see below that this relation is a simple linear transformation.

To see, in fact, that under the condition $\rho^2 = -4(|s| + k)^2$, $k = 1, 2, 3, \ldots$, one obtains the spinor representation out of Eq. (6-1), we put

$$M = s + \frac{i}{2}\rho - 1, \qquad N = -s + \frac{i}{2}\rho - 1 \tag{6-11}$$

Then by Eqs. (6-9), (6-10) and (6-11), one obtains $\gamma = M - \alpha$ and $\delta = N - \beta$. Accordingly, the polynomial $p(u)$ of Eq. (6-8) can now be written as

$$p(u) = \sum_{\alpha=0}^{M} \sum_{\beta=0}^{N} a_{\alpha\beta} u_{21}^{\alpha} \bar{u}_{21}^{\beta} u_{22}^{M-\alpha} \bar{u}_{22}^{N-\beta} \tag{6-12}$$

which is homogeneous in the variables u_{21} and u_{22} for degree M, and in $\bar{u}_{21}$ and $\bar{u}_{22}$ of degree N. The polynomial (6-12) also satisfies the condition $p(\gamma u) = e^{i(M-N)\psi/2}p(u)$, where the matrix $\gamma \in SU_2$ is given by Eq. (3-41). Comparing Eq. (6-12) with Eq. (3-55b), we see that the quantity $a_{\alpha\beta}$ is just $\pi^{1/2}p_{rs}$. Hence $a_{\alpha\beta}$ is related to 2-component spinors, by Eq. (3-26), by

$$a_{\alpha\beta} = \pi^{1/2} \binom{M}{\alpha} \binom{N}{\beta} \phi_{A_1 \cdots A_M \dot{X}_1 \cdots \dot{X}_N} \tag{6-13}$$

with $A_1 + \cdots + A_M = \alpha$ and $\dot{X}_1 + \cdots + \dot{X}_N = \beta$.

Relation between Spinors and ϕ_m^j

We are now in a position to find out the connection between spinors and the generalized Fourier transform ϕ_m^j in the finite-dimensional case. Since the polynomial $p(u)$ belongs to the space $L_2^{2s}(SU_2)$, one can expand it into a generalized Fourier series,

$$p(u) = \sum_{j=|s|}^{\infty} (2j + 1) \sum_{m=-j}^{j} \phi_m^j D_{sm}^j(u) \tag{6-14a}$$

where ϕ_m^j is related to $p(u)$ by

$$\phi_m^j = \int p(u) \bar{D}_{sm}^j(u)\, du. \tag{6-14b}$$

Using the expansion (6-12) for $p(u)$ in Eq. (6-14b), one obtains

$$\phi_m^j = \sum_{\alpha=0}^{M} \sum_{\beta=0}^{N} \tilde{C}_{m\alpha\beta}^{jMN} a_{\alpha\beta} \tag{6-15}$$

where the six-index symbols $\tilde{C}_{m\alpha\beta}^{jMN}$ are some numerical coefficients given by

$$\tilde{C}_{m\alpha\beta}^{jMN} = \int \bar{D}_{sm}^j(u) u_{21}^{\alpha} \bar{u}_{21}^{\beta} u_{22}^{M-\alpha} \bar{u}_{22}^{N-\beta}\, du \tag{6-16}$$

In terms of the 2-component spinors, by Eq. (6-13), one obtains

$$\phi_m^j = \sum_{\alpha=0}^{M} \sum_{\beta=0}^{N} C_{m\alpha\beta}^{jMN} \phi_{A_1 \cdots A_M \dot{X}_1 \cdots \dot{X}_N} \tag{6-17}$$

where the new 6-index symbol $C_{m\alpha\beta}^{jMN}$ is related to $\tilde{C}_{m\alpha\beta}^{jMN}$ by

$$C_{m\alpha\beta}^{jMN} = \pi^{1/2}\binom{M}{\alpha}\binom{N}{\beta}\tilde{C}_{m\alpha\beta}^{jMN} \tag{6-18}$$

Here $A_1 + \cdots + A_M = \alpha$ and $\dot{X}_1 + \cdots + \dot{X}_N = \beta$.

The generalized Fourier transform ϕ_m^j is, therefore, related to the spinors $\phi_{A_1 \cdots A_M \dot{X}_1 \cdots \dot{X}_N}$ via a linear transformation, given explicitly by Eqs. (6-18), (6-16) as an integral over the group SU_2.

Invariant Bilinear Functionals[9]

In Sections 4-1 and 4-2 we introduced linear spaces of homogeneous functions $D(\chi)$ and wrote down a representation formula, Eq. (4-36), for the group $SL(2, C)$. We now continue the discussion of the theory of representations in the space $D(\chi)$.

We start with the idea of an invariant functional for a pair of representations $D^{(1)}(g)$ and $D^{(2)}(g)$ of the group $SL(2, C)$ on the two linear infinite-dimensional spaces R_1 and R_2, respectively. Let $B(\phi, \psi)$ be a *continuous bilinear functional* whose arguments ϕ and ψ may be arbitrary elements of the spaces R_1 and R_2. This means one has $B(a\phi_1 + b\phi_2, \psi) = aB(\phi_1, \psi) + bB(\phi_2, \psi)$ and $B(\phi, a\psi_1 + b\psi_2) = aB(\phi, \psi_1) + bB(\phi, \psi_2)$, along with the fact that $B(\phi, \psi)$ is a continuous function of ϕ and ψ in the direct sum of the spaces R_1 and R_2. The bilinear functional $B(\phi, \psi)$ is then called *invariant under the representations* $D^{(1)}(g)$ *and* $D^{(2)}(g)$ if for arbitrary ϕ of R_1 and ψ of R_2, and any g of $SL(2, C)$, one has $B(D^{(1)}(g)\phi, D^{(2)}(g)\psi) = B(\phi, \psi)$.

Consider now the two representations of the group $SL(2, C)$,

$$D(g; \chi_1)\phi(z) = (bz + d)^{n_1 - 1}(\bar{b}\bar{z} + \bar{d})^{n_2 - 1}\phi\left(\frac{az + c}{bz + d}\right) \tag{6-19a}$$

$$D(g; \chi_2)\psi(z) = (bz + d)^{m_1 - 1}(\bar{b}\bar{z} + \bar{d})^{m_2 - 1}\psi\left(\frac{az + c}{bz + d}\right) \tag{6-19b}$$

where $\chi_1 = (n_1, n_2)$ and $\chi_2 = (m_1, m_2)$, realized in the spaces of homogeneous functions $D(\chi_1)$ and $D(\chi_2)$, respectively. One then wishes to find a bilinear functional that is invariant with respect to the operators $D(g; \chi_1)$ and $D(g; \chi_2)$. To this end one associates with each matrix

$$g = \begin{pmatrix} a & b \\ c & d \end{pmatrix}; \qquad ad - bc = 1$$

of the group $SL(2, C)$ the Möbius transformation $w = (az + c)/(bz + d)$ in the complex plane. Matrix multiplication then corresponds to multiplication of these transformations, and both e and $-e$, where e is the 2×2 unit matrix, are associated with the identity transformation. Every such transformation of the complex

[9] For more details see, e.g., I. M. Gelfand, M. I. Graev, and N. Ya Vilenkin, *Generalized Functions, Vol. 5: Integral Geometry and Representation Theory*, Academic Press, New York, and London, 1966.

plane can be obtained from the following three types of transformations (Problem 6.3):

(1) Parallel translation,

$$z \to z + z_0; \quad g = \begin{pmatrix} 1 & 0 \\ z_0 & 1 \end{pmatrix} \tag{6-20a}$$

(2) Dilation,

$$z \to \alpha^2 z; \quad g = \begin{pmatrix} \alpha & 0 \\ 0 & \alpha^{-1} \end{pmatrix} \tag{6-20b}$$

(3) Inversion

$$z \to -1/z; \quad g = \begin{pmatrix} 0 & 1 \\ -1 & 0 \end{pmatrix} \tag{6-20c}$$

Hence it is enough to discuss invariance properties of a bilinear functional by checking its invariance under each one of these subgroups. The associated representation formulae to these transformations, corresponding to parallel translation, dilation, and inversion, respectively, are then given by

$$\begin{aligned} D(g; \chi) f(z) &= f(z + z_0) \\ D(g; \chi) f(z) &= \alpha^{1-n_1} \bar{\alpha}^{1-n_2} f(\alpha^2 z) \\ D(g; \chi) f(z) &= z^{n_1 - 1} \bar{z}^{n_2 - 1} f(-1/z) \end{aligned} \tag{6-21}$$

It follows that if $s_1 = -\frac{1}{2}(n_1 + m_1)$ and $s_2 = -\frac{1}{2}(n_2 + m_2)$ are *not* simultaneously negative integers or zero, then every bilinear functional invariant under parallel translations and dilations is given, for every pair of functions ϕ and ψ, by[10, 11]

$$B(\phi, \psi) = \left(\frac{i}{2}\right)^2 \int (z_1 - z_2)^{s_1 - 1} (\bar{z}_1 - \bar{z}_2)^{s_2 - 1} \times \phi(z_1) \psi(z_2) \, dz_1 \, d\bar{z}_1 \, dz_2 \, d\bar{z}_2 \tag{6-22a}$$

On the other hand, if s_1 and s_2 are simultaneously negative integers or zero, the invariant bilinear functional is given by (Problem 6.4)

$$B(\phi, \psi) = \frac{1}{2} \int \phi^{(-s_1, s_2)}(z) \psi(z) \, dz \, d\bar{z}, \tag{6-22b}$$

where use has been made of the notation $\phi^{(n_1, n_2)}(z) = \partial^{n_1 + n_2} \phi(z) / \partial z^{n_1} \, \partial \bar{z}^{n_2}$.

[10] For Re $(s_1 + s_2) < 0$, the integral is to be understood in the sense of its regularization. For instance if $-m < \text{Re}\,(s_1 + s_2) < -m + 1$, this regularization is defined by

$$B(\phi, \psi) = \left(\frac{i}{2}\right)^2 \int z^{s_1 - 1} \bar{z}^{s_2 - 1} \int \psi(z_1) \left\{ \phi(z + z_1) - \sum_{i+j=0}^{m-1} \frac{\phi^{(i,j)}(z_1) z^i \bar{z}^j}{i!j!} \right\} dz \, d\bar{z} \, dz_1 \, d\bar{z}_1$$

[11] Meanwhile we restrict our discussion to those functions $\phi(z)$ and $\psi(z)$ that belong to the space of infinitely differentiable functions of bounded support.

The requirement that the bilinear functionals (6-22) be also invariant under inversion leads in the first case (i.e., when s_1 and s_2 are not both negative integers or zero) to the condition that $\chi_1 = (n_1, n_2) = \chi_2 = (m_1, m_2)$, or $n_k = m_k = -s_k$, for $k = 1, 2$. The bilinear functional (6-22a) then has the form (Problem 6.5)

$$B(\phi, \psi) = \left(\frac{i}{2}\right)^2 \int (z_1 - z_2)^{-n_1-1}(\bar{z}_1 - \bar{z}_2)^{-n_2-1} \times \phi(z_1)\psi(z_2)\, dz_1\, d\bar{z}_1\, dz_2\, d\bar{z}_2 \tag{6-23}$$

In the second case (i.e. when s_1 and s_2 are negative integers or zero) the requirement that the bilinear functional (6-22b) be invariant under inversion also is satisfied if one of the following conditions is fulfilled: (a) $n_1 = m_1$, $n_2 = m_2$; $n_1, n_2 = 0, 1, 2, \ldots$; (b) $n_1 = -m_1$, $n_2 = -m_2$; (c) $n_1 = m_1$, $n_2 = -m_2$, $n_1 = 1, 2, \ldots$; (d) $n_1 = -m_1$, $n_2 = m_2$, $n_2 = 1, 2, \ldots$. If such an invariant bilinear functional exists it is given, in each of these cases, by:

$$B(\phi, \psi) = \frac{i}{2} \int \phi^{(n_1, n_2)}(z)\psi(z)\, dz\, d\bar{z} \tag{6-24a}$$

$$B(\phi, \psi) = \frac{i}{2} \int \phi(z)\psi(z)\, dz\, d\bar{z} \tag{6-24b}$$

$$B(\phi, \psi) = \frac{i}{2} \int \phi^{(n_1, 0)}(z)\psi(z)\, dz\, d\bar{z} \tag{6-24c}$$

$$B(\phi, \psi) = \frac{i}{2} \int \phi^{(0, n_2)}(z)\psi(z)\, dz\, d\bar{z} \tag{6-24d}$$

We have thus not only found the values of the parameters n_k and m_k for which an invariant bilinear functional may exist, but also have found the form of this functional when it does exist. It is possible to show that these conditions are not only necessary but also *sufficient* (Problem 6.6), where now the functions ϕ and ψ belong to the space of homogeneous functions $D(\chi)$.

Intertwining Operators

We now discuss the conditions on $\chi_1 = (n_1, n_2)$ and $\chi_2 = (m_1, m_2)$ such that the two representations

$$D(g; \chi_1)\phi(z) = (bz + d)^{n_1-1}(\bar{b}\bar{z} + \bar{d})^{n_2-1}\phi\left(\frac{az + c}{bz + d}\right) \tag{6-25a}$$

$$D(g; \chi_2)\psi(z) = (bz + d)^{m_1-1}(\bar{b}\bar{z} + \bar{d})^{m_2-1}\psi\left(\frac{az + c}{bz + d}\right) \tag{6-25b}$$

of the group $SL(2, C)$ are equivalent.

Definition Suppose for the two representation operators $D(g; \chi_1)$ and $D(g; \chi_2)$ there exists a continuous linear mapping $A \neq 0$ of the space $D(\chi_1)$

into the space $D(\chi_2)$, which is not necessarily one-to-one or onto all of $D(\chi_2)$, such that

$$AD(g; \chi_1) = D(g; \chi_2)A \tag{6-26}$$

The mapping A is then said to *intertwine* $D(\chi_1)$ and $D(\chi_2)$. If A is an isomorphism, that is if A is a bicontinuous one-to-one mapping of $D(\chi_1)$ onto $D(\chi_2)$, then the representations $D(g; \chi_1)$ and $D(g; \chi_2)$ are called *equivalent*.

One can verify that the above definition is equivalent to the one given in Section 6-1.

We now establish a relationship between intertwining operators and invariant bilinear functionals.

Lemma 6.1 *The operator A intertwines the representation operators $D(g; \chi_1)$ and $D(g; \chi_2)$, namely $AD(g; \chi_1) = D(g; \chi_2)A$, if and only if the bilinear functional $B(\phi, \psi)$, defined by*

$$B(\phi, \psi) = (\psi, A\phi) \tag{6-27}$$

where

$$(\psi, \phi) = \frac{i}{2} \int \psi(z)\phi(z) \, dz \, d\bar{z} \tag{6-28}$$

is invariant under the representations $D(g; \chi_1)$ and $D(g; -\chi_2)$.

In fact, to see that the condition of intertwining $AD(g; \chi_1) = D(g; \chi_2)A$ yields the invariance of the bilinear functional (6-27) and vice versa, one writes these conditions in equivalent ways. The condition $AD(g; \chi_1) = D(g; \chi_2)A$ is equivalent to the statement that the equation

$$(D(g; -\chi_2)\psi, AD(g; \chi_1)\phi) = (D(g; -\chi_2)\psi, D(g; \chi_2)A\phi) \tag{6-29a}$$

holds for all ϕ of $D(\chi_1)$ and ψ of $D(-\chi_2)$. The invariance of the bilinear functional $B(\phi, \psi)$ of Eq. (6-27), on the other hand, can be written as

$$(D(g; -\chi_2)\psi, AD(g; \chi_1)\phi) = (\psi, A\phi) \tag{6-29b}$$

But Eq. (2-29a) is the same as Eq. (6-29b) since the left-hand sides of these equations are identical, whereas the right-hand sides are equal because (ψ, ϕ) is invariant under $D(g; -\chi_2)$ and $D(g; \chi_2)$.

We have found all possible χ_1 and χ_2 for which an invariant bilinear functional may exist. Consequently, to find out the conditions under which the intertwining operator A may exist we need only to replace χ_2 by $-\chi_2$ in the above results. Hence one arrives at the following results: An operator $A \neq 0$ mapping the space $D(\chi_1)$ continuously into the space $D(\chi_2)$, where $\chi_1 = (n_1, n_2)$ and $\chi_2 = (m_1, m_2)$, and intertwining the representations operators, namely

$AD(g; \chi_1) = D(g; \chi_2)A$, exists if and only if one of the following four conditions is satisfied:[12]

(1) $\chi_1 = \chi_2$. The operator A is of the form

$$A\phi(z) = c\phi(z) \tag{6-30}$$

Hence A is a multiple of the unit operator,[13]

(2) $\chi_1 = (n_1, n_2)$, and $\chi_2 = -\chi_1 = (-n_1, -n_2)$, where n_1 and n_2 are not both non-negative integers. The operator A is then of the form

$$A\phi(z) = c\frac{i}{2}\int (z - z_1)^{-n_1-1}(\bar{z} - \bar{z}_1)^{-n_2-1}\phi(z_1)\, dz_1\, d\bar{z}_1 \tag{6-31}$$

(3) $\chi_1 = -\chi_2$, where n_1 and n_2 are non-negative integers. The operator A is then of the form

$$A\phi(z) = c\frac{\partial^{n_1+n_2}\phi(z)}{\partial z^{n_1}\, \partial \bar{z}^{n_2}} \tag{6-32}$$

(4) $\chi_1 = (n_1, n_2)$, $\chi_2 = (-n_1, n_2)$, where n_1 is a positive integer. The operator A is then of the form

$$A\phi(z) = c\frac{\partial^{n_1}\phi(z)}{\partial z^{n_1}} \tag{6-33a}$$

(4′) $\chi_1 = (n_1, n_2)$, $\chi_2 = (n_1, -n_2)$, where n_2 is a positive integer. The operator A is then of the form

$$A\phi(z) = c\frac{\partial^{n_2}\phi(z)}{\partial \bar{z}^{n_2}} \tag{6-33b}$$

In the above equations c is a constant.

6-3 UNITARY REPRESENTATIONS CASE

Before we find the conditions under which a representation in the space of homogeneous functions, $D(\chi)$, is unitary we discuss the problem of equivalence of representations.

[12] These are the same four cases given above, but are now renumbered, for convenience, since χ_2 here is replaced by $-\chi_2$.

[13] Note that when $\chi_1 = \chi_2$, i.e., $D(\chi_1) = D(\chi_2)$, means that any continuous linear operator in the space $D(\chi)$ that commutes with all the operators of $D(g; \chi)$ is a multiple of the unit operator. Such a representation $D(g; \chi)$ is called an *operator irreducible*.

Equivalence of Representations

In Section 6-2 we have already defined equivalence of representations with the help of intertwining operators. Two representations $D(g; \chi_1)$ and $D(g; \chi_2)$ in the two spaces of homogeneous functions $D(\chi_1)$ and $D(\chi_2)$, respectively, are called equivalent if there exists an operator A which is a bicontinuous one-to-one mapping of $D(\chi_1)$ onto $D(\chi_2)$ such that

$$AD(g; \chi_1) = D(g; \chi_2)A \tag{6-34}$$

Since we have found in Section 6-2 the necessary and sufficient conditions on χ_1 and χ_2 for the existence of such an operator A, we can now find what additional conditions should be imposed on the parameters χ_1 and χ_2 if A is to be a bicontinuous one-to-one mapping of the spaces $D(\chi_1)$ and $D(\chi_2)$. Hence one can go through the four cases discussed above. In fact the first case is of no interest since if $\chi_1 = \chi_2$ the space of homogeneous functions $D(\chi_1)$ and $D(\chi_2)$ simply coincide. Hence we are left with cases (2), (3), (4) and (4′). One can then prove the following:

Lemma 6.2 *Two representations $D(g; \chi_1)$ and $D(g; \chi_2)$, where $\chi_1 = (n_1, n_2)$ and $\chi_2 = (m_1, m_2)$ with $\chi_1 \neq \chi_2$, are equivalent if and only if $n_1 = -m_1$ and $n_2 = -m_2$, where n_1 and n_2 are not integers of the same sign. When the two representations $D(g; \chi_1)$ and $D(g; \chi_2)$ are equivalent, then the bicontinuous one-to-one mapping A of the space $D(\chi_1)$ onto $D(\chi_2)$ such that $AD(g; \chi_1) = D(g; \chi_2)A$, is given by*[14]

$$A\phi(z) = c\frac{i}{2}\int (z - z_1)^{-n_1-1}(\bar{z} - \bar{z}_1)^{-n_2-1}\phi(z_1)\, dz_1\, d\bar{z}_1 \tag{6-35}$$

Indeed, let us first assume that $\chi_1 = (n_1, n_2)$ and $\chi_2 = -\chi_1 = (-n_1, -n_2)$ where n_1 and n_2 are not simultaneously non-negative integers. It was already shown by Eq. (6-31) that in this case the operator A is of the form (6-35) (case 2 discussed above). It then follows that the representations $D(g; \chi_1)$ and $D(g; -\chi_1)$ are equivalent. For consider the operator A mapping $D(\chi_1)$ into $D(-\chi_1)$ and satisfying $AD(g; \chi_1) = D(g; -\chi_1)A$. Consider also the operator A_1 mapping $D(-\chi_1)$ into $D(\chi_1)$ and intertwining the representations, namely satisfying $A_1 D(g; -\chi_1) = D(g; \chi_1)A_1$. Then it is possible to show (Problem 6.9) that $A_1 A = AA_1 = \mu 1$, where 1 is the unit operator, and μ is given by

$$\mu = (-1)^{n_1-n_2} 4\pi^2(n_1 + n_2 + |n_1 - n_2|)^{-1}\,(-n_1 - n_2 + |n_1 - n_2|)^{-1} \tag{6-36}$$

Consequently, $\mu \neq 0$ and the operators A and A_1 are isomorphisms of the spaces $D(\chi_1)$ and $D(\chi_2)$. This proves the equivalence of $D(g; \chi_1)$ and $D(g; -\chi_1)$ for n_1 and n_2 not being nonzero integers of the same sign.

[14] For Re $(n_1 + n_2) > 0$ the integral in Eq. (6-35) is to be understood in the sense of its regularization. For more details see I. M. Gelfand, M. I. Graev, and N. Ya. Vilenkin, *op. cit.*

We now consider the case for which $\chi_1 = -\chi_2$ with n_1 and n_2 nonzero integers of the same sign (Case 3 discussed above). [If $n_1 = n_2 = 0$ then $D(\chi_1)$ coincides with $D(\chi_2)$.] Then one has here $D(g; \chi_1)$ and $D(g; \chi_2)$ are not equivalent. This is so since the operator A in this case is of the form given by Eq. (6-32) which is singular since it annihilates all polynomials in the space $D(\chi_1)$ of the form

$$\phi(z) = \sum_{j=0}^{n_1-1} \sum_{k=0}^{n_2-1} a_{jk} z^j \bar{z}^k$$

The only two cases left are those for which $\chi_1 = (n_1, n_2), \chi_2 = (-n_1, n_2)$ for positive integer n_1, and $\chi_1 = (n_1, n_2), \chi_2 = (n_1, -n_2)$ for positive integer n_2 (cases 4 and 4′ discussed above). In both of these two cases the representations $D(g; \chi_1)$ and $D(g; \chi_2)$ are not equivalent (Problem 6.10).

We are now in a position to discuss unitarity conditions for the representations in the spaces of homogeneous functions $D(\chi)$.

Unitary Representations

Under certain conditions one can define in the space $D(\chi)$ of homogeneous functions with $\chi = (n_1, n_2)$, a Hermitian bilinear functional (ϕ, ψ) which is invariant under the representation operators, i.e., $(D(g; \chi)\phi, D(g; \chi)\psi) = (\phi, \psi)$ for all $\phi(z)$ and $\psi(z)$ of the space $D(\chi)$ and any element g of the group $SL(2, C)$. When the Hermitian bilinear functional (ϕ, ψ) is positive definite it could then be used as the scalar product in the space $D(\chi)$ and hence the representation will be unitary (Appendix C).

In Chapters 4 and 5 the unitary representations of the group $SL(2, C)$ were discussed and found. They were the representation of the principal series and the representation of the complementary series. We shall see below that the first case corresponds to $n_1 = -\bar{n}_2$ for which the scalar product is given by

$$(\phi, \psi) = \frac{i}{2} \int \phi(z)\bar{\psi}(z)\, dz\, d\bar{z} \tag{6-37}$$

whereas the second case corresponds to $n_1 = n_2 = \sigma - 1$, where σ is a real number such that $0 < \sigma < 2$, for which case the scalar product is given by

$$\langle \phi, \psi \rangle = \frac{i}{\Gamma\left(\frac{\sigma}{\tau}\right)} \left(\frac{i}{2}\right)^2 \int |z_1 - z_2|^{\sigma-2} \phi(z_1)\bar{\psi}(z_2)\, dz_1\, d\bar{z}_1\, dz_2\, d\bar{z}_2 \tag{6-38}$$

Invariant Hermitian Functionals on $D(\chi)$

Let us find the conditions for the existence of an invariant, not necessarily positive definite for the time being, Hermitian functional on the space $D(\chi)$. To this end we associate with every Hermitian functional (ϕ, ψ), defined on the space $D(\chi)$, with $\chi = (n_1, n_2)$, the invariant bilinear functional

$$(\phi, \psi) = B(\phi, \bar{\psi}) \tag{6-39}$$

This invariant bilinear functional is defined for $\phi(z) \in D(\chi)$ and $\psi(z) \in D(\chi^\dagger)$, where $\chi^\dagger = (\bar{n}_2, \bar{n}_1)$ for $\chi = (n_1, n_2)$. The Hermitian invariant form (ϕ, ψ) is then invariant under the operator $D(g; \chi)$ if and only if $B(\phi, \bar{\psi}) = (\phi, \psi)$ is invariant under the representation operators $D(g; \chi)$ and $D(g; \chi^\dagger)$. This is so since

$$\begin{aligned} B(D(g; \chi)\phi, D(g; \chi^\dagger)\bar{\psi}) &= (D(g; \chi)\phi, \overline{D(g; \chi^\dagger)\bar{\psi}}) \\ &= (D(g; \chi)\phi, D(g; \chi)\psi) \end{aligned}$$

Consequently, the invariance of the Hermitian form under the representation operator $D(g; \chi)$ is the same as the invariance of the bilinear functionals under the operators $D(g; \chi)$ and $D(g; \chi^\dagger)$. This latter invariance was discussed in detail above. All one needs is to identify χ_1 with χ and χ_2 with $\chi^\dagger$. Hence one immediately obtains:

Lemma 6.3 *An invariant Hermitian functional exists in the space of homogeneous functions* $D(\chi)$, *where* $\chi = (n_1, n_2)$, *if and only if either (1)* $n_1 = -\bar{n}_2$ *; or (2)* $n_1 = \bar{n}_2$.

In the first case, one has $n_1 = -\bar{n}_2$, or $n_1 = \frac{1}{2}(n + i\rho)$, $n_2 = \frac{1}{2}(-n + i\rho)$, where n is an integer and ρ is any real number. The Hermitian functional in this case is given by

$$(\phi, \psi) = \frac{i}{2} \int \phi(z)\bar{\psi}(z)\, dz\, d\bar{z}. \tag{6-40}$$

The second case $(n_1 = \bar{n}_2)$ implies, since $n_1 - n_2$ should be an integer, that $n_1 = n_2 = \tau$ is real. The invariant Hermitian functional is then given by

$$\langle \phi, \psi \rangle = \left(\frac{i}{2}\right)^2 \int |z_1 - z_2|^{-2\tau - 2} \phi(z_1)\bar{\psi}(z_2)\, dz_1\, d\bar{z}_1\, dz_2\, d\bar{z}_2 \tag{6-41a}$$

for $\tau \neq 0, 1, \ldots$, and by

$$\langle \phi, \psi \rangle = \frac{i}{2} \int \phi^{(q,\, q)}(z)\bar{\psi}(z)\, dz\, d\bar{z} \tag{6-41b}$$

for $\tau = q$ which is a non-negative integer. The last two formulae may be unified into

$$\langle \phi, \psi \rangle = \frac{1}{\Gamma(-\tau)}\left(\frac{i}{2}\right)^2 \int |z_1 - z_2|^{-2\tau - 2} \phi(z_1)\bar{\psi}(z_2)\, dz_1\, d\bar{z}_1\, dz_2\, d\bar{z}_2 \tag{6-42}$$

by using the identity

$$\left.\frac{|z|^{-2\tau - 2}}{\Gamma(-\tau)}\right|_{\tau = q} = \frac{(-1)^q \pi}{q!} \delta^{(q,\, q)}(z)$$

known from the theory of distributions.

Positive Definite Hermitian Functionals

We now turn to the problem of positive definiteness of the Hermitian functionals discussed above.

In the first case, for which $\chi = (\frac{1}{2}[-n + i\rho], \frac{1}{2}[-n + i\rho])$, the Hermitian functional (6-40) is obviously positive definite. We now find the restrictions on the real parameter τ under which the Hermitian functional (6-42) becomes positive definite.

One first notices that, since the representations $D(g; \chi)$ and $D(g; -\chi)$ are equivalent, one can restrict the choice of τ to $-\infty < \tau \leqslant 0$. Replacing the variable τ into $\sigma = -2\tau$, thus $0 \leqslant \sigma < \infty$. The invariant Hermitian form (6-42) will then have the form

$$\langle \phi, \psi \rangle = \frac{1}{\Gamma\left(\frac{\sigma}{2}\right)}\left(\frac{i}{2}\right)^2 \int |z_1 - z_2|^{\sigma - 2}\phi(z_1)\bar{\psi}(z_2)\, dz_1\, d\bar{z}_1\, dz_2\, d\bar{z}_2 \qquad (6\text{-}43)$$

with $0 \leqslant \sigma < \infty$. But the integral (6-43) was already discussed in Section 5-1, Eq. (5-12). It was found there that the condition for positive definiteness is satisfied only for $0 < \sigma < 2$.

Summarizing our results one obtains:

Lemma 6.4 *A representation $D(g; \chi)$ defined on the space of homogeneous functions $D(\chi)$, where $\chi = (n_1, n_2)$, will have a Hermitian positive definite invariant form in either one of the two cases: (1) $\chi = (\frac{1}{2}[n + i\rho], \frac{1}{2}[-n + i\rho])$, where n is an integer and ρ is a real number. In this case the Hermitian functional has the form*

$$(\phi, \psi) = \frac{i}{2}\int \phi(z)\bar{\psi}(z)\, dz\, d\bar{z}.$$

(2) $\chi = (-\sigma/2, -\sigma/2)$, where $0 < \sigma < 2$. In this case the Hermitian functional has the form

$$\langle \phi, \psi \rangle = \frac{1}{\Gamma\left(\frac{\sigma}{2}\right)}\left(\frac{i}{2}\right)^2 \int |z_1 - z_2|^{\sigma - 2}\phi(z_1)\bar{\psi}(z_2)\, dz_1\, dz_2\, d\bar{z}_1\, d\bar{z}_2$$

Representations in the first case are the representations of the *principal series.* Representations in the second case are the representations of the *complementary series.* Both series were extensively discussed in Chapter 4 and 5.

Unitary Representations on a Hilbert Space

The positive definite invariant Hermitian functional can be taken as the scalar product in the space $D(\chi)$. The space $D(\chi)$ can then be completed with respect to the norm $|\phi| = (\phi, \phi)^{1/2}$ to obtain a Hilbert space H in which $D(\chi)$ forms a subset which is dense everywhere. The representation operators $D(g; \chi)$ can be extended uniquely to unitary operators on the Hilbert space H, and are again denoted by $D(g; \chi)$. They satisfy $(D(g; \chi)\phi, D(g; \chi)\psi) = (\phi, \psi)$, along with the requirements that $D(e; \chi) = 1$ and $D(g_1 g_2; \chi) = D(g_1; \chi)D(g_2; \chi)$, where e is the unit element of $SL(2, C)$ and 1 is the unit operator on $D(\chi)$. Hence $D(g; \chi)$ provides a unitary representation of the group $SL(2, C)$. One can show that there exists no closed proper subspace in H which is invariant under the operator $D(g; \chi)$.

6-4 HARMONIC ANALYSIS ON THE GROUP $SL(2, C)$

In Section 4-1 we discussed the concept of Fourier transforms on the group SU_2. The purpose of this section is to obtain analogous results for the group $SL(2, C)$.[15] We will see below that a regular representation of the group $SL(2, C)$ is decomposed into the direct sum of representations of the principal series. Because the representations of the principal series are not one-dimensional but infinite-dimensional the corresponding Plancherel formula and the corresponding equation (4-20) will contain the trace of the operators of these representations just as the group SU_2 does (see Problem 4.3). On the other hand, since the representations of the principal series depend on discrete index as well as on continuous index, the obtained Plancherel's equation will contain summation as well as integration.

Fourier Transform on the Group $SL(2, C)$

The analogs of the exponential $e^{i\lambda x}$ in Fourier theory for the group $SL(2, C)$ is a solution of the functional equation $f(g_1 g_2) = f(g_1)f(g_2)$ since the ordinary exponentials $e^{i\lambda x}$ are solutions of the functional equation $f(x_1 + x_2) = f(x_1)f(x_2)$. The only *scalar* solution to the equation $f(g_1 g_2) = f(g_1)f(g_2)$ is $f(g) = 1$. Since every solution to this equation, along with the condition $f(e) = 1$, is a representation, hence we look for an operator solution to it. We have already studied representations of the group $SL(2, C)$ in varieties of spaces. If we consider the spaces of homogeneous functions $D(\chi)$ (see Section 4-1) as the spaces where these operators are defined, then these operators are $D(g; \chi)$, where χ is a pair of complex numbers, $\chi = (n_1, n_2)$, whose difference is an integer and satisfy [compare Eq. (4-36)]

$$D(g; \chi)\phi(z) = (bz + d)^{n_1 - 1}(\bar{b}\bar{z} + \bar{d})^{n_2 - 1}\phi\left(\frac{az + c}{bz + d}\right) \tag{6-44}$$

[15] Section 6-4 is a brief review of the subject. For more details on the Lorentz group and harmonic analysis see W. Rühl, *The Lorent Group and Harmonic Analysis*, Benjamin, New York, 1970, Chapter 4; I. M. Gelfand, M. I. Graev and N. Ya. Vilenkin, *Generalized Functions, Vol. 5: Integral Geometry and Representation Theory*, Academic Press, New York and London, 1966, Chapter IV; M. A. Naimark, *Linear Representations of the Lorentz Group*, Pergamon Press, New York, 1964, Section 14.

Here a, b, c, d are complex matrix elements of $g \in SL(2, C)$,

$$g = \begin{pmatrix} a & b \\ c & d \end{pmatrix}; \qquad ad - bc = 1 \tag{6-45}$$

In this sense $D(g; \chi)$, g, and χ play the roles of $e^{i\lambda x}$, x, and $i\lambda$, respectively.[16]

Definition The *operator* function

$$F(\chi) = \int f(g) D(g; \chi) \, dg \tag{6-46}$$

is called the Fourier transform of the function $f(g)$ defined on the group $SL(2, C)$.[17] The function $F(\chi)$ is an operator in the space $D(\chi)$ whose action is defined by

$$F(\chi)\phi(z) = \int f(g) D(g; \chi) \phi(z) \, dg \tag{6-47}$$

It can be shown that if $f(g)$ is a continuous rapidly decreasing function, the integral on the right-hand side of Eq. (6-47) converges for all $\phi(z)$ of the space $D(\chi)$. Moreover, $F(\chi)\phi(z)$ belongs to the space $D(\chi)$ and $F(\chi)$ is continuous in the topology of $D(\chi)$.

Properties of Fourier Transform on $SL(2, C)$

We start with finding the Fourier transform $F(\chi)$ under the translation $g \to gg_0$ of $f(g)$. The Fourier transform of $f(gg_0)$ is the operator function

$$\int f(gg_0) D(g; \chi) \, dg$$

Since this is an invariant integral (see Section 4-3), we can make the substitution $g \to gg_0^{-1}$, thus we obtain

$$\begin{aligned} \int f(gg_0) D(g; \chi) \, dg &= \int f(g) D(gg_0^{-1}; \chi) \, dg \\ &= \int f(g) D(g; \chi) D(g_0^{-1}; \chi) dg \\ &= \int f(g) D(g; \chi) \, dg D^{-1}(g_0; \chi) \end{aligned}$$

[16] Solutions of equations of the form $D(g_1 g_2) = D(g_1)(g_2)$ give rise to special functions that are widely known in mathematical literature. These special functions are matrix elements of the operators $D(g)$. The only exception, so far, are the Lamé and Mathieu functions.

[17] Functions defined on the group $SL(2, C)$ were discussed in detail in Section 4-3.

As a result one obtains

$$\int f(gg_0)D(g; \chi)\,dg = F(\chi)D^{-1}(g_0; \chi) \tag{6-48}$$

In the same way one obtains

$$\int f(g_0^{-1}g)D(g; \chi)\,dg = D(g_0; \chi)F(\chi) \tag{6-49}$$

We now prove the following lemma.

Lemma 6.5 *The Fourier transform of any rapidly decreasing continuous function $f(g)$ is an integral operator given by*

$$F(\chi)\phi(z_1) = \frac{i}{2}\int K(z_1, z_2; \chi)\phi(z_2)\,dz_2\,d\bar{z}_2 \tag{6-50}$$

$K(z_1, z_2; \chi)$ is a function of the two variables z_1 and z_2, which is called a *kernel.*

In fact $F(\chi)\phi(z_1)$ is defined by

$$F(\chi)\phi(z_1) = \int f(g)D(g; \chi)\phi(z_1)\,dg \tag{6-51}$$

for any $\phi(z)$ of the space $D(\chi)$. If we use in (6-51) the expression for $D(g; \chi)\phi(z_1)$, given by

$$D(g; \chi)\phi(z_1) = (\beta z_1 + \delta)^{n_1 - 1}(\bar{\beta}\bar{z}_1 + \bar{\delta})^{n_2 - 1}\phi\left(\frac{\alpha z_1 + \gamma}{\beta z_1 + \delta}\right)$$

and the expression for dg in terms of the three variables α, β, and δ given by Eq. (4-63), we obtain

$$F(\chi)\phi(z_1) = \left(\frac{i}{2}\right)^3 \int f(\alpha, \beta, \delta)(\beta z_1 + \delta)^{n_1 - 1}(\bar{\beta}\bar{z}_1 + \bar{\delta})^{n_2 - 1} \times \phi\left(\frac{\alpha z_1 + \gamma}{\beta z_1 + \delta}\right)|\beta|^{-2}\,d\alpha\,d\bar{\alpha}\,d\beta\,d\bar{\beta}\,d\delta\,d\bar{\delta} \tag{6-52}$$

The integral above can be written in a somewhat different way if we change variables according to $\lambda = \beta z_1 + \delta$, and put $z_2 = (\alpha z_1 + \gamma)/(\beta z_1 + \delta)$. Then

$\delta = \lambda - \beta z_1$, $\alpha = \lambda^{-1} + \beta z_2$, and $\gamma = (\alpha\delta - 1)\beta^{-1}$. The Jacobian of the transformation of the coordinates gives

$$\left| \frac{\partial(\alpha, \beta, \delta)}{\partial(z_2, \beta, \lambda)} \right| = \beta$$

Using these results in Eq. (6-52), one obtains

$$F(\chi)\phi(z_1) = \left(\frac{i}{2}\right)^3 \int f(\lambda^{-1} + \beta z_2, \beta, \lambda - \beta z_1)\phi(z_2) \times \lambda^{n_1 - 1}\bar{\lambda}^{n_2 - 1}\, d\lambda\, d\bar{\lambda}\, d\beta\, d\bar{\beta}\, dz_2\, d\bar{z}_2 \tag{6-53}$$

Equation (6-53) shows that $F(\chi)\phi(z_1)$ is of the form given by Eq. (6-50) and that the kernel $K(z_1, z_2; \chi)$ is given by

$$K(z_1, z_2; \chi) = \left(\frac{i}{2}\right)^2 \int f(\lambda^{-1} + \beta z_2, \beta, \lambda - \beta z_1) \times \lambda^{n_1 - 1}\bar{\lambda}^{n_2 - 1}\, d\lambda\, d\bar{\lambda}\, d\beta\, d\bar{\beta} \tag{6-54}$$

Finally we assert without a proof that if the function $f(g)$ is an infinitely differentiable function, decreasing rapidly with all of its derivatives, then the operator

$$F(\chi) = \int f(g) D(g; \chi)\, dg \tag{6-55}$$

has a trace which is given by

$$\operatorname{Tr} F(\chi) = \frac{i}{2} \int K(z, z; \chi)\, dz \tag{6-56}$$

In terms of the function $f(g)$ the trace can be written as

$$\operatorname{Tr} F(\chi) = \int f(g)(\lambda^{n_1}\bar{\lambda}^{n_2} + \lambda^{-n_1}\bar{\lambda}^{-n_2})\,|\lambda - \bar{\lambda}|^{-2}\, dg \tag{6-57}$$

where λ and λ^{-1} are the eigenvalues of the matrix g of the group $SL(2, C)$.

Inverse Fourier Transform

Let $F(\chi)$ be the Fourier transform of a square integrable function $f(g) = f(\alpha, \beta, \delta)$ defined on the group $SL(2, C)$. Then $F(\chi)$ is defined on the "real axis", that is for those values of $\chi = (n_1, n_2)$ corresponding to unitary representations of the principal series. Hence the points of the "real axis" are characterized by $n_2 = -\bar{n}_1$, or $n_1 = \frac{1}{2}(n + i\rho)$ and $n_2 = \frac{1}{2}(-n + i\rho)$, where ρ is a real number and n is an integer.

We would like first to write down the function $f(g)$ in terms of $F(\chi)$ for values of χ on the "real axis". Alternatively, this could also be done by expressing the function $f(g)$ in terms of the kernel function $K(z_1, z_2; \chi)$ of the operator $F(\chi)$, with χ defined on the "real axis". Since the kernel function was given by Eq. (6-54), our problem reduces to inverting Eq. (6-54).

To this end we define the function $\phi(z_1, z_2; \chi)$ by

$$\phi(z_1, z_2; \lambda) = \frac{i}{2} \int f(\lambda^{-1} + \beta z_2, \beta, \lambda - \beta z_1)\, d\lambda\, d\bar{\lambda} \tag{6-58}$$

Hence the kernel function can now be written as

$$K(z_1, z_2; \chi) = \frac{i}{2} \int \phi(z_1, z_2; \lambda) \lambda^{n_1 - 1} \bar{\lambda}^{n_2 - 1}\, d\lambda\, d\bar{\lambda} \tag{6-59}$$

The inversion of Eq. (6-54) can then be achieved by first inverting Eq. (6-59), thus writing $\phi(z_1, z_2; \lambda)$ in terms of $K(z_1, z_2; \chi)$, and then inverting Eq. (6-58), that is expressing $f(g)$ in terms of $\phi(z_1, z_2; \lambda)$.

The inverse of Eq. (6-59) is straightforward. The result is (Problem 6.11):

$$\phi(z_1, z_2; \lambda) = \frac{1}{(2\pi)^2} \int_{\chi_0} K(z_1, z_2; \chi) \lambda^{-n_1} \bar{\lambda}^{-n_2}\, d\chi \tag{6-60}$$

where χ_0 is the "real axis", that is the set of points $\chi = (\frac{1}{2}[n + i\rho], \frac{1}{2}[-n + i\rho])$, where ρ is real and n is an integer. The integration with respect to χ is understood as integration over the variable ρ and summation over n. The inverse of Eq. (6-58) is more cumbersome. The result is

$$f(\alpha, \beta, \delta) = -\frac{i}{4\pi^2} \frac{1}{|\beta|^2} \int \phi_{\lambda\bar{\lambda}} \left(\frac{\lambda - \delta}{\beta}, \frac{\alpha - \lambda^{-1}}{\beta}; \lambda \right) d\lambda\, d\bar{\lambda} \tag{6-61}$$

where $\phi_{\lambda\bar{\lambda}} = \partial^2 \phi / \partial\lambda\, \partial\bar{\lambda}$.

We are now in a position to express $f(g) = f(\alpha, \beta, \delta)$ in terms of the kernel function of its Fourier transform. Accordingly, we differentiate Eq. (6-60) twice with respect to λ and $\bar{\lambda}$ and using the result in Eq. (6-61) we obtain

$$\begin{aligned} f(\alpha, \beta, \delta) = -\frac{i}{16\pi^4} \int_{\chi_0} \int n_1 n_2 K\left(z, \frac{\alpha z + \gamma}{\beta z + \delta}; \chi \right) \\ \times (\beta z + \delta)^{-n_1 - 1} (\bar{\beta} \bar{z} + \bar{\delta})^{-n_2 - 1}\, dz\, d\bar{z}\, d\chi \end{aligned} \tag{6-62}$$

Here the variable $\chi = (n_1, n_2) = (\frac{1}{2}[n + i\rho], \frac{1}{2}[-n + i\rho])$, where n is an integer and

ρ is real, and integration over χ means integration over ρ and summation over n. Taking these facts into account, one thus obtains

$$f(\alpha, \beta, \delta) = \frac{i}{64\pi^4} \sum_{n=-\infty}^{\infty} \int_{-\infty}^{\infty} (n^2 + \rho^2)\, d\rho \times \int K\left(z, \frac{\alpha z + \gamma}{\beta z + \delta}; \frac{n + i\rho}{2}, \frac{-n + i\rho}{2}\right) \times (\beta z + \delta)^{-(1/2)(n+i\rho)-1} (\bar{\beta}\bar{z} + \bar{\delta})^{(1/2)(n-i\rho)-1}\, dz\, d\bar{z} \tag{6-63}$$

We finally write $f(g)$ in terms of its Fourier transform $F(\chi)$. Since

$$K\left(z_1, \frac{\alpha z_2 + \gamma}{\beta z_2 + \delta}; \chi\right)(\beta z_2 + \delta)^{-n_1-1}(\bar{\beta}\bar{z}_2 + \bar{\delta})^{-n_2-1}$$

is the kernel of $F(\chi)D^{-1}(g; \chi)$, it follows that

$$\frac{i}{2} \int K\left(z_1, \frac{\alpha z + \gamma}{\beta z + \delta}; \chi\right)(\beta z + \delta)^{-n_1-1}(\bar{\beta}\bar{z} + \bar{\delta})^{-n_2-1}\, dz\, d\bar{z} = \mathrm{Tr}\, \{F(\chi)D^{-1}(g; \chi)\} \tag{6-64}$$

Using this result in Eq. (6-62) gives

$$f(g) = -\frac{1}{8\pi^4} \int_{\chi_0} \mathrm{Tr}\, \{F(\chi)D^{-1}(g; \chi)\} c(\chi)\, d\chi \tag{6-65}$$

where $c(\chi) = n_1 n_2$.

Plancherel's Theorem for $SL(2, C)$

The mapping $f(g) \to K(z_1, z_2; \chi)$ can now be made to be isometric by choosing a suitable scalar product in the space of functions $f(g)$ and the space of kernels. To this end we define the scalar product for the functions on the group $SL(2, C)$ by

$$(f_1, f_2) = \int f_1(g) \bar{f}_2(g)\, dg \tag{6-66}$$

whereas that for the kernels by

$$(K_1, K_2) = \frac{1}{32\pi^4} \int_{\chi_0} K_1(z_1, z_2; \chi) \bar{K}_2(z_1, z_2; \chi) \times c(\chi)\, dz_1\, d\bar{z}_1\, dz_2\, d\bar{z}_2\, d\chi \tag{6-67}$$

where $c(\chi) = n_1 n_2 = -|n_1|^2$, and the integral over χ means integration over ρ and summation over n.

Lemma 6.6 *Let $f(g)$ be a square integrable function and $K(z_1, z_2; \chi)$ the kernel of its Fourier transform. Then $(K, K) < \infty$ and*

$$(f, f) = (K, K) \tag{6-68}$$

Equation (6-68) is the analog of Plancherel's theorem for square integrable functions defined over the group $SL(2, C)$. The proof of Eq. (6-68) is left for the reader. The explicit form of Eq. (6-68) is given by

$$\int |f(g)|^2 \, dg = \frac{1}{32\pi^4} \int_{\chi o} \int |K(z_1, z_2; \chi)|^2 c(\chi) \, dz_1 \, d\bar{z}_1 \, dz_2 \, d\bar{z}_2 \, d\chi \tag{6-69}$$

In terms of its Fourier transform, it is given by

$$\int |f(g)|^2 \, dg = -\frac{1}{8\pi^4} \int_{\chi o} c(\chi) \, \mathrm{Tr} \, \{F(\chi) F^*(\chi)\} \, d\chi \tag{6-70}$$

Here $F^* = \bar{F}'$, where $F'(\chi)$ is the conjugate of $F(\chi)$, the integral over χ is to be understood as the integral over ρ and summation over n, and $\mathrm{Tr}\,(FF^*)$ denotes the trace of the operator FF^*.[18] It will be noticed that the integral does not involve all the irreducible unitary representations, but only those of the principal series.

Decomposition of the Regular Representation

We conclude this chapter by discussing the Fourier integral and the decomposition of the regular representation of the Lorentz group into irreducible representations.

The right regular representation of the group $SL(2, C)$ is realized on the Hilbert space of functions $f(g)$ on the group, such that the integral

$$|f|^2 = \int |f(g)|^2 \, dg \tag{6-71}$$

converges. With any element g_0 of the group $SL(2, C)$ we associate the operator $D(g_0; R)$, called the *right translation operator*, which transforms $f(g)$ into

$$D(g_0; R) f(g) = f(g g_0) \tag{6-72}$$

The operator $D(g_0; R)$ is a representation since $D(e; R) = 1$ and $D(g_1; R)D(g_2; R) = D(g_1 g_2; R)$. This representation, which is unitary, is called the *regular representation*, and can be decomposed into irreducible representations.

The decomposition of the regular representation of the group $SL(2, C)$ into

[18] The operator $F(\chi)$ is a Hilbert-Schmidt operator.

irreducible unitary representations $D(g; \chi)$ of the principal series can be seen if one recalls that under the translation $f(g) \to f(gg_0)$ the kernel transforms according to

$$K(z_1, z_2; \chi) \to (\beta_0 z + \delta_0)^{-n_1-1}(\bar{\beta}_0 \bar{z} + \bar{\delta}_0)^{-n_2-1} K\left(z_1, \frac{\alpha_0 z + \gamma_0}{\beta_0 z + \delta_0}; \chi\right) \quad (6\text{-}73)$$

where α_0, β_0, and δ_0 are the elements of the matrix g_0 of the group $SL(2, C)$,

$$g_0 = \begin{pmatrix} \alpha_0 & \beta_0 \\ \gamma_0 & \delta_0 \end{pmatrix}$$

Accordingly, as a function of z the kernel transforms under such translations according to the representation $D(g_0; -\chi)$ of the principal series. Hence the expansion of the functions on the group $SL(2, C)$ in the Fourier integral describes the decomposition of its regular representation $D(g; R)$ by a direct integral into irreducible representations of the principal series. To each $\chi = (n_1, n_2)$ there corresponds a set of component spaces depending on the other parameter z_1, on each of which the same representation $D(g; -\chi)$ is induced.

Finally we remark that the correspondence $f(g) \to K(z_1, z; \chi)$ defines also the decomposition of the *left* regular representation $D(g; L)$ of the group $SL(2, C)$ into irreducible components. The left regular representation is defined by

$$D(g_0; L)f(g) = f(g_0^{-1}g) \quad (6\text{-}74)$$

The corresponding equation to Eq. (6-73) is now given by

$$K(z_1, z; \chi) \to (\beta_0 z_1 + \delta_0)^{n_1-1}(\bar{\beta}_0 \bar{z}_1 + \bar{\delta}_0)^{n_2-1} K\left(\frac{\alpha_0 z_1 + \gamma_0}{\beta_0 z_1 + \delta_0}, z; \chi\right) \quad (6\text{-}75)$$

Hence under the left translations the kernel $K(z_1, z; \chi)$ transforms as a function of z_1 under the unitary representation $D(g; \chi)$ of the principal series. Consequently, the Fourier transform of the function $f(g)$ describes also the decomposition of the left regular representation into its irreducible components.

PROBLEMS

6.1 Verify that the operators $D(g)$ defined by Eqs. (6-1) are unitary if and only if ρ is real.

6.2 Show that every matrix g of $SL(2, C)$ can be represented in the form $g = u_1 \varepsilon u_2$, where u_1 and u_2 are unitary matrices of SU_2, and ε is defined by

$$\varepsilon = \begin{pmatrix} \varepsilon_{22}^{-1} & 0 \\ 0 & \varepsilon_{22} \end{pmatrix}$$

where ε_{22} is real.

6.3 Prove that every linear-fractional transformation $w = (az + c)/(bz + d)$ of the complex plane can be obtained by combining the following three types of transformations: (1) Parallel translation, for which $a = 1$, $b = 0$, $c = z_0$, $d = 1$; (2) Dilation, for which $a = \alpha$, $b = 0$, $c = 0$, $d = \alpha^{-1}$; and (3) Inversion, for which $a = 0$, $b = 1$, $c = -1$, $d = 0$.

6.4 Prove Eqs. (6-22).

6.5 Prove Eqs. (6-23) and (6-24).

6.6 Prove the sufficiency conditions of Eqs. (6-23) and (6-24).

6.7 Verify Eqs. (6-30)–(6-33).

6.8 Show that Eq. (6-32) can be considered as a special case of Eq. (6-31) by choosing the constant c such that $c^{-1} = \Gamma(-\frac{1}{2}n_1 - \frac{1}{2}n_2 + \frac{1}{2}|n_1 - n_2|)$ and remembering that the generalized function

$$\frac{(z_1 - z)^{-n_1-1}(\bar{z}_1 - \bar{z})^{-n_2-1}}{\Gamma(-\frac{1}{2}n_1 - \frac{1}{2}n_2 + \frac{1}{2}|n_1 - n_2|)}$$

becomes $\delta^{(n_1, n_2)}(z_1 - z)$ for non-negative integers n_1 and n_2.

6.9 Verify Eq. (6-36).

6.10 Show that the representations $D(g; \chi_1)$ and $D(g; \chi_2)$ are not equivalent in cases 4 and 4′.

6.11 Prove Eq. (6-60).

6.12 Prove Eq. (6-68).

CHAPTER

SEVEN

ELEMENTS OF GENERAL RELATIVITY THEORY

In the previous five chapters all finite- and infinite-dimensional representations of the Lorentz group were found. We now turn to the description of the gravitational field in which the Lorentz group plays an important role. This description will be given in Chapters 8 to 12. The present chapter is intended as an introductory presentation of classical general relativity theory.

We begin the discussion with a brief review of Riemannian geometry, followed by a description of the physical foundations of general relativity. These are the principles of equivalence and general covariance. The gravitational field equations are then derived in a tensorial form. Exact and approximate solutions of Einstein's field equations are subsequently given. Applications of these solutions are then made in connection with experimental verifications of general relativity. Finally, the equations of motion of material bodies are discussed. These include the Einstein–Infeld–Hoffmann equation of motion for post-Newtonian accuracy.

7-1 RIEMANNIAN GEOMETRY

Transformation of Coordinates

Any four independent variables x^μ, where the greek letter takes the values 0, 1, 2, 3, may be considered as the coordinates of a four-dimensional space V_4. Each set of values of x^μ defines a point of V_4. Let there be another set of coordinates x'^μ related to the first set x^ν by[1]

$$x'^\mu = f^\mu(x^\nu) \tag{7-1}$$

[1] L. P. Eisenhart, *Riemannian Geometry*, Princeton, New Jersey, 1949.

where f^μ are four independent real functions of x^ν. A necessary and sufficient condition that f^μ be independent is that their Jacobian

$$\left|\frac{\partial f^\mu}{\partial x^\nu}\right| = \begin{vmatrix} \dfrac{\partial f^0}{\partial x^0} & \cdots & \dfrac{\partial f^3}{\partial x^0} \\ \vdots & & \\ \dfrac{\partial f^0}{\partial x^3} & \cdots & \dfrac{\partial f^3}{\partial x^3} \end{vmatrix} \tag{7-2}$$

does not vanish identically. Equation (7-1) defines a *transformation of coordinates* in the space V_4. Since the Jacobian is different from zero, one can also write x^μ in terms of x'^ν as

$$x^\mu = g^\mu(x'^\nu) \tag{7-3}$$

A direction at a point P in the space V_4 is determined by the differential dx^μ. The same direction is determined in another set of coordinates x'^μ by the differential dx'^μ. The two differentials are related, using Eq. (7-1), by

$$dx'^\mu = (\partial x'^\mu/\partial x^\nu)\, dx^\nu = (\partial f^\mu/\partial x^\nu)\, dx^\nu \tag{7-4}$$

Here the Einstein summation convention is used, according to which repeated Greek indices are summed over the values 0, 1, 2, 3.

Contravariant Vectors

Let two sets of functions V^μ and V'^μ be related by

$$V'^\mu = (\partial x'^\mu/\partial x^\nu) V^\nu \tag{7-5}$$

similar to the way the differentials dx'^μ and dx^μ are related. V^μ and V'^μ are then called the *components* of a *contravariant vector* in the coordinate systems x^μ and x'^μ, respectively. Hence any four functions of the x's in one coordinate system can be taken as the components of a contravariant vector whose components in any other coordinate system are given by Eq. (7-5).[2]

Invariants. Covariant Vectors

Two functions $f(x)$ and $f'(x')$ define an *invariant* if they are reducible to each other by a coordinate transformation.

[2] A contravariant vector determines a direction at each point of the space V_4. Let V^μ be the components of a contravariant vector and let dx^μ be a displacement in the direction of V^μ. Then $dx^0/V^0 = \cdots = dx^3/V^3$. This set of equations admits three independent $f^k(x^\mu) = c^k$, where $k = 0, 1, 2$, and the c's are arbitrary constants and the matrix $\partial f^k/\partial c^\mu$ is of rank three. The functions f^k are solutions of the partial differential equation $V^\nu\, \partial f^k/\partial x^\nu = 0$. Hence using the transformation laws (7-1) and (7-3) one obtains $V'^k = 0$ for $k = 0, 1, 2$, and $V'^3 \neq 0$. Hence a system of coordinates can be chosen in terms of which all components but one of a given contravariant vector are equal to zero.

Let f be a function of the coordinates. Then

$$\partial f/\partial x'^{\mu} = (\partial f/\partial x^{\nu})(\partial x^{\nu}/\partial x'^{\mu}) \tag{7-6}$$

Two sets of functions V_μ and V'_μ are called the components of a *covariant vector* in the systems x and x', respectively, if they are related by the transformation law of the form (7-6),

$$V'_\mu = (\partial x^{\nu}/x'^{\mu})V_\nu \tag{7-7}$$

For example, if f is a scalar function, then $\partial f/\partial x^\mu$ is a covariant vector. It is called the *gradient* of f. The product $V^\mu W_\mu$ is an invariant if V is a contravariant vector and W is a covariant vector. Conversely, if the quantity $V^\mu W_\mu$ is an invariant and either V^μ or W_μ are arbitrary vectors, then the other set is a vector.

Tensors

Tensors of any order are defined by generalizing Eqs. (7-5) and (7-7). Thus the equation[3]

$$T'^{\mu_1\cdots\mu_m}_{\nu_1\cdots\nu_n} = \frac{\partial x'^{\mu_1}}{\partial x^{\rho_1}}\cdots\frac{\partial x'^{\mu_m}}{\partial x^{\rho_m}}\frac{\partial x^{\sigma_1}}{\partial x'^{\nu_1}}\cdots\frac{\partial x^{\sigma_n}}{\partial x'^{\nu_n}}T^{\rho_1\cdots\rho_m}_{\sigma_1\cdots\sigma_n} \tag{7-8}$$

defines a mixed tensor of order $m + n$, contravariant of the mth order and covariant of the nth order. If the Kronecker delta function is taken as the components of a mixed tensor of the second order in one set of coordinates, for example, then it defines the components of a tensor in any set of coordinates. An invariant is a tensor of zero order and a vector is a tensor of order one. When the relative position of two indices, either contravariant or covariant, is immaterial, the tensor is called *symmetric* with respect to these indices. When the relative position of two indices of a tensor is interchanged and the tensor obtained differs only in sign from the original one, the tensor is called *skew-symmetric* with respect to these indices. The process by means of which from a mixed tensor of order r one obtains a tensor of order $r - 2$ is called *contraction.*

Let $g_{\mu\nu}$ be the components of the *metric tensor*, i.e., a symmetric covariant tensor, which is a function of coordinates, and let $g = \det g_{\mu\nu}$. The quantity $g^{\mu\nu}$, denoting the cofactor of $g_{\mu\nu}$ divided by g, is a symmetric contravariant tensor and satisfies

$$g^{\mu\rho}g_{\nu\rho} = \delta^\mu_\nu \tag{7-9}$$

[3] Certain other quantities transform according to the law

$$T'^{\mu\cdots}_{\alpha\cdots} = J^N\frac{\partial x'^{\mu}}{\partial x^{\rho}}\frac{\partial x^{\beta}}{\partial x'^{\alpha}}\cdots T^{\rho\cdots}_{\beta\cdots}$$

Here J is the Jacobian determinant $|\partial x^\alpha/\partial x'^\beta|$. The superscript N is the power to which J is raised. $T^{\mu\cdots}_{\nu\cdots}$ is called a *tensor density* of weight N. For example, if g' denotes det $g'_{\mu\nu}$ then $g' = J^2 g$, where $g = \det g_{\mu\nu}$. Hence one has for the four-dimensional elements in two coordinate systems the equality:

$$(-g)^{1/2}\,d^4x = (-g')^{1/2}\,d^4x'$$

The element of length is defined by means of a quadratic differential form $ds^2 = g_{\mu\nu}\, dx^\mu\, dx^\nu$. By means of the tensors $g_{\mu\nu}$ and $g^{\mu\nu}$ one can lower or raise tensor indices:

$$T^{\mu}{}_{\nu\rho} = g^{\mu\sigma} T_{\sigma\nu\rho} \tag{7-10}$$

$$T_{\alpha}{}^{\beta\gamma} = g_{\alpha\rho}\, T^{\rho\beta\gamma}$$

Christoffel Symbols

From the two tensors $g_{\mu\nu}$ and $g^{\mu\nu}$ one can define the two functions

$$\Gamma_{\alpha\rho\sigma} = \frac{1}{2}\left(\frac{\partial g_{\rho\alpha}}{\partial x^\sigma} + \frac{\partial g_{\sigma\alpha}}{\partial x^\rho} - \frac{\partial g_{\rho\sigma}}{\partial x^\alpha}\right) \tag{7-11}$$

$$\Gamma^{\mu}_{\rho\sigma} = g^{\mu\alpha}\Gamma_{\alpha\rho\sigma} \tag{7-12}$$

They are symmetric in ρ and σ, and are called the *Christoffel symbols* of the *first* and *second* kind, respectively. Both Christoffel symbols are not components of tensors. By starting with the differential transformation law for $g_{\mu\nu}$ it is not too difficult to show that $\Gamma_{\alpha\rho\sigma}$ transforms according to the following relation (see Problem 7.1):

$$\Gamma'_{\nu\mu\alpha} = \frac{\partial x^\beta}{\partial x'^\mu}\frac{\partial x^\gamma}{\partial x'^\alpha}\frac{\partial x^\delta}{\partial x'^\nu}\Gamma_{\delta\beta\gamma} + g_{\beta\gamma}\frac{\partial x^\beta}{\partial x'^\nu}\frac{\partial^2 x^\gamma}{\partial x'^\mu\,\partial x'^\alpha} \tag{7-13}$$

Making use of the transformation law for $g^{\alpha\beta}$ then leads to the transformation law of $\Gamma^{\delta}_{\beta\nu}$ as

$$\Gamma'^{\delta}_{\beta\nu} = \frac{\partial x'^\delta}{\partial x^\alpha}\frac{\partial x^\mu}{\partial x'^\beta}\frac{\partial x^\sigma}{\partial x'^\nu}\Gamma^{\alpha}_{\mu\sigma} + \frac{\partial x'^\delta}{\partial x^\sigma}\frac{\partial^2 x^\sigma}{\partial x'^\beta\,\partial x'^\nu} \tag{7-14}$$

From Eq. (7-11) we obtain

$$\Gamma^{\mu}_{\alpha\mu} = \frac{1}{2} g^{\mu\nu}\frac{\partial g_{\mu\nu}}{\partial x^\alpha} \tag{7-15}$$

This equation can be rewritten in terms of the determinant g of $g_{\mu\nu}$. The rule for expansion of a determinant leads to the formula

$$\partial g/\partial g_{\mu\nu} = \Delta^{\mu\nu} \tag{7-16}$$

where $\Delta^{\mu\nu}$ is the cofactor of the element $g_{\mu\nu}$. From the law for obtaining the inverse of a determinant, and from the definition of $g^{\mu\nu}$, Eq. (7-16) may be written as

$$\partial g/\partial g_{\mu\nu} = g g^{\mu\nu}$$

and consequently

$$dg = g g^{\mu\nu}\, dg_{\mu\nu} = -g g_{\mu\nu}\, dg^{\mu\nu}$$

Hence we have

$$\partial_\alpha g = g g^{\mu\nu}\, \partial_\alpha g_{\mu\nu} = -g g_{\mu\nu}\, \partial_\alpha g^{\mu\nu} \tag{7-17}$$

The use of Eq. (7-17) enables us to write Eq. (7-15) in the form

$$\Gamma^\mu_{\alpha\mu} = \partial_\alpha(\ln \sqrt{(-g)}) \tag{7-18}$$

Covariant Differentiation

We have seen that the derivatives of an invariant are components of a covariant vector. This is the only case for a general system of coordinates in which the derivative of a tensor is a tensor. However, there are expressions involving first derivatives which are components of a tensor. To see this we proceed as follows.

Let V^μ and V'^ν be a contravariant vector in two coordinate systems x and x'. Then

$$V^\mu = V'^\nu(\partial x^\mu/\partial x'^\nu)$$

Differentiating this equation with respect to x^α and using Eq. (7-14) gives (see Problem 7.2):

$$\frac{\partial V^\mu}{\partial x^\alpha} = \left(\frac{\partial V'^\rho}{\partial x'^\nu} + V'^\sigma \Gamma'^\rho_{\sigma\nu}\right)\frac{\partial x'^\nu}{\partial x^\alpha}\frac{\partial x^\mu}{\partial x'^\rho} - V^\rho \Gamma^\mu_{\rho\alpha} \tag{7-19}$$

Hence if we define a *covariant derivative* of V^μ by

$$\nabla_\alpha V^\mu = \partial_\alpha V^\mu + \Gamma^\mu_{\rho\alpha} V^\rho \tag{7-20}$$

the above equation can be written as

$$\nabla_\alpha V^\mu = \nabla_\nu V'^\rho \frac{\partial x'^\nu}{\partial x^\alpha}\frac{\partial x^\mu}{\partial x'^\rho}$$

Therefore $\nabla_\alpha V^\mu$ is a mixed tensor of second order.

In the same way one shows that the covariant derivative of a covariant vector V_μ is given by:

$$\nabla_\alpha V_\mu = \partial_\alpha V_\mu - \Gamma^\rho_{\mu\alpha} V_\rho \tag{7-21}$$

From the above equation one has for the *curl* of a vector V_μ:

$$\nabla_\beta V_\alpha - \nabla_\alpha V_\beta = \partial_\beta V_\alpha - \partial_\alpha V_\beta \tag{7-22}$$

Hence a necessary and sufficient condition that the first covariant derivative of a covariant vector be symmetric is that the vector be a gradient.

It is easily seen, using the law of covariant differentiation of tensors (see Problem 7.3), that

$$\begin{aligned} \nabla_\rho g^{\mu\nu} &= 0 \\ \nabla_\rho g_{\mu\nu} &= 0 \\ \nabla_\rho \delta^\mu_\nu &= 0 \end{aligned} \tag{7-23}$$

Other properties of covariant differentiation can be established (see Problem 7.4).

Riemann and Ricci Tensors

If we differentiate covariantly the tensor $\nabla_\alpha V_\mu$, given by Eq. (7-21), we obtain

$$(\nabla_\gamma \nabla_\beta - \nabla_\beta \nabla_\gamma) V_\alpha = R^\delta{}_{\alpha\beta\gamma} V_\delta \tag{7-24}$$

where $R^\delta{}_{\alpha\beta\gamma}$ is called the *Riemann tensor* and is given by

$$R^\delta{}_{\alpha\beta\gamma} = \partial_\beta \Gamma^\delta_{\alpha\gamma} - \partial_\gamma \Gamma^\delta_{\alpha\beta} + \Gamma^\mu_{\alpha\gamma} \Gamma^\delta_{\mu\beta} - \Gamma^\mu_{\alpha\beta} \Gamma^\delta_{\mu\gamma} \tag{7-25}$$

A generalization of Eq. (7-24) to an arbitrary tensor can be made (see Problem 7.5). One can show that in order that there can exist a coordinate system in which the first covariant derivatives reduce to ordinary ones at every point in space, it is necessary and sufficient that the Riemann tensor be zero and that the coordinates be those in which the metric is constant.

One notices that the Riemann tensor satisfies

$$\begin{gathered} R_{\alpha\beta\gamma\delta} = -R_{\beta\alpha\gamma\delta} = -R_{\alpha\beta\delta\gamma} = R_{\gamma\delta\alpha\beta} \\ R_{\alpha\beta\gamma\delta} + R_{\alpha\gamma\delta\beta} + R_{\alpha\delta\beta\gamma} = 0 \end{gathered} \tag{7-26}$$

Moreover, counting the number of components, one finds that in a four-dimensional space the Riemann tensor has 20 components.

From the Riemann tensor one can define the *Ricci tensor* and the *Ricci scalar* by

$$R_{\mu\nu} = R^\alpha{}_{\mu\alpha\nu} = \frac{1}{\sqrt{(-g)}} (\sqrt{(-g)} \Gamma^\alpha_{\mu\nu})_{,\,\alpha} - (\ln \sqrt{(-g)})_{,\,\mu\nu} - \Gamma^\alpha_{\mu\beta} \Gamma^\beta_{\nu\alpha} \tag{7-27}$$

$$R = R^\mu{}_\mu \tag{7-28}$$

respectively. Here a comma denotes partial differentiation, $f_{,\,\alpha} = \partial_\alpha f$. The *Einstein tensor* is then defined by

$$G_{\mu\nu} = R_{\mu\nu} - \tfrac{1}{2} g_{\mu\nu} R \tag{7-29}$$

The last important tensor constructed from the Riemann tensor is the *Weyl conformal tensor*:

$$\begin{aligned} C_{\rho\sigma\mu\nu} = R_{\rho\sigma\mu\nu} &- \tfrac{1}{2}(g_{\rho\mu} R_{\nu\sigma} - g_{\rho\nu} R_{\mu\sigma} - g_{\sigma\mu} R_{\nu\rho} + g_{\sigma\nu} R_{\mu\rho}) \\ &- \tfrac{1}{6}(g_{\rho\nu} g_{\mu\sigma} - g_{\rho\mu} g_{\nu\sigma}) R \end{aligned} \tag{7-30}$$

It has the special property that

$$C^\rho{}_{\mu\rho\nu} = 0$$

Furthermore, if the Weyl tensor vanishes everywhere, then the metric is *conformally flat*;[4] that is, there exists a mapping such that $g_{\mu\nu}$ can be diagonalized,

[4] Two spaces V and $\tilde{V}$ are called conformal spaces if their metric tensors $g_{\mu\nu}$ and $\tilde{g}_{\mu\nu}$ are related by $\tilde{g}_{\mu\nu} = e^\beta g_{\mu\nu}$, where β is a function of the coordinates.

with $\pm\beta(x)$ appearing in the diagonal positions, and where $\beta(x)$ is some function. This follows from the fact that the Weyl tensor can be expressed entirely in terms of the density $\tilde{g}_{\mu\nu} = (-g)^{-1/4} g_{\mu\nu}$ and its inverse, and is equal to the Riemann tensor formed by replacing $g_{\mu\nu}$ by $\tilde{g}_{\mu\nu}$, $R_{\alpha\beta\gamma\delta}(\tilde{g}_{\mu\nu}) = C_{\alpha\beta\gamma\delta}(g_{\mu\nu})$. Consequently, the vanishing of the Weyl tensor implies the vanishing of $R_{\alpha\beta\gamma\delta}(\tilde{g}_{\mu\nu})$, which in turn implies that there exists a mapping such that $\tilde{g}_{\mu\nu}$ is everywhere diagonal, with ± 1 appearing along the diagonal. Only g is arbitrary and $\pm g^{1/4}(x)$ appears along the diagonal of $g_{\mu\nu}$.

Geodesics

The differential equations of the curves of extremal length are called *geodesic equations*. To find their equations we seek the relations which must be satisfied to give a stationary value to the integral $\int \mathrm{d}s$. Hence we have to find the solution of the variational problem

$$\delta \int L\,\mathrm{d}s = 0$$

where the Lagrangian L is given by

$$L = \left(g_{\mu\nu}\frac{\mathrm{d}x^\mu}{\mathrm{d}s}\frac{\mathrm{d}x^\nu}{\mathrm{d}s}\right)^{1/2} \tag{7-31}$$

Accordingly we have

$$\delta \int L\,\mathrm{d}s = \int \left[\frac{\partial L}{\partial x^\mu}\,\delta x^\mu + \frac{\partial L}{\partial(\mathrm{d}x^\mu/\mathrm{d}s)}\,\delta\left(\frac{\mathrm{d}x^\mu}{\mathrm{d}s}\right)\right]\mathrm{d}s$$

The second term of the integrand may be written as the two terms

$$\frac{\mathrm{d}}{\mathrm{d}s}\left[\frac{\partial L}{\partial(\mathrm{d}x^\mu/\mathrm{d}s)}\,\delta x^\mu\right] - \frac{\mathrm{d}}{\mathrm{d}s}\left[\frac{\partial L}{\partial(\mathrm{d}x^\mu/\mathrm{d}s)}\right]\delta x^\mu$$

On integration, the first of these expressions contributes nothing since the variations are assumed to vanish at the end points of the curve. As expected, the equation obtained is the usual Langrange equation:

$$\frac{\mathrm{d}}{\mathrm{d}s}\frac{\partial L}{\partial(\mathrm{d}x^\mu/\mathrm{d}s)} - \frac{\partial L}{\partial x^\mu} = 0 \tag{7-32}$$

A simple calculation then gives, using the Lagrangian given by Eq. (7-31),

$$\frac{\mathrm{d}^2 x^\mu}{\mathrm{d}s^2} + \Gamma^\mu_{\alpha\beta}\frac{\mathrm{d}x^\alpha}{\mathrm{d}s}\frac{\mathrm{d}x^\beta}{\mathrm{d}s} = 0 \tag{7-33}$$

Bianchi Identities

A study of Eq. (7-14) shows that it is always possible to choose a coordinate system in which all the Christoffel symbols vanish at a point. For, suppose the Christoffel symbols do not vanish at a point A. One can then carry out the coordinate transformation

$$x'^{\alpha} = x^{\alpha} - x^{\alpha}_{A} + \tfrac{1}{2}\,\Gamma^{\alpha}_{\beta\gamma}(A)(x^{\beta} - x^{\beta}_{A})(x^{\gamma} - x^{\gamma}_{A})$$

where the subscript A indicates to the value at the point A. By Eq. (7-14) one finds that the Christoffel symbols in the new coordinate system vanish at the point A.

A coordinate system for which the Christoffel symbols vanish at a point is called *geodesic*.[5]

If we choose a geodesic coordinate system at a point A, then at A one has

$$\nabla_{\nu} R^{\mu}{}_{\delta\beta\gamma} = \partial_{\beta}\,\partial_{\nu}\Gamma_{\delta\gamma} - \partial_{\gamma}\,\partial_{\nu}\Gamma^{\mu}_{\delta\beta}$$

Consequently, at the point A one has:

$$\nabla_{\nu} R^{\mu}{}_{\delta\beta\gamma} + \nabla_{\gamma} R^{\mu}{}_{\delta\nu\beta} + \nabla_{\beta} R^{\mu}{}_{\delta\gamma\nu} = 0 \tag{7-34}$$

Since the terms of this equation are components of a tensor, this equation holds for any coordinate system and at each point. Hence Eq. (7-34) is an identity throughout the space. It is known as the *Bianchi identities*.

Multiplication of Eq. (7-34) by $g^{\delta\beta}\,\delta^{\gamma}_{\mu}$ gives

$$g^{\delta\beta}(\nabla_{\nu} R^{\gamma}{}_{\delta\beta\gamma} + \nabla_{\gamma} R^{\gamma}{}_{\delta\nu\beta} + \nabla_{\beta} R^{\gamma}{}_{\delta\gamma\nu}) = 0$$

Using the symmetry properties of the Riemann tensor, the last equation becomes:

$$\nabla_{\nu}(R_{\gamma}{}^{\nu} - \tfrac{1}{2}\,\delta^{\nu}_{\gamma} R) = \nabla_{\nu} G_{\gamma}{}^{\nu} = 0 \tag{7-35}$$

Equation (7-35) is called the *contracted Bianchi identity*.

After having developed the mathematical tools to describe general relativity theory, we now turn to the physical foundations of the theory.

7-2 PRINCIPLE OF EQUIVALENCE

Null Experiments. Eötvös Experiment

One of the most interesting *null experiments* in physics is due to Eötvös,[6] first performed in 1890 and recently repeated by Dicke.[7] The experiment showed, in great precision, that all bodies fall with the same acceleration. The roots of the

[5] It is also possible to transform away the Christoffel symbols along a given curve. See E. Fermi, *Atti Accad. Nazl. Lincei* **21,** 21 and 51 (1922).

[6] R. V. Eötvös, *Math. Natur. Ber. Ungarn* **8,** 65 (1890); *Beibl. Ann. Physick* **15,** 688 (1891); *Ann. Physik* **59,** 354 (1896); R. V. Eötvös, D. Pekar and E. Fekete, *Ann. Physik* **68,** 11 (1922).

[7] P. G. Roll, R. Krotkov, and R. H. Dicke, *Ann. Phys. (N.Y.)* **26,** 442 (1964).

experiment go back to Newton and Galileo, who demonstrated experimentally that the gravitational acceleration of a body is independent of its composition.

The importance of the Eötvös experiment is in the fact that the null result of the experiment is a *necessary* condition for the theory of general relativity to be valid.[8]

Eötvös employed a static torsion balance, balancing a component of the Earth's gravitational pull on a weight against the centrifugal force field of the Earth acting on the weight. He employed a horizontal torsion beam, 40 cm long, suspended by a fine wire. From the ends of the torsion beam were suspended, two masses of different compositions, one lower than the other. A lack of exact proportionality between the inertial and gravitational masses of the two bodies would then lead to a torque tending to rotate the balance. There appears to be no need for the one mass to be suspended lower than the other.

The experiment of Eötvös showed, with an accuracy of a few parts in 10^9, that inertial and gravitational masses are equal.[9]

In the experiment performed by Dicke, the gravitational acceleration toward the Sun of small gold and aluminium weights were compared and found to be equal with an accuracy of about one part in 10^{11}. Hence the necessary condition to be satisfied for the validity of general relativity theory seems to be rather satisfactorily met.

The question therefore arises as to what extent is this experiment also a sufficient condition to be satisfied in order that general relativity theory be valid.

It has been emphasized by Dicke that gold and aluminium differ from each other rather greatly in several important aspects. First, the neutron to proton ratio is quite different in the two elements, varying from 1.08 in aluminium to 1.50 in gold. Second, the electrons in aluminium move with nonrelativistic velocities, but in gold the k-shell electrons have a 15 per cent increase in their masses as a result of their relativistic velocities. Third, the electromagnetic negative contribution to the binding energy of the nucleus varies as z^2 and represents $\frac{1}{2}$ per cent of the total mass of a gold atom, whereas it is negligible in aluminium. Fourth, the virtual pair field and other fields would be expected to be different in the two atoms. We thus conclude that the physical aspects of gold and aluminium differ substantially, and consequently the equality of their accelerations represents an important condition to be satisfied by any theory of gravitation.

Since the accuracy of the Eötvös experiment is great, the question arises as to whether it implies that the equivalence principle is very nearly valid. We show below that this is true in a limited sense; certain aspects of the equivalence principle are not supported in the slightest by the Eötvös experiment.

In order to understand the limited conclusions to be drawn from the Eötvös

[8] R. H. Dicke, in *Relativity, Groups, and Topology*, (C. DeWitt *et al.*, Eds), Gordon and Breach, New York, 1964; S. Weinberg, *Gravitation and Cosmology*, Wiley, New York, 1973.

[9] Infeld has suggested that intertial mass is different from gravitational mass in general relativity theory and that they are equal only in the lowest approximation. See L. Infeld, *Ann. Phys.* (*N.Y.*) **6**, 341 (1959).

experiment, we first consider the significance of the equivalence principle for relativity.

Strong and Weak Principle of Equivalence

It is convenient to make a distinction between the *strong* equivalence principle, upon which Einstein's general relativity theory is based, and the *weak* equivalence principle, supported by the Eötvös experiment.

The strong equivalence principle might be stated as the assumption that *in a freely falling, non-rotating, laboratory the local laws of physics take on some standard form, including a standard numerical content, independent of the position of the laboratory in space*. It is implicit in this statement that the effects of gradients in the gravitational field strength are negligibly small, i.e., tidal interaction effects are negligible.

The weak principle of equivalence says much less, and it states only that *the local gravitational acceleration is substantially independent of the composition and structure of the matter being accelerated.*

The significance of the strong form of the principle of equivalence of general relativity theory appears to be its requirement that the laws of physics, expressed in a coordinate system, locally inertial, shall take on some standard form and have some standard numerical content. As a laboratory, neither rotating nor accelerating, in gravity-free space provides a particular example of such a coordinate system, the assumption that physical laws are correctly described in gravity-free space by the usual Lorentz-invariant formalism implies that for the more general situation, where gravitation is present, the formalism should reduce locally to this standard Lorentz invariant form.

The numerical content of the locally observed laws of physics is contained in the dimensionless physical constants appearing in the formulation of physical laws. These include the ratios of the masses of elementary particles, the various coupling constants of the theory, such as the fine structure constant, and the ratios of the masses of elementary particles to the characteristic fundamental gravitational mass $(hc/G)^{1/2}$ where h is Plank's constant, c is the speed of light and G is Newton's gravitational constant. Hence one of the assumptions of the strong equivalence principle is that these dimensionless constants are truly constants, i.e., coordinate independent.

It is evident that the weak principle of equivalence is supported directly and strongly by the Eötvös experiment. But does the Eötvös experiment also support the strong principle of equivalence?

Let us consider the implication of the Eötvös experiment for the constancy of the dimensionless physical constants, hence for the strong equivalence principle. This may be discussed by an example. Consider the question of whether the fine structure constant is really constant, hence independent of position. If we make the assumptions of energy conservation and mass energy equivalence then the fine structure constant could not vary appreciably with position.

To see this let us first assume the contrary. Consequently, the internal energy of the atom is a function of position, since the electrostatic self energy of the nucleus is proportional to $z(z-1)\alpha$, where z is the atomic number and α is the fine structure constant.

Let us assume now a closed cycle in which the atom is first slowly lifted, while in its ground state, from the floor of the laboratory to the ceiling. In the raised position the atom is taken apart, being broken down either into elementary particles or into two or more atoms of smaller masses. These fragments are then gently lowered to the ground floor and reassembled into the original atom. Now it is easily seen that if the internal energy of the primary atom is a function of height, but not that of the fragments, the atom has an additional (anomalous) weight equal to the negative gradient of its internal energy. This is a necessary requirement for energy conservation. More generally, if the energies of the fragments are also variable, but by different amounts, the anomalous forces will be different for the fragments. It should be remembered that only energy conservation is required to exhibit these anomalous forces. With the additional assumption that the inertial masses of the atoms and elementary particles are equal to their internal energies, anomalous gravitational acceleration must appear contrary to the null result of the Eötvös experiment. The expected anomalous acceleration is equal to $\delta a = -c^2 \nabla m/m$.

A strong test of the constancy of the fine structure constant is also provided by the electrostatic contribution to the binding energy of the nucleus. The equality of the gravitational acceleration toward the Sun of aluminium and gold, to an accuracy of one part in 10^{11}, implies that the fine structure constant could vary with position relative to the Sun by only an extremely small amount. The fractional gradient of the fine structure constant due to the presence of the Sun would not exceed 5×10^{-31} per cm, or else the gravitational acceleration of gold relative to the sun would differ by more than one part in 10^{11} from that of aluminium. This is a very severe limit to the constancy of the fine structure constant, for one might expect that if there was a cosmological effect leading to a variation of the fine structure constant, the effect of the Sun's presence would be[10, 11] of the order of $\delta\alpha \sim -\alpha^2 GM/rc^2$, where $\delta\alpha$ is the change in the fine structure constant due to the presence of the Sun, and M and r are the mass of and distance to the Sun. If there were a variation in the fine structure constant as large as that given above, there would be an anomalous gravitational acceleration 5×10^6 times as large as the limit set by the Eötvös experiment.[12]

The conclusion from the Eötvös experiment about the constancy of the physical constants is, accordingly, that all the dimensionless physical constants differing from unity by no more than a few powers of 10, are constants or are very nearly constants.[13]

[10] R. H. Dicke, *Science, N.Y.* **129**, 621 (1959).

[11] L. Landau, in *Niels Bohr and the Development of Physics* (W. Pauli, Ed.), McGraw-Hill, New York, 1955.

[12] This conclusion depends on the assumption that the inertial mass is equivalent to the energy, and the assumption of energy conservation.

[13] There are two physical constants which differ from unity by many powers of 10, for which this argument is without validity. These are the Fermi and the gravitational constants, whose values are, respectively, about 10^{-13} and 10^{-40}.

Negative Mass

It has been suggested that the gravitational mass of an antiparticle, such as the positron, might be negative.[14] Schiff has shown that an argument based on the Eötvös experiment and the standard Lorentz invariance can be used to exclude this possibility.[15] The argument is based on the fact that in the virtual photon field of the nucleus there exist also virtual electron–positron pairs associated with the vacuum polarization. If the positrons in the pairs field were to tend to fall, up not down, there would be an anomalous weight for the atom, substantially greater for large atomic number than small. Because of the null result of the Eötvös experiment, it is concluded that positrons and other antiparticles fall down, not up.

In conclusion, not only is the weak principle of equivalence supported directly and strongly by the Eötvös experiment but, except for the question of the invariance of the gravitational and Fermi interaction constants, the strong principle is also supported by it.

7-3 PRINCIPLE OF GENERAL COVARIANCE

We have seen in the preceding section that a gravitational field can be considered locally equivalent to an accelerated frame. This implies that the special theory of relativity cannot be valid in an extended region where gravitational fields are present. A curved space-time is needed and all laws of nature should be covariant under the most general coordinate transformations.

The original formulation of general relativity by Einstein was based on two principles: (1) the principle of equivalence (discussed in detail in the last section); and (2) the principle of general covariance.[16]

The principle of general covariance is often stated in one of the following forms, which are not exactly equivalent:[17]

(1) All coordinate systems are equally good for stating the laws of physics, and they should be treated on the same footing.
(2) The equations of physics should have tensorial forms.
(3) The equations of physics should have the same form in all coordinate systems.

According to the principle of general covariance, the coordinates become nothing more than a bookkeeping system to label the events. The principle is a valuable guide to deducing correct equations.

[14] For a discussion on the dynamics of negative mass see H. Bondi, *Rev. Mod. Phys.* **29**, 3 and 423 (1957).

[15] L. I. Schiff, *Phys. Rev. Letters* **1**, 254 (1958); *Proc. Natl. Acad. Sci.* **45**, 69 (1959).

[16] A. Einstein, *Ann. Physik* **49**, 769 (1916).

[17] A. Trautman, *Lectures on General Relativity*, Prentice-Hall, Englewood Cliffs, N.J., 1965.

It has been pointed out by Kretschmann[18] that any space-time physical law can be written in a covariant form and hence the principle of general covariance has no necessary physical consequences, and Einstein[19] concurred with this view.

In spite of Einstein's acceptance of Kretschmann's objection, it appears that the principle of general covariance was introduced by Einstein as a generalization of the principle of special relativity[20] and he often referred to it as the principle of general relativity. In fact the principle of equivalence (which necessarily leads to the introduction of a curved space-time), plus the assumption of general covariance, is most of what is needed to generate Einstein's theory of general relativity. They lead directly to the idea that gravitation can be explained by means of Riemannian geometry. This is done in the next section.

7-4 GRAVITATIONAL FIELD EQUATIONS

We have seen in Section 7-1 that the Riemannian geometry is characterized by a geometrical metric, i.e., a symmetric, tensor $g_{\mu\nu}$ from which one can construct other quantities. General relativity theory identifies this tensor as the gravitational potential. Hence in general relativity there are ten components to the gravitational potential, as compared with the single potential function in the Newtonian theory of gravitation.

Einstein's Field Equations

In trying to arrive at the desired gravitational field equations that the metric tensor has to satisfy, we are guided by the requirement that, in an appropriate limit, the theory should be reduced to the Newtonian gravitational theory. In the latter theory, the gravitational potential ϕ is determined by the Poisson equation:

$$\nabla^2\phi = 4\pi G\rho \tag{7-36}$$

where G ($= 6.67 \times 10^{-8}$ cm^3 gm^{-1} sec^{-2}) is the Newton gravitational constant and ρ is the mass density of matter. Hence $g_{\mu\nu}$ should satisfy a second order differential equation, and it should also be related to the energy–momentum tensor $T_{\mu\nu}$ linearly. Such equations are[21]

$$R_{\mu\nu} - \tfrac{1}{2}g_{\mu\nu}R = \kappa T_{\mu\nu} \tag{7-37}$$

where κ is some constant to be determined. But the contracted Bianchi identities,

[18] E. Kretschmann, *Ann. Physik* **53**, 575 (1917).

[19] A. Einstein, *Ann. Physik* **55**, 241 (1918).

[20] The principle of special relativity states that the laws of physics are valid in all inertial frames of references.

[21] In cosmology theory, one sometimes adds an additional term, $\lambda g_{\mu\nu}$, to the left-hand side of Eq. (7-37). The constant λ is known as a cosmological constant.

Eq. (7-35), show that the covariant divergence of the left-hand side of Eq. (7-37) vanishes. Hence

$$\nabla_\nu T_\mu{}^\nu = 0 \tag{7-38}$$

which expresses the covariant conservation of energy and momentum. The constant κ can be determined by going to the limit of weak gravitational field (see Problem 7.10). Its value is $\kappa = 8\pi G/c^4$. The constant κ is known as Einstein's gravitational constant.

Deduction of Einstein's Equations from Variational Principle

We start with the action integral

$$I = \int \sqrt{(-g)}(L_G - 2\kappa L_F)\, \mathrm{d}^4x \tag{7-39}$$

and demand its variation to be zero. Here L_G and L_F are the Lagrangians for the gravitational and other fields, respectively. We take $L_G = R$, where R is the Ricci scalar, $R = R_{\mu\nu}g^{\mu\nu}$.

The first part of the integral (7-39) gives

$$\delta \int \sqrt{(-g)}R\, \mathrm{d}^4x = \int \sqrt{(-g)}g^{\mu\nu}\, \delta R_{\mu\nu}\, \mathrm{d}^4x + \int R_{\mu\nu}\, \delta(\sqrt{(-g)}g^{\mu\nu})\, \mathrm{d}^4x \tag{7-40}$$

To find $\delta R_{\mu\nu}$ we note that in a geodesic coordinate system one has[22]

$$\delta R_{\mu\nu} = \nabla_\alpha(\delta\Gamma^\alpha_{\mu\nu}) - \nabla_\nu(\delta\Gamma^\alpha_{\mu\alpha})$$

But the latter is a tensorial equation. Hence it is valid in all coordinate systems. Consequently, the first integral on the right-hand side of Eq. (7-40) can be written as

$$\int \sqrt{(-g)}g^{\mu\nu}\, \delta R_{\mu\nu}\, \mathrm{d}^4x = \int \sqrt{(-g)}\nabla_\alpha(g^{\mu\nu}\, \delta\Gamma^\alpha_{\mu\nu} - g^{\mu\alpha}\, \delta\Gamma^\beta_{\mu\beta})\, \mathrm{d}^4x$$

and hence (by Problem 7.7) is equal to

$$\int \partial_\alpha[\sqrt{(-g)}(g^{\mu\nu}\, \delta\Gamma^\alpha_{\mu\nu} - g^{\mu\alpha}\, \delta\Gamma^\beta_{\mu\beta})]\, \mathrm{d}^4x$$

This integral, however, vanishes since by Gauss' theorem it is equal to a surface integral which is equal to zero in consequence of the vanishing of the variations on the boundary.

The second integral on the right-hand side of Eq. (7-40) gives, by Eq. (7-17),

$$\int R_{\mu\nu}\, \delta(\sqrt{(-g)}g^{\mu\nu})\, \mathrm{d}^4x = \int \sqrt{(-g)}(R_{\mu\nu} - \tfrac{1}{2}g_{\mu\nu}R)\, \delta g^{\mu\nu}\, \mathrm{d}^4x \tag{7-41}$$

[22] A. Palatini, *Rend. circ. mat. Palermo* **43**, 203 (1919).

The second part of the integral (7-39) leads to (see Problem 7.11)

$$\delta \int \sqrt{(-g)} L_F \, d^4x = -\tfrac{1}{2} \int \sqrt{(-g)} T_{\mu\nu} \, \delta g^{\mu\nu} \, d^4x \tag{7-42}$$

where $T_{\mu\nu}$ is the energy–momentum tensor and is given by

$$T_{\mu\nu} = \frac{-2}{\sqrt{(-g)}} \left[\left(\frac{\partial(\sqrt{(-g)} L_F)}{\partial g^{\mu\nu}_{,\alpha}} \right)_{,\alpha} - \frac{\partial(\sqrt{(-g)} L_F)}{\partial g^{\mu\nu}} \right] \tag{7-43}$$

and a comma denotes partial differentiation, $f_{,\alpha} = \partial_\alpha f$. Combining Eqs. (7-39), (7-41) and (7-42) then leads to the field equations (7-37):

$$R_{\mu\nu} - \tfrac{1}{2} g_{\mu\nu} R = \kappa T_{\mu\nu}$$

Maxwell's Equations in Curved Space

The Lagrangian density for the electromagnetic field is given by

$$\mathfrak{L} = -(1/16\pi) F_{\alpha\beta} F^{\alpha\beta} + (1/c) J^\alpha A_\alpha + \mathfrak{L}_e \tag{7-44}$$

The field $F_{\mu\nu}$ is related to the potential A_α by

$$F_{\mu\nu} = \partial_\nu A_\mu - \partial_\mu A_\nu \tag{7-45}$$

where J^α is the four-current density, and $\mathfrak{L}_e$ is the Lagrangian density of the charged particles.

Maxwell's equations are then given by

$$\partial_\nu F^{\mu\nu} = (4\pi/c) J^\mu \tag{7-46}$$

$$\partial_\gamma F_{\alpha\beta} + \partial_\beta F_{\gamma\alpha} + \partial_\alpha F_{\beta\gamma} = 0 \tag{7-47}$$

It is easily seen that generalizations of these equations to curved space are achieved by the following equations:

$$\nabla_\nu F^{\mu\nu} = (4\pi/c) J^\mu \tag{7-48}$$

$$F_{\mu\nu} = \nabla_\nu A_\mu - \nabla_\mu A_\nu \tag{7-49}$$

for Eqs. (7-46) and (7-45), whereas

$$\nabla_\delta [\varepsilon^{\alpha\beta\gamma\delta} (-g)^{-1/2} F_{\alpha\beta}] = 0 \tag{7-50}$$

generalizes Eq. (7-47). It will be noted that Eqs. (7-49) and (7-50) are identical to Eqs. (7-45) and (7-47), respectively. In Eq. (7-50) $\varepsilon^{\alpha\beta\gamma\delta}$ is the totally skew-symmetric tensor density of weight $+1$ with values $+1$ and -1, depending on whether $\alpha\beta\gamma\delta$ is an even or an odd permutation of 0123 and zero otherwise.

Stationary and Static Gravitational Fields

A gravitational field is called *stationary* if it admits a time-like Killing vector field ξ^μ, i.e.,

$$\nabla_\mu \xi_\nu + \nabla_\nu \xi_\mu = 0 \tag{7-51}$$

where $\xi_\mu = g_{\mu\nu}\xi^\nu$ and $\xi^2 = \xi_\mu \xi^\mu > 0$. Equation (7-51) is called *Killing's equation*, and can be written in the equivalent form

$$\mathscr{L}_\xi g_{\alpha\beta} = \xi^\sigma \partial_\sigma g_{\alpha\beta} + g_{\beta\sigma}\, \partial_\alpha \xi^\sigma + g_{\alpha\sigma}\, \partial_\beta \xi^\sigma = 0 \tag{7-52}$$

If a space possesses a time-like Killing vector field ξ^μ, then there always exists a coordinate system in which $\xi^\mu = (1, 0, 0, 0)$, and therefore

$$\partial_0 g_{\alpha\beta} = \partial_t g_{\alpha\beta} = 0 \tag{7-53}$$

by Eq. (7-52), where $t = x^0$ is the time-like coordinate.

A special case of a stationary space is one for which the trajectories of the vector ξ^μ are orthogonal to a family of hypersurfaces. Such a space is called *static*.

In a static space there exists a coordinate system, which is said to be adapted to the Killing vector field ξ^μ, in which

$$\partial_t g_{\mu\nu} = 0, \qquad g_{0k} = 0 \tag{7-54}$$

where $k = 1, 2, 3$.

For suppose $\xi_\mu = \chi\, \partial_\mu \sigma$. Then $\xi_{[\alpha}\, \partial_\beta\, \xi_{\gamma]} = 0$, or $\xi_{[\alpha}\, \nabla_\beta\, \xi_{\gamma]} = 0$, where the square bracket indicates an antisymmetrization of the indices (for example $T_{[\alpha\beta\gamma]} = (1/3!)$ $(T_{\alpha\beta\gamma} + T_{\beta\gamma\alpha} + T_{\gamma\alpha\beta} - T_{\beta\alpha\gamma} - T_{\alpha\gamma\beta} - T_{\gamma\beta\alpha})$). Using Killing's equation (7-51), the latter equation can be written as

$$\xi_\alpha \nabla_\beta \xi_\gamma + \xi_\beta \nabla_\gamma \xi_\alpha + \xi_\gamma \nabla_\alpha \xi_\beta = 0$$

Multiplying now this equation by ξ^γ, putting $\xi^2 = \xi_\gamma \xi^\gamma$, and again using Killing's equation (7-51), gives

$$(\xi_\alpha \nabla_\beta - \xi_\beta \nabla_\alpha)\xi^2 + \xi^2(\nabla_\alpha \xi_\beta - \nabla_\beta \xi_\alpha) = 0$$

Hence one obtains the equation

$$\partial_\alpha(\xi_\beta/\xi^2) = \partial_\beta(\xi_\alpha/\xi^2)$$

whose solution is $\xi_\alpha = \xi^2\, \partial_\alpha \sigma$, and therefore $\chi = \xi^2$.

If we choose now a coordinate system in which $\xi^\alpha = \delta^\alpha_0$, then the last equation gives $g_{\alpha 0} = g_{\alpha\beta}\, \delta^\beta_0 = g_{\alpha\beta}\, \xi^\beta = \xi_\alpha = \xi^2\, \partial_\alpha \sigma$. But $\xi^2 = g_{\alpha\beta}\, \xi^\alpha \xi^\beta = g_{\alpha\beta}\, \delta^\alpha_0\, \delta^\beta_0 = g_{00}$. Hence we obtain

$$g_{\alpha 0} = g_{00}\, \partial_\alpha \sigma \tag{7-55}$$

For $\alpha = 0$, Eq. (7-55) gives $\partial_0 \sigma = 1$, and consequently $\sigma = x^0 + f(x^1, x^2, x^3)$, where f is an arbitrary function of the coordinates x^k.

If we now define the new coordinate system $x'^0 = x^0 + f(x'^k)$, $x'^j = x^j$, then the components of the metric tensor g'_{0k} in the new coordinate system are given by

$$g'_{0k} = \frac{\partial x^\beta}{\partial x'^0}\frac{\partial x^\gamma}{\partial x'^k} g_{\beta\gamma} = g_{0k} - g_{00}\,\partial_k \sigma$$

and therefore

$$g'_{0k} = 0$$

by Eq. (7-55). This shows Eqs. (7-54) to be valid for a static field.

In the next section we find solutions of the Einstein field equations some of which will be static.

7-5 SOLUTIONS OF EINSTEIN'S FIELD EQUATIONS

In spite of the nonlinearity of the Einstein field equations, there are numerous exact solutions to these equations. Moreover, there are other solutions which are not exact but approximate. Detailed exact solutions will be given in Chapter 11 using special methods. Here we give some of the known elementary ones.

Schwarzschild Solution

The simplest of all exact solutions to Einstein's field equations is that of Schwarzschild.[23] The solution is *spherical symmetric* and *static*. Such a field can be produced by a spherically symmetric distribution and motion of matter. It follows that the requirement of spherical symmetry alone is sufficient to yield a static solution.

The spherical symmetry of the metric means that the expression for the interval $\mathrm{d}s = (g_{\mu\nu}\,\mathrm{d}x^\mu\,\mathrm{d}x^\nu)^{1/2}$ must be the same for all points located at the same distance from the centre. In flat space their distance is equal to the radius vector, and the metric is given by (c is taken as equal to 1):

$$\mathrm{d}s^2 = \mathrm{d}t^2 - \mathrm{d}r^2 - r^2(\mathrm{d}\theta^2 + \sin^2\theta\,\mathrm{d}\phi^2) \tag{7-56}$$

In a non-Euclidean space, such as the one we have in the presence of a gravitational field, there is no quantity which has all the properties of the flat space radius vector, such as that it is equal both to the distance from the centre and to the length of the circumference divided by 2π. Therefore, the choice of a radius vector is here arbitrary.

When a mass with spherical symmetry is introduced at the origin, the flat space line element (7-56) must be modified but in a way that retains spherical symmetry. The most general spherically symmetric expression for $\mathrm{d}s^2$ is

$$\mathrm{d}s^2 = a(r,t)\,\mathrm{d}t^2 + b(r,t)\,\mathrm{d}r^2 + c(r,t)\,\mathrm{d}r\,\mathrm{d}t + \mathrm{d}(r,t)(\mathrm{d}\theta^2 + \sin^2\theta\,\mathrm{d}\phi^2) \tag{7-57}$$

[23] K. Schwarzschild, *S.B. Preuss. Akad. Wiss.*, 189 (1916).

Because of the arbitrariness in the choice of the coordinate system in general relativity theory, we can perform a coordinate transformation which does not destroy the spherical symmetry of ds^2. Hence we can choose new coordinates r' and t' given by some two functions $r' = r'(r, t)$ and $t' = t'(r, t)$.

Making use of these transformations, we can choose the new coordinates so that the coefficient $c(r, t)$ of the mixed term $dr\, dt$ vanishes and the coefficient $d(r, t)$ of the angular part to be $-r'^2$, in the metric (7-57). The latter condition implies that the radius vector is now defined in such a way that the circumference of a circle whose centre is at the origin of the coordinates is equal to $2\pi r$. It is convenient to express the functions $a(r, t)$ and $b(r, t)$ in exponential forms, e^ν and $-e^\lambda$, respectively, where ν and λ are functions of the new coordinates r' and t'. Consequently, the line element (7-57) will have the form

$$ds^2 = e^\nu\, dt^2 - e^\lambda\, dr^2 - r^2(d\theta^2 + \sin^2\theta\, d\phi^2) \tag{7-58}$$

where we have dropped the primes from the new coordinates r' and t', and the speed of light c is taken as equal to 1.

We now denote the coordinates t, r, θ, ϕ by x^0, x^1, x^2, x^3, respectively. Hence the nonvanishing components of the covariant metric tensor are given by:

$$g_{\mu\nu} = \begin{pmatrix} e^\nu & & & 0 \\ & -e^\lambda & & \\ & & -r^2 & \\ 0 & & & -r^2\sin^2\theta \end{pmatrix} \tag{7-59a}$$

whereas those of the contravariant metric tensor are:

$$g^{\mu\nu} = \begin{pmatrix} e^{-\nu} & & & 0 \\ & -e^{-\lambda} & & \\ & & -r^{-2} & \\ 0 & & & -r^{-2}\sin^{-2}\theta \end{pmatrix} \tag{7-59b}$$

To find out the differential equations that the functions ν and λ have to satisfy, according to Einstein's equations, we first need to calculate the Christoffel symbols associated with the metric (7-59). The nonvanishing components are:

$$\begin{gathered} \Gamma^0_{00} = \dot{\nu}/2,\ \Gamma^0_{10} = \nu'/2,\ \Gamma^0_{11} = (\dot{\lambda}/2)\, e^{\lambda-\nu} \\ \Gamma^1_{00} = (\nu'/2)\, e^{\nu-\lambda},\ \Gamma^1_{10} = \dot{\lambda}/2,\ \Gamma^1_{11} = \lambda'/2 \\ \Gamma^1_{22} = -r\, e^{-\lambda},\ \Gamma^1_{33} = -r\sin^2\theta\, e^{-\lambda},\ \Gamma^2_{12} = 1/r \\ \Gamma^2_{33} = -\sin\theta\cos\theta,\ \Gamma^3_{13} = 1/r,\ \Gamma^3_{23} = \cot\theta \end{gathered} \tag{7-60}$$

where dots and primes denote differentiation with respect to t and r, respectively.

With these Christoffel symbols, we compute the following expressions for the nonvanishing components of the Einstein tensor $G_\nu{}^\mu = R_\nu{}^\mu - \frac{1}{2}\,\delta_\nu^\mu R$:

$$G_0{}^0 = -\mathrm{e}^{-\lambda}\left(\frac{1}{r^2} - \frac{\lambda'}{r}\right) + \frac{1}{r^2}$$

$$G_0{}^1 = -\mathrm{e}^{-\lambda}\frac{\dot{\lambda}}{r} \tag{7-61}$$

$$G_1{}^1 = -\mathrm{e}^{-\lambda}\left(\frac{\nu'}{r} + \frac{1}{r^2}\right) + \frac{1}{r^2}$$

$$G_2{}^2 = G_3{}^3 = -\frac{1}{2}\,\mathrm{e}^{-\lambda}\left(\nu'' + \frac{\nu'^2}{2} + \frac{\nu' - \lambda'}{r} - \frac{\nu'\lambda'}{2}\right) + \frac{1}{2}\,\mathrm{e}^{-\nu}\left(\ddot{\lambda} + \frac{\dot{\lambda}^2}{2} - \frac{\dot{\lambda}\dot{\nu}}{2}\right).$$

All other components vanish identically.

The gravitational field equations can now be integrated exactly for the spherical symmetric field in vacuum, i.e., outside the masses producing the field. Setting Eqs. (7-61) equal to zero leads to the independent equations:

$$\mathrm{e}^{-\lambda}\left(\frac{\nu'}{r} + \frac{1}{r^2}\right) - \frac{1}{r^2} = 0 \tag{7-62a}$$

$$\mathrm{e}^{-\lambda}\left(\frac{\lambda'}{r} - \frac{1}{r^2}\right) + \frac{1}{r^2} = 0 \tag{7-62b}$$

$$\dot{\lambda} = 0 \tag{7-62c}$$

From Eq. (7-62a) and (7-62b) we find $\nu' + \lambda' = 0$, so that $\nu + \lambda = f(t)$, where $f(t)$ is a function of t only. If we perform now the coordinate transformation $x^0 = h(x'^0)$, $x^k = x'^k$, then $g'_{00} = \dot{h}^2 g_{00}$. Such a transformation amounts to adding to the function ν an arbitrary function of time, while leaving unaffected the other components of the metric. Hence we can choose the function h so that $\nu + \lambda = 0$. Consequently, we see, by Eq. (7-62c), that both ν and λ are time-independent. In other words the spherically symmetric gravitational field in vacuum is automatically static.

Equations (7-62) can now be integrated. They give:

$$\mathrm{e}^{-\lambda} = \mathrm{e}^{\nu} = 1 - \frac{K}{r}$$

where K is an integration constant. We see that for $r \to \infty$, $\mathrm{e}^{-\lambda} = \mathrm{e}^{\nu} = 1$, i.e., far from the gravitational bodies, the metric reduces to that of the flat space (7-56). The constant K can easily be determined from the requirement that Newton's law of motion be obtained at large distances from the central mass. From the geodesic equation it follows that the radial acceleration of a small test mass at rest with respect to the central mass is (see Problem 7.10):

$$-\Gamma^1_{00} = -\frac{1}{2}\left(1 - \frac{K}{r}\right)\frac{K}{r^2} \to -\frac{K}{2r^2}$$

Comparing this expression with the Newtonian value $-Gm/r^2$ gives $K = 2Gm$, where m is the central mass and G is the Newton constant.[24]

We therefore obtain for the spherically symmetric metric the form:

$$g_{\mu\nu} = \begin{pmatrix} 1 - 2Gm/r & & & \\ & -(1 - 2Gm/r)^{-1} & & 0 \\ & 0 & -r^2 & \\ & & & -r^2 \sin^2 \theta \end{pmatrix} \tag{7-63}$$

It is known as the *Schwarzschild solution* and describes the most general spherically symmetric solution to the Einstein field equations in a region of space where the energy–momentum tensor $T^{\mu\nu}$ vanishes. Although $g_{\mu\nu}$ goes to the flat space metric when r goes to infinity, it was *not* necessary to require this asymptotic behavior to obtain the solution. Finally, we remark that all spherically symmetric solutions of the Einstein field equations in vacuum which satisfy the boundary conditions at infinity are equivalent to the Schwarzschild field, i.e., their time-dependence can be eliminated by a suitable coordinate transformation. This result is due to Birkhoff.[25]

Finally, it is convenient to introduce Cartesian coordinates by means of the coordinate transformation

$$\begin{aligned} x^1 &= r \sin \theta \cos \phi \\ x^2 &= r \sin \theta \sin \phi \\ x^3 &= r \cos \theta \end{aligned}$$

In terms of these coordinates, the Schwarzschild metric (7-63) will then have the form

$$\begin{aligned} g_{00} &= 1 - \frac{2Gm}{r} \\ g_{0r} &= 0 \\ g_{rs} &= -\delta_{rs} - \frac{2Gm/r}{1 - 2Gm/r} \frac{x^r x^s}{r^2} \end{aligned} \tag{7-63a}$$

Maximal Extension of the Schwarzschild Metric

The surface for which $r = 2Gm$ in the Schwarzschild metric (7-63) is often called the *Schwarzschild singularity*. At this surface the component of the metric g_{11} diverges. This singularity, however, is different from that at $r = 0$ of the same

[24] The constant $2Gm$, or $2Gm/c^2$ in units where c is not taken as equal to 1, is often called the *Schwarzschild radius* of the mass m. For example, the Schwarzschild radius for the Sun is 2.95 km, that for the Earth is 8.9 mm, and that for an electron is 13.5×10^{-56} cm.

[25] G. Birkhoff, *Relativity and Modern Physics*, Harvard University Press, Cambridge, Mass., 1923.

metric. It has been shown by Kasner,[26] Lemaitre,[27] Einstein and Rosen,[28] Robertson,[29] Synge,[30] Ehlers,[31] Finkelstein,[32] and Fronsdal[33] that the singularity at $r = 2Gm$ is not an intrinsic feature of the Schwarschild metric but rather a property of the coordinate system used to express the metric. An indication of this effect is that the determinant of the metric, $g = -r^4 \sin^2 \theta$, is regular at $r = 2Gm$. So is the scalar $R_{\alpha\beta\gamma\delta} R^{\alpha\beta\gamma\delta} = 48G^2m^2/r^6$. It follows that the topology of the space–time manifold of the Schwarzschild metric is *not* equivalent to that of the Euclidean four-dimensional space.

An exhibition of these properties was given by Kruskal[34] who presented a particularly simple transformation of the Schwarzschild metric into new coordinates, whereby the singularity at $r = 2Gm$ is removed and the maximal singularity-free extension is clearly exhibited. A manifold with a metric geometry imposed upon it is said to be *maximal* if every geodesic line emanating from an arbitrary point of the manifold has an infinite length in both directions, or terminates on a physical singularity (that cannot be removed by a suitable coordinate transformation). If all geodesics emanating from a point have infinite length in both directions, the manifold is said to be *complete*. It thus follows that a manifold that is maximal but not complete possesses singular points. The Kruskal manifold is maximal but not complete.

Kruskal introduced a new spherically symmetric coordinate system in which radial light rays everywhere have the slope $dx^1/dx^0 = \pm 1$. The metric will then have the form

$$g_{\mu\nu} = \begin{pmatrix} f^2 & & & 0 \\ & -f^2 & & \\ & & -r^2 & \\ 0 & & & -r^2 \sin^2 \theta \end{pmatrix} \tag{7-64}$$

where $x^0 = v$, $x^1 = u$, $x^2 = \theta$, and $x^3 = \phi$. By identifying Eqs. (7-63) and (7-64), and requiring the function f to depend on r alone and to remain finite and nonzero for $v = u = 0$, one finds the following essentially unique equations of transformation between the exterior of the "spherical singularity", $r > 2Gm$, and the quadrant $u > |v|$ in the plane of the new variables:

$$\begin{aligned} v &= [(r/2Gm) - 1]^{1/2} \exp(r/4Gm) \sinh(t/4Gm) \\ u &= [(r/2Gm) - 1]^{1/2} \exp(r/4Gm) \cosh(t/4Gm) \end{aligned} \tag{7-65}$$

[26] E. Kasner, *Am. J. Math.* **43,** 130 (1921).
[27] G. Lemaitre, *Ann. Soc. Sci. Bruxelles* A53, 51 (1933).
[28] A. Einstein and N. Rosen, *Phys. Rev.* **48,** 73 (1935).
[29] H. P. Robertson, lecture in Toronto, 1939 (unpublished), cited by J. L. Synge.
[30] J. L. Synge, *Proc. Roy. Irish Acad.* A53, 83 (1950).
[31] J. Ehlers, thesis, Hamburg, 1957 (unpublished).
[32] D. Finkelstein, *Phys. Rev.* **110,** 965 (1958).
[33] C. Fronsdal, *Phys. Rev.* **116,** 778 (1959).
[34] M. D. Kruskal, *Phys. Rev.* **119,** 1743 (1960).

The inverse transformation is given by

$$[(r/2Gm) - 1] \exp(r/2Gm) = u^2 - v^2$$
$$t/4Gm = \operatorname{arctanh}(v/u) \tag{7-66}$$

and the function f is defined by

$$f^2 = (32Gm^3/r) \exp(-r/2Gm)$$
$$= \text{a transcendental function of } (u^2 - v^2) \tag{7-67}$$

The new coordinates give an analytic extension, K, of the limited region of space-time, S, which is described without singularity by the Schwarzschild coordinates with $r > 2Gm$. The metric in the extended region joins on smoothly, and without singularity, to the metric at the boundary of the region S at $r = 2Gm$. The extended region K, moreover, is the *maximal* possible singularity-free extension of the region S. This may be seen by direct examination of the geodesics; every geodesic, followed in whichever direction, either runs into the "barrier" of intrinsic singularities at $r = 0$ ($v^2 - u^2 = 1$), or is continuable infinitely.

The maximal extension K has a *non-Euclidean* topology (see Fig. 7.1). It therefore belongs to the class of topologies considered by Einstein and Rosen, Wheeler,[35] and Misner and Wheeler.[36] It presents a "bridge" between two otherwise Euclidean spaces. It may also be interpreted as describing the "throat of a wormhole" connecting two distant regions in one Euclidean space (when the

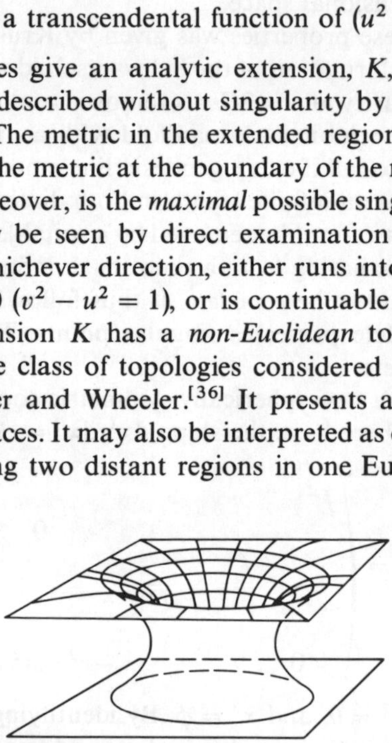

(a)

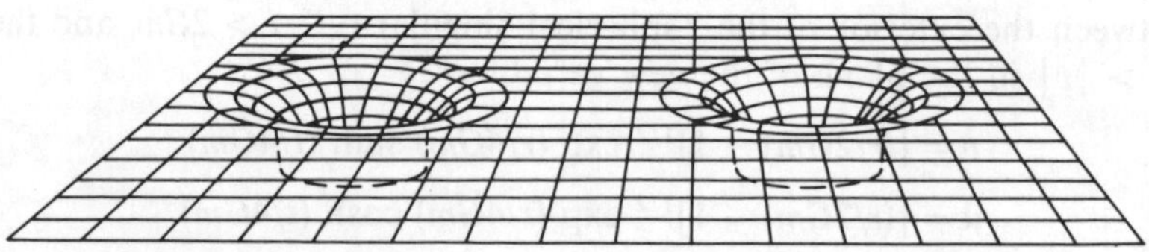

(b)

Figure 7.1 Two interpretations of the three-dimensional "maximally extended Schwarzschild metric" at the time $t = 0$. (*a*) A connection or bridge in the sense of Einstein and Rosen between *two* otherwise Euclidean spaces. (*b*) A wormhole in the sense of Wheeler connecting two regions in *one* Euclidean space, in the limiting case where these regions are extremely far apart compared to the dimensions of the throat of the wormhole.

[35] J. A. Wheeler, *Phys. Rev.* **97**, 511 (1955).

[36] C. W. Misner and J. A. Wheeler, *Ann. Phys. (N.Y.)* **2**, 525 (1957).

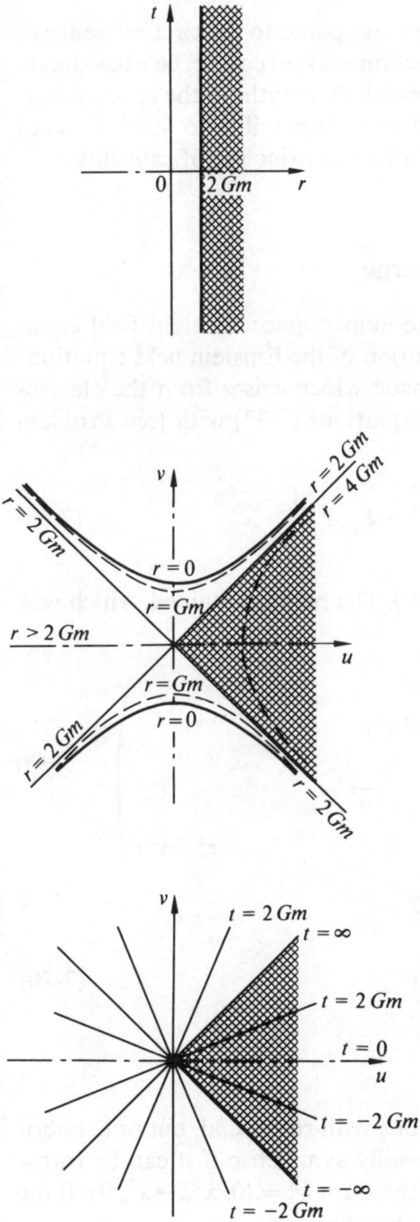

Figure 7.2 The Kruskal diagram. Corresponding regions of the (r, t) and (u, v) planes. In the latter, curves of constant r are hyperbolas asymptotic to the lines $r = 2Gm$, while t is constant on straight lines through the origin. The exterior of the singular sphere, $r > 2Gm$, corresponds to the region $|v| < u$ (the hatched areas). The whole line $r = 2Gm$ in the (r, t) plane corresponds to the origin $u = v = 0$, while two one-dimensional families of ideal limit points with $r \to 2Gm$ and $t \to \pm\infty$ correspond to the remaining boundary points $u = |v| > 0$.

In the (u, v) plane the metric is entirely regular not only in the hatched area but in the entire area between the two branches of the hyperbola $r = 0$. This comprises two images of the exterior of the spherical singularity and two of its interior. (The expressions in Eqs. (7-65) and (7-66) are valid in the right-hand quadrant $u > |v|$. To obtain formulae valid in the left-hand quadrant replace u and v by their negatives everywhere. To obtain formulae valid in the upper or lower quadrant replace u by $\pm v$, v by $\pm u$, and $r/2Gm - 1$ by its negative everywhere. Note that the formula for r and the final formula for t remain invariant under these substitutions). The purely radial ($d\theta = d\phi = 0$) null geodesics are lines inclined at 45°. The points with $r = 2Gm$ have no local topological distinction, but rather a global one: if a test particle crosses $r = 2Gm$ into the interior (where r is time-like and t space-like), it can never get back out but must inevitably hit the irremovable singularity $r = 0$.

separation of the wormhole mouths is very large compared to the circumference of the throat). The length of the wormhole connection may, of course, be exceedingly short compared to the distance between the wormhole mouths in the approximating Euclidean space. However, (see Fig. 7.2), it is impossible to send a signal through the throat in such a way as to contradict the principle of causality.

Gravitational Field of a Point Electric Charge

The Schwarzschild metric is a solution of the empty-space Einstein field equations. The field due to a point charge is a solution of the Einstein field equations with a non-vanishing energy–momentum tensor which arises from the electromagnetic field. Hence we must use the field equations (7-37) with (see Problem 7.13)

$$T_{\mu\nu} = \frac{1}{4\pi}\left\{\frac{1}{4} g_{\mu\nu} F_{\alpha\beta} F^{\alpha\beta} - F_{\mu\alpha} F_{\nu}^{\ \alpha}\right\} \tag{7-68}$$

along with Maxwell's equations (7-48) to (7-50). The metric obtained, which was first given by Reissner,[37] is given by

$$g_{\mu\nu} = \begin{pmatrix} 1 - 2Gm/r + Ge^2/r^2 & & & \\ & -(1 - 2Gm/r + Ge^2/r^2)^{-1} & & 0 \\ & & -r^2 & \\ 0 & & & -r^2 \sin^2\theta \end{pmatrix} \tag{7-69}$$

The electromagnetic potential is then given by

$$A_\mu = \left(\frac{e}{r}, \mathbf{0}\right) \tag{7-70}$$

Solution with Rotational Symmetry

Weyl[38] and Levi-Civita[39] found static solutions with rotational, but not spherical, symmetry. A gravitational field is rotationally symmetric if it can be transformed so that there exists a Killing vector of the form $\xi^\mu = (0, x^2, -x^1, 0)$. If the field is also static then it can be transformed into the form

$$ds^2 = e^{2\psi}\, dt^2 - e^{2\gamma - 2\psi}(d\rho^2 + dz^2) - \rho^2 e^{-2\psi}\, d\phi^2 \tag{7-71}$$

[37] H. Reissner, *Ann. Physik* **50,** 106 (1916).
[38] H. Weyl, *Ann. Physik* **54,** 117 (1917); **59,** 185 (1919); G. Beck and H. Weyl, *Math. Z.* **13,** 142 (1921).
[39] T. Levi-Civita, *Rend. Acc. Lincei,* Several Notes (1918–1919).

where ρ, z, and ϕ can be interpreted as cylindrical polar coordinates and t as the time. In empty space the Einstein field equations then yield

$$\frac{\partial^2\psi}{\partial\rho^2} + \frac{1}{\rho}\frac{\partial\psi}{\partial\rho} + \frac{\partial^2\psi}{\partial z^2} = 0 \tag{7-72a}$$

$$\frac{\partial\gamma}{\partial\rho} = \rho\left[\left(\frac{\partial\psi}{\partial\rho}\right)^2 - \left(\frac{\partial\psi}{\partial z}\right)^2\right] \tag{7-72b}$$

$$\frac{\partial\gamma}{\partial z} = 2\rho\,\frac{\partial\psi}{\partial\rho}\,\frac{\partial\psi}{\partial z} \tag{7-72c}$$

Equation (7-72a) is the Laplace equation in cylindrical polar coordinates for a function with rotational symmetry. If one chooses a suitable solution to Eq. (7-72a) for ψ, one can then solve Eqs. (7-72b) and (7-72c) for γ, since these equations are compatible on the basis of Eq. (7-72a).

It is convenient to introduce ellipsoidal coordinates λ, μ given by $\lambda = (r_1 + r_2)/2m$, $\mu = (r_1 - r_2)/2m$, where $r_1^2 = \rho^2 + (z+m)^2$, $r_2^2 = \rho^2 + (z-m)^2$, so that $\lambda \geqslant 1$, $-1 \leqslant \mu \leqslant +1$. If we take the new coordinates as independent variables, the field equations (7-72a)–(7-72c) are replaced by

$$\frac{\partial}{\partial\lambda}\left[(\lambda^2-1)\frac{\partial\psi}{\partial\lambda}\right] + \frac{\partial}{\partial\mu}\left[(1-\mu^2)\frac{\partial\psi}{\partial\mu}\right] = 0 \tag{7-73a}$$

$$\frac{\partial\gamma}{\partial\lambda} = \frac{1-\mu^2}{\lambda^2-\mu^2}\left\{\lambda(\lambda^2-1)\left(\frac{\partial\psi}{\partial\lambda}\right)^2 - \lambda(1-\mu^2)\left(\frac{\partial\psi}{\partial\mu}\right)^2 - 2\mu(\lambda^2-1)\,\frac{\partial\psi}{\partial\lambda}\,\frac{\partial\psi}{\partial\mu}\right\} \tag{7-73b}$$

$$\frac{\partial\gamma}{\partial\mu} = \frac{\lambda^2-1}{\lambda^2-\mu^2}\left\{\mu(\lambda^2-1)\left(\frac{\partial\psi}{\partial\lambda}\right)^2 - \mu(1-\mu^2)\left(\frac{\partial\psi}{\partial\mu}\right)^2 + 2\lambda(1-\mu^2)\,\frac{\partial\psi}{\partial\lambda}\,\frac{\partial\psi}{\partial\mu}\right\} \tag{7-73c}$$

Equations (7-73) were solved by Erez and Rosen[40] by separating variables. Thus, if we write $\psi = \Lambda(\lambda)M(\mu)$, we obtain the equations

$$\frac{\mathrm{d}}{\mathrm{d}\lambda}\left[(\lambda^2-1)\frac{\mathrm{d}\Lambda}{\mathrm{d}\lambda}\right] - a\Lambda = 0$$

$$\frac{\mathrm{d}}{\mathrm{d}\mu}\left[(1-\mu^2)\,\frac{\mathrm{d}M}{\mathrm{d}\mu}\right] + aM = 0$$

where a is a separation constant. To get a well-behaved solution for M we take $a = l(l+1)$, where $l = 0, 1, 2, \ldots$. The solution of Eq. (7-73a), which is well-behaved at infinity, is then given by

$$\psi_l = P_l(\mu)Q_l(\lambda) \tag{7-74}$$

[40] G. Erez and N. Rosen, *Bull. Res. Counc. Israel* **8F**, 47 (1959).

where $P_l(\mu)$ is the Legendre polynomial and $Q_l(\lambda)$ is the second Legendre function. In particular one has

$$\psi_0 = \frac{1}{2} \ln \frac{\lambda - 1}{\lambda + 1}$$

which gives the Schwarzschild solution if one goes from λ, μ back to r, θ.

The general solution of Eq. (7-73a) is a linear combination of the solutions (7-74). For example, a solution of the form

$$\psi = \psi_0 + q_l \psi_l \tag{7-75}$$

where $l \neq 0$ and q_l is an arbitrary constant, can be considered as a generalization of the Schwarzschild solution describing the gravitational field of a particle which possesses a mass-multipole of order l in addition to its mass.

Particle with Quadrupole Moment

The simplest case, which is nontrivial from the physical point of view, is that for which $l = 2$. One finds in this case (dropping the subscript from q_2):

$$\psi = \frac{1}{2}\left\{\left[1 + \frac{1}{4} q(3\lambda^2 - 1)(3\mu^2 - 1)\right] \ln \frac{\lambda - 1}{\lambda + 1} + \frac{3}{2} q\lambda(3\mu^2 - 1)\right\}$$

$$\begin{aligned}
\gamma = \frac{9}{64} q^2 \Big\{ & (9\lambda^4 - 10\lambda^2 + 1) \ln^2 \frac{\lambda - 1}{\lambda + 1} + (36\lambda^3 - 28\lambda) \ln \frac{\lambda - 1}{\lambda + 1} \\
& + 36\lambda^2 - 16\Big\}\mu^4 + \Big\{\frac{9}{32} q^2(-5\lambda^4 + 6\lambda^2 - 1) \ln^2 \frac{\lambda - 1}{\lambda + 1} \\
& + \left[\frac{3}{2} q\lambda + \frac{9}{32} q^2\left(-20\lambda^3 + \frac{52}{3}\lambda\right)\right] \ln \frac{\lambda - 1}{\lambda + 1} + 3q \\
& + \frac{9}{32} q^2\left(-20\lambda^2 + \frac{32}{3}\right)\Big\}\mu^2 + \left(\frac{1}{2} q^2 + q + \frac{1}{2}\right) \ln \frac{\lambda^2 - 1}{\lambda^2 - \mu^2} \\
& + \frac{9}{64} q^2(\lambda^4 - 2\lambda^2 + 1) \ln^2 \frac{\lambda - 1}{\lambda + 1} + \left\{\frac{1}{16} q^2(9\lambda^2 - 15\lambda) - \frac{3}{2} q\lambda\right\} \ln \frac{\lambda - 1}{\lambda + 1} \\
& + \frac{9}{16} q^2\left(\lambda^2 - \frac{3}{4}\right) + 3q
\end{aligned}$$

The constant of integration in γ has been chosen so that $\gamma \to 0$ as $\lambda \to \infty$, i.e., the space is Euclidean at infinity.

To study the behaviour of the solution at large distances from the origin, we note that the line element (7-71) is given in terms of the variables λ, μ by

$$\begin{aligned}
ds^2 = e^{2\psi} \, dt^2 - m^2 \, e^{2\gamma - 2\psi}(\lambda^2 - \mu^2)\left(\frac{d\lambda^2}{\lambda^2 - 1} + \frac{d\mu^2}{1 - \mu^2}\right) \\
- m^2 \, e^{-2\psi}(\lambda^2 - 1)(1 - \mu^2) \, d\phi^2
\end{aligned} \tag{7-71a}$$

If we now carry out the transformation $\lambda = (r/m) - 1$, $\mu = \cos\theta$, we obtain

$$ds^2 = e^{2\psi}\,dt^2 - e^{2\gamma - 2\psi}\left\{\left(1 + \frac{m^2 \sin^2\theta}{r^2 - 2mr}\right) dr^2 + (r^2 - 2mr + m^2 \sin^2\theta)\,d\theta^2\right\} - e^{-2\psi}(r^2 - 2mr)\sin^2\theta\,d\phi^2 \quad (7\text{-}71b)$$

Denoting the coordinates t, r, θ, ϕ by x^0, x^1, x^2, x^3, and expanding in inverse powers of r, one finds that for large values of r,

$$g_{00} = 1 + 2\left\{-\frac{m}{r} - \frac{Q}{r^3}P_2(\cos\theta) - \frac{9Qm}{r^4}P_2(\cos\theta) + \frac{9}{14}\frac{Qm^2}{r^5}P_2(\cos\theta) + \frac{1}{r^6}\left[-\frac{50}{14}Qm^3P_2(\cos\theta) + \tfrac{1}{2}Q^2(P_2(\cos\theta))^2\right] + \cdots\right\},$$

where we have introduced the quadrupole moment Q given by $q = 15Q/2m^3$. The other components of the metric tensor can be similarly expanded. For example[41]

$$g_{11} = -1 - \frac{2m}{r} - \left(\frac{m}{r}\right)^2\left[4 + \frac{9}{5}q^2 - 2(q + q^2)\sin^2\theta\right] - \left(\frac{m}{r}\right)^3\left[8 - \frac{16}{3}q + \frac{36}{5}q^2 - \left(\frac{38}{5}q + 4q^2\right)\sin^2\theta\right] + \cdots$$

Cylindrical Gravitational Waves

Einstein's gravitational field equations have some solutions which describe wave phenomenae. Exact solutions which describe gravitational waves were first given by Einstein and Rosen.[42] These solutions describe cylindrical waves. However, spherical and plane waves were also found. In the following we deal with some special cases of cylindrical gravitational waves.[43]

41 In the more general case given by Eq. (7-75), if one expands the Legendre function in inverse powers of the argument

$$Q_l(\lambda) = \frac{b_l}{\lambda^{l+1}} + \cdots; \qquad b_l = \frac{1}{2^l}\int_0^1 (1 - x^2)^l\,dx$$

one finds that

$$g_{00} = 1 - 2\left[\frac{m}{r} + \frac{Q_l}{r^{l+1}}P_l(\cos\theta) + \cdots\right]$$

for large values of r. Here we introduced the multiple-moment Q_l by the relation $Q_l = -q_l b_l m^{l+1}$. We see that the first terms in the expansion of g_{00} correspond to the Newtonian potential of a mass plus a multiple of order l. To get the other components of the metric tensor one has to integrate Eqs. (7-73b) and (7-73c) for γ. The calculation gets progressively more tedious as l increases.

42 A. Einstein and N. Rosen, *J. Franklin Inst.* **223**, 43 (1937).

43 N. Rosen, *Bull. Res. Counc. Israel* **3**, 328 (1954).

To discuss cylindrical waves one modifies the static, axial-symmetric, gravitational field of Weyl and Levi-Civita discussed above. What we do is essentially to interchange the roles of z and t in the line element (7-71). Thus, we take for the line element, instead of Eq. (7-71),

$$ds^2 = e^{2\gamma - 2\psi}(dt^2 - d\rho^2) - e^{-2\psi}\rho^2 \, d\phi^2 - e^{2\psi} \, dz^2. \tag{7-76}$$

The Einstein equations then give

$$\frac{\partial^2 \psi}{\partial \rho^2} + \frac{1}{\rho}\frac{\partial \psi}{\partial \rho} - \frac{\partial^2 \psi}{\partial t^2} = 0 \tag{7-77a}$$

$$\frac{\partial \gamma}{\partial \rho} = \rho \left[\left(\frac{\partial \psi}{\partial \rho} \right)^2 + \left(\frac{\partial \psi}{\partial t} \right)^2 \right] \tag{7-77b}$$

$$\frac{\partial \gamma}{\partial t} = 2\rho \frac{\partial \psi}{\partial \rho}\frac{\partial \psi}{\partial t} \tag{7-77c}$$

Let us consider periodic solutions of the wave equation (7-77a) representing monochromatic waves, i.e., waves having a sinusoidal dependence on t. The solutions of this type are of the form

$$\psi = AJ_0(\omega\rho)\cos(\omega t + \alpha) + BN_0(\omega\rho)\cos(\omega t + \beta) \tag{7-78}$$

where J_0 and N_0 are Bessel functions of the first and second kind, respectively, of order zero,[44] and the frequency ω and the other constants A, B, α and β are arbitrary.

As a particular case, let us take a standing wave described by the solution

$$\psi = AJ_0(\omega\rho)\cos \omega t. \tag{7-79a}$$

Substituting into Eqs. (7-77b) and (7-77c) we get

$$\frac{\partial \gamma}{\partial \rho} = A^2\omega^2\rho\{[J_0'(\omega\rho)]^2\cos^2 \omega t + [J_0(\omega\rho)]^2\sin^2 \omega t\}$$

$$\frac{\partial \gamma}{\partial t} = -A^2\omega^2\rho J_0(\omega\rho)J_0'(\omega\rho)\sin 2\omega t$$

Integrating these equations we get

$$\begin{aligned}\gamma = {} & \tfrac{1}{2}A^2\omega\rho J_0(\omega\rho)J_0'(\omega\rho)\cos 2\omega t \\ & + \tfrac{1}{2}A^2\omega^2\rho^2\{[J_0'(\omega\rho)]^2 - J_0(\omega\rho)J_0''(\omega\rho)\}\end{aligned} \tag{7-79b}$$

Hence both ψ and γ are periodic functions of t.

A solution of this form, free from singularities, would be suitable to describing a situation in which standing waves are set up by reflection at the surface of a large sphere with centre at the origin. If in Eqs. (7-79) we replace $J_0(\omega\rho)$ by $N_0(\omega\rho)$ we obtain a solution with a singularity at the origin. This might be interpreted as

44 The Bessel functions of the second kind are the same as those denoted by Y_0 by G. N. Watson, *A Treatise on the Theory of Bessel Functions*, Cambridge, 1966.

describing a standing cylindrical gravitational wave with matter present along the z-axis.

Since, for large values of ρ, the asymptotic expansions of the Bessel functions are given by

$$J_0(\omega\rho) \simeq (2/\pi\omega\rho)^{1/2} \cos\left(\omega\rho - \frac{\pi}{4}\right)$$

$$N_0(\omega\rho) \simeq (2/\pi\omega\rho)^{1/2} \sin\left(\omega\rho - \frac{\pi}{4}\right)$$

we obtain an outgoing wave if we take

$$\psi = AJ_0(\omega\rho) \cos \omega t + AN_0(\omega\rho) \sin \omega t \tag{7-80a}$$

since the asymptotic expansion then has the form

$$\psi \simeq A(2/\pi\omega\rho)^{1/2} \cos\left(\omega\rho - \omega t - \frac{\pi}{4}\right).$$

Substituting the expression (7-80a) into Eqs. (7-77b) and (7-77c) and carrying out the integration, one obtains

$$\begin{aligned}\gamma = \tfrac{1}{2}A^2\omega\rho\{&J_0(\omega\rho)J_0'(\omega\rho) + N_0(\omega\rho)N_0'(\omega\rho)\\ &+ \omega\rho[(J_0(\omega\rho))^2 + (J_0'(\omega\rho))^2 + (N_0(\omega\rho))^2 + (N_0'(\omega\rho))^2]\\ &+ [J_0(\omega\rho)J_0'(\omega\rho) - N_0(\omega\rho)N_0'(\omega\rho)] \cos 2\omega t\\ &+ [J_0(\omega\rho)N_0'(\omega\rho) + N_0(\omega\rho)J_0'(\omega\rho)] \sin 2\omega t\} - (2/\pi)A^2\omega t.\end{aligned} \tag{7-80b}$$

It should be pointed out that in the present case the solution for γ contains an *aperiodic* term in t. The continuous transfer of gravitational energy by such a wave brings about a permanent change in the metric tensor. However, a wave of this kind would have to be excluded on physical grounds; since the wave carries away energy from the matter located along the z-axis, there must be a change in the motion of the latter in the course of time in consequence of which the solution for ψ cannot remain periodic in t.

7-6 EXPERIMENTAL TESTS OF GENERAL RELATIVITY

Up to a few years ago, general relativity was verified by three tests: the red shift, the deflection of light, and the planetary orbit effect on the planets. The first two tests could also be explained, in fact, without the use of the Einstein field equations.[45] However, this picture has been changed recently.

45 L. I. Schiff, *Am. J. Phys.* **28**, 340 (1960).

Gravitational Red Shift

Consider the clocks at rest at two points 1 and 2. The rate of change of times at these points are then given by $ds(1) = \sqrt{(g_{00}(1))}\, dt$ and $ds(2) = \sqrt{(g_{00}(2))}\, dt$. The relation between the rates of similar clocks in a gravitational field is therefore given by $(g_{00}(2)/g_{00}(1))^{1/2}$. The frequency of an atom, ν_0, located at point 1, when seen by an observer at point 2 is, hence, given by

$$\nu = \nu_0(g_{00}(1)/g_{00}(2))^{1/2} \tag{7-81}$$

For a gravitational field like the Schwarzschild field, one therefore obtains for the frequency shift per unit frequency:

$$\frac{\Delta\nu}{\nu_0} = \frac{\nu - \nu_0}{\nu_0} \simeq -Gm\left(\frac{1}{r_1} - \frac{1}{r_2}\right) \tag{7-81a}$$

to first order in Gm/r. If we take r_1 to be the observed radius of the Sun and r_2 the radius of the Earth's orbit (thus neglecting completely the Earth's gravitational field), then $\Delta\nu/\nu_0 = -2.12 \times 10^{-6}$. This frequency shift is usually referred to as the *gravitational red shift*.

The gravitational red shift was tested for the Sun and for white dwarfs,[46] and it was suggested that it be tested by atomic clocks.[47] However, the red shift was also observed directly using the Mössbauer effect by Pound and Rebka,[48] and by Cranshaw, Schiffer, and Whitehead.[49] The latter employed Fe^{57} and a total height difference of 12.5 metres. A red shift 0.96 ± 0.45 times the predicted value was observed by them. Pound and Rebka's result is more precise. They obtained a red shift 1.05 ± 0.10 times the predicted value.

Effects on Planetary Motion

We assume that test particles move along geodesics in the gravitational field (see next section), and that planets are small masses as compared with the mass of the Sun, thus behaving like test particles. Consequently, to find the equation of motion of a planet moving in the gravitational field of the Sun one has to write the geodesic equation in the Schwarzschild field. In fact one does not need the exact solution (7-63a) but its first approximation,

$$\begin{aligned} g_{00} &= 1 - 2Gm/r \\ g_{0r} &= 0 \\ g_{rs} &= -\delta_{rs} - 2Gmx^r x^s/r^3 \end{aligned} \tag{7-82}$$

[46] B. Bertotti, D. Brill, and R. Krotkov, Experiments on gravitation, in *Gravitation: An Introduction to Current Research* (L. Witten, Ed.), Wiley, New York, 1962.

[47] S. F. Singer, *Phys. Rev.* **104,** 11 (1956); C. Moller, *Nuovo Cim. Suppl.* **6,** 381, (1957).

[48] R. V. Pound and G. A. Rabka, Jr., *Phys. Rev. Letters,* **4,** 337 (1960).

[49] T. F. Cranshaw, S. P. Schiffer, and A. B. Whitehead, *Phys. Rev. Letters* **4,** 163 (1960).

Using the approximate metric (7-82) in the geodesic equation (7-33) gives (see Problem 7.15)

$$\ddot{\mathbf{x}} - Gm\nabla\frac{1}{r} = Gm\left\{2(\dot{\mathbf{x}}^2)\nabla\frac{1}{r} - 2Gm\frac{1}{r}\nabla\frac{1}{r} - 2\left(\dot{\mathbf{x}}\cdot\nabla\frac{1}{r}\right)\dot{\mathbf{x}} + \frac{3}{r^5}(\mathbf{x}\cdot\dot{\mathbf{x}})^2\mathbf{x}\right\} \tag{7-83}$$

where we have used three-dimensional notation and a dot denotes differentiation with respect to t. Multiplying Eq. (7-83), vectorially, by $\mathbf{x}$ gives

$$\mathbf{x}\times\ddot{\mathbf{x}} = -2Gm\left(\dot{\mathbf{x}}\cdot\nabla\frac{1}{r}\right)(\mathbf{x}\times\dot{\mathbf{x}})$$

thus leading to the first integral

$$\mathbf{x}\times\dot{\mathbf{x}} = \mathbf{J}\,\mathrm{e}^{-2Gm/r}$$

where $\mathbf{J}$ is a *constant* vector, the angular momentum per mass unit. Hence the radius vector $\mathbf{x}$ moves in a plane perpendicular to the vector $\mathbf{J}$. Introducing in this plane polar coordinates r, ϕ to describe the motion of the planet, the equation of motion (7-83), consequently, decomposes into

$$\ddot{r} - r\dot{\phi}^2 + \frac{Gm}{r^2} = \frac{Gm}{r^2}\left\{3\dot{r}^2 - 2r^2\dot{\phi}^2 + 2\frac{Gm}{r}\right\} \tag{7-84a}$$

$$r^2\dot{\phi} = J\,\mathrm{e}^{-2Gm/r} \tag{7-84b}$$

where J is the magnitude of the vector $\mathbf{J}$.

Introducing the new variable $u = 1/r$ one can rewrite Eqs. (7-84) in terms of $u(\phi)$:

$$u'' + u - \frac{Gm}{J^2} = Gm\left(-u'^2 + 2u^2 + 2\frac{Gm}{J^2}u\right) \tag{7-85}$$

Here a prime denotes a derivation with respect to the angle ϕ.

Let us try a solution of the form

$$u = b(1 + \varepsilon\cos\alpha\phi)$$

Here ε is the eccentricity and α is some parameter to be determined and whose value in the usual nonrelativistic mechanics is unity.[50] The other constant b is related to J in the nonrelativistic mechanics by $Gm/J^2 = b$. Using the above solution in Eq. (7-85) and equating coefficients of $\cos\alpha\phi$ gives

$$\alpha^2 = 1 - 2Gm(2b + Gm/J^2)$$

Substituting for Gm/J^2 its nonrelativistic value b then gives $\alpha^2 = 1 - 6Gmb$, or, to a first approximation in Gm,

$$\alpha = 1 - 3Gmb$$

[50] H. Goldstein, *Classical Mechanics*, Addison-Wesley, Reading, Mass., 1965.

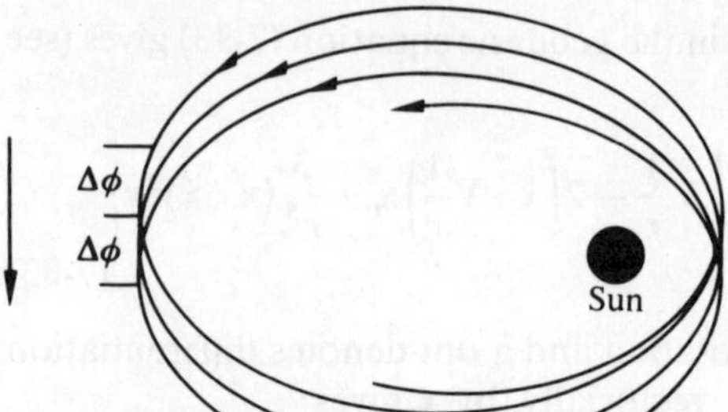

Figure 7.3 Elliptic orbit with perihelion advance.

Successive perihelia occur when

$$(1 - 3Gmb)(2\pi + \Delta\phi) = 2\pi$$

Consequently, there will be an advance in the perihelion of the orbit per revolution (see Fig. 7.3) given by $\Delta\phi = 6\pi Gmb$, or $\Delta\phi = 6\pi Gm/a(1 - \varepsilon^2)$ if we make use of the nonrelativistic value of the constant b, and where a is the semimajor axis of the orbit. Reinstating now c, the velocity of light, finally gives for the perihelion advance

$$\Delta\phi = \frac{6\pi Gm}{c^2 a(1 - \varepsilon^2)} \tag{7-86}$$

in radians per revolution. We list below the calculated values of $\Delta\phi$ per century for four planets:

Planet	$\Delta\phi$
Mercury	43.03″
Venus	8.60″
Earth	3.80″
Mars	1.35″

The astronomical observations for the planet Mercury give 43.11 ± 0.45 sec per century,[51] in good agreement with the calculated value.

Deflection of Light

To discuss the deflection of light in the gravitational field we must again solve the geodesic equation, but now with the conditions $ds = 0$. Using the approximate solution (7-82) then gives for $g_{\mu\nu}\, dx^\mu\, dx^\nu = 0$

$$(1 + 2Gm/r)[\dot{\mathbf{x}} \cdot \dot{\mathbf{x}} + (2Gm/r^3)(\mathbf{x} \cdot \dot{\mathbf{x}})^2] = 1$$

Using polar coordinates r, ϕ, consequently, gives to the first approximation in Gm

$$\dot{r}^2 + r^2\dot{\phi}^2 + 4Gm\dot{r}^2/r + 2Gmr\dot{\phi}^2 = 1$$

[51] R. L. Duncombe, *Astron. J.* **61**, 174 (1956).

Again changing variables into $u(\phi) = 1/r$, and using Eq. (7-84b), gives

$$u'^2 + u^2 + 2Gmu(2u'^2 + u^2) = J^{-2}\, e^{4Gmu}$$

Differentiation of this equation with respect to ϕ gives

$$u'' + u + Gm(2u'^2 + 4uu'' + 3u^2) = 2GmJ^{-2} \tag{7-87}$$

to the first approximation in Gm.

To solve Eq. (7-87) we note that in the lowest approximation one has

$$u'^2 \simeq J^{-2} - u^2$$

$$u'' \simeq -u$$

Using these values in Eq. (7-87) gives

$$u'' + u = 3Gmu^2 \tag{7-88}$$

for the orbit of the light ray. In the lowest approximation u satisfies $u'' + u = 0$, whose solution is a straight line

$$u = \frac{\cos\phi}{R}$$

where R is a constant. This shows that $r = 1/u$ has a minimum value R at $\phi = 0$. Substituting into the right-hand side of Eq. (7-88) then gives

$$u'' + u = 3\frac{Gm}{R^2}\cos^2\phi \tag{7-88a}$$

The solution of this equation is

$$u = \frac{\cos\phi}{R} + \frac{Gm}{R^2}(1 + \sin^2\phi)$$

Introducing new Cartesian coordinates $x = r\cos\phi$ and $y = r\sin\phi$, the above equation gives

$$x = R - \frac{Gm}{R}\frac{x^2 + 2y^2}{\sqrt{(x^2 + y^2)}}$$

For large values of $|y|$ this equation becomes

$$x \simeq R - \frac{2Gm}{R}|y|$$

Hence, asymptotically, the orbit of the light ray is a straight line in space. This result is expected, since far away from the central mass the space is flat. The angle $\Delta\phi$ between the two asymptotes is, however, equal to

$$\Delta\phi = 4\frac{Gm}{c^2R} \tag{7-89}$$

in units in which c is different from unity.

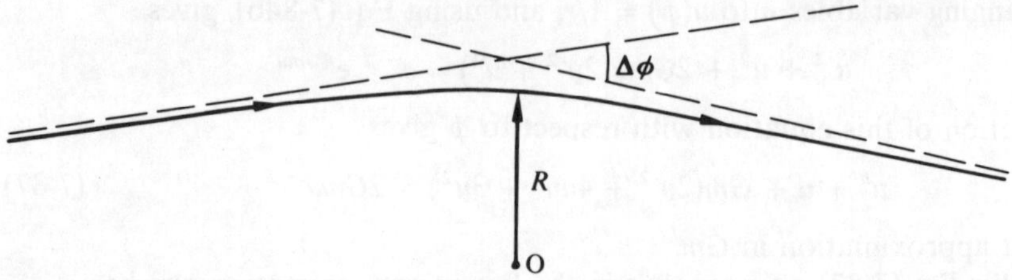

Figure 7.4 Bending of a light ray in the Schwarzschild field.

The angle $\Delta\phi$ represents the angle of *deflection* of a light ray in passing through the Schwarzschild field (see Fig. 7.4). For a light ray just grazing the Sun Eq. (7-89) gives $\Delta\phi = 1.75$ sec. Observations indeed confirm this result; one of the latest results gives 1.75 ± 0.10 sec.[52]

Gravitational Radiation Experiments

Weber[53] has developed methods to detect gravitational waves (see Sec. 7-5) that Einstein's gravitational field equations predict. The experiment involves detectors at opposite ends of a 1000 km baseline. Sudden increases in detector output were observed by him roughly once in several days, coincident within the resolution time of 0.25 seconds. The statistics rule out an accidental origin and experiments rule out seismic and electromagnetic effects.

Weber's apparatus[54] measures the Fourier transform of the Riemann tensor (see Sec. 7-1). The method uses the fact that the distance η^μ between two neighbouring test particles, which follow geodesics, satisfies the *geodesic deviation* equation[55]

$$\frac{\delta^2 \eta^\mu}{\delta s^2} + R^\mu{}_{\alpha\nu\beta}\,\lambda^\alpha \eta^\nu \lambda^\beta = 0$$

where λ^α is the tangent vector to one of the geodesics, and $\delta/\delta s = \lambda^\alpha \nabla_\alpha$ is a directional covariant derivative. Weber measures the strains of a large aluminium cylinder, having mass of the order of 10^6 grams, by means of a piezoelectric crystal attached to the cylinder which transforms the mechanical movement into an electric current. The detector was developed for operation in the vicinity of 1662 cycle/sec. A high frequency source was developed for dynamic gravitational fields and the detector was tested by doing a communication experiment with high frequency Coulomb fields.[56]

[52] B. Bertotti, D. Brill, and R. Krothov, *op. cit.*

[53] J. Weber, Gravitational Radiation Experiments, in *Relativity* (M. Carmeli, S. I. Fickler, and L. Witten, Eds), Plenum Press, New York, 1970.

[54] J. Weber, *General Relativity and Gravitational Waves*, Interscience Publishers, New York, 1961.

[55] J. L. Synge and A. Schild, *Tensor Calculus*, University of Toronto Press, 1949.

[56] J. Sinsky and J. Weber, *Phys. Rev. Letters* **18**, 795 (1967).

Radar Experiment

Shapiro[57] has designed a radar experiment to test general relativity by measuring the effect of solar gravity on time delays of round-trip travel times of radar pulses transmitted from the Earth toward an inner planet, i.e., Venus or Mercury. The experiment is based on the phenomenon that electromagnetic waves "slow down" in the gravitational field. Within the framework of general relativity there should be an anomalous delay of 200 microseconds in the arrival time of a radar echo from Mercury, positioned on the far side of the Sun near the limb. For example, if we calculate the proper time τ at $r = r_2$ for a radial round-trip travel $r_2 \to r_1 \to r_2$, with $r_2 > r_1$, of a radar pulse in the Schwarszchild field, and subtract from τ the corresponding value τ_0 when the spherical mass $m = 0$, we find

$$\Delta\tau = \frac{4Gm}{c^3}\left(\ln\frac{r_2}{r_1} - \frac{r_2 - r_1}{r_2}\right) + 0(m^2)$$

In general one finds

$$\Delta\tau \simeq \frac{4Gm}{c^3}\ln\left(\frac{r_e + r_p + R}{r_e + r_p - R}\right)$$

where r_e is the Earth–Sun distance, r_p the planet–Sun distance, and R the Earth–planet distance.

Shapiro[58] found that the retardation of radar signals are 1.02 ± 0.05 times the corresponding effect predicted by general relativity.

Low-Temperature Experiments

Schiff[59] has proposed an experiment to check the equations of motion in general relativity by means of a gyroscope, which is forced to go around the Earth either in a stationary laboratory fixed to the Earth or a satellite. The unique experiment is made possible by complete use of a low-temperature environment, and the properties of superconductors, including the use of zero magnetic fields and ultrasensitive magnetometry.[60] Schiff has calculated, using results obtained by Papapetrou[61] for the motion of spinning bodies in general relativity, that a perfect gyroscope subject to no external torques will experience an anomalous precession with respect to the fixed stars as it travels around the Earth.

[57] I. I. Shapiro, *Phys. Rev. Letters* **13,** 789 (1964). For lightlike behavior of motion of test particles see J. Jaffe and I. I. Shapiro, *Phys. Rev.* D**6**, 405 (1972); M. Carmeli, *Nuovo Cimento Letters* **3,** 379 (1972).

[58] I. I. Shapiro, *General Relativity and Gravitation* **3,** 135 (1972).

[59] L. I. Schiff, *Proc. Natl. Acad. Sci.* **46,** 871 (1960).

[60] C. W. F. Everitt, W. M. Fairbank, and W. O. Hamilton, General Relativity Experiments Using Low Temperature Techniques, in *Relativity* (M. Carmeli, S. I. Fickler, and L. Witten, Eds), Plenum Press, New York, 1970.

[61] A. Papapetrou, *Proc. R. Soc. Lond.* (A)**209,** 248 (1951).

7-7 EQUATIONS OF MOTION

Geodesic Postulate

In the last section it was assumed that the planet's motion around the Sun is described by the geodesic equation (7-33). The assumption that the equations of motion of a test particle, moving in a gravitational field, are given by the geodesic equation is known as the *geodesic postulate* and was suggested by Einstein[62] in his first article on the general theory of relativity.

It was eleven years later when Einstein and Grommer[63] showed that the geodesic postulate need not be assumed, but that it rather follows from the gravitational field equations; this is a consequence of the non-linearity of the field equations along with the fact that they satisfy the four contracted Bianchi identities (see Section 7-1). The discovery of Einstein and Grommer is considered to be one of the most important achievements, and one of the most attractive features of the general theory of relativity since its publication. Later on Infeld and Schild[64] showed that the equations of motion of a test particle are given by the geodesic equation in the *external gravitational* field. This result, however, does not differ from the geodesic postulate because, by definition, a test particle has no self-field.[65]

Equations of Motion as a Consequence of Field Equations

In order to establish the relation between the Einstein field equations and the equations of motion one proceeds as follows. We have seen in Section 7-4 that because of the contracted Bianchi identity it follows that the energy–momentum tensor $T^{\mu\nu}$ satisfies a generally covariant conservation law of the form given by Eq. (7-38). Consequently, one obtains for the energy–momentum tensor *density* $\mathscr{T}^{\mu\nu}$

$$\nabla_\nu \mathscr{T}^{\mu\nu} \equiv \partial_\nu \mathscr{T}^{\mu\nu} + \Gamma^\mu_{\alpha\beta} \mathscr{T}^{\alpha\beta} = 0 \tag{7-90}$$

where $\mathscr{T}^{\mu\nu} \equiv (-g)^{1/2} T^{\mu\nu}$.

For a system of N particles of finite masses, represented as singularities of the gravitational field, $\mathscr{T}^{\mu\nu}$ may be taken in the form

$$\mathscr{T}^{\mu\nu} = \sum_{A=1}^{N} m_A v^\mu_A v^\nu_A \, \delta_A(\mathbf{x} - \mathbf{z}_A) \tag{7-91}$$

Here z^μ_A are the coordinates of the Ath particle. (Roman capital indices, $A, B, \ldots$, run from 1 to N. For these indices the summation convention will be suspended).

[62] A. Einstein, *Ann. Physik* **49**, 761 (1916); English translation in *The Principle of Relativity*, Dover, New York, 1923.

[63] A. Einstein and J. Grommer, *Sitzungsber. Ber. Akad. Wiss.* **2**, 235 (1927).

[64] L. Infeld and A. Schild, *Rev. Mod. Phys.* **21**, 408 (1949).

[65] The significance of the geodesic postulate for a finite mass was subsequently discussed by M. Carmeli, *Phys. Letters* **11**, 24 (1964).

$v^\mu = \dot{z}^\mu = dz^\mu/dt(v^0_A = \dot{z}^0_A = 1)$, and δ is the three-dimensional Dirac delta function satisfying the following conditions:

$$\delta(\mathbf{x}) = 0 \quad \text{for} \quad \mathbf{x} \neq 0$$

$$\int \delta(\mathbf{x} - \mathbf{z})\, d^3x = 1$$

$$\int f(\mathbf{x})\, \delta(\mathbf{x} - \mathbf{z})\, d^3x = f(\mathbf{z}),$$

for any continuous function $f(\mathbf{x})$ in the neighbourhood of $\mathbf{z}$. In Eq. (7-91), m_A is a function of time which may be called the *inertial mass* of the Ath particle.

If we put the energy–momentum tensor density (7-91) into Eq. (7-90) and integrate over the three-dimensional region surrounding the first singularity, we obtain

$$dp^\mu/dt = \int F^\mu\, \delta(\mathbf{x} - \mathbf{z})\, d^3x \tag{7-92}$$

where

$$p^\mu = mv^\mu$$

$$F^\mu = -m\Gamma^\mu_{\alpha\beta} v^\alpha v^\beta$$

and we have put, for simplicity, $m = m_1$, $z^\mu = z^\mu_1$, $v^\mu = v^\mu_1$, and $\delta(\mathbf{x} - \mathbf{z}) = \delta_1(\mathbf{x} - \mathbf{z}_1)$.

Self-Action Terms

Equation (7-92) may be interpreted as an "exact equation of motion" of the first particle. However, since the Christoffel symbols are singular at the location of the particle, the equation contains infinite self-action terms. However, these terms can be removed as follows.[66]

Putting Eq. (7-91) into Eq. (7-90) we obtain

$$\partial_0 \left[\sum_{A=1}^{N} m_A v^\mu_A\, \delta_A \right] + \partial_n \left[\sum_{A=1}^{N} m_A v^\mu_A v^n_A\, \delta_A \right] + \sum_{A=1}^{N} m_A \Gamma^\mu_{\alpha\beta} v^\alpha_A v^\beta_A\, \delta_A = 0 \tag{7-90a}$$

where Latin indices run from 1 to 3. The first term on the left-hand side of Eq. (7-90) can be written as

$$\partial_0 \left[\sum_{A=1}^{N} m_A v^\mu_A\, \delta_A \right] = \sum_{A=1}^{N} \partial_0 (m_A v^\mu_A)\, \delta_A + \sum_{A=1}^{N} m_A v^\mu_A\, \partial_0\, \delta_A$$

with

$$\partial_0\, \delta_A = \partial_0\, \delta_A(x^s - z^s_A) = -\partial_n\, \delta_A v^n_A$$

[66] M. Carmeli, *Phys. Rev.* **140,** B1441 (1965).

Using the above results in Eq. (7-90a), we obtain

$$\sum_{A=1}^{N} \{d(m_A v_A^\mu)/\,dt + m_A \Gamma^\mu_{\alpha\beta} v_A^\alpha v_A^\beta\}\, \delta_A = 0 \tag{7-90b}$$

Equation (7-90b), which is identical with Eq. (7-90), is satisfied for any space–time point, since otherwise the Bianchi identity or the Einstein field equations will not be satisfied.

Let us now examine the behaviour of Eq. (7-90b) in the infinitesimal neighbourhood of the first singularity, which we assume not to contain any other singularity. In this region $\delta_B(\mathbf{x} - \mathbf{z}_B) = 0$ for $B = 2, 3, \ldots, N$. Hence Eq. (7-90b) gives for the conservation law near the first singularity

$$\{d(mv^\mu)/dt + m\Gamma^\mu_{\alpha\beta} v^\alpha v^\beta\}\, \delta(\mathbf{x} - \mathbf{z}) = 0 \tag{7-93}$$

Let us further assume that the Christoffel symbols near the first singularity can be expanded into a power series in the infinitesimal distance r, defined by $r^2 = (x^s - z^s)(x^s - z^s)$, where $z^s = z^s_1$, in the vicinity of the first particle. Then we have

$$\Gamma^\mu_{\alpha\beta} = {}_{-k}\Gamma^\mu_{\alpha\beta} + {}_{-k+1}\Gamma^\mu_{\alpha\beta} + \cdots + {}_0\Gamma^\mu_{\alpha\beta} + \cdots$$

where the indices written as subscripts on the left of a function indicate its behaviour with respect to r, and k is a positive integer. For example ${}_0\Gamma^\mu_{\alpha\beta}$ is that part of the Christoffel symbol which varies as r^0, i.e., is finite at the location of the first particle. When one uses spherical coordinates r, θ, and ϕ, one can write

$${}_{-k}\Gamma^\mu_{\alpha\beta} = r^{-k} A^\mu_{\alpha\beta}(\theta, \phi)$$

$${}_{-k+1}\Gamma^\mu_{\alpha\beta} = r^{-k+1} B^\mu_{\alpha\beta}(\theta, \phi)$$

$$\cdot \quad \cdot \quad \cdot \quad \cdot \quad \cdot \quad \cdot$$

$${}_0\Gamma^\mu_{\alpha\beta} = D^\mu_{\alpha\beta}(\theta, \phi), \text{ etc.}$$

Terms like ${}_1\Gamma^\mu_{\alpha\beta}$, ${}_2\Gamma^\mu_{\alpha\beta}$, etc., however, need not be taken into account when one puts the above expansion into Eq. (7-93) since $r^j\, \delta(\mathbf{x} - \mathbf{z}) = 0$ for any positive integer j. If we denote now $mA^\mu_{\alpha\beta} v^\alpha v^\beta, \ldots$ by $A^\mu, \ldots$ we can write Eq. (7-93) in the form

$$\{r^{-k}A^\mu + r^{-k+1}B^\mu + \cdots + r^{-1}C^\mu + D^\mu_1\}\, \delta(\mathbf{x} - \mathbf{z}) = 0 \tag{7-93a}$$

where we have used the notation $D^\mu_1 = d(mv^\mu)/\,dt + D^\mu$.

In order to get rid of terms proportional to negative powers of r in Eq. (7-93a) we proceed as follows. Multiplying Eq. (7-93a) by r^k and using $r^j\, \delta(\mathbf{x} - \mathbf{z}) = 0$ we obtain

$$A^\mu(\theta, \phi)\, \delta(\mathbf{r}) = 0$$

the integration of which over the three-dimensional region yields, using spherical coordinates,

$$\iint A^\mu(\theta, \phi) \sin\theta\, d\theta\, d\phi \int r^2\, \delta(\mathbf{r})\, dr = 0$$

From the property of the delta-function

$$\int \delta(\mathbf{r})\, \mathrm{d}^3x = \iint \sin\theta\, \mathrm{d}\theta\, \mathrm{d}\phi \int \delta(\mathbf{r}) r^2\, \mathrm{d}r = 1$$

one obtains $\int \delta(\mathbf{r}) r^2\, \mathrm{d}r = (4\pi)^{-1}$. Hence we obtain

$$\iint A^\mu(\theta, \phi) \sin\theta\, \mathrm{d}\theta\, \mathrm{d}\phi = 0 \tag{7-94a}$$

independent of the value of the variable r. Thus the angular distribution of $A^\mu(\theta, \phi)$ is such that its average equals zero. However, not only does the above equation hold, but also (s is any finite positive integer)

$$a(r) = r^{-s} \iint A^\mu(\theta, \phi) \sin\theta\, \mathrm{d}\theta\, \mathrm{d}\phi = 0$$

for small values of r as well as when r tends to zero, as can be verified by using L'Hospital's theorem, for example. It follows then that $a(r)$ is a function of r whose value is zero for any small r, including $r = 0$. Using the property of the delta-function we obtain

$$\int r^2\, \delta(\mathbf{r}) f(r)\, \mathrm{d}r = (4\pi)^{-1} f(0)$$

for any continuous function of r. Since $a(r)$ is certainly continuous, one obtains

$$\int r^2\, \delta(\mathbf{r}) a(r)\, \mathrm{d}r = 0 \tag{7-95a}$$

Hence when one integrates Eq. (7-93a) over the three-dimensional space, there will be no contribution from the first term.

In order to show that the second term of Eq. (7-93a) will also not contribute to the three-dimensional integration of the same equation, we multiply it by r^{k-1}. We obtain now, after neglecting terms that do not contribute,

$$\{r^{-1} A^\mu(\theta, \phi) + B^\mu(\theta, \phi)\}\, \delta(\mathbf{r}) = 0$$

Integration of this equation, again using spherical coordinates, shows that the first term will not contribute anything because of Eq. (7-95a), and we are left with

$$\iint B^\mu(\theta, \phi) \sin\theta\, \mathrm{d}\theta\, \mathrm{d}\phi \int r^2\, \delta(\mathbf{r})\, \mathrm{d}r = 0$$

Hence

$$\iint B^\mu(\theta, \phi) \sin\theta\, \mathrm{d}\theta\, \mathrm{d}\phi = 0 \tag{7-94b}$$

independent of r. From this equation one obtains another one, analogous to Eq. (7-95a) but with B^μ instead of A^μ:

$$\int r^2\, \delta(\mathbf{r}) b(r)\, \mathrm{d}r = 0 \tag{7-95b}$$

with

$$b(r) = r^{-s} \iint B^{\mu}(\theta, \phi) \sin\theta \, d\theta \, d\phi = 0$$

Proceeding in this way, one verifies that the angular distribution of all functions $A^{\mu}(\theta, \phi)$, $B^{\mu}(\theta, \phi)$, etc., is such that they all satisfy equations like Eqs. (7-94). From this it is clear that one obtains

$$\int D_1^{\mu}(\theta, \phi) \, \delta(\mathbf{r}) \, d^3x = 0$$

which gives

$$dp^{\mu}/dt + mv^{\alpha}v^{\beta} \int {}_0\Gamma^{\mu}_{\alpha\beta} \, \delta(\mathbf{r}) \, d^3x = 0 \tag{7-92a}$$

or equivalently

$$\dot{v}^k + v^{\alpha}v^{\beta} \int ({}_0\Gamma^k_{\alpha\beta} - v^k \, {}_0\Gamma^0_{\alpha\beta}) \, \delta(\mathbf{r}) \, d^3x = 0 \tag{7-92b}$$

Equation (7-92a) is the "exact equation of motion".

Einstein–Infeld–Hoffmann Method

Having found the law of motion (7-92a), one can now proceed to find the equation of motion of two finite masses, each moving in the field produced by both of them. In the following we find such an equation of motion in the case for which the particles' velocities are much smaller than the speed of light. Moreover, we will confine ourselves to an accuracy of post-Newtonian. This means the equation of motion obtained will contain the Newtonian equation as a limit, but is a first generalization of it. Such an equation was first obtained by Einstein, Infeld, and Hoffmann.[67] To obtain this equation we solve the field equations and formulate the equations of motion explicitly by means of an approximation method, the Einstein–Infeld–Hoffmann (EIH) method, to be described below.

Let us assume a function ϕ developed in a power series in the parameter $\lambda = 1/c$, where c is the speed of light. One then has

$$\phi = {}_0\phi + {}_1\phi + {}_2\phi + \cdots$$

The indices written as left subscripts indicate the order of λ absorbed by the ϕ's.

If a function $\phi(x)$ varies rapidly in space but slowly with x^0, then we are justified in not treating all its derivatives in the same manner. The derivatives with respect to x^0 will be of a higher order than the space derivatives. We thus write

$$\partial_0({}_l\phi) = {}_{l+1}\psi$$

[67] A. Einstein, L. Infeld, and B. Hoffmann, *Ann. Math.* **39**, 65 (1938); A. Einstein and L. Infeld, *Canad. J. Math.* **1**, 209 (1949).

That is, differentiation with respect to x^0 raises the order by one. Thus if the coordinates z^s of a particle are considered to be of order zero, $\dot{z}^s$ will be of order one, and $\ddot{z}^s$ of order two. Using now the Newtonian approximation mass × acceleration = mass × mass/(distance)^2, we see the mass is of order two. In all the power developments we take into account only even or only odd powers of $1/c$.[68]

Thus, because of the order with which we start m and $\dot{z}^s$, we have

$$\mathcal{T}^{00} = {}_2\mathcal{T}^{00} + {}_4\mathcal{T}^{00} + \cdots$$
$$\mathcal{T}^{0n} = {}_3\mathcal{T}^{0n} + {}_5\mathcal{T}^{0n} + \cdots$$
$$\mathcal{T}^{mn} = {}_4\mathcal{T}^{mn} + {}_6\mathcal{T}^{mn} + \cdots$$

As to the metric tensor, we write

$$g_{\mu\nu} = \eta_{\mu\nu} + h_{\mu\nu}\,,\; g^{\mu\nu} = \eta^{\mu\nu} + h^{\mu\nu}$$

The gravitational field equations (7-37) can be written as

$$\sqrt{(-g)}R_{\alpha\beta} = \kappa(\mathcal{T}_{\alpha\beta} - \tfrac{1}{2}g_{\alpha\beta}\mathcal{T}) \tag{7-96}$$

where $\mathcal{T} = \mathcal{T}_{\mu\nu}g^{\mu\nu}$, and $R_{\alpha\beta}$ is given by Eq. (7-27). From the right-hand side of the field equations it follows that R_{00} and R_{mn} (when $m = n$) start with order two, R_{mn} (when $m \neq n$) start with order four, while R_{0m} starts with order three. The lowest order expressions of the left-hand side are

$$\begin{aligned} R_{00} &\simeq \tfrac{1}{2}h_{00,\,ss} \\ R_{0m} &\simeq \tfrac{1}{2}(h_{0m,\,ss} - h_{0s,\,ms} - h_{ms,\,0s} + h_{ss,\,0m}) \\ R_{mn} &\simeq \tfrac{1}{2}(h_{mn,\,ss} - h_{ms,\,ns} - h_{ns,\,ms} - h_{00,\,mn} + h_{ss,\,mn}) \end{aligned} \tag{7-97}$$

where a comma denotes a partial derivative, $\phi_{,s} = \partial_s\phi$. Hence we have

$$h_{00} = {}_2h_{00} + {}_4h_{00} + \cdots$$
$$h_{0m} = {}_3h_{0m} + {}_5h_{0m} + \cdots$$
$$h_{mn} = {}_2h_{mn} + {}_4h_{mn} + \cdots$$

Newtonian Equation of Motion

We now find the equation of motion in the lowest (Newtonian) approximation. We do it in such a way as to make the generalization to the post-Newtonian approximation as simple as possible.

[68] The expansion of the metric tensor, etc., in a power series in c^{-2} (such as $\phi = {}_0\phi + {}_2\phi + \ldots$, or $\phi = {}_1\phi + {}_3\phi + \cdots$) corresponds to the choice of the symmetric Green function, thus excluding radiation.

Because of Eqs. (7-96) and (7-97), the field equations of the lowest order are in h_{00},

$$\tfrac{1}{2}\,{}_2h_{00,\,ss} = \kappa({}_2\mathcal{T}^{00} - \tfrac{1}{2}\,{}_2\mathcal{T}^{00}) = (\kappa/2)_2\,\mathcal{T}^{00} = (\kappa/2)\sum_{A=1}^{2} \mu_A\,\delta_A$$

where, for simplicity, we have put $\mu_A = {}_2m_A$. Hence the equation obtained is

$${}_2h_{00,\,ss} = \kappa \sum_{A=1}^{2} \mu_A\,\delta_A$$

The solution of this equation that represents two masses is

$${}_2h_{00} = -2G\sum_{A=1}^{2} \mu_A r_A^{-1}$$

where $r_A^2 = (x^s - z_A^s)(x^s - z_A^s)$. Using ${}_2h_{00}$ in the equation of motion (7-92b), we obtain in the lowest (second) order for the equation of motion of the first particle

$$\ddot{z}_1^k - G\int \partial_k(\mu_2 r_2^{-1})\,\delta(\mathbf{x} - \mathbf{z}_1)\,\mathrm{d}^3x = 0$$

This gives

$$\ddot{z}_1^k = G(\partial/\partial z_1^k)(\mu_2/z) \tag{7-98}$$

where $z^2 = (z_1^s - z_2^s)(z_1^s - z_2^s)$. Equation (7-98) is, of course, the Newton equatin of motion.

Einstein–Infeld–Hoffmann Equation

To find the equation of motion up to the fourth order, we must know besides ${}_2h_{00}$ the functions ${}_4h_{00}$, ${}_3h_{0n}$, and ${}_2h_{mn}$. The second and third functions are easy to find. The left-hand side of the corresponding equations is written out in Eq. (7-97), whereas the right-hand side is given by Eq. (7-96) and it is $-\kappa\sum \mu_A \dot{z}_A^m\,\delta_A$ for the $0m$ component, and $(\kappa/2)\,\delta_{mn}\sum \mu_A\,\delta_A$ for the mn component. Therefore, for the ${}_2h_{mn}$ we have the equation

$${}_2h_{mn,\,ss} - {}_2h_{ms,\,ns} - {}_2h_{ns,\,ms} + {}_2h_{ss,\,mn} - {}_2h_{00,\,mn} = \delta_{mn}\,{}_2h_{00,\,ss}$$

whose solution is

$${}_2h_{mn} = \delta_{mn}\,{}_2h_{00}$$

The equation for ${}_3h_{0n}$ is

$${}_3h_{0n,\,ss} - {}_3h_{0s,\,ns} - {}_2h_{ns,\,0s} + {}_2h_{ss,\,n0} = -2\kappa\sum \mu_A \dot{z}_A^n\,\delta_A$$

Using the value of ${}_2h_{mn}$ in terms of the ${}_2h_{00}$ found above, we obtain

$${}_3h_{0n,\,ss} - {}_3h_{0s,\,ns} + 2\,{}_2h_{00,\,n0} = -2\kappa\sum_{A=1}^{2} \mu_A \dot{z}_A^n\,\delta_A$$

The solution of this equation is

$$ {}_{3}h_{0n} = 4G \sum_{A=1}^{2} \mu_A \dot{z}_A^n r_A^{-1}. $$

Calculation of ${}_{4}h_{00}$ is somewhat more complicated. The relevant part of ${}_{4}h_{00}$, for two masses, that contributes to the equation of motion of the first particle, is

$$ {}_{4}h_{00} \simeq G\{2G\mu_2^2 r_2^{-2} - 3\mu_2 \dot{z}_2^s \dot{z}_2^s r_2^{-1} - \mu_2 r_{2,\,00} + 2G\mu_1\mu_2(zr_2)^{-1}\} \qquad (7\text{-}99) $$

Using these values for ${}_{4}h_{00}$, ${}_{3}h_{0n}$, and ${}_{2}h_{mn}$ in the equation of motion (7-92b) gives, for the two-body problem (Problem 7.16):[69]

$$ \begin{aligned} \ddot{z}_1^n - \mu_2\, \partial(1/z)/\partial z_1^n \\ = \mu_2\{[\dot{z}_1^s\dot{z}_1^s + \tfrac{3}{2}\dot{z}_2^s\dot{z}_2^s - 4\dot{z}_1^s\dot{z}_2^s - 4\mu_2/z - 5\mu_1/z]\,\partial(1/z)/\partial z_1^n \\ + [4\dot{z}_1^s(\dot{z}_2^n - \dot{z}_1^n) + 3\dot{z}_1^n\dot{z}_2^s - 4\dot{z}_2^n\dot{z}_2^s]\,\partial(1/z)/\partial z_1^s + \tfrac{1}{2}\dot{z}_2^s\dot{z}_2^r\,\partial^3 z/\partial z_1^s\,\partial z_1^r\,\partial z_1^n\} \end{aligned} \qquad (7\text{-}100) $$

In Eq. (7-100) the Newtonian gravitational constant G was taken as equal to 1. The equation of motion for the second particle is obtained by replacing μ_1, μ_2, z_1, z_2 by μ_2, μ_1, z_2, z_1, respectively.

Equation (7-100) is known as the Einstein–Infeld–Hoffmann equation of motion, and is a generalization of the Newton equation.[70] The essential relativistic correction may be obtained by fixing one of the particles. Writing M for μ_2, neglecting μ_1 and $\dot{z}_2^s$, and using an obvious three-dimensional vector notation, Eq. (7-100) simplifies to

$$ \ddot{\mathbf{z}} - M\nabla\left(\frac{1}{z}\right) = M\left\{\left(\dot{\mathbf{z}}\cdot\dot{\mathbf{z}} - \frac{4M}{z}\right)\nabla\frac{1}{z} - 4\dot{\mathbf{z}}\left(\dot{\mathbf{z}}\cdot\nabla\frac{1}{z}\right)\right\} \qquad (7\text{-}101) $$

where $\mathbf{z}$ denotes the three-vector z_1^s.

PROBLEMS

7.1 Find the transformation laws (7-13) and (7-14) of the Christoffel symbols of the first and second kinds.

7.2 Prove Eq. (7-19).

[69] L. Infeld, *Rev. Mod. Phys.* **29,** 398 (1957).

[70] For the problem of motion and gravitational radiation see M. Carmeli, *Nuovo Cim.* (X) **37,** 842 (1965); W. L. Burke and K. S. Thorne, Gravitational Radiation Damping, in *Relativity* (M. Carmeli, S. I. Fickler and L. Witten, Eds.), Plenum Press, New York, 1970.

7.3 Show that the covariant derivatives of the tensors $T_{\alpha\beta}$, $T^{\alpha\beta}$, and $T^{\alpha}{}_{\beta}$ are given by

$$\nabla_\gamma T_{\alpha\beta} = \frac{\partial T_{\alpha\beta}}{\partial x^\gamma} - \Gamma^\delta_{\beta\gamma} T_{\alpha\delta} - \Gamma^\delta_{\alpha\gamma} T_{\delta\beta}$$

$$\nabla_\gamma T^{\alpha\beta} = \frac{\partial T^{\alpha\beta}}{\partial x^\gamma} + \Gamma^\alpha_{\delta\gamma} T^{\delta\beta} + \Gamma^\beta_{\delta\gamma} T^{\alpha\delta}$$

$$\nabla_\gamma T^\alpha{}_\beta = \frac{\partial T^\alpha{}_\beta}{\partial x^\gamma} + \Gamma^\alpha_{\delta\gamma} T^\delta{}_\beta - \Gamma^\delta_{\beta\gamma} T^\alpha{}_\delta$$

From this find the general rule for covariant differentiation.

7.4 Show that covariant differentiation of the sum, difference, outer and inner product of tensors obeys the usual rules of ordinary differentiation.

7.5 Generalize Eq. (7-24) for a tensor $T_{\mu\nu}$.

7.6 If $T_{\alpha\beta}$ is the curl of a covariant vector, show that

$$\nabla_\gamma T_{\alpha\beta} + \nabla_\alpha T_{\beta\gamma} + \nabla_\beta T_{\gamma\alpha} = 0$$

and that this is equivalent to

$$\partial_\gamma T_{\alpha\beta} + \partial_\alpha T_{\beta\gamma} + \partial_\beta T_{\gamma\alpha} = 0$$

7.7 Show that the divergence $\nabla_\mu V^\mu$ of the vector V^μ is given by

$$\nabla_\mu V^\mu = \frac{1}{\surd(-g)} \frac{\partial}{\partial x^\mu} (V^\mu \surd(-g))$$

Also show that for a skew-symmetric tensor $F^{\alpha\beta}$ the covariant divergence is

$$\nabla_\beta F^{\alpha\beta} = \frac{1}{\surd(-g)} \frac{\partial}{\partial x^\beta} (F^{\alpha\beta} \surd(-g))$$

7.8 Find the expression for the Riemann tensor $R_{\alpha\beta\gamma\delta}$. From it prove Eqs. (7-26).

7.9 Discuss the constancy of the weak and gravitational coupling constants.

7.10 Use the geodesic equations, Eq. (7-33), to determine the force per unit mass on a body at rest, and show that it is given by $F^i = -c^2\Gamma^i_{00}$ where $i = 1, 2, 3$. In the weak field approximation $g^{i\alpha}$ are very close to the Lorentz metric, and for a time-independent metric $F^i = c^2\Gamma^i_{00} \simeq (c^2/2)\, \partial_i g_{00}$. Show that in the weak field case Eq. (7-37) reduces to the Poisson equation (7-36) where $g_{00} \simeq 1 - 2\phi$. From this show that the constant κ in Eq. (7-37) is given by $\kappa = 8\pi G/c^4$.

7.11 Prove Eqs. (7-42) and (7-43).

7.12 Derive the gravitational field equations (7-37) using calculus of variation by treating both $g_{\mu\nu}$ and $\Gamma^\mu_{\alpha\beta}$ as independent variants, and obtain thereby equations that determine both objects. Such a procedure is known as the Palatini formalism. The procedure is analogous to the one employed in deriving the electromagnetic field equations from a variational principle where both the field $F^{\mu\nu}$ and the potential A_μ are variants of an action principle. [See A. Einstein, *S.B. preuss. Akad. Wiss.*, 414 (1925).]

7.13 Find the energy–momentum tensor $T_{\mu\nu}$ for (1) a system of neutral particles of inertial mass M (function of time); (2) the electromagnetic field; and (3) a scalar field ϕ. Show that they are given by:

$$(1)\quad T^{\mu\nu} = \sum M \dot{z}^\mu \dot{z}^\nu\, \delta(\mathbf{x} - \mathbf{z})$$

$$(2)\quad T_{\mu\nu} = \frac{1}{4\pi}\left\{\frac{1}{4} g_{\mu\nu} F_{\alpha\beta} F^{\alpha\beta} - F_{\mu\alpha} F_\nu{}^\alpha\right\}$$

$$(3)\quad T_{\mu\nu} = \partial_\mu\phi\,\partial_\nu\phi - \frac{1}{2} g_{\mu\nu}(\nabla^\alpha\phi\nabla_\alpha\phi - m^2\phi^2)$$

7.14 Show that if one takes as the solutions of Eqs. (7-73a) to (7-73c)

$$\psi = \frac{1}{2} \ln \frac{\lambda - 1}{\lambda + 1}, \qquad \gamma = \frac{1}{2} \ln \frac{\lambda^2 - 1}{\lambda^2 - \mu^2}$$

and then carries out the transformation $\lambda = (r/m) - 1$, $\mu = \cos\theta$, one obtains the usual form of the Schwarzschild solution. [See G. Erez and N. Rosen, *Bull. Res. Counc. Israel* **8F**, 47 (1959)].

7.15 Use the approximate metric (7-82) into the geodesic equation (7-33) to show that the equation obtained is (7-83).

7.16 Prove Eqs. (7-99) and (7-100).

CHAPTER

EIGHT

SPINORS IN GENERAL RELATIVITY

In Chapter 3 spinors were introduced in connection with the finite-dimensional representations of the group $SL(2, C)$. In this chapter we use them in the description of physical quantities within the framework of general relativity theory. Thus the connection between spinors and tensors, defined in a Riemannian space, will be given and accordingly we define the Maxwell, Weyl, and Riemann spinors. The classifications of the Maxwell and the Weyl spinors are consequently outlined.

8-1 CONNECTION BETWEEN SPINORS AND TENSORS

In Chapter 3 it was shown how 2-component spinors are associated with the finite-dimensional representations of the group $SL(2, C)$ when the representation is realized in the space of polynomials. In particular, it was shown that spinors appear (up to factorial terms) as the coefficients of the polynomials of the space in which the representation is realized. Furthermore, it was shown that their transformation law under the group translation provides another form for the representation.

We now use 2-component spinors in the description of the gravitational field. Accordingly, these quantities become functions of space–time when they are applied in physics. To this end one associates, with each tensor describing a physical quantity in general relativity, a spinor.

Spinors in Riemannian Space

2-component spinors are introduced in a Riemannian space at each space–time point, in a tangent two-dimensional complex space. The correspondence between tensors and spinors is then obtained by means of mixed-indices quantities.[1] They are four 2 × 2 Hermitian matrices, denoted by $\tilde{\sigma}^{\mu}{}_{AB'}$. Greek letters are used for tensor indices running over 0, 1, 2, 3, and Roman capitals for spinor indices taking the values 0, 1. Primed indices refer to the complex conjugate. These indices replace the dotted indices of spinors used in previous chapters. The Hermiticity of the matrices $\tilde{\sigma}^{\mu}$ means $\tilde{\sigma}^{\mu}{}_{AB'} = \bar{\tilde{\sigma}}^{\mu}{}_{B'A}$. When a locally Cartesian coordinate frame is used, the $\tilde{\sigma}^{\mu}$ matrices may be taken (apart from a factor) as the unit matrix (for $\tilde{\sigma}^{0}$) and the three Pauli matrices (for $\tilde{\sigma}^{k}$). Other sets may be obtained from these by coordinate transformations. We will not need an explicit knowledge of any one set of $\tilde{\sigma}^{\mu}$.

The four matrices $\tilde{\sigma}^{\mu}$ satisfy the relation

$$g_{\mu\nu}\tilde{\sigma}^{\mu}{}_{AB'}\tilde{\sigma}^{\nu}{}_{CD'} = \varepsilon_{AC}\varepsilon_{B'D'} \tag{8-1}$$

where $g_{\mu\nu}$ is the *geometrical* metric tensor, and ε_{AC} and $\varepsilon_{B'D'}$, along with ε^{AC} and $\varepsilon^{B'D'}$, are the skew-symmetric Levi–Civita symbols, given by

$$\varepsilon_{AB} = \begin{pmatrix} 0 & 1 \\ -1 & 0 \end{pmatrix} \tag{8-2}$$

Raising or lowering a spinor index is made by means of the above symbols ε, with the following conventions:

$$\begin{aligned} \xi^{A} &= \varepsilon^{AB}\xi_{B}, \qquad & \xi_{A} &= \xi^{B}\varepsilon_{BA} \\ \eta^{A'} &= \varepsilon^{A'B'}\eta_{B'}, \qquad & \eta_{A'} &= \eta^{B'}\varepsilon_{B'A'} \end{aligned} \tag{8-3}$$

Equivalence of Spinors and Tensors

The spinor equivalent of a tensor is a quantity which has an unprimed and a primed spinor index for each tensor index. The spinor representing the tensor $T^{\alpha\beta}{}_{\gamma}$, for example, is

$$T^{AB'CD'}{}_{EF'} = \tilde{\sigma}_{\alpha}{}^{AB'}\tilde{\sigma}_{\beta}{}^{CD'}\tilde{\sigma}^{\gamma}{}_{EF'}T^{\alpha\beta}{}_{\gamma} \tag{8-4}$$

The tensor representing the spinor $S^{AB'}{}_{CD'}$, on the other hand, is given by

$$S^{\alpha}{}_{\beta} = \tilde{\sigma}^{\alpha}{}_{AB'}\tilde{\sigma}_{\beta}{}^{CD'}S^{AB'}{}_{CD'} \tag{8-5}$$

Greek indices are lowered and raised, as usual, by the metric tensor $g_{\alpha\beta}$ and its inverse, $g^{\alpha\beta}$. The spinor expressions for the metric tensor are given by

$$g_{AB'CD'} = \varepsilon_{AC}\varepsilon_{B'D'}, \qquad g^{AB'CD'} = \varepsilon^{AC}\varepsilon^{B'D'} \tag{8-6}$$

When taking the complex conjugate of a spinor, unprimed indices become primed, and vice versa. The complex conjugate of the spinor $S^{AB'}$, for example, is

[1] L. Infeld and B. L. van der Waerden, *Sb. preuss Akad. Wiss., Phys.-mat. Kl.* 380 (1933); W. L. Bade and J. Jehle, *Rev. Mod. Phys.* **25**, 714 (1953).

$\bar{S}^{A'B}$. Accordingly, the condition for the vector S^α to be real is that its spinor equivalent be Hermitian:

$$S^{AB'} = \bar{S}^{B'A} \tag{8-7}$$

Covariant Derivative of Spinors

The covariant derivative, $\nabla_\mu \xi_A$, of a spinor ξ_A is

$$\nabla_\mu \xi_A = \partial_\mu \xi_A - \Gamma^B{}_{A\mu} \xi_B \tag{8-8}$$

where $\Gamma^B{}_{A\mu}$ is the spinor affine connection. The corresponding quantity $\bar{\Gamma}^{B'}{}_{A'\mu}$ deals with the spinor $\bar{\xi}_{A'}$. The spinor affine connection is fixed by the requirement that the covariant derivatives of $\tilde{\sigma}^\mu{}_{AB'}$, ε_{AB}, and $\varepsilon_{A'B'}$ shall all vanish:[2]

$$\begin{gathered} \nabla_\alpha \tilde{\sigma}^\mu{}_{AB'} = 0 \\ \nabla_\alpha \varepsilon_{AB} = 0 \\ \nabla_\alpha \varepsilon_{A'B'} = 0 \end{gathered} \tag{8-9}$$

8-2 MAXWELL, WEYL, AND RIEMANN SPINORS

We now find the spinors describing the electromagnetic and gravitational fields. They are obtained, using the procedure outlined in the last section, by associating a spinor with each of the Maxwell, Weyl, and Riemann tensors. However, since these tensors have some special symmetry properties, the corresponding spinor equivalents are simplified.

The Electromagnetic Field

Let $F_{\mu\nu}$ describe the Maxwell tensor, i.e., a *real skew-symmetric* tensor with two indices. Let us denote the spinor equivalent of this tensor by $F_{AB'CD'}$. It obviously satisfies

$$F_{AB'CD'} = -F_{CD'AB'} \tag{8-10}$$

and, as a result, one obtains the identity

$$F_{AB'CD'} = \tfrac{1}{2}(F_{AB'CD'} - F_{CB'AD'}) + \tfrac{1}{2}(F_{CB'AD'} - F_{CD'AB'})$$

Accordingly, one obtains (see Problem 8.2):

$$F_{AB'CD'} = \tfrac{1}{2}(\varepsilon_{AC} F_{GB'}{}^{G}{}_{D'} + \varepsilon_{B'D'} F_{CG'A}{}^{G'}) \tag{8-11}$$

The last equation can be simplified. If one denotes $\frac{1}{2}F_{CG'A}{}^{G'}$ by ϕ_{AC}, then

$$\phi_{AC} = \tfrac{1}{2}F_{CG'A}{}^{G'} = -\tfrac{1}{2}F_A{}^{G'}{}_{CG'} = \tfrac{1}{2}F_{AG'C}{}^{G'} = \phi_{CA}$$

[2] R. Penrose, *Ann. Phys. (N.Y.)* **10,** 171 (1960).

by the antisymmetry property of F. Hence the spinor ϕ_{AC} is symmetric. Taking the complex conjugate of ϕ_{BD}, on the other hand, gives

$$\bar{\phi}_{B'D'} = \overline{\tfrac{1}{2}F_{BP'D}{}^{P'}} = \tfrac{1}{2}\bar{F}_{B'PD'}{}^{P} = \tfrac{1}{2}F_{PB'}{}^{P}{}_{D'}$$

where the last equality was a consequence of the Hermiticity property of F. Using the above two equations in Eq. (8-11) we obtain the rather simple decomposition of the spinor $F_{AB'CD'}$:

$$F_{AB'CD'} = \varepsilon_{AC}\bar{\phi}_{B'D'} + \phi_{AC}\varepsilon_{B'D'} \tag{8-12}$$

We thus see that the Maxwell tensor is equivalent to a symmetric spinor with two indices. In other words, the six real components of the skew-symmetric tensor $F_{\mu\nu}$ are equivalent to the three complex components of the symmetric spinor ϕ_{AB}. In the following ϕ_{AB} will be referred to as the *Maxwell spinor*.

The Gravitational Field

Let now $R_{\mu\nu\rho\sigma}$ be the Riemann tensor, i.e., a real four-index tensor having the symmetry properties (see Chapter 7):[3]

$$\begin{gathered} R_{\mu\nu\rho\sigma} = -R_{\nu\mu\rho\sigma} = -R_{\mu\nu\sigma\rho} \\ R_{\mu\nu\rho\sigma} + R_{\mu\rho\sigma\nu} + R_{\mu\sigma\nu\rho} = 0 \end{gathered} \tag{8-13}$$

Following the procedure outlined above for the Maxwell tensor, one obtains for the *Riemann spinor*:[4]

$$\begin{aligned} -R_{AB'CD'EF'GH'} = {} & \chi_{ACEG}\varepsilon_{B'D'}\varepsilon_{F'H'} + \phi_{ACF'H'}\varepsilon_{B'D'}\varepsilon_{EG} \\ & + \varepsilon_{AC}\bar{\phi}_{B'D'EG}\varepsilon_{F'H'} + \varepsilon_{AC}\varepsilon_{EG}\bar{\chi}_{B'D'F'H'} \end{aligned} \tag{8-14}$$

where

$$\chi_{ACEG} = -\tfrac{1}{4}R_{AB'C}{}^{B'}{}_{EF'G}{}^{F'}$$

and

$$\phi_{ACF'H'} = -\tfrac{1}{4}R_{AB'C}{}^{B'}{}_{EF'}{}^{E}{}_{H'}$$

The two spinors χ_{ABCD} and $\phi_{ABC'D'}$ uniquely define the curvature spinors. From the symmetries of the Riemann tensor, Eqs. (8-13), it follows that the spinors χ_{ABCD} and $\phi_{ABC'D'}$ have the following symmetry properties:

$$\chi_{ABCD} = \chi_{BACD} = \chi_{ABDC} = \chi_{CDAB} \tag{8-15}$$

and

$$\phi_{ABC'D'} = \phi_{BAC'D'} = \phi_{ABD'C'} = \bar{\phi}_{C'D'AB} \tag{8-16}$$

[3] Note that as a consequence of the Eqs. (8-13) one has $R_{\mu\nu\rho\sigma} = R_{\rho\sigma\mu\nu}$.
[4] L. Witten, *Phys. Rev.* **113**, 357 (1959).

The Weyl Spinor

The spinor χ_{ABCD} may further be decomposed. To this end one writes it as

$$\chi_{ABCD} = \tfrac{1}{3}(\chi_{ABCD} + \chi_{ACBD} + \chi_{ADBC}) + \tfrac{1}{3}(\chi_{ABCD} - \chi_{ACBD}) + \tfrac{1}{3}(\chi_{ABCD} - \chi_{ADBC})$$

and hence, using Eq. (8-15), one can write

$$\chi_{ABCD} = \psi_{ABCD} + \tfrac{1}{3}\varepsilon_{BC}\chi_{AE}{}^{E}{}_{D} + \tfrac{1}{3}\varepsilon_{BD}\chi_{AEC}{}^{E} \tag{8-17}$$

where we have used the notation

$$\psi_{ABCD} = \tfrac{1}{3}(\chi_{ABCD} + \chi_{ACBD} + \chi_{ADBC}) \tag{8-18}$$

But the expression $\chi_{AE}{}^{E}{}_{D}$ is skew-symmetric in the indices A, D since

$$\chi_{AE}{}^{E}{}_{D} = \chi^{E}{}_{DAE} = \chi_{D}{}^{E}{}_{EA} = -\chi_{DE}{}^{E}{}_{A}$$

Therefore

$$\chi_{AE}{}^{E}{}_{D} = \tfrac{1}{2}(\chi_{AE}{}^{E}{}_{D} - \chi_{DE}{}^{E}{}_{A}) = \tfrac{1}{2}\varepsilon_{AD}\chi_{CE}{}^{EC}$$

Accordingly, we obtain for Eq. (8-17)

$$\chi_{ABCD} = \psi_{ABCD} + \tfrac{1}{6}(\varepsilon_{BC}\varepsilon_{AD} + \varepsilon_{BD}\varepsilon_{AC})\lambda \tag{8-19}$$

where we have denoted $\chi_{AB}{}^{AB}$ by λ. The spinor ψ_{ABCD} is, of course, completely symmetric in its four indices. It corresponds uniquely to the Weyl conformal tensor $C_{\mu\nu\rho\sigma}$ (see Chapter 7). It will be referred to as the *Weyl spinor*.[5] Moreover, using the second of Eqs. (8-13), one can show that λ is real (see Problem 8.6).

One thus obtains for the Riemann spinor, Eq. (8-14), the following decomposition:

$$\begin{aligned} -R_{AB'CD'EF'GH'} = {} & \psi_{ACEG}\varepsilon_{B'D'}\varepsilon_{F'H'} + \varepsilon_{AC}\varepsilon_{EG}\bar{\psi}_{B'D'F'H'} \\ & + \frac{\lambda}{6}\left\{\left(\varepsilon_{CE}\varepsilon_{AG} + \varepsilon_{CG}\varepsilon_{AE}\right)\varepsilon_{B'D'}\varepsilon_{F'H'}\right. \\ & \left. + \varepsilon_{AC}\varepsilon_{EG}\left(\varepsilon_{D'F'}\varepsilon_{B'H'} + \varepsilon_{D'H'}\varepsilon_{B'F'}\right)\right\} \\ & + \phi_{ACF'H'}\varepsilon_{B'D'}\varepsilon_{EG} + \varepsilon_{AC}\varepsilon_{F'H'}\bar{\phi}_{B'D'EG} \end{aligned} \tag{8-20}$$

Counting components, one finds five complex components for ψ_{ABCD}, three real and three complex components for $\phi_{ABC'D'}$, and one real λ. Their sum, is thus, equivalent to the twenty real components of the Riemann tensor.

[5] From the Weyl spinor ψ_{ABCD} one can construct the tensor $T_{\alpha\beta\gamma\delta}$ whose spinor equivalent is given by $T_{AB'CD'EF'GH'} = \psi_{ACEG}\bar{\psi}_{B'D'F'H'}$. It is known as the Robinson-Bel tensor (sometimes, gravitational density or super energy). The tensor $T_{\alpha\beta\gamma\delta}$ is symmetric in all its indices, has vanishing traces, and vanishing covariant devergences when $\phi_{ABC'D'} = 0$. See L. Bel, *C. R. Acad. Sci., Paris* **247**, 1094 (1958) and **248**, 1297 (1959).

Ricci's and Einstein's Spinors

To conclude this section we find below the spinors that correspond to the *Ricci* and *Einstein* tensors.

The Ricci tensor $R_{\mu\nu} = R^{\sigma}{}_{\mu\sigma\nu}$ has the spinor form

$$\begin{aligned} R_{AB'CD'} &= R^{EF'}{}_{AB'EF'CD'} \\ &= -\lambda\varepsilon_{AC}\varepsilon_{B'D'} + 2\phi_{ACB'D'} \end{aligned} \tag{8-21}$$

The Ricci scalar (scalar curvature) $R = R^{\alpha}{}_{\alpha}$ is given by

$$R = R^{AB'}{}_{AB'} = -4\lambda \tag{8-22}$$

Hence the spinor $\phi_{ABC'D'}$ represents the *trace-free part of the Ricci tensor*. The Einstein tensor $G_{\mu\nu} = R_{\mu\nu} - \frac{1}{2}g_{\mu\nu}R$, using Eqs. (8-6), (8-21) and (8-22), therefore, takes the spinorial form

$$G_{AB'CD'} = \lambda\varepsilon_{AC}\varepsilon_{B'D'} + 2\phi_{ACB'D'} \tag{8-23}$$

8-3 CLASSIFICATION OF MAXWELL SPINOR

We are now in a position to classify both the electromagnetic and gravitational fields. The classification of the gravitational field has been of great interest in relation to the study of gravitational radiation and in understanding some general features of exact solutions of the field equations.

In this section we will discuss the classification of the Maxwell tensor; the gravitational field will be discussed in the next section. The discussion of the electromagnetic field will be made in such a way as to emphasize the analogy to the classification of the Weyl tensor for the gravitational field case.[6]

Complex 3-space

Let $F_{\mu\nu}$ be the Maxwell tensor and let $^*F_{\mu\nu}$ be its dual (see Problem 8.4). Let us also define the tensor $F^{+}{}_{\mu\nu}$ by

$$F^{+}{}_{\mu\nu} = F_{\mu\nu} + i^*F_{\mu\nu} \tag{8-24}$$

so that $^*F^{+}{}_{\mu\nu} = -iF^{+}{}_{\mu\nu}$. The spinor equivalent of the tensor $F_{\mu\nu}$ was found in the previous section and is given by Eq. (8-12), whereas that of the tensor $F^{+}{}_{\mu\nu}$ is given by

$$F^{+}{}_{AB'CD'} = 2\phi_{AC}\varepsilon_{B'D'} \tag{8-25}$$

where ϕ_{AC} is the Maxwell spinor.

[6] G. Ludwig, *Am. J. Phys.* **37**, 1225 (1969).

Classification of the electromagnetic field can be made by classifying ϕ_{AB}. To this end one studies the *eigenspinors* and *eigenvalues* of the spinorial equation

$$\phi^A{}_B\alpha^B = \lambda\alpha^A \tag{8-26}$$

To study this equation one introduces a *basis* in spin space. Let the two spinors of the basis be denoted by l_A and n_A, satisfying the normalization condition $l_A n^A = 1$. This basis induces another basis, given by

$$\xi_{0AB} = n_A n_B, \qquad \xi_{1AB} = -2l_{(A}n_{B)}, \qquad \xi_{2AB} = l_A l_B \tag{8-27}$$

in the 3-dimensional space, E_3, of bispinors. Here $l_{(A}n_{B)} = \frac{1}{2}(l_A n_B + l_B n_A)$. This means a bispinor ϕ_{AB} can be written in terms of the basis (8-27) as

$$\phi_{AB} = \sum_{m=0}^{2} \phi_m \xi_{mAB} \tag{8-28}$$

where ϕ_0, ϕ_1, and ϕ_2 are called *dyad* components of the bispinor and correspond to the six real components of the tensor $F_{\mu\nu}$. The spin frame l_A, n_A induces other bases in E_3, such as the one given by

$$\begin{aligned} \eta_{0AB} &= 2^{1/2} i l_{(A}n_{B)} \\ \eta_{1AB} &= 2^{-1/2}(l_A l_B + n_A n_B) \\ \eta_{2AB} &= 2^{-1/2} i (l_A l_B - n_A n_B) \end{aligned} \tag{8-29}$$

This basis satisfies the orthogonality relation

$$\eta_{mAB}\eta_n{}^{AB} = \delta_{mn} \tag{8-30}$$

In terms of this last basis ϕ_{AB} can now be written as

$$\phi_{AB} = \sum_{m=0}^{2} \chi_m \eta_{mAB} \tag{8-31}$$

The two sets of three components χ and ϕ are then related by

$$\begin{aligned} \chi_0 &= 2^{1/2} i \phi_1 \\ \chi_1 &= 2^{-1/2}(\phi_0 + \phi_2) \\ \chi_2 &= 2^{-1/2} i (\phi_0 - \phi_2) \end{aligned} \tag{8-32}$$

Classification

In terms of the dyad components ϕ_m, the eigenvalue equation (8-26) becomes

$$\Phi\alpha = \lambda\alpha \tag{8-26a}$$

where Φ is a 2×2 matrix given by

$$\Phi = \begin{pmatrix} \phi_1 & \phi_2 \\ -\phi_0 & -\phi_1 \end{pmatrix} \tag{8-33a}$$

and α is a column matrix given by

$$\alpha = \begin{pmatrix} \alpha^0 \\ \alpha^1 \end{pmatrix} \tag{8-33b}$$

where α^a are the dyad components of α^A, i.e., $\alpha^a = \zeta^a{}_A \alpha^A$, and we have denoted $\zeta_0{}^A = l^A$ and $\zeta_1{}^A = n^A$. The two eigenvalues of Eq. (8-26a) are $\lambda = \pm(\phi_1{}^2 - \phi_0 \phi_2)^{1/2}$. One, therefore, has two cases: (1) $\phi_1{}^2 - \phi_0\phi_2 \neq 0$, in which case there are two different eigenspinors. The spinor ϕ_{AB} is called *algebraically general* or *non-null*; and (2) $\phi_1{}^2 - \phi_0\phi_2 = 0$, in which case $\lambda = 0$ and there is only one eigenspinor. The spinor ϕ_{AB} is then called *algebraically special* or *null*.[7]

Change of Spin Frame

Let us introduce another basis l'_A, n'_A in spin space that is related to the original basis l_A, n_A by

$$\begin{pmatrix} l'_A \\ n'_A \end{pmatrix} = \begin{pmatrix} a & b \\ c & d \end{pmatrix} \begin{pmatrix} l_A \\ n_A \end{pmatrix} \tag{8-34}$$

Here a, b, c, and d are complex numbers satisfying $ad - bc = 1$. Thus the matrix

$$g = \begin{pmatrix} a & b \\ c & d \end{pmatrix} \tag{8-35}$$

is an element of the group $SL(2, C)$. We can now write ϕ_{AB} in terms of the new basis,

$$\phi_{AB} = \sum_{m=0}^{2} \phi'_m \xi'_{mAB} \tag{8-36}$$

where ξ'_{mAB} is the induced basis in E_3 and is given in terms of l'_A and n'_A in accordance with Eq. (8-27). The dyad components ϕ'_m can then be obtained in terms of ϕ_n by

$$\Phi' = (g^t)^{-1} \Phi g^t \tag{8-37}$$

or by

$$\begin{pmatrix} \phi'_0 \\ \phi'_1 \\ \phi'_2 \end{pmatrix} = \begin{pmatrix} a^2 & 2ab & b^2 \\ ac & bc + ad & bd \\ c^2 & 2cd & d^2 \end{pmatrix} \begin{pmatrix} \phi_0 \\ \phi_1 \\ \phi_2 \end{pmatrix} \tag{8-38}$$

[7] A null bispinor could also be defined as one which is orthogonal to itself, or has a zero inner product in E_3, where inner products of two spinors ϕ_{AB} and ϕ'_{AB} are defined by $\sum \chi_m \chi'_m = \phi_{AB}\phi'^{AB}$ and χ_m and χ'_m are the dyad components in the orthogonal basis η_{mAB}.

The corresponding transformation law for χ's are obtained, using Eqs. (8-32) and (8-38):

$$\begin{pmatrix} \chi'_0 \\ \chi'_1 \\ \chi'_2 \end{pmatrix} = \begin{pmatrix} ad + bc & i(ac + db) & ac - bd \\ -i(ab + cd) & \frac{1}{2}(a^2 + b^2 + c^2 + d^2) & -\frac{i}{2}(a^2 - b^2 + c^2 - d^2) \\ ab - cd & \frac{i}{2}(a^2 + b^2 - c^2 - d^2) & \frac{1}{2}(a^2 - b^2 - c^2 + d^2) \end{pmatrix} \begin{pmatrix} \chi_0 \\ \chi_1 \\ \chi_2 \end{pmatrix} \tag{8-39}$$

Let us denote the two square 3×3 complex matrices in Eqs. (8-38) and (8-39) by Q and P, respectively. They give *three-dimensional representations* for the proper, orthochronous, homogeneous Lorentz group discussed in Chapter 2. The matrix in Eq. (8-39) is *orthogonal*, $P^{-1} = P^t$, with determinant unity,[8] whereas that in Eq. (8-38) satisfies the relation

$$\sum_{k=0}^{2} \frac{(-1)^k}{k!(2-k)!} Q_{kj} Q_{(2-k),\,l} = \frac{(-1)^l}{l!(2-l)!} \delta_{l,\,(2-j)}$$

The spin frame transformation (8-34) also induces a proper, orthochronous, Lorentz transformation on the null tetrad in the curved space constructed from the two spinors l_A and n_A. The null tetrad induced by the two spinors l^A and n^A is given by:[9]

$$\begin{aligned} l^\mu &= \tilde{\sigma}^\mu{}_{AB'} l^A \bar{l}^{B'} \\ m^\mu &= \tilde{\sigma}^\mu{}_{AB'} l^A \bar{n}^{B'} \\ \bar{m}^\mu &= \tilde{\sigma}^\mu{}_{AB'} n^A \bar{l}^{B'} \\ n^\mu &= \tilde{\sigma}^\mu{}_{AB'} n^A \bar{n}^{B'} \end{aligned} \tag{8-40}$$

Accordingly, a change of a null tetrad in the curved space is represented in the space E_3 by a proper orthogonal matrix, provided the basis in E_3 is chosen to be orthogonal as in Eq. (8-30).

As has been shown in Chapter 4 that the matrix $g \in SL(2, C)$, given in Eqs. (8-34) and (8-35), can be written as a product of three matrices of the form

$$g_1(z) = \begin{pmatrix} 1 & 0 \\ z & 1 \end{pmatrix}, \quad g_2(z) = \begin{pmatrix} z & 0 \\ 0 & z^{-1} \end{pmatrix}, \quad g_3(z) = \begin{pmatrix} 1 & z \\ 0 & 1 \end{pmatrix} \tag{8-41}$$

[8] For applications of the 3×3 complex orthogonal matrix representation of the homogeneous Lorentz group to other physical problems, see B. Kursunoglu, *Modern Quantum Theory*, W. H. Freeman, San Francisco, 1962.

[9] The null tetrad l^μ, m^μ, $\bar{m}^\mu$, and n^μ satisfies the normalization conditions:

$$\begin{aligned} l_\mu n^\mu &= -m_\mu \bar{m}^\mu = 1 \\ l_\mu l^\mu &= n_\mu n^\mu = m_\mu m^\mu = \bar{m}_\mu \bar{m}^\mu \\ &= l_\mu m^\mu = l_\mu \bar{m}^\mu = n_\mu m^\mu = n_\mu \bar{m}^\mu = 0 \end{aligned}$$

It will be noted that l^μ and n^μ are real, whereas m^μ is complex.

where z is a complex number. The transformation $g_1(z)$ leaves the spinor l_A, and hence the null vector l_μ, invariant. It is called a one-(complex) parameter *null rotation* about l_μ. The transformation $g_3(z)$ is also a one-parameter null rotation, but about the vector n_μ. The transformation $g_2(z)$ corresponds to an ordinary Lorentz transformation (boost) in the $l_\mu - n_\mu$ plane, along with a spatial rotation in the $m_\mu - \bar{m}_\mu$ plane.[10]

The matrices $Q_1(z)$, $Q_2(z)$ and $Q_3(z)$ obtained from Eq. (8-38), corresponding to the three matrices $g_1(z)$, $g_2(z)$, and $g_3(z)$ of the group $SL(2, C)$, can be obtained by putting the appropriate values in Q. The transformations of $\boldsymbol{\phi} = (\phi_0, \phi_1, \phi_2)$ under $Q_1(z)$, $Q_2(z)$, and $Q_3(z)$ are then given by:

$$\begin{aligned} \phi'_0 &= \phi_0 \\ \phi'_1 &= z\phi_0 + \phi_1 \\ \phi'_2 &= z^2\phi_0 + 2z\phi_1 + \phi_2 \end{aligned} \tag{8-42}$$

$$\begin{aligned} \phi'_0 &= z^2\phi_0 \\ \phi'_1 &= \phi_1 \\ \phi'_2 &= z^{-2}\phi_2 \end{aligned} \tag{8-43}$$

and

$$\begin{aligned} \phi'_0 &= \phi_0 + 2z\phi_1 + z^2\phi_2 \\ \phi'_1 &= \phi_1 + z\phi_2 \\ \phi'_2 &= \phi_2 \end{aligned} \tag{8-44}$$

respectively.

Invariants

The matrix Φ given by Eqs. (8-26a) and (8-33) can also be written, using Eq. (8-32), in terms of the components of the 3-vector $\boldsymbol{\chi} = (\chi_0, \chi_1, \chi_2)$ as:

$$\Phi = (i/\sqrt{2})X \tag{8-45}$$

$$X = \begin{pmatrix} -\chi_0 & \chi_2 - i\chi_1 \\ \chi_2 + i\chi_1 & \chi_0 \end{pmatrix} \tag{8-46}$$

Under a change of basis, the trace of the matrix X must be invariant. But $\operatorname{Tr} X = 0$, and thus it does not yield an interesting invariant. However, $\operatorname{Tr} X^2 =$

[10] The matrices P, Eq. (8-39), corresponding to the three matrices $g_1(z)$, $g_2(z)$, and $g_3(z)$, can be obtained by putting the appropriate values in P. One can verify that the determinants of $P_1(z)$, $P_2(z)$, and $P_3(z)$ are all $+1$. Hence, the determinant of P is also $+1$.

$2\chi_m \chi_m = 2\chi \cdot \chi = 2\phi_{AB}\phi^{AB}$ is an "obvious" invariant. In fact, from $X^2 = \chi \cdot \chi I$, where I is the 2×2 unit matrix, it follows that

$$\begin{aligned} \text{Tr } X^{2n-1} &= 0 \\ \text{Tr } X^{2n} &= 2(\chi \cdot \chi)^n \end{aligned} \tag{8-47}$$

for any natural number n.

We thus see that the invariant $\chi \cdot \chi$ plays an important role in the classification of the bivector. If $\chi \cdot \chi$ vanishes, the bivector is null; otherwise it is non-null.

Canonical Forms

There are two canonical forms which correspond to the two types of bivectors.

If the bivector is *null* one can always choose a spin frame l'_A, n'_A in such a way that the direction of $n'_A n'_B$ in E_3 coincides with that of the given null bispinor. To see this we proceed as follows. Let $\boldsymbol{\phi} = (\phi_0, \phi_1, \phi_2)$, with $\chi \cdot \chi = 2(\phi_0 \phi_2 - \phi_1{}^2) = 0$, be the components of the bispinor in the basis (8-27). Without loss of generality one can assume that $\phi_0 \neq 0$. (A null rotation $g_3(z)$ about n_μ, of the form (8-44), could always make it so.) Under a null rotation around l_μ, the components of $\boldsymbol{\phi}$ transform according to Eq. (8-42). ϕ'_2 is a quadratic polynomial in z, whereas ϕ'_1 is proportional to the derivative of ϕ'_2 with respect to z. The condition $\phi_0 \phi_2 - \phi_1{}^2 = 0$ yields a double root for ϕ'_2 given by $z = -\phi_1/\phi_0$. Choosing this root for z makes both ϕ'_2 and ϕ'_1 vanish simultaneously. Accordingly, in the new frame, $\phi_{AB} = \phi'_0 n'_A n'_B$ and $\boldsymbol{\phi}' = (\phi'_0, 0, 0)$. The matrix Φ and the eigenspinor α of Eq. (8-33) will have the forms:

$$\begin{pmatrix} 0 & 0 \\ -\phi'_0 & 0 \end{pmatrix} \tag{8-48}$$

and

$$\begin{pmatrix} 0 \\ 1 \end{pmatrix} \tag{8-49}$$

i.e., n'_A. It will be noted that it is equally possible to make the direction $l'_A l'_B$ coincides with the given bispinor by making a null rotation around n_α instead.

If the given bispinor is non-null, we can make ϕ'_2, but not ϕ'_1, to vanish by choosing z to be one of the roots of the quadratic form ϕ'_2. Applying a null rotation around n_α with appropriate value for z will leave ϕ''_1 as the only non-zero component of $\boldsymbol{\phi}''$. Hence $\phi_{AB} = -2\phi''_1 l''_{(A} n''_{B)}$. The matrix Φ, Eq. (8-33a), is then given by

$$\begin{pmatrix} \phi''_1 & 0 \\ 0 & -\phi''_1 \end{pmatrix} \tag{8-50}$$

whereas the eigenspinors Eq. (8-33b), will be given by

$$\begin{pmatrix}1\\0\end{pmatrix} \text{ and } \begin{pmatrix}0\\1\end{pmatrix} \tag{8-51}$$

i.e., by the new basis spinors l''_A and n''_A.

Spinor Method

The classification of the Maxwell spinor ϕ_{AB} could also be made by decomposing it into the symmetrized product of spinors with one index. This is done as follows.

Let ζ^A be an arbitrary spinor, and consider the expression $\phi_{AB}\zeta^A\zeta^B$. This is a homogeneous polynomial of second degree in ζ^0 and ζ^1. This polynomial can be factored into two linear factors, thus writing identically

$$\phi_{AB}\zeta^A\zeta^B = (\alpha_A\zeta^A)(\beta_B\zeta^B)$$

or

$$\{\phi_{AB} - \alpha_{(A}\beta_{B)}\}\zeta^A\zeta^B = 0$$

Therefore, since ζ^A is arbitrary, one obtains a decomposition of ϕ_{AB},

$$\phi_{AB} = \alpha_{(A}\beta_{B)} \tag{8-52}$$

which is called the *canonical decomposition* of ϕ_{AB}.

The spinors α_A and β_B are determined up to a (complex) scalar factor. They are called *principal spinors*, and each of them, in turn, determines a real null *direction*. They need not be distinct. As a result, the decomposition (8-52) determines at least one and at most two real null directions, called the *principal null directions* of ϕ_{AB}. Classification of ϕ_{AB} may be based on counting the multiplicities of principal null directions. If α_A and β_B coincide, the bispinor ϕ_{AB} is null, otherwise it is general. This classification coincides with our previous discussion.

Tensor Method

Finally, we briefly mention the tensor method of classification. For every skew-symmetric tensor there exist two null directions $\zeta_\mu \neq 0$, which may or may not coincide, satisfying the equation

$$F^+{}_{\mu[\nu}\zeta_{\rho]}\zeta^\mu = 0 \tag{8-53}$$

If the directions coincide, $F_{\mu\nu}$ is null, otherwise it is non-null. Equation (8-53) is equivalent to the spinor equation

$$\phi_{AB}\zeta^A\zeta^B = 0$$

Our previous discussion shows the tensor method to be equivalent to previous methods.

8-4 CLASSIFICATION OF WEYL SPINOR

In the previous section bivectors were discussed as a preliminary to discussing the Weyl spinor. Moreover, bivectors occur as eigenvectors of the Weyl tensor. In this section we discuss the Weyl tensor.

The Weyl tensor $C_{\alpha\beta\gamma\delta}$ has the same symmetry properties of the Riemann tensor, Eqs. (8-13). In addition, it satisfies

$$C^{\rho}{}_{\alpha\rho\beta} = 0 \tag{8-54}$$

These identities reduce the number of independent components of $C_{\alpha\beta\gamma\delta}$ to ten.

In the last section the spinor equivalent of $C_{\alpha\beta\gamma\delta}$ was found to be a symmetric spinor of four indices, ψ_{ABCD},

$$-C_{AB'CD'EF'GH'} = \varepsilon_{AC}\varepsilon_{EG}\bar{\psi}_{B'D'F'H'} + \psi_{ACEG}\varepsilon_{B'D'}\varepsilon_{F'H'} \tag{8-55}$$

Corresponding to the Weyl tensor $C_{\alpha\beta\gamma\delta}$ one can define the tensor $C^{+}{}_{\alpha\beta\gamma\delta}$ by

$$C^{+}{}_{\alpha\beta\gamma\delta} = C_{\alpha\beta\gamma\delta} + i^{*}C_{\alpha\beta\gamma\delta} \tag{8-56}$$

where * denotes the (left- or right-hand) dual. The spinor equivalent to (8-56) is given by

$$-C^{+}{}_{AB'CD'EF'GH'} = 2\varepsilon_{B'D'}\varepsilon_{F'H'}\psi_{ACEG} \tag{8-57}$$

Complex 5-space

In order to classify the Weyl tensor we classify the Weyl spinor ψ_{ABCD} in terms of its eigenvalues and eigenspinors. The characteristic equation now is:

$$\psi_{ABCD}\phi^{CD} = \lambda\phi_{AB} \tag{8-58}$$

The basis l_A, n_A in spinorial space induces the basis

$$\begin{gathered}\xi_{0\,ABCD} = n_A n_B n_C n_D, \qquad \xi_{1\,ABCD} = -4l_{(A}n_B n_C n_{D)} \\ \xi_{2\,ABCD} = 6l_{(A}l_B n_C n_{D)}, \qquad \xi_{3\,ABCD} = -4l_{(A}l_B l_C n_{D)} \\ \xi_{4\,ABCD} = l_A l_B l_C l_D\end{gathered} \tag{8-59}$$

in the 5-dimensional complex, E_5, of completely symmetric four-spinors. The Weyl spinor can now be written in terms of the basis (8-59) as

$$\psi_{ABCD} = \sum_{n=0}^{4} \psi_n \xi_{n\,ABCD} \tag{8-60}$$

where ψ_n, with $n = 0, 1, \ldots, 4$, are the dyad components of the Weyl spinor, and correspond to the ten real components of the Weyl tensor.

Since E_5 is a subspace of $E_3 \times E_3$, one can expand ψ_{ABCD} in terms of the basis $\eta_{m\,AB}\eta_{n\,CD}$ of $E_3 \times E_3$:

$$\psi_{ABCD} = \sum_{m,\,n=0}^{2} \psi_{mn}\eta_{m\,AB}\eta_{n\,CD} \tag{8-61}$$

One can then write the coefficients ψ_{mn} in terms of the dyad components

$\psi_0, \ldots, \psi_4$ by use of Eqs. (8-29) and (8-59), to obtain a symmetric and trace-free matrix:

$$\psi_{mn} = \begin{pmatrix} -2\psi_2 & i(\psi_1+\psi_3) & (\psi_3-\psi_1) \\ i(\psi_1+\psi_3) & \frac{1}{2}(2\psi_2+\psi_0+\psi_4) & \frac{i}{2}(\psi_0-\psi_4) \\ (\psi_3-\psi_1) & \frac{i}{2}(\psi_0-\psi_4) & \frac{1}{2}(2\psi_2-\psi_0-\psi_4) \end{pmatrix} \tag{8-62}$$

We have seen that the Weyl tensor can be regarded as a vector in a five-dimensional space. The space E_5 has properties similar to E_3 discussed in the last section. It will be useful to introduce an orthonormal basis in E_5. Such a basis is provided by the following five, completely symmetric, four-spinors:

$$\begin{aligned} \eta_{0\,ABCD} &= (1/\sqrt{2})(l_A l_B l_C l_D + n_A n_B n_C n_D) \\ \eta_{1\,ABCD} &= i\sqrt{2}(l_{(A} l_B l_C n_{D)} + l_{(A} n_B n_C n_{D)}) \\ \eta_{2\,ABCD} &= \sqrt{6} l_{(A} l_B n_C n_{D)} \\ \eta_{3\,ABCD} &= \sqrt{2}(l_{(A} l_B l_C n_{D)} - l_{(A} n_B n_C n_{D)}) \\ \eta_{4\,ABCD} &= (i/\sqrt{2})(l_A l_B l_C l_D - n_A n_B n_C n_D) \end{aligned} \tag{8-63}$$

As can be easily verified, they satisfy

$$\eta_{m\,ABCD}\,\eta_n{}^{ABCD} = \delta_{mn}; \qquad m, n = 0, \ldots, 4 \tag{8-64}$$

and an arbitrary element of the space E_5 can be written as linear combination of them:

$$\psi_{ABCD} = 2 \sum_{m=0}^{4} \chi_m \eta_{m\,ABCD} \tag{8-65}$$

in analogy to Eq. (8-31) for the Maxwell spinor, and where the factor 2 has been introduced for convenience. The components χ_m can then be expressed in terms of the components ψ_n of Eq. (8-60).[11] We find:

$$\begin{aligned} \chi_0 &= 2^{-3/2}(\psi_0+\psi_4) \qquad & \chi_1 &= 2^{-1/2} i(\psi_1+\psi_3) \\ \chi_2 &= (3/2)^{1/2}\psi_2 \qquad & \chi_3 &= 2^{-1/2}(\psi_1-\psi_3) \\ & & \chi_4 &= 2^{-3/2} i(\psi_0-\psi_4) \end{aligned} \tag{8-66}$$

[11] Just as in the 3-dimensional case for the Maxwell spinor, one can introduce an inner product in E_5 as follows: If two Weyl tensors $C_{\alpha\beta\gamma\delta}$ and $C'_{\alpha\beta\gamma\delta}$ have components χ_m and χ'_n, respectively, in the basis (8-63), then their inner product is defined by

$$4\chi \cdot \chi' = 4 \sum_{m=0}^{4} \chi_m \chi'_m = \psi_{ABCD}\psi'^{ABCD} = (1/16) C^{+}{}_{\alpha\beta\gamma\delta} C'^{+\alpha\beta\gamma\delta}$$

where ψ_{ABCD} and ψ'_{ABCD} are the Weyl spinors associated with the Weyl tensors $C_{\alpha\beta\gamma\delta}$ and $C'_{\alpha\beta\gamma\delta}$, respectively. Two Weyl tensors or, equivalently, Weyl spinors are orthogonal if their inner product vanishes. A Weyl tensor which is orthogonal to itself is called *null*. A *unit* Weyl tensor is one for which the self-inner product is unity. As for bivectors, the existence of inner product allows the introduction of the notion of direction in E_5 in the usual manner.

Classification

In terms of the matrix Ψ of Eq. (8-62), the eigenvalue equation (8-58) can be written as

$$\Psi\chi = \lambda\chi \tag{8-67}$$

where χ is the column matrix whose elements are χ_m, $m = 0, 1, 2$, and χ_m are the components of ϕ_{AB} in the orthonormal basis η_{mAB}.

The Weyl spinor can now be classified according to the possible numbers of eigenvalues and eigenvectors of the matrix Ψ, Eq. (8-62). The maximum number of eigenvalues for the matrix Ψ is three. Corresponding to every eigenvalue there is at least one eigenvector. Accordingly, we obtain the following table:

Distinct eigenvectors		3			2		1
Distinct eigenvalues	3	2	1	2	1		1
Petrov type	I	D	0	II	N		III

Remark 1. In the following it will be shown that if there is only one distinct eigenvalue, then that eigenvalue is necessarily zero. Therefore, if there were three linearly independent eigenvectors corresponding to it, every vector of E_3 would also be an eigenvector. This is possible if and only if the Weyl spinor is identically zero.

Remark 2. Type I is also known as *algebraically general*, the others are known as *algebraically special*.[12]

Change of Frame

A change of the basis according to the transformation (8-34) induces changes in the various field components. Comparing Eqs. (8-31) and (8-61) shows that if the law of transformation of the vector χ is given by (8-39) then the law of transformation of the matrix Ψ should be given by

$$\Psi' = P\Psi P^t \tag{8-68}$$

One then can obtain the law of transformation for the dyad components $\psi_0, \ldots, \psi_4$ which is found to be

$$\begin{pmatrix} \psi'_0 \\ \psi'_1 \\ \psi'_2 \\ \psi'_3 \\ \psi'_4 \end{pmatrix} = \begin{pmatrix} a^4 & 4a^3b & 6a^2b^2 & 4ab^3 & b^4 \\ a^3c & a^2(3bc+ad) & 3ab(ad+bc) & b^2(3ad+bc) & b^3d \\ a^2c^2 & 2ac(ad+bc) & 1+6abcd & 2bd(ad+bc) & b^2d^2 \\ ac^3 & c^2(3ad+bc) & 3cd(ad+bc) & d^2(3bc+ad) & bd^3 \\ c^4 & 4c^3d & 6c^2d^2 & 4cd^3 & d^4 \end{pmatrix} \begin{pmatrix} \psi_0 \\ \psi_1 \\ \psi_2 \\ \psi_3 \\ \psi_4 \end{pmatrix} \tag{8-69}$$

[12] A Weyl spinor of type N is often called null.

Using Eq. (8-66) one finds the transformation law for the components χ_m, $m = 0, 1, \ldots, 4$. The result can be written in the form

$$\chi' = R\chi \tag{8-70}$$

The 5×5 complex matrix R is a function of the complex variables a, b, c, and d appearing as elements of the matrix $g \in SL(2, C)$. One can show that the matrix R is orthogonal and has a determinant unity (see Problem 8.11). The 5×5 matrices in Eqs. (8-69) and (8-70) give five-dimensional representations of the proper, orthochronous, homogeneous Lorentz group.

The transformation (8-69) can be applied for specific cases when the matrix $g \in SL(2, C)$ of Eq. (8-35) is taken as $g_1(z)$, $g_2(z)$, and $g_3(z)$. Under a null rotation $g_1(z)$ around l_μ, the dyad components $\psi_0, \ldots, \psi_4$ of the Weyl spinor transform into

$$\begin{aligned}
\psi'_0 &= \psi_0 \\
\psi'_1 &= z\psi_0 + \psi_1 \\
\psi'_2 &= z^2\psi_0 + 2z\psi_1 + \psi_2 \\
\psi'_3 &= z^3\psi_0 + 3z^2\psi_1 + 3z\psi_2 + \psi_3 \\
\psi'_4 &= z^4\psi_0 + 4z^3\psi_1 + 6z^2\psi_2 + 4z\psi_3 + \psi_4
\end{aligned} \tag{8-71}$$

$g_2(z)$ induces the transformation

$$\begin{aligned}
\psi'_0 &= z^4\psi_0 \\
\psi'_1 &= z^2\psi_1 \\
\psi'_2 &= \psi_2 \\
\psi'_3 &= z^{-2}\psi_3 \\
\psi'_4 &= z^{-4}\psi_4
\end{aligned} \tag{8-72}$$

whereas $g_3(z)$ induces the null rotation around n_α:

$$\begin{aligned}
\psi'_0 &= \psi_0 + 4z\psi_1 + 6z^2\psi_2 + 4z^3\psi_3 + z^4\psi_4 \\
\psi'_1 &= \psi_1 + 3z\psi_2 + 3z^2\psi_3 + z^3\psi_4 \\
\psi'_2 &= \psi_2 + 2z\psi_3 + z^2\psi_4 \\
\psi'_3 &= \psi_3 + z\psi_4 \\
\psi'_4 &= \psi_4
\end{aligned} \tag{8-73}$$

Invariants

By writing the components $\psi_0, \ldots, \psi_4$ in terms of $\chi_0, \ldots, \chi_4$ as expressed by Eq. (8-66), the matrix Ψ of Eq. (8-62) may be written in terms of the components of the Weyl tensor in the orthonormal basis as:

$$\Psi = 2^{1/2}\begin{pmatrix} -(2\chi_2/\sqrt{3}) & \chi_1 & -\chi_3 \\ \chi_1 & (\chi_2/\sqrt{3})+\chi_0 & \chi_4 \\ -\chi_3 & \chi_4 & (\chi_2/\sqrt{3})-\chi_0 \end{pmatrix} \tag{8-74}$$

As can be verified from the transformation law (8-68), the characteristics of the symmetric and traceless matrix (8-74) are independent of the spin frame.

Since the matrix (8-74) has a zero trace, we consider the invariant Tr Ψ^2 which is equal to

$$\operatorname{Tr}\Psi^2 = 4\chi\cdot\chi = 4\sum_{m=0}^{4}\chi_m\chi_m = \psi_{ABCD}\psi^{ABCD} \tag{8-75}$$

If the invariant Tr Ψ^2 vanishes, the Weyl tensor is null. Otherwise, it is non-null. Since an orthogonal transformation in E_5 does not necessarily represent a change of spin frame, there is another invariant. It is Tr Ψ^3,

$$\operatorname{Tr}\Psi^3 = \psi^{AB}{}_{CD}\psi^{CD}{}_{EF}\psi^{EF}{}_{AB} \tag{8-76}$$

Now the eigenvalues of Ψ satisfy the equation $|\Psi - \lambda I| = 0$, where I is the 3×3 unit matrix. This equation gives the cubic equation in λ:

$$f(\lambda) = \lambda^3 - \tfrac{1}{2}A\lambda - \tfrac{1}{3}B = 0 \tag{8-77}$$

where $A = 4\chi\cdot\chi$ and $B = 3\det\Psi$. By the Cayley–Hamilton theorem,

$$\Psi^3 - \tfrac{1}{2}A\Psi - \tfrac{1}{3}BI = 0$$

and hence Tr $\Psi^3 = B$. One also easily verifies that Tr Ψ^n, where $n = 4, 5, \ldots$, can be expressed in terms of A and B, and therefore there are no further independent invariants.

Let λ_1, λ_2 and λ_3 be the eigenvalues of Ψ (which may or may not be distinct). From Eq. (8-77) one then obtains

$$\begin{gathered}\lambda_1+\lambda_2+\lambda_3 = 0\\ -2(\lambda_1\lambda_2+\lambda_2\lambda_3+\lambda_3\lambda_1) = A\\ 3\lambda_1\lambda_2\lambda_3 = B\end{gathered} \tag{8-78}$$

Accordingly, if $\lambda_1 = \lambda_2 = \lambda_3$, then $\lambda_1 = \lambda_2 = \lambda_3 = 0$, and hence the two invariants A and B, vanish. This is the case of gravitational fields of types N and III. A gravitational field is of type II or D if two of the eigenvalues, let us say λ_1 and λ_2, are equal, $\lambda_1 = \lambda_2$, and $\lambda_3 \neq \lambda_1$. Equations (8-78) then show that $\lambda_1 = (A/6)^{1/2} = \lambda_2$, and $\lambda_3 = -(2A/3)^{1/2}$, and that $A^3 = 6B^2 \neq 0$. A Weyl tensor is of type I if and only if $\lambda_1 \neq \lambda_2 \neq \lambda_3$, and hence $A^3 \neq 6B^2$.

The classification given above is invariant under a change of frame. This is so since if χ is an eigenvector of Ψ with eigenvalue λ then, because of the transformation law (8-68), $P\chi$ is an eigenvector of Ψ' with the same eigenvalue. Conversely, if χ' is an eigenvector of Ψ', then $P^t\chi'$ is an eigenvector of Ψ with the same eigenvalue. One can also show that if two Weyl tensors can be transformed to each other by a change of a basis, then they must be of the same type and have the same eigenvalues. The converse is also correct: If Ψ and Ψ' are of the same type and have the same eigenvalues, they can be transformed one into the other. This result enables us to put a matrix Ψ, corresponding to a non-zero Weyl tensor, in one of five canonical forms by choosing the spin frame in an appropriate way. Thus, every element of the space E_5 can be put into one of five standard forms.

Canonical Forms

We can assume without loss of generality that $\psi_0 \neq 0$, since otherwise a transformation of the type (8-73) will allow us to make ψ_0 non-zero. Using Eq. (8-71) we now consider ψ'_4 as a quartic in z, ψ'_3 a cubic in z, etc., and notice that $\psi'_3, \psi'_2, \psi'_1$, and ψ'_0 are proportional to the first, second, third, and fourth derivative, respectively, of ψ'_4 with respect to z. If ψ'_4 has a double root we make ψ'_3 vanish simultaneously with ψ'_4 by choosing z to be this root. If ψ'_4 has a triple root we make $\psi'_4, \psi'_3, \psi'_2$ vanish simultaneously by choosing z to be this triple root. If ψ'_4 has a quadruple root, choosing z as this root will make $\psi'_4, \psi'_3, \psi'_2, \psi'_1$ zero. One then finds that a necessary and sufficient condition for ψ'_4 to have a quadruple or a triple root is $A = B = 0$. For one or two double roots the condition is $A^3 = 6B^2 \neq 0$, for no multiple roots it is $A^3 \neq 6B^2$.

After the transformation (8-71) has been performed, let us drop the primes from the components of the Weyl spinor. Equation (8-71) can now be followed up by a transformation of the type (8-73) describing a null rotation about n_α. If the quartic in (8-71) has a quadruple root, allowing us to make ψ_4, ψ_3, ψ_2 and ψ_1 vanish, no further transformation is necessary. The Weyl spinor is in the standard form

$$(\psi_0, 0, 0, 0, 0) \text{ with } \psi_0 \neq 0 \tag{8-79}$$

If the quartic had a triple root, so that $\psi_4 = \psi_3 = \psi_2 = 0$, $\psi_1 \neq 0$, then $\psi'_4 = \psi'_3 = \psi'_2 = 0$, $\psi'_1 = \psi_1$ and ψ'_0 can be made to vanish by choosing $z = -\psi_0/4\psi_1$, yielding the standard form

$$(0, \psi_1, 0, 0, 0) \text{ with } \psi_1 \neq 0 \tag{8-80}$$

If the quartic had a double root, so that $\psi_4 = \psi_3 = 0$, $\psi_2 \neq 0$, then $\psi'_4 = \psi'_3 = 0$, $\psi'_2 = \psi_2$. ψ'_0 is a quadratic in z and can be made to vanish by choosing z to be one of its roots. If this root is a double root ψ'_1 will vanish also for this choice of z, otherwise it will not. The former case occurs if the quartic had two double roots, the latter if it had one double root and two single ones. To show this is easy, but tedious. Thus we get, dropping the primes, the standard forms

$$(0, 0, \psi_2, 0, 0) \text{ with } \psi_2 \neq 0 \tag{8-81}$$

and

$$(0, \psi_1, \psi_2, 0, 0) \text{ with } \psi_1 \neq 0, \psi_2 \neq 0 \tag{8-82}$$

If the quartic has only single roots, then $A^3 \neq 6B^2$. This is also the condition that the quartic in (8-73) has no repeated roots. Hence only one of $\psi'_0, \psi'_1, \psi'_2, \psi'_3$ can be made to vanish by an appropriate choice of z. Thus we see that we can find a spin frame l_A, n_A which induces a corresponding basis (8-59) in E_5 such that the components of the given type I Weyl spinor take on the standard form,

$$(0, \psi_1, \psi_2, \psi_3, 0) \tag{8-83}$$

where ψ_1, ψ_2, ψ_3 are all non-zero and satisfy the condition

$$A^3 - 6B^2 = \tfrac{1}{2}\psi_1\psi_3(9\psi_2{}^2 - 16\psi_1\psi_3) \neq 0$$

The Weyl spinors (8-79) to (8-82) are, respectively, of types N, III, D, II, as can be seen by finding the corresponding matrices from (8-62) and calculating the eigenvectors and eigenvalues. The results are shown in the following table. Note that the algebraically special types are characterized by the existence of a null eigenbispinor. All the eigenbispinors of a type I Weyl spinor are non-null, as is best shown from the canonical form of the corresponding matrix (given in the sequel).

	Eigenvalues	Eigenbispinors	Type
(8-79)	0, 0, 0	$l_{(A}n_{B)}$, $n_A n_B$	N
(8-80)	0, 0, 0	$n_A n_B$	III
(8-81)	ψ_2, ψ_2	$l_A l_B$, $n_A n_B$	D
	$-2\psi_2$	$l_{(A}n_{B)}$	
(8-82)	ψ_2, ψ_2	$n_A n_B$	II
	$-2\psi_2$	$l_{(A}n_{B)} - \dfrac{1}{3}\dfrac{\psi_1}{\psi_2} n_A n_B$	

For each type of Weyl spinor we have found a standard form. For example, a Weyl spinor of type D can be put into the form (8-81) by an appropriate choice of spin-frame. Our method is essentially the matrix method; the matrix, however, was obtained using spinors. Corresponding to each type there is a canonical form of the matrix which is obtained by choosing the spin-frame appropriately. These canonical forms are listed below with the components of the Weyl spinor in the basis (8-59) also given:

$$\begin{pmatrix} 0 & 0 & 0 \\ 0 & 1 & i \\ 0 & i & -1 \end{pmatrix} \quad \text{for type N, } (2, 0, 0, 0, 0)$$

$$\begin{pmatrix} 0 & 1 & 0 \\ 1 & 0 & i \\ 0 & i & 0 \end{pmatrix} \quad \text{for type III, } \left(1, -\frac{i}{2}, 0, -\frac{i}{2}, -1\right)$$

$$\begin{pmatrix} -2\lambda_2 & 0 & 0 \\ 0 & \lambda_2 + 1 & i \\ 0 & i & \lambda_2 - 1 \end{pmatrix} \quad \text{for type II, } (2, 0, \lambda_2, 0, 0)$$

$$\begin{pmatrix} -2\lambda_1 & 0 & 0 \\ 0 & \lambda_1 & 0 \\ 0 & 0 & \lambda_1 \end{pmatrix} \quad \text{for type D, } (0, 0, \lambda_1, 0, 0)$$

$$\begin{pmatrix} -(\lambda_1 + \lambda_2) & 0 & 0 \\ 0 & \lambda_1 & 0 \\ 0 & 0 & \lambda_2 \end{pmatrix} \quad \text{for type I, } \left(\frac{\lambda_1 - \lambda_2}{2}, 0, \frac{\lambda_1 + \lambda_2}{2}, 0, \frac{\lambda_1 - \lambda_2}{2}\right)$$

λ_1 and λ_2 are eigenvalues.

The standard forms for each type can be obtained from the appropriate canonical form by a spin-frame transformation and vice versa. For instance, the type III Weyl spinor $(1, -i/2, 0, -i/2, -1)$ can be transformed into $(0, i/2, 0, 0, 0)$ by performing transformation (8-71) with $z = i$ followed by (8-73) with $z = i/2$. Conversely, performing (8-73) with $z = -\psi_1$ followed by (8-71) with $z = 1/2\psi_1$, followed by (8-72) with $z^2 = i/2\psi_1$ transforms $(0, \psi_1, 0, 0, 0)$ into $(1, -i/2, 0, -i/2, -1)$.

Spinor Method

The spinor method of classifying a Weyl tensor is analogous to that of classifying a bivector. The expression $\psi_{ABCD}\zeta^A\zeta^B\zeta^C\zeta^D$ can be written as a quartic polynomial in $C = \zeta^0/\zeta^1$:

$$\begin{aligned} \psi_{ABCD}\zeta^A\zeta^B\zeta^C\zeta^D &= (\zeta^1)^4[\psi_{0000}C^4 + 4\psi_{1000}C^3 + 6\psi_{1100}C^2 \\ &\quad + 4\psi_{0111}C + \psi_{1111}] \\ &= (\zeta^1)^4[\psi_0 C^4 + 4\psi_1 C^3 + 6\psi_2 C^2 + 4\psi_3 C + \psi_4] \end{aligned} \tag{8-84}$$

where the dyad components are taken with respect to some spin-frame. This quartic polynomial can be factored:

$$\begin{aligned} \psi_{ABCD}\zeta^A\zeta^B\zeta^C\zeta^D &= (\zeta^1)^4(\alpha_0 C + \alpha_1)(\beta_0 C + \beta_1)(\gamma_0 C + \gamma_1)(\delta_0 C + \delta_1) \\ &= \alpha_A\zeta^A\beta_B\zeta^B\gamma_C\zeta^C\,\delta_D\zeta^D \end{aligned}$$

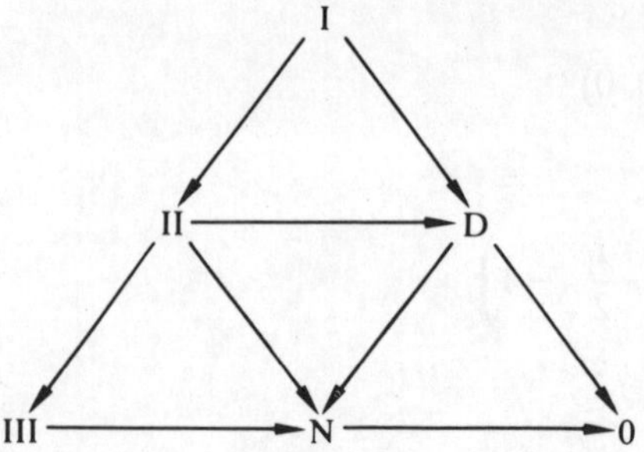

Figure 8.1 The Penrose diagram for classification of the gravitational field. An arrow A → B indicates that the type B field is obtained from the type A field.

where the spinors α_A, β_A, γ_A, δ_A are determined up to a constant of proportionality. Since ψ_{ABCD} is symmetric,

$$\psi_{ABCD} = \alpha_{(A}\beta_B\gamma_C\delta_{D)} \tag{8-85}$$

The directions of the null vectors corresponding to these spinors are called principal null directions. If two or more coincide, the Weyl spinor is said to be algebraically special; if they are distinct it is algebraically general. The Weyl spinor is said to be of type N if all four spinors coincide; of type III if three of the spinors coincide; of type II if two coincide with the remaining two distinct; of type D if the spinors coincide in pairs; of type I if the spinors are all distinct. This is usually expressed in the Penrose diagram (see Fig. 8.1), where A → B indicates that type B is obtained form type A by the confluence of two principal null directions. Type 0, a vanishing Weyl spinor, has been included here for completeness. That this way of classifying Weyl spinors is equivalent to the matrix method becomes clear when we note that the quartic polynomial appearing in (8-84) is precisely the one that was under discussion in the previous section.

Tensor Method

The tensor method depends on the fact that for every Weyl tensor there exist four null directions $\zeta_\mu \neq 0$, some of which may coincide, satisfying the equation

$$\zeta_{[\alpha}C^{+}{}_{\beta]\gamma\delta[\rho}\zeta_{\sigma]}\zeta^\gamma\zeta^\delta = 0 \tag{8-86}$$

A Weyl tensor is of Petrov type I if the four directions are distinct; of type II if two coincide with the remaining pair distinct; of type D if they coincide in pairs; of type III if three directions coincide; of type N if all four directions coincide. Equation (8-86) is equivalent to the spinor equation

$$\psi_{ABCD}\zeta^A\zeta^B\zeta^C\zeta^D = 0.$$

Our discussion following Eq. (8-84) shows that the tensor method is equivalent to the matrix and spinor methods.

PROBLEMS

8.1 Use Eq. (8-1) to show that the geometrical metric can be written as

$$g^{\mu\nu} = \tilde{\sigma}^{\mu}{}_{AB'}\tilde{\sigma}^{\nu AB'}$$

Also show that

$$\tilde{\sigma}_{\mu}{}^{A}{}_{C'}\tilde{\sigma}_{\nu}{}^{BC'} + \tilde{\sigma}_{\nu}{}^{A}{}_{C'}\tilde{\sigma}_{\mu}{}^{BC'} = g_{\mu\nu}\varepsilon^{AB}$$

8.2 Use the identity

$$\varepsilon_{AB}\varepsilon_{CD} + \varepsilon_{AC}\varepsilon_{DB} + \varepsilon_{AD}\varepsilon_{BC} = 0$$

to show that an arbitrary spinor with two indices, ξ^{AB}, will satisfy the equation

$$\xi_{AB} - \xi_{BA} = \varepsilon_{AB}\xi_{C}{}^{C}$$

where $\xi_{C}{}^{C} = \varepsilon^{CD}\xi_{CD}$.

8.3 Show that the spinor equivalent to the tensor

$$\varepsilon^{\alpha\beta}_{\mu\nu} = (-g)^{1/2}\varepsilon_{\rho\sigma\mu\nu}g^{\alpha\rho}g^{\beta\sigma}$$

is given by

$$\varepsilon^{AB'CD'}_{EF'GH'} = i(\delta^{A}_{E}\delta^{C}_{G}\delta^{B'}_{H'}\delta^{D'}_{F'} - \delta^{A}_{G}\delta^{C}_{E}\delta^{B'}_{F'}\delta^{D'}_{H'})$$

[See R. Penrose, *Ann. Phys.* (*N.Y.*) **10,** 171 (1960).]

8.4 The tensor $*F_{\mu\nu}$, defined by

$$*F_{\mu\nu} = \tfrac{1}{2}(-g)^{1/2}F^{\rho\sigma}\varepsilon_{\mu\nu\rho\sigma}$$

is called the *dual* to the tensor $F_{\mu\nu}$. Show that if the spinor equivalent of $F_{\mu\nu}$ is given by Eq. (8-12), then its dual can be given as

$$*F_{AB'CD'} = i(\varepsilon_{AC}\bar{\phi}_{B'D'} - \phi_{AC}\varepsilon_{B'D'})$$

8.5 Prove Eqs. (8-14) to (8-16).

8.6 Show that the function $\lambda = \chi_{AB}{}^{AB}$ is real.

8.7 The *right dual* tensor of the Riemann tensor is defined by

$$S_{\mu\nu\rho\sigma} = \tfrac{1}{2}(-g)^{1/2}R_{\mu\nu}{}^{\alpha\beta}\varepsilon_{\alpha\beta\rho\sigma}$$

Find its spinor equivalent in terms of the spinors χ_{ABCD} and $\phi_{ABC'D'}$.

8.8 Show that the Bianchi identity,

$$\nabla_{\alpha}R_{\mu\nu\beta\gamma} + \nabla_{\beta}R_{\mu\nu\gamma\alpha} + \nabla_{\gamma}R_{\mu\nu\alpha\beta} = 0$$

is equivalent to the equation

$$\nabla_{DE'}\chi_{ABC}{}^{D} = \nabla_{CF'}\phi_{ABE'}{}^{F'}$$

[R. Penrose, *Ann. Phys.* (*N.Y.*) **10,** 171 (1960).]

8.9 Show that the eigenvalue equation (8-26) corresponds to the tensor equation

$$F^{+}_{\mu\nu}\alpha^{\nu} = -2\lambda\alpha_{\mu}$$

or equivalently to

$$F_{\mu\nu}\alpha^{\nu} = -2(\text{Re }\lambda)\alpha_{\mu}$$

where the vector α_{μ} is null and given by $\alpha^{\mu} = \tilde{\sigma}^{\mu}_{AB'}\alpha^{A}\bar{\alpha}^{B'}$.

8.10 Show that the eigenvalue equation (8-58) corresponds to the tensor equation

$$C^{+}{}_{\alpha\beta\gamma\delta}F^{+\gamma\delta} = 4\lambda F^{+}{}_{\alpha\beta}$$

or the equation

$$C_{\alpha\beta\gamma\delta}F^{+\gamma\delta} = 2\lambda F^{+}{}_{\alpha\beta}$$

where $F^{+}{}_{\alpha\beta}$ and ϕ_{AB} are related by Eq. (8-25).

8.11 Find the matrix R of Eq. (8-70); show that it is orthogonal and its determinant equals $+1$.

8.12 Show that

$$\begin{aligned}\mathrm{Tr}\,\Psi^3 &= 3\det\Psi\\ &= 3(\chi_0\chi_1^2 + 2\chi_0^2\chi_2/\sqrt{3} - \chi_0\chi_3^2\\ &\quad -\chi_1^2\chi_2/\sqrt{3} - 2\chi_1\chi_3\chi_4 - 2\chi_2^3/3\sqrt{3}\\ &\quad -\chi_2\chi_3^2/\sqrt{3} + 2\chi_2\chi_4^2/\sqrt{3})/\sqrt{2}\end{aligned}$$

8.13 Show that the three scalar components of the Maxwell field ϕ_0, ϕ_1, and ϕ_2 given by Eq. (8-28) can be obtained directly from the Maxwell tensor $F_{\mu\nu}$ by use of the tetrad of vectors (8-40). Show that

$$\phi_0 = F_{\mu\nu}l^\mu m^\nu$$

$$\phi_1 = \tfrac{1}{2}F_{\mu\nu}(l^\mu n^\nu + \bar{m}^\mu m^\nu)$$

$$\phi_2 = F_{\mu\nu}\bar{m}^\mu n^\nu$$

8.14 Show that the five scalar components of the Weyl spinor $\psi_0, \ldots, \psi_4$ can be obtained directly from the Weyl tensor $C_{\alpha\beta\gamma\delta}$ by

$$\psi_0 = -C_{\mu\nu\rho\sigma}l^\mu m^\nu l^\rho m^\sigma$$

$$\psi_1 = -C_{\mu\nu\rho\sigma}l^\mu n^\nu l^\rho m^\sigma$$

$$\psi_2 = -\tfrac{1}{2}C_{\mu\nu\rho\sigma}(l^\mu n^\nu l^\rho n^\sigma - l^\mu n^\nu m^\rho \bar{m}^\sigma)$$

$$\psi_3 = -C_{\mu\nu\rho\sigma}\bar{m}^\mu n^\nu l^\rho n^\sigma$$

$$\psi_4 = -C_{\mu\nu\rho\sigma}\bar{m}^\mu n^\nu \bar{m}^\rho n^\sigma$$

8.15 Show that Maxwell's field equations with a magnetic monopole can be written in the form

$$\nabla_\nu F^{\mu\nu} = ej^\mu$$

$$\nabla_\nu {}^*F^{\mu\nu} = gj^\mu$$

where $^*F$ is the dual to F (see Problem 8.4). Use now the following dual rotations

$$\tilde{F}_{\mu\nu} = F_{\mu\nu}\cos\alpha + {}^*F_{\mu\nu}\sin\alpha$$

$$h = e\cos\alpha + g\sin\alpha$$

$$0 = -e\sin\alpha + g\cos\alpha$$

in order to show that Maxwell's equations can now be written as

$$\nabla_\nu \tilde{F}^{\mu\nu} = hj^\mu$$

$$\nabla_\nu {}^*\tilde{F}^{\mu\nu} = 0.$$

Show that the above duality transformation of the Maxwell field is equivalent to

$$\tilde{\phi}_{AB} = \phi_{AB}\, e^{-i\alpha}$$

where ϕ_{AB} is the Maxwell spinor. [See R. Penrose, *Ann. Phys.* **10**, 171 (1960)].

CHAPTER

NINE

$SL(2, C)$ GAUGE THEORY OF THE GRAVITATIONAL FIELD: THE NEWMAN–PENROSE EQUATIONS

After discussing the use of spinors in general relativity theory in the last chapter, we are now in a position to formulate the gravitational field as a gauge theory. The theory of isotopic spin and gauge fields is hence discussed. This includes the description of the electromagnetic and the Yang–Mills fields. The method of Utiyama, generalizing the Yang–Mills field into gravitation by use of the homogeneous Lorentz group instead of the group SU_2, is consequently described. Finally the group $SL(2, C)$ is employed. A scheme is given according to which the spin coefficients of the gravitational field and the Riemann tensor are presented in the form of linear combinations of the infinitesimal generators of the group $SL(2, C)$ similar to the way Yang and Mills write their dynamical variables in terms of the generators of the group SU_2. In this approach the spin coefficients take the role of the Yang–Mills potentials and the Riemann tensor takes the role of the fields. The gravitational field equations of general relativity are then derived using the variational principle. The field equations obtained are the familiar ones known as the Newman–Penrose equations.

9-1 ISOTOPIC SPIN AND GAUGE FIELDS

In ordinary gauge invariance of a charged field which is described by a complex wave function ψ, a change of *gauge*[1] means a change of *phase factor* $\psi \rightarrow \psi'$, $\psi' = (\exp i\alpha)\psi$, a change that is devoid of any physical consequences. Since ψ

[1] H. Weyl, *The Theory of Groups and Quantum Mechanics*, Dover, New York, 1931.

depends on space–time points, the relative phase factor of ψ at two different points is completely arbitrary and α is, accordingly, a function of space–time. In other words, the arbitrariness in choosing the phase factor is local in character.

To preserve invariance it is then necessary to counteract the variation of the phase α with space–time coordinates by introducing the electromagnetic potentials $A_\mu(x)$ which change under a gauge transformation as

$$A'_\mu = A_\mu + \frac{1}{e}\frac{\partial\alpha}{\partial x^\mu}$$

and to replace the derivative of ψ by a "covariant derivative" with the combination $(\partial_\mu - ieA_\mu)\psi$.

Isotopic Spin

Historically, an *isotopic spin* parameter was first introduced by Heisenberg[2] in 1932 to describe the two charge states, namely neutron and proton, of a nucleon. The idea that the neutron and proton correspond to two states of the same particle was suggested at the time by the fact that their masses are nearly equal, and that the light stable even nuclei contain equal numbers of them. Later on it was pointed out that the p–p and n–p interactions are approximately equal in the ^1S state.[3, 4] and consequently it was assumed that the equality holds also in the other states available to both the n–p and p–p systems. Under such an assumption one arrives at the concept of a *total* isotopic spin[5] which is conserved in nucleon–nucleon interactions. Experiments on the energy levels of light nuclei strongly suggest that this assumption is indeed correct.[6] This implies that all strong interactions, such as the pion–nucleon interaction, should also satisfy the same conservation law. This, and the fact that there are three charge states for the pion can be coupled to the nucleon field singly, lead to the conclusion that pions have isotopic spin unity. A verification of this conclusion was found in experiments which compare the differential cross-section of the process $n + p \to \pi^0 + d$ with that of the previously measured process $p + p \to \pi^+ + d$.[7]

Conservation of Isotopic Spin and Invariance

The conservation of isotopic spin is identical with the requirement that all interactions be invariant under isotopic spin rotation, when electromagnetic interactions are neglected. This means that the orientation of the isotopic spin has no physical

[2] W. Heisenberg, *Z. Phys.* **77,** 1 (1932).

[3] G. Breit, E. Condon, and R. Present, *Phys. Rev.* **50,** 825 (1936).

[4] J. Schwinger [*Phys. Rev.* **78,** 135 (1950)] pointed out that the small difference may be attributed to magnetic interactions.

[5] The total isotopic spin was first introduced by E. Wigner, *Phys. Rev.* **51,** 106 (1937); B. Cassen and E. Condon, *Phys. Rev.* **50,** 846 (1936).

[6] T. Lauritsen, *Ann. Rev. Nucl. Sci.* **1,** 67 (1952); D. R. Inglis, *Rev. Mod. Phys.* **25,** 390 (1953).

[7] R. H. Hildebrand, *Phys. Rev.* **89,** 1090 (1953).

significance. Differentiation between a neutron and a proton is then an arbitrary process. This arbitrariness is subject to the limitation that once one chooses what to call a proton and what to call a neutron at one space–time point, one is then not free to make any other choices at other space–time points. It also seems not to be consistent with the localized field concept which underlies the usual physical theories.

Isotopic Spin and Gauge Fields

The possibility of requiring that all interactions be invariant under *independent* rotations of the isotopic spin at all space–time points, so that the relative orientation of the isotopic spin at two space–time points becomes physically meaningless, was accordingly explored by Yang and Mills.[8] They introduced *isotopic gauge* as an arbitrary way of choosing the orientation of the isotopic spin axes at all space–time points, in analogy with the electromagnetic gauge which represents an arbitrary way of choosing the complex phase factor of a charged field at all space–time points. This suggests that all physical processes, which do not involve the electromagnetic field, be invariant under the isotopic gauge transformation $\psi \to \psi'$, $\psi' = S^{-1}\psi$, where S represents a space–time dependent isotopic spin rotation which is a 2×2 unitary matrix with determinant unity, i.e., an element of the group SU_2 discussed in Chapter 1.

In an entirely similar manner to what is done in electrodynamics, Yang and Mills introduced a B-field in the case of the isotopic gauge transformation to counteract the dependence of the matrix S on the space–time coordinates. Accordingly, and in analogy with the electromagnetic case, all derivatives of the wave function ψ describing a field with isotopic spin $\frac{1}{2}$ should appear as "covariant derivatives" of the form $(\partial_\mu - iB_\mu)\psi$, where B_μ are four 2×2 Hermitian matrices. The field equations satisfied by the twelve independent components of the B-field, which is called the **b** field, and their interaction with any field having an isotopic spin, are fixed just as in the electromagnetic case.

Isotopic Gauge Transformation

Under an isotopic gauge transformation, a two-components wave function ψ describing a field with isotopic spin $\frac{1}{2}$ transforms according to

$$\psi = S\psi' \tag{9-1}$$

Invariance then requires that the covariant derivative expression transforms as $S(\partial_\mu - iB'_\mu)\psi' = (\partial_\mu - iB_\mu)\psi$. When combined with Eq. (9-1), we obtain the isotopic gauge transformation of the 2×2 potential matrix B_μ:

$$B'_\mu = S^{-1}B_\mu S + iS^{-1}\,\partial_\mu S \tag{9-2}$$

[8] C. N. Yang and R. L. Mills, *Phys. Rev.* **96,** 191 (1954).

In analogy to the procedure of obtaining gauge invariant field strengths in the electromagnetic case, Yang and Mills define their field as

$$F_{\mu\nu} = \partial_\nu B_\mu - \partial_\mu B_\nu + i[B_\mu, B_\nu] \tag{9-3}$$

where the commutator $[B_\mu, B_\nu] = B_\mu B_\nu - B_\nu B_\mu$. Under the transformation (9-1) the 2×2 field matrix (9-3) transforms as

$$F'_{\mu\nu} = S^{-1} F_{\mu\nu} S \tag{9-4}$$

Now Eq. (9-2) is valid for any S and its corresponding B_μ. Furthermore, the matrix $S^{-1}\ \partial S/\partial x^\mu$ appearing in Eq. (9-2) is a linear combination of the isotopic spin "angular momentum" matrices T^i, $i = 1, 2, 3$, corresponding to the isotopic spin on the field ψ under consideration. Here $T^i = \frac{1}{2}\sigma^i$, where σ^i are the three Pauli spin matrices given by Eq. (1-5). Accordingly, the matrix B_μ itself must also contain a linear combination of the matrices T^i; any part of B_μ in addition to this, denote it by $\tilde{B}_\mu$, is a scalar or tensor combination of the T's, and must transform by the homogenous part of (9-2), $\tilde{B}'_\mu = S^{-1}\tilde{B}_\mu S$. Such a field is extraneous and was allowed by the very general form we took for the B-field, but is irrevelevant to the question of isotopic gauge. Therefore, the relevant part of the B-field can be written as a linear combination of the matrices T^i:

$$B_\mu = 2\mathbf{b}_\mu \cdot \mathbf{T} = \mathbf{b}_\mu \boldsymbol{\sigma} \tag{9-5}$$

where bold-face letters denote 3-component vectors in the isotopic space.

The isotopic-gauge covariant field matrices $F_{\mu\nu}$ can also be expressed as a linear combination of the T's. One obtains

$$F_{\mu\nu} = 2\mathbf{f}_{\mu\nu} \cdot \mathbf{T} = \mathbf{f}_{\mu\nu} \cdot \boldsymbol{\sigma} \tag{9-6}$$

where

$$\mathbf{f}_{\mu\nu} = \frac{\partial \mathbf{b}_\mu}{\partial x^\nu} - \frac{\partial \mathbf{b}_\nu}{\partial x^\mu} + 2\mathbf{b}_\mu \times \mathbf{b}_\nu \tag{9-7}$$

One notices that $\mathbf{f}_{\mu\nu}$ transforms like a vector under an isotopic gauge transformation. The corresponding transformation of $\mathbf{b}_\mu$ is cumbersome. Under *infinitesimal* isotopic gauge transformations, $S = 1 - i\boldsymbol{\sigma} \cdot \boldsymbol{\delta\omega}$. Then

$$\mathbf{b}'_\mu = \mathbf{b}_\mu - 2\mathbf{b}_\mu \times \boldsymbol{\delta\omega} + \partial\boldsymbol{\delta\omega}/\partial x^\mu \tag{9-8}$$

Field Equations

In analogy to the electromagnetic case one can write down an isotopic gauge invariant Lagrangian density:

$$\mathcal{L} = -\tfrac{1}{8}\mathrm{Tr}\, F_{\mu\nu} F^{\mu\nu} = -\tfrac{1}{4}\mathbf{f}_{\mu\nu} \cdot \mathbf{f}^{\mu\nu} \tag{9-9}$$

One can also include a field with isotopic spin $\frac{1}{2}$ to obtain the following total Lagrangian density:

$$\mathcal{L} = -\tfrac{1}{8}\,\mathrm{Tr}\, F_{\mu\nu} F^{\mu\nu} - \bar{\psi}\gamma_\mu(\partial_\mu - iB_\mu)\psi - m\bar{\psi}\psi \tag{9-10}$$

The equations of motion obtained from the Lagrangian (9-10) are:[9]

$$\partial \mathbf{f}_{\mu\nu}/\partial x^{\nu} - 2(\mathbf{b}_{\nu} \times \mathbf{f}_{\mu\nu}) + \mathbf{J}_{\mu} = 0 \tag{9-11}$$

$$\gamma_{\mu}(\partial_{\mu} - i\boldsymbol{\sigma} \cdot \mathbf{b}_{\mu})\psi + m\psi = 0 \tag{9-12}$$

where

$$\mathbf{J}_{\mu} = i\bar{\psi}\gamma_{\mu}\,\boldsymbol{\sigma}\psi \tag{9-13}$$

Since the divergence of $\mathbf{J}_{\mu}$ does not vanish, one may define

$$\mathscr{J}_{\mu} = \mathbf{J}_{\mu} - 2\mathbf{b}_{\nu} \times \mathbf{f}_{\mu\nu} \tag{9-14}$$

which leads to the *equation of continuity*.

$$\partial \mathscr{J}_{\mu}/\partial x^{\mu} = 0 \tag{9-15}$$

Equation (9-15) guarantees that the total isotopic spin

$$T = \int \mathscr{J}_0 \, \mathrm{d}^3 x \tag{9-16}$$

is independent of time and Lorentz transformation.

Nonlinearity of the Field Equations

Equation (9-14) shows that the isotopic spin arises from both the spin-$\frac{1}{2}$ field $\mathbf{J}_{\mu}$ and from the $\mathbf{b}_{\mu}$ field itself. This fact makes the field equations for the B-field *nonlinear*, even in the absence of the spin-$\frac{1}{2}$ field. The situation here is different from that of the electromagnetic case whose field is chargeless, and hence satisfies linear equations.

Internal Holonomy Group of Gauge Fields

Group-theoretical considerations concerning the gauge theory discussed above involved so far the gauge group SU_2, only. In addition to the possibility of generalizing the gauge group to other groups,[10] there exists another group which is defined by the potential B_{μ}.[11] One arrives at this group by the observation that the potential can be used to define "parallel displacement" of multiplets ψ at neighbouring points in the same way as the Christoffel symbols define parallelity of vectors in Riemann space (see Chap. 7). By making a parallel displacement of

[9] The equations of motion (9-12) and (9-13) can be completed by the supplementary condition $\partial \mathbf{b}_{\mu}/\partial x^{\mu} = 0$ which serves to eliminate the scalar part of the field in $\mathbf{b}_{\mu}$. This imposes a condition on the possible isotopic gauge transformations. The infinitesimal isotopic gauge transformation $S = 1 - i\boldsymbol{\sigma} \cdot \delta\boldsymbol{\omega}$ must satisfy the condition:

$$2\mathbf{b}_{\mu} \times \partial\, \delta\boldsymbol{\omega}/\partial x^{\mu} + \partial^2\, \delta\boldsymbol{\omega}/\partial x^{\mu}\, \partial x_{\nu} = 0$$

This equation is the analog of the equation $\partial^2 \alpha/\partial x^{\mu}\, \partial x_{\nu} = 0$ that must be satisfied by the gauge transformation $A'_{\mu} = A_{\mu} + e^{-1}(\partial\alpha/\partial x^{\mu})$ of the electromagnetic field.

[10] S. L. Glashow and M. Gell-Mann, *Ann. Phys. (N.Y.)* **15,** 437 (1961).

[11] H. G. Loos, *J. Math. Phys.* **8,** 2114 (1967).

multiplets around a closed curve in space–time, one obtains a linear transformation of multiplets. Doing this for all closed curves passing through a given point x^μ results in a continuous set of linear multiplet transformations at x^μ. This set turns out to be a Lie group, called the *internal holonomy* group, in analogy to the ordinary holonomy group in the Riemann space.[12] To see this we proceed as follows.

We call the multiplets $\psi(x^\mu)$ and $\psi(x^\mu + dx^\mu)$ equivalent if

$$\psi(x^\mu + \mathrm{d}x^\mu) - \psi(x^\mu) = B_\alpha(x^\mu)\psi(x^\mu)\,\mathrm{d}x^\alpha \tag{9-17}$$

This equivalence relation, which is invariant under the gauge transformation (9-1) because of Eq. (9-2), can be used to execute an equivalence displacement of multiplets ψ along a curve in space–time. The question is whether such equivalence displacement is path-dependent. If it is not, we have

$$\nabla_\kappa \psi = \partial_\kappa \psi - iB_\kappa \psi = 0 \tag{9-18}$$

which leads to the integrability condition

$$\nabla_{[\lambda}\nabla_{\kappa]}\psi = -\tfrac{1}{2}F_{\lambda\kappa}\psi = 0 \tag{9-19}$$

Since this is true everywhere for linearly independent internal vectors ψ, it follows that $F_{\lambda\kappa} = 0$. Accordingly, in order to have nonvanishing gauge fields the equivalence transport of ψ must be *path-dependent*. In the same way one can show that if the field $F_{\mu\nu}$ vanishes everywhere, then the potential B_μ can be transformed away by a gauge transformation (see Problem 9.5).

Now let C be a closed, piecewise, continuously differentiable, and has a sense of circumscription. Taking a multiplet ψ around C by equivalence displacement results in a linear transformation $H(C)$:

$$\psi' = H(C)\psi \tag{9-20}$$

Doing this for all closed curves C through a point x one gets a set $\mathcal{H}$ of linear internal transformations. The inverse of $H(C)$ is produced by equivalence displacement around C in the opposite direction. The composition $H(C_2)H(C_1)$ is the element $H(C_1 + C_2)$, where $C_1 + C_2$ describes the loop consisting of C_1 and C_2. Hence $\mathcal{H}$ is a group which is a subgroup of the full complex linear group $GL(n, c)$. It is a connected Lie group. This is the *internal holonomy group.* It follows that the internal holonomy groups at different points are isometric (Problem 9.6).

Finally, an interesting result is obtained if we consider the relation between the internal homonomy group $\mathcal{H}$ and the gauge group $\mathcal{G}$. We will leave it to the reader (Problem 9.7) to show that compatibility of the groups $\mathcal{H}$ and $\mathcal{G}$ requires that either (1) $\mathcal{H}$ be a subgroup of $\mathcal{G}$, or (2) $\mathcal{G}$ be an invariant subgroup of $\mathcal{H}$.

[12] J. F. Schell, *J. Math. Phys.* **2**, 202 (1961); J. N. Goldberg and R. P. Kerr, *ibid.*, 327 and 332 (1961).

9-2 LORENTZ INVARIANCE AND THE GRAVITATIONAL FIELD

In the previous section we saw that the existence of the electromagnetic field and the Yang–Mills field can be related to invariance properties. Thus, if the Lagrangian density is invariant under phase transformations $\psi \to (\exp i\alpha)\psi$, and if we wish to make it invariant under the general gauge transformations for which α is a function of x, then it is necessary to introduce a new field A_μ which transforms according to $A_\mu \to A_\mu + e^{-1}\,\partial_\mu \alpha$, and to replace the derivative of ψ by a "covariant derivative" $(\partial_\mu - ieA_\mu)\psi$. A similar argument was applied by Yang and Mills to isotopic spin rotations, to yield a triplet of vector fields. It is thus an attractive idea to relate the existence of the gravitational field to Lorentz invariance.

Homogeneous Lorentz Group and the Gravitational Field

Utiyama[13] has proposed a method which leads to the introduction of 24 field variables $A^{ij}{}_\mu$ by considering the homogeneous Lorentz transformations of the Lorentz group specified by six parameters. One starts by assuring that the action integral

$$I = \int L(\psi^A, \psi^A{}_{,k})\, \mathrm{d}^4x \tag{9-21}$$

where $\psi_{,k} = \partial_k \psi$, is invariant under Lorentz transformations. Besides the x-system one introduces an arbitrary system of curvilinear coordinates u^μ. We will use Latin and Greek indices to represent quantities defined with respect to the x-system (local Lorentz frame) and the u-system respectively. The square of the invariant length of the infinitesimal line element is given by

$$\mathrm{d}s^2 = \eta_{ik}\, \mathrm{d}x^i\, \mathrm{d}x^k = g_{\mu\nu}\, \mathrm{d}u^\mu\, \mathrm{d}u^\nu$$

where η_{ik} is the Minkowskian metric and $g_{\mu\nu}(u) = (\partial x^i/\partial u^\mu)(\partial x^k/\partial u^\nu)\eta_{ik}$. Defining the functions

$$h^k{}_\mu(u) = \partial x^k/\partial u^\mu, \qquad h_k{}^\mu(u) = \partial u^\mu/\partial x^k \tag{9-22}$$

then gives $\eta_{kl}\, h^k{}_\mu\, h^l{}_\nu = g_{\mu\nu}(u)$, $g_{\mu\nu}\, h_k{}^\mu h_l{}^\nu = \eta_{kl}$, $h_k{}^\mu h^l{}_\mu = \delta^l_k$, $h_k{}^\mu h^k{}_\nu = \delta^\mu_\nu$, and $\det g_{\mu\nu} = g = -h^2 = -(\det h^k{}_\mu)^2$. Raising or lowering both kind of indices is made by means of the metrics $g^{\mu\nu}$, $g_{\mu\nu}$ or η_{kl} and η^{kl}. Under a Lorentz transformation $x^k \to x^k + \varepsilon^k{}_l\, x^l$, where $\varepsilon^{kl} = -\varepsilon^{lk}$ are infinitesimal parameters, one has $h_k{}^\mu \to h_k{}^\mu + \delta h_k{}^\mu$, with $\delta h_k{}^\mu = -\varepsilon^l{}_k\, h_l{}^\mu$. Using the h functions we can transform a world tensor into a corresponding local tensor defined with respect to the local frame, and vice versa. For example, $\psi^k(u) = h^k{}_\mu(u)\psi^\mu(u)$ and $\psi^\mu(u) = h_k{}^\mu(u)\psi^k(u)$, where $\psi^k(u) = \psi^k(x(u))$.

Accordingly, we can rewrite the action integral as

$$I = \int \mathfrak{L}\,(\psi^A(u), \psi^A{}_{,\mu}(u), h^k{}_\mu(u))\, \mathrm{d}^4u \tag{9-23}$$

where $\mathfrak{L} = L(\psi^A(u), h_k{}^\mu(u)\psi^A{}_{,\mu}(u))h$, and $\psi^A{}_{,\mu} = \partial\psi^A(u)/\partial u^\mu$.

[13] R. Utiyama, *Phys. Rev.* **101**, 1597 (1956).

Invariance of the Action Integral

The action integral I is invariant under: (1) the Lorentz transformation which yields

$$\delta h^k{}_\mu = \varepsilon^k{}_l h^l{}_\mu \tag{9-24a}$$

$$\delta\psi^A = \tfrac{1}{2} T_{(kl)}{}^A{}_B \psi^B \varepsilon^{kl} \tag{9-24b}$$

where u^μ is unchanged, and $T_{(kl)}{}^A{}_B$ is the AB matrix element of the infinitesimal generator of the Lorentz group. The matrix $T_{(kl)}$ satisfies

$$[T_{(kl)}, T_{(mn)}] = \tfrac{1}{2} f_{kl}{}^{ab}{}_{mn} T_{(ab)}, \qquad T_{(kl)} = -T_{(lk)}$$

(2) the general point transformation

$$u^\mu \to u^\mu + \lambda^\mu(u)$$

where $\lambda^\mu(u)$ is an arbitrary function of u, which yields

$$\delta h^k{}_\mu = -(\partial\lambda^\nu/\partial u^\mu) h^k{}_\nu \tag{9-25a}$$

$$\delta\psi^A(u) = \psi'^A(u') - \psi^A(u) = 0 \tag{9-25b}$$

$$\delta\psi^A{}_{,\mu} = -(\partial\lambda^\nu/\partial u^\mu)\psi^A{}_{,\nu} \tag{9-25c}$$

In the following the set $h^k{}_\mu$ shall be considered as 16 independent given functions.

Generalized Lorentz Transformation

We now generalize the Lorentz transformation into one in which ε^{ik} are replaced by arbitrary functions $\varepsilon^{ik}(u)$. Under this "generalized Lorentz transformation" we assume that ψ^A and $h^k{}_\mu$ transform as

$$\delta\psi^A = \tfrac{1}{2} A^{kl}{}_\mu T_{(kl)}{}^A{}_B \psi^B \tag{9-26a}$$

$$\delta h^k{}_\mu = \varepsilon^k{}_l(u) h^l{}_\mu \tag{9-26b}$$

Then in order that I remains invariant under (9-26), it is necessary to introduce a new field $A^{kl}{}_\mu(u) = -A^{lk}{}_\mu(u)$ with the following transformation law:

$$\delta A^{kl}{}_\mu = \varepsilon^k{}_m A^{ml}{}_\mu + \varepsilon^l{}_m A^{km}{}_\mu + \partial\varepsilon^{kl}/\partial u^\mu \tag{9-27}$$

The new Lagrangian density is then given by

$$\mathfrak{L}(\psi^A, \nabla_\mu \psi^A, h^k{}_\mu) = hL(\psi^A, h_k{}^\mu \nabla_\mu \psi^A) \tag{9-28}$$

where

$$\nabla_\mu \psi^A = \partial\psi^A/\partial u^\mu - \tfrac{1}{2} A^{kl}{}_\mu T_{(kl)}{}^A{}_B \psi^B \tag{9-29}$$

We now take as our basic space–time, some Riemannian space whose metric is $g_{\mu\nu}(u) = h^k{}_\mu h_{k\nu}$ and whose affine connection is $\Gamma^\lambda{}_{\mu\nu} = \tfrac{1}{2} g^{\lambda\sigma}(\partial g_{\sigma\mu}/\partial u^\nu + \partial g_{\nu\sigma}/\partial u^\mu - \partial g_{\mu\nu}/\partial u^\sigma)$. In order to obtain the relationship between $A^{kl}{}_\mu$ and $h^k{}_\mu$, let us take for ψ^A the local tensor ψ^{kl}. Then from Eq. (9-20) we obtain

$$\nabla_\mu \psi^{kl} = \partial\psi^{kl}/\partial u^\mu - A^{km}{}_\mu \psi_m{}^l - A^{lm}{}_\mu \psi^k{}_m$$

Accordingly, replacing ψ^{kl} by $\psi^{kv} = h_m{}^v \psi^{km}$, we obtain

$$\nabla_\mu \psi^{kv} = \partial\psi^{kv}/\partial u^\mu - A^{km}{}_\mu \psi_m{}^v + \Gamma'{}_{\rho\mu}{}^v \psi^{k\rho} \tag{9-30}$$

where

$$\Gamma'{}_{\mu\nu}{}^\rho = h_l{}^\rho(\partial h^l{}_\nu/\partial u^\mu) - h_k{}^\rho h_{l\nu} A^{kl}{}_\mu \tag{9-31}$$

Equation (9-30) shows that the covariant derivative obtained here is the usual one. where for Greek indices the Γ' appear instead of the usual affinity Γ, and for Latin indices the $A^{kl}{}_\mu$ must be inserted instead of Γ. The relation (9-30) can be generalized (see Problem 9.8). One also finds, under the *ad hoc* assumption that Γ' is symmetric in its lower two indices (see Problem 9.9), that

$$\Gamma'^\rho_{\mu\nu} = \frac{1}{2} g^{\rho\sigma}\left(\frac{\partial g_{\sigma\mu}}{\partial u^\nu} + \frac{\partial g_{\nu\sigma}}{\partial u^\mu} - \frac{\partial g_{\mu\nu}}{\partial u^\sigma}\right) = \Gamma^\rho_{\mu\nu} \tag{9-32}$$

and

$$h_l{}^\rho \frac{\partial h^l{}_\nu}{\partial u^\mu} - A^\rho{}_{\nu\mu} = \Gamma^\rho{}_{\nu\mu} \tag{9-33}$$

where $A^\rho{}_{\nu\mu} = h_k{}^\rho h_{l\nu} A^{kl}{}_\mu$.

Accordingly, we have obtained a general expression for the covariant derivative. For example, if we take for the field ψ^A the spinor field ψ, we obtain

$$\nabla_\mu \psi = \partial\psi/\partial x^\mu - (i/4)A^{kl}{}_\mu[\gamma_k, \gamma_l]\psi$$

where γ_k are the usual Dirac γ matrices.

Free Field Case

Let us now consider the Lagrangian density $\mathfrak{L}_0$ for the free field, i.e., the case without the multiplet ψ^A. The Lagrangian density is a function of the functions h and A, $\mathfrak{L}_0(h^k{}_\mu, A^{kl}{}_\mu, \partial A^{kl}{}_\mu/\partial u^\nu)$. Since $\mathfrak{L}_0$ should be invariant under the "generalized Lorentz transformation", it follows that $\mathfrak{L}_0$ must depend on the field A through the form $\mathfrak{L}_0(h^k{}_\mu, F^{kl}{}_{\mu\nu})$, where F is defined by

$$\begin{aligned} F^{kl}{}_{\mu\nu} &= \frac{\partial A^{kl}{}_\nu}{\partial u^\mu} - \frac{\partial A^{kl}{}_\mu}{\partial u^\nu} - \frac{1}{4} f_{ab}{}^{kl}{}_{mn}(A^{ab}{}_\mu A^{mn}{}_\nu - A^{ab}{}_\nu A^{mn}{}_\mu) \\ &= \frac{\partial A^{kl}{}_\nu}{\partial u^\mu} - \frac{\partial A^{kl}{}_\mu}{\partial u^\nu} + A^{kb}{}_\mu A^l{}_{b\nu} - A^{kb}{}_\nu A^l{}_{b\mu} \end{aligned} \tag{9-34}$$

One can then show that (Problem 9.10):

$$F^{kl}{}_{\mu\nu} = h^{l\lambda} h^k{}_\alpha R^\alpha{}_{\lambda\mu\nu} \tag{9-35}$$

where $R^\alpha{}_{\lambda\mu\nu}$ is the Riemann tensor.

The total Langrangian density is given by $\mathfrak{L}_t = \mathfrak{L}(\psi^A, \nabla_\mu \psi^A, h^k{}_\mu) + \mathfrak{L}_0(h^k{}_\mu, F^{kl}{}_{\mu\nu})$. The field equations for ψ and h are given by

$$\frac{\delta \mathfrak{L}}{\delta \psi^A} = 0, \qquad \frac{\delta \mathfrak{L}_t}{\delta h^i{}_\mu} = 0$$

The field equations of gravitation are usually obtained from a particular Lagrangian density, $\mathfrak{L}_0 = hR$, where R is defined by $R = g^{\mu\nu}R_{\mu\nu} = h_l{}^\mu h_k{}^\nu F^{kl}{}_{\mu\nu}$, and $R_{\mu\nu} = R^\lambda{}_{\mu\lambda\nu}$. Taking the variation with respect to h gives

$$\frac{\delta\mathfrak{L}_0}{\delta h^i{}_\mu} + \frac{\delta\mathfrak{L}}{\delta h^i{}_\mu} = 0$$

One obtains

$$\frac{\delta\mathfrak{L}_0}{\delta h^i{}_\mu}\,\delta h^i{}_\mu = \frac{\partial\mathfrak{L}_0}{\delta g_{\rho\mu}}\,h_{i\rho}\,\delta h^i{}_\mu + \frac{\partial}{\partial u^\mu}\left\{\frac{\delta\mathfrak{L}_0}{\partial g_{\rho\sigma,\,\mu}}\,h_{i\rho}\,\delta h^i{}_\sigma\right\}$$

where

$$\frac{\delta\mathfrak{L}_0}{\delta g_{\rho\sigma}} = -h\left(R^{\rho\sigma} - \frac{1}{2}g^{\rho\sigma}R\right)$$

Accordingly, the action principal leads to the field equations

$$h(R^{\rho\sigma} - \tfrac{1}{2}g^{\rho\sigma}R) = \mathscr{T}^{\rho\sigma} \tag{9-36}$$

where $\mathscr{T}^{\rho\sigma} = \mathscr{T}^\rho{}_i\,h^{i\sigma}$, and $\mathscr{T}^\rho{}_i$ is given by

$$\mathscr{T}^\rho{}_i = \delta\mathfrak{L}/\delta h^i{}_\rho$$

Here $\mathscr{T}^{\rho\sigma}$ is the symmetric energy–momentum tensor density of the original field ψ. The symmetry character of $\mathscr{T}^{\rho\sigma}$ can be proved (see Problem 9.11).

Poincaré Invariance and the Gravitational Field

Kibble[14] has pointed out that Utiyama's method, discussed above, is a rather unsatisfactory procedure since it is the purpose of the method to supply an argument for introducing the gravitational field variables, including the metric and the affine connections. To overcome these difficulties and make possible the introduction of the vierbein components $h^i{}_\mu$ as well as the local affine connections $A^{ij}{}_\mu$ as new field variables analogous to the electromagnetic potentials A_μ, Kibble extended Utiyama's discussion and considered the 10-parameter inhomogeneous Lorentz group (Poincaré group) instead of the homogeneous 6-parameter Lorentz group. He showed it is then unnecessary to introduce *a priori* curvilinear coordinates or a Riemannian metric, and that the new field variables introduced as a consequence of the argument include the vierbein components $h_k{}^\mu$ as well as the local affine connection $A^{ij}{}_\mu$. The extended transformations for which the 10 parameters become arbitrary functions of position may be interpreted as general coordinate transformations and rotations of the $h_k{}^\mu$ field. The Lagrangian density proposed, then yields the free space field equations $R_{\mu\nu} = 0$, but when matter is presented the resultant equations show that there is a difference from the theory of general relativity which arises from the fact that $A^{ij}{}_\mu$ appear in the matter field

14 T. W. B. Kibble, *J. Math. Phys.* **2**, 212 (1961).

Lagrangian. As a consequence this means that, although the covariant derivative of the metric vanishes, the affine connection $\Gamma^{\lambda}_{\alpha\beta}$ is nonsymmetric.

We will not discuss in details this Poincaré invariant theory. Instead, we will return and formulate the gravitational field equations in a *SL*(2, *C*) invariant way so as to exhibit the gauge aspects of the theory. This is done in the next section.

9-3 *SL*(2, *C*) INVARIANCE AND THE GRAVITATIONAL FIELD

In the previous section we discussed the theories of Utiyama and Kibble of applying the Yang–Mills method in order to relate the gravitational field to a generalized gauge field associated with the Lorentz group, where one starts with flat space and introduces at each point a curved space–time. On the other hand, we saw in Section 8-1 how spinors are introduced in a Riemannian space, at each space–time point in a tangent two-dimensional complex space. The two procedures are therefore the opposite of each other. It is thus an attractive idea to relate the two approaches, the one that is based on the Yang–Mills method and the other that is based on spinor formalism, to the gravitational field. In this section it is shown how the theory of general relativity, given in the spinorial form, can be recast into a Yang–Mills-type theory by use of the group *SL*(2, *C*). To be sure, we will not follow the prescription of Utiyama, thus not starting with a Dirac field and going into a curved space–time since, as has been pointed out by Weinberg,[15] this is a somewhat arbitrary procedure. Instead, we will reverse Utiyama's procedure since we start with the curved space–time and subsequently introduce at each space–time point a tangent space in which a complex three-dimensional linear space is introduced. Another difference exists between the present theory and that of gauge fields since in the latter case it is the spin affinities that are considered as potentials, whereas the potential matrices here will be defined differently [see Eq. (9-38) below]. Obviously, spin affinities are not space–time vectors whereas the potentials to be defined here are.

Spin Frame Gauge

The gravitational field dynamical variables of general relativity can be divided into three sets: (1) the Riemann tensor, decomposed into its irreducible components (the Weyl tensor, the trace-free parts of the Ricci tensor, and the Ricci scalar); (2) the spin coefficients (to be introduced in the sequel); and (3) a tetrad system (see Chapter 8) of vectors (from which one obtains the metric tensor). They are connected by three sets of first-order partial differential equations which describe the gravitational field.

We will represent the spin coefficients and the components of the Riemann tensor in the form of linear combinations of the infinitesimal generators of the group *SL*(2, *C*). This representation is very similar to the way Yang and Mills write their dynamical variables in terms of the Pauli spin matrices. The spin

[15] S. Weinberg, *Phys. Rev.* **138,** B988 (1965).

coefficients take the role of the Yang–Mills-like potentials, whereas the Riemann tensor components take the role of the fields.

There is an essential difference, however, between this representation and that of Yang and Mills. The group underlying the symmetry here is $SL(2, C)$ whereas in the Yang–Mills case it is SU_2. The group $SL(2, C)$ seems to fit in with general relativity in a remarkable and natural way, just as 2-component spinors do. This is not an unexpected result since spinors, as we have seen in Chapter 3, describe the finite-dimensional representation of the group $SL(2, C)$.

We start by introducing at each point of space–time two 2-component spinors $\zeta_a{}^A$, where $a = 0, 1$, to define a spin frame. Each one of these two spinors might be considered as a complex wave function describing a spin $\frac{1}{2}$ particle, but one assumes nothing as to whether they satisfy any dynamical wave equations. As in Chapter 8 the two spinors $\zeta_a{}^A$ are normalized such as $\zeta_a{}^B \varepsilon_{BA} \zeta_b{}^A = \zeta_{aA} \zeta_b{}^A = \varepsilon_{ab}$, where, as usual, ε's are the skew-symmetric Levi–Civita symbols defined by $\varepsilon_{01} = 1$. Such a frame has already been discussed in Chapter 8 where the two spinors were denoted by l^A and n^A.

A spin frame gauge can be defined[16] as an arbitrary way of choosing the orientation of the spin frame axes at all space–time points, in analogy with the isotopic gauge which is an arbitrary way of choosing the orientation of the isotopic spin axes at all points. One then demands that all physical processes be invariant under the spin frame transformation

$$\zeta = S\zeta' \tag{9-37}$$

where ζ is a 2×2 complex matrix whose elements are $\zeta_a{}^A$, and S represents a spin frame rotation which is a 2×2 unimodular complex matrix whose elements $S_a{}^b$ are functions of space–time.

An arbitrary spinor $G^{AB'}$ can now be written in terms of the spin frame, $G^{AB'} = G^{ab'} \zeta_a{}^A \bar{\zeta}_{b'}{}^{B'}$, where $G^{ab'}$ are the *dyad components* of the spinor $G^{AB'}$ and are given by $G^{ab'} = G_{AB'} \zeta^{aA} \bar{\zeta}^{b'B'}$. As before, lower-case indices are used for dyad components. These indices behave the same way algebraically as ordinary spinor indices except when covariant differentiation is applied in which case no term involving an affine connection appears for them. By the same token, the quantity $\nabla_\mu \xi^A$, obtained by taking the covariant derivative of a spinor ξ^A, can also be written in terms of the spin frame as $\nabla_\mu \xi^A = B^b{}_\mu \zeta_b{}^A$, where $B^b{}_\mu$, with $b = 0, 1$, are some space–time vectors. In particular the last formulae applies to the two spinors $\zeta_a{}^A$ defining the spin frame. This gives $\nabla_\mu \zeta_a{}^A = B_a{}^b{}_\mu \zeta_b{}^A$, where $B_a{}^b{}_\mu$, with $a, b = 0, 1$, are some vectors.

Potentials and Fields

In the Yang–Mills theory it is the spinor affine connections which are considered as potentials. However, these quantities are not space–time vectors, as has been shown in Chapter 7, in the Riemannian space, and alternative quantities have to

[16] M. Carmeli, *Ann. Phys.* (*N.Y.*) **71**, 603 (1972).

be found. Fortunately, such quantities are available. For instance, one can take the vectors $B_{a\ \mu}^{\ b}$, obtained above from the covariant derivatives of $\zeta_a^{\ A}$, as the potentials. For convenience one rewrites the relation $\nabla_\mu \zeta_a^{\ A} = B_{a\ \mu}^{\ b} \zeta_b^{\ A}$ as

$$\nabla_\mu \zeta = B_\mu \zeta \tag{9-38}$$

where B_μ and ζ are 2×2 complex matrices whose elements are $B_{a\ \mu}^{\ b}$ and $\zeta_a^{\ B}$, respectively. The normalization condition that the two spinors $\zeta_a^{\ A}$ have to satisfy then implies that the matrix B_μ be traceless and the matrix ζ be unimodular.

The commutator of the covariant derivatives, $(\nabla_\nu \nabla_\mu - \nabla_\mu \nabla_\nu)$, when applied on ζ gives $F_{\mu\nu} \zeta$, where

$$F_{\mu\nu} = \partial_\nu B_\mu - \partial_\mu B_\nu + [B_\mu, B_\nu] \tag{9-39}$$

is a 2×2 complex traceless matrix whose elements are skew-symmetric tensors. The commutator $[B_\mu, B_\nu] = B_\mu B_\nu - B_\nu B_\mu$. Hence the relation between the F- and B-matrices is similar to that of the Yang–Mills field, Eq. (9-3), but with the exception that the potentials are now defined by Eq. (9-38) rather than taken as the spinor affine connections as is done in that case. Furthermore, under a change of the spin frame (9-37) one easily finds that B_μ and $F_{\mu\nu}$ transform into

$$B'_\mu = S^{-1} B_\mu S - S^{-1} \partial_\mu S \tag{9-40}$$

$$F'_{\mu\nu} = S^{-1} F_{\mu\nu} S \tag{9-41}$$

similarily to those of the Yang–Mills field, Eqs. (9-2) and (9-4), when subjected under an isotopic gauge transformation.

The matrix B_μ defines 12 complex functions, whereas the matrix $F_{\mu\nu}$ defines 18 complex functions. The latter is equivalent to the 20 real components of the Riemann tensor plus the 16 real components of the tetrad field $\sigma^\mu_{\ ab'}$ [see Eq. (9-46) below].

Spin Coefficients as Potentials

Since the matrices B_μ and $F_{\mu\nu}$ are traceless, it follows that they both can be written as linear combinations of the infinitesimal generators of the group $SL(2, C)$ similar to the way Yang and Mills write their dynamical variables in terms of the Pauli spin matrices. The infinitesimal generators of the group $SL(2, C)$ are three traceless matrices that can be chosen as[17]

$$g_1 = \begin{pmatrix} 0 & 0 \\ 1 & 0 \end{pmatrix}, \quad g_2 = \begin{pmatrix} 1 & 0 \\ 0 & -1 \end{pmatrix}, \quad g_3 = \begin{pmatrix} 0 & 1 \\ 0 & 0 \end{pmatrix} \tag{9-42}$$

[17] Our matrices are related to those of Gelfand, Graev, and Vilenkin by $g_1 = a_-$, $g_2 = 2a_0$, and $g_3 = a_+$. See I. M. Gelfand, M. I. Graev, and N. Ya. Vilenkin, *Integral Geometry and Representation Theory*, Academic Press, New York, 1966.

The matrices g_1, g_2, and g_3 are tangent vectors to the one-parameter subgroups

$$g_1(z) = \begin{pmatrix} 1 & 0 \\ z & 1 \end{pmatrix}, \qquad g_2(z) = \begin{pmatrix} e^z & 0 \\ 0 & e^{-z} \end{pmatrix}, \qquad g_3(z) = \begin{pmatrix} 1 & z \\ 0 & 1 \end{pmatrix} \tag{9-43}$$

where z is a complex parameter, and satisfy

$$g_m(z_1 + z_2) = g_m(z_1)g_m(z_2); \qquad \text{(no summation on } m)$$

for $m = 1, 2, 3$. The matrices (9-43) are identical to those appearing in Eq. (8-41) [but with a slight change in the notation of $g_2(z)$], and every matrix of the group $SL(2, C)$ can be presented uniquely as product of them. The infinitesimal generators g_1, g_2, and g_3 are obtained from $g_1(z)$, $g_2(z)$, and $g_3(z)$, as usual, by

$$g_m = [\mathrm{d}g_m(z)/\mathrm{d}z]_{z=0}$$

and conversely, the matrices $g_1(z)$, $g_2(z)$, and $g_3(z)$ can be expressed in terms of the infinitesimal generators g_1, g_2, and g_3 by

$$g_m(z) = \exp(zg_m)$$

Accordingly, one can write

$$B_\mu = \mathbf{b}_\mu \cdot \mathbf{g} \tag{9-44}$$

$$F_{\mu\nu} = \mathbf{f}_{\mu\nu} \cdot \mathbf{g} \tag{9-45}$$

where $\mathbf{g} = (g_1, g_2, g_3)$, and $\mathbf{b}_\mu$ and $\mathbf{f}_{\mu\nu}$ are vectors in the complex 3-dimensional space of $SL(2, C)$.

We now define another set of Hermitian matrices related to the matrix $\tilde{\sigma}^\mu$ (see Sec. 8-1) by

$$\sigma^\mu = \zeta\tilde{\sigma}^\mu\zeta^\dagger \tag{9-46}$$

where $\zeta^\dagger$ is the Hermitian conjugate of ζ. Contrary to $\tilde{\sigma}^\mu$ whose covariant derivative vanishes by definition, the covariant derivative of the matrix σ^μ does not vanish and one has, using Eq. (9-38),

$$\nabla_\alpha \sigma^\mu = B_\alpha \sigma^\mu + \sigma^\mu B_\alpha^\dagger \tag{9-47}$$

The geometrical metric can then be written as $g^{\mu\nu} = \tilde{\sigma}^\mu{}_{AB'}\tilde{\sigma}^{\nu AB'} = \sigma^\mu{}_{ab'}\sigma^{\nu ab'}$. The elements of the matrix σ^μ define a null tetrad of vectors where $\sigma^\mu{}_{00'}$ and $\sigma^\mu{}_{11'}$ are real, whereas $\sigma^\mu{}_{01'}$ and $\sigma^\mu{}_{10'}$ are complex, conjugate to each other. Moreover, they satisfy the orthogonality relation $\sigma^\mu{}_{ab'}\sigma_{\mu cd'} = \varepsilon_{ac}\varepsilon_{b'd'}$. These are the same null vectors l^μ, m^μ, $\bar{m}^\mu$, and n^μ introduced in Chapter 8, where $\sigma^\mu{}_{00'} = l^\mu$, $\sigma^\mu{}_{01'} = m^\mu$, $\sigma^\mu{}_{10'} = \bar{m}^\mu$, and $\sigma^\mu{}_{11'} = n^\mu$.

The three sets of matrices B_μ, $F_{\mu\nu}$, and σ^α describe all the dynamical variables of the gravitational field. From the B_μ and $F_{\mu\nu}$ one can obtain two new sets of matrices which are just new representations of the B and F matrices:

$$B_{ab'} = \sigma^\mu{}_{ab'}B_\mu \tag{9-48}$$

$$F_{ab'cd'} = \sigma^\mu{}_{ab'}\sigma^\nu{}_{cd'}F_{\mu\nu} \tag{9-49}$$

Again one may write the latter matrices as linear combinations of $\mathbf{g}$:

$$B_{cd'} = \mathbf{b}_{cd'} \cdot \mathbf{g} \tag{9-50}$$

$$F_{ab'cd'} = \mathbf{f}_{ab'cd'} \cdot \mathbf{g} \tag{9-51}$$

where the new four 3-vectors $\mathbf{b}_{cd'}$ and the six 3-vectors $\mathbf{f}_{ab'cd'}$ are related to $\mathbf{b}_\mu$ and $\mathbf{f}_{\mu\nu}$ of Eqs. (9-44) and (9-45) by

$$\mathbf{b}_{cd'} = \sigma^\mu{}_{cd'}\,\mathbf{b}_\mu\,, \qquad \mathbf{f}_{ab'cd'} = \sigma^\mu{}_{ab'}\sigma^\nu{}_{cd'}\,\mathbf{f}_{\mu\nu}$$

The four 3-vectors $\mathbf{b}_{cd'}$ in the complex $SL(2, C)$ space will be denoted by

$$\begin{aligned} \mathbf{b}_{00'} &= (\pi, \varepsilon, -\kappa), \qquad \mathbf{b}_{01'} = (\mu, \beta, -\sigma) \\ \mathbf{b}_{10'} &= (\lambda, \alpha, -\rho), \qquad \mathbf{b}_{11'} = (\nu, \gamma, -\tau) \end{aligned} \tag{9-52}$$

Using Eq. (9-50), we see that the four matrices $B_{cd'}$ will then have the form

$$\begin{aligned} B_{00'} &= \begin{pmatrix} \varepsilon & -\kappa \\ \pi & -\varepsilon \end{pmatrix}, \qquad B_{01'} = \begin{pmatrix} \beta & -\sigma \\ \mu & -\beta \end{pmatrix} \\ B_{10'} &= \begin{pmatrix} \alpha & -\rho \\ \lambda & -\alpha \end{pmatrix}, \qquad B_{11'} = \begin{pmatrix} \gamma & -\tau \\ \nu & -\gamma \end{pmatrix} \end{aligned} \tag{9-53}$$

when the representation (9-42) is used for $\mathbf{g}$.

The twelve complex functions ε, κ, π, etc., were first introduced by Newman and Penrose[18] and are known in general relativity theory as spin coefficients. From the point of view of Yang–Mills field theory these same quantities are the potentials the field of which is given by the F-matrices according to Eq. (9-39).

Symmetry of $F_{ab'cd'}$

It is convenient to introduce another matrix, $\tilde{B}_\mu$, connected to the matrix B_μ by a similarity transformation

$$\zeta \tilde{B}_\mu = B_\mu \zeta \tag{9-54}$$

The new matrix then satisfies

$$\nabla_\mu \zeta = \zeta \tilde{B}_\mu \tag{9-55}$$

The matrix elements of $\tilde{B}_\mu$ and B_μ are related as follows. If $B_a{}^b{}_\mu$ is the ab element of the matrix B_μ, then $B_A{}^B{}_\mu$ is the AB element of the matrix $\tilde{B}_\mu$. This fact can easily be seen by writing the matrix elements of both sides of Eq. (9-54). The left-hand side gives

$$(\zeta \tilde{B}_\mu)_e{}^F = \zeta_e{}^D B_D{}^F{}_\mu$$

[18] E. T. Newman and R. Penrose, *J. Math. Phys.* **3**, 566 (1962).

whereas the right-hand side gives

$$(B_\mu \zeta)_e{}^F = B_e{}^d{}_\mu \zeta_d{}^F$$

As can be easily seen, both of these expressions are equal to $B_e{}^F{}_\mu$. Hence, while the matrix element indices of B_μ are dyad components, the matrix element indices of $\tilde{B}_\mu$ are spinorial.

In the same way we can define another matrix $\tilde{F}_{\mu\nu}$,

$$\zeta \tilde{F}_{\mu\nu} = F_{\mu\nu} \zeta \tag{9-56}$$

which satisfies

$$(\nabla_\nu \nabla_\mu - \nabla_\mu \nabla_\nu)\zeta = \zeta \tilde{F}_{\mu\nu} \tag{9-57}$$

and whose explicit expression is given by

$$\tilde{F}_{\mu\nu} = \nabla_\nu \tilde{B}_\mu - \nabla_\mu \tilde{B}_\nu - [\tilde{B}_\mu, \tilde{B}_\nu] \tag{9-58}$$

Similar to the potential matrix B_μ, the matrix elements of $F_{\mu\nu}$ and $\tilde{F}_{\mu\nu}$ will be $F_a{}^b{}_{\mu\nu}$ and $F_A{}^B{}_{\mu\nu}$, respectively.

To find the $SL(2, C)$ structure of the matrices $F_{ab'cd'}$ we proceed as follows.[19] Let ξ^P be an arbitrary spinor. Then

$$\begin{aligned}(\nabla_\nu \nabla_\mu - \nabla_\mu \nabla_\nu)\xi^P &= (\nabla_\nu \nabla_\mu - \nabla_\mu \nabla_\nu)\xi^g \zeta_g{}^P \\ &= \xi^g (\nabla_\nu \nabla_\mu - \nabla_\mu \nabla_\nu)\zeta_g{}^P\end{aligned}$$

Now, using Eq. (9-57), we obtain

$$(\nabla_\nu \nabla_\mu - \nabla_\mu \nabla_\nu)\xi^P = \xi^q \zeta_g{}^H F_H{}^P{}_{\mu\nu}$$

Hence we have

$$(\nabla_\nu \nabla_\mu - \nabla_\mu \nabla_\nu)\xi_Q = F_{PQ\mu\nu}\xi^P$$

or equivalently,

$$(\nabla_{AC'} \nabla_{BD'} - \nabla_{BD'} \nabla_{AC'})\xi_Q = F_{PQBD'AC'}\xi^P \tag{9-59}$$

By decomposing the commutator of differentiation on the left-hand side of Eq. (9-59), we obtain (see Problem 9.12):

$$\tfrac{1}{2}\varepsilon_{C'D'}(\nabla_{AF'}\nabla_B{}^{F'} + \nabla_{BF'}\nabla_A{}^{F'})\xi_Q + \tfrac{1}{2}\varepsilon_{AB}(\nabla_{EC'}\nabla^E{}_{D'} + \nabla_{ED'}\nabla^E{}_{C'})\xi_Q = F_{PQBD'AC'}\xi^P \tag{9-60}$$

But the left-hand side of Eq. (9-60) is equal (see Problem 9.13) to:

$$\varepsilon_{C'D'}[\psi_{ABQP} - \Lambda(\varepsilon_{PA}\varepsilon_{BQ} + \varepsilon_{PB}\varepsilon_{AQ})]\xi^P + \varepsilon_{AB}\phi_{QPC'D'}\xi^P \tag{9-61}$$

where ψ_{ABCD} is the totally symmetric spinor which represents the Weyl spinor, $\phi_{QPC'D'}$ represents the trace-free part of the Ricci spinor having the symmetry

$$\phi_{QPC'D'} = \phi_{PQC'D'} = \phi_{QPD'C'} = \bar{\phi}_{C'D'QP}$$

and $\Lambda = -R/24$, where R is the Ricci scalar.

[19] M. Carmeli, *J. Math. Phys.* **11**, 2728 (1970).

Accordingly, we obtain

$$F_{PQBD'AC'} = \varepsilon_{C'D'}[\psi_{ABQP} - \Lambda(\varepsilon_{PA}\varepsilon_{BQ} + \varepsilon_{PB}\varepsilon_{AQ})] + \varepsilon_{AB}\phi_{QPC'D'}$$

and the same relation holds for lower-case indices:

$$F_p{}^q{}_{bd'ac'} = \varepsilon_{c'd'}[\psi_p{}^q{}_{ab} - \Lambda(\varepsilon_{pa}\ \delta_b{}^q + \varepsilon_{pb}\ \delta_a{}^q)] + \varepsilon_{ab}\phi_p{}^q{}_{c'd'}$$

Using the standard notation

$$\psi_{0000} = \psi_0, \qquad \psi_{0001} = \psi_1, \qquad \psi_{0011} = \psi_2$$
$$\psi_{0111} = \psi_3, \qquad \psi_{1111} = \psi_4$$

and

$$\phi_{000'0'} = \phi_{00}, \qquad \phi_{010'0'} = \phi_{10}, \qquad \phi_{110'0'} = \phi_{20}$$
$$\phi_{000'1'} = \phi_{01}, \qquad \phi_{010'1'} = \phi_{11}, \qquad \phi_{110'1'} = \phi_{21}$$
$$\phi_{001'1'} = \phi_{02}, \qquad \phi_{011'1'} = \phi_{12}, \qquad \phi_{111'1'} = \phi_{22}$$

we finally obtain for the 3-vector $\mathbf{f}_{ab'cd'}$:

$$\begin{aligned}
\mathbf{f}_{01'00'} &= (\psi_2 + 2\Lambda, \psi_1, -\psi_0)\\
\mathbf{f}_{11'10'} &= (\psi_4, \psi_3, -\psi_2 - 2\Lambda)\\
\mathbf{f}_{10'00'} &= (\phi_{20}, \phi_{10}, -\phi_{00})\\
\mathbf{f}_{11'01'} &= (\phi_{22}, \phi_{12}, -\phi_{02})\\
\mathbf{f}_{11'00'} &= (\psi_3 + \phi_{21}, \psi_2 + \phi_{11} - \Lambda, -\psi_1 - \phi_{01})\\
\mathbf{f}_{10'01'} &= (-\psi_3 + \phi_{21}, -\psi_2 + \phi_{11} + \Lambda, \psi_1 - \phi_{01})
\end{aligned} \tag{9-62}$$

and for the six matrices $F_{ab'cd'}$:

$$\begin{aligned}
F_{01'00'} &= \begin{pmatrix} \psi_1 & -\psi_0 \\ \psi_2 + 2\Lambda & -\psi_1 \end{pmatrix}\\
F_{11'10'} &= \begin{pmatrix} \psi_3 & -\psi_2 - 2\Lambda \\ \psi_4 & -\psi_3 \end{pmatrix}\\
F_{10'00'} &= \begin{pmatrix} \phi_{10} & -\phi_{00} \\ \phi_{20} & -\phi_{10} \end{pmatrix}\\
F_{11'01'} &= \begin{pmatrix} \phi_{12} & -\phi_{02} \\ \phi_{22} & -\phi_{12} \end{pmatrix}\\
F_{11'00'} &= \begin{pmatrix} \psi_2 + \phi_{11} - \Lambda & -\psi_1 - \phi_{01} \\ \psi_3 + \phi_{21} & -\psi_2 - \phi_{11} + \Lambda \end{pmatrix}\\
F_{10'01'} &= \begin{pmatrix} -\psi_2 + \phi_{11} + \Lambda & \psi_1 - \phi_{01} \\ -\psi_3 + \phi_{21} & \psi_2 - \phi_{11} - \Lambda \end{pmatrix}
\end{aligned} \tag{9-63}$$

In Appendix E the transformation laws under the group $SL(2, C)$ of the tetrad of null vectors, the spin coefficients, and the components of the Riemann tensor will be given explicitly. Notice that $\psi_0, \ldots, \psi_4$ are complex, Λ is real, and ϕ_{mn} satisfy $\phi_{mn} = \bar{\phi}_{nm}$.

9-4 GRAVITATIONAL FIELD EQUATIONS

Having defined the gravitational field dynamical variables, given by the components of the Riemann tensor (9-63), the spin coefficients (9-53), and the four null vectors $\sigma^{\mu}{}_{ab'}$, we are now in a position to find out the gravitational field equations that relate these quantities. We encounter a different situation from that of the Yang–Mills case. In the latter case the field equations are obtained from a Lagrangian density that is postulated. In the present case we have the Einstein equations which relate the Einstein tensor to matter. This latter condition, is imposed on the field equations already at our disposal, including the identities, which consequently ceased to be identities and become part of the field equations. The procedure of using the identities as part of the field equations is well known in general relativity.[20] Nevertheless, we will see that two out of three sets of these same equations can be derived from a Lagrangian density.

Identities

The matrices $F_{\mu\nu}$ and $\tilde{F}_{\mu\nu}$ defined in the previous section satisfy some identities which can be found as follows.

We calculate the expression

$$\nabla_\alpha F_{\beta\gamma} + \nabla_\beta F_{\gamma\alpha} + \nabla_\gamma F_{\alpha\beta} \tag{9-64}$$

The curl part of F does not contribute to (9-64) as can easily be verified. The second part of F, the commutator $[B_\mu, B_\nu]$, contributes to (9-64) the expression

$$[(\nabla_\gamma B_\alpha - \nabla_\alpha B_\gamma), B_\beta] + [(\nabla_\alpha B_\beta - \nabla_\beta B_\alpha), B_\gamma] + [(\nabla_\beta B_\gamma - \nabla_\gamma B_\beta), B_\alpha]$$

Now add to this expression the term

$$[[B_\alpha, B_\gamma], B_\beta] + [[B_\beta, B_\alpha], B_\gamma] + [[B_\gamma, B_\beta], B_\alpha]$$

which is identically zero, we obtain

$$\nabla_\alpha F_{\beta\gamma} + \nabla_\beta F_{\gamma\alpha} + \nabla_\gamma F_{\alpha\beta} = [B_\alpha, F_{\beta\gamma}] + [B_\beta, F_{\gamma\alpha}] + [B_\gamma, F_{\alpha\beta}] \tag{9-65}$$

Note that the covariant derivatives in Eq. (9-65) can be replaced by partial derivatives without affecting that equation.

To find the identity the $\tilde{F}_{\mu\nu}$ satisfy, we express its covariant derivative in terms of those of $F_{\mu\nu}$. Since, by Eq. (9-56), $\tilde{F}_{\beta\gamma} = \zeta^{-1} F_{\beta\gamma} \zeta$, one obtains

$$\nabla_\alpha \tilde{F}_{\beta\gamma} = (\nabla_\alpha \zeta^{-1}) F_{\beta\gamma} \zeta + \zeta^{-1} (\nabla_\alpha F_{\beta\gamma}) \zeta + \zeta^{-1} F_{\beta\gamma} \nabla_\alpha \zeta$$

[20] M. Carmeli, *Nuovo Cim. Lett.* **4**, 40 (1970); *Nuovo Cim.* **7A**, 9 (1972).

Using Eq. (9-38), and the fact that $\nabla_\alpha \zeta^{-1} = -\zeta^{-1}(\nabla_\alpha \zeta)\zeta^{-1}$, one obtains

$$\nabla_\alpha \tilde{F}_{\beta\gamma} = \zeta^{-1}(\nabla_\alpha F_{\beta\gamma} - [B_\alpha, F_{\beta\gamma}])\zeta$$

Using this equation in Eq. (9-65) we obtain the identity that the matrix $\tilde{F}_{\alpha\beta}$ has to satisfy:

$$\nabla_\alpha \tilde{F}_{\beta\gamma} + \nabla_\beta \tilde{F}_{\gamma\alpha} + \nabla_\gamma \tilde{F}_{\alpha\beta} = 0 \tag{9-66}$$

Field Equations

In Section 9-3 we have defined the matrix $F_{\mu\nu}$ in terms of the matrix B_μ by Eq. (9-39), and in this section we showed that these matrices satisfy the identity (9-65).

By contracting Eqs. (9-39) and (9-65) with $\sigma^\mu{}_{ab'}\sigma^\nu{}_{cd'}$ and $\sigma^\alpha{}_{ab'}\sigma^\beta{}_{cd'}\sigma^\gamma{}_{ef'}$, respectively, and using Eq. (9-47), one obtains two sets of first-order partial differential equations that now connect the four matrices $B_{ab'}$ and the six matrices $F_{cd'ef'}$. A supplementary set of equations which connect the matrices σ^μ with the matrices $B_{ab'}$ and which is called the *metric equation* is, furthermore, obtained from Eq. (9-47).

Multiplying Eq. (9-39) by $\sigma^\mu{}_{ab'}\sigma^\nu{}_{cd'}$ and using Eq. (9-48) we obtain

$$\partial_{cd'} B_{ab'} - \partial_{ab'} B_{cd'} - (\nabla_{cd'}\sigma^\mu{}_{ab'} - \nabla_{ab'}\sigma^\mu{}_{cd'})B_\mu + [B_{ab'}, B_{cd'}] = F_{ab'cd'} \tag{9-67}$$

where the two differential operators ∂ and ∇ are defined by $\partial_{ab'} = \sigma^\mu{}_{ab'}\,\partial_\mu$ and $\nabla_{ab'} = \sigma^\mu{}_{ab'}\,\nabla_\mu$.

The third and fourth terms in Eq. (9-67) can be found using Eq. (9-47). Contracting the latter with $\sigma^\alpha{}_{cd'}$ we obtain

$$\nabla_{cd'}\sigma^\mu = B_{cd'}\sigma^\mu + \sigma^\mu B^\dagger_\alpha \bar{\sigma}^\alpha{}_{d'c}$$

where use has been made of the Hermicity property of $\sigma^\alpha{}_{cd'}$. Hence we obtain

$$\nabla_{cd'}\sigma^\mu = B_{cd'}\sigma^\mu + \sigma^\mu B^\dagger{}_{d'c}. \tag{9-68}$$

Here the four matrices $B^\dagger{}_{d'c}$ are Hermitian conjugate to the matrices $B_{dc'}$ given in Eq. (9-53). For completeness, and for the convenience of the reader, we list them below:

$$\begin{aligned} B^\dagger{}_{0'0} &= \begin{pmatrix} \bar{\varepsilon} & \bar{\pi} \\ -\bar{\kappa} & -\bar{\varepsilon} \end{pmatrix}, & B^\dagger{}_{0'1} &= \begin{pmatrix} \bar{\beta} & \bar{\mu} \\ -\bar{\sigma} & -\bar{\beta} \end{pmatrix} \\ B^\dagger{}_{1'0} &= \begin{pmatrix} \bar{\alpha} & \bar{\lambda} \\ -\bar{\rho} & -\bar{\alpha} \end{pmatrix}, & B^\dagger{}_{1'1} &= \begin{pmatrix} \bar{\gamma} & \bar{\nu} \\ -\bar{\tau} & -\bar{\gamma} \end{pmatrix} \end{aligned} \tag{9-69}$$

Accordingly we obtain from Eq. (9-68)

$$\nabla_{cd'}\sigma^\mu{}_{ab'} = (B_{cd'}\sigma^\mu)_{ab'} + (\sigma^\mu B^\dagger{}_{d'c})_{ab'}$$

where $(\quad)_{ab'}$ is the ab' element of the matrix $(\quad)$. Writing this last equation in terms of matrix elements we obtain

$$\nabla_{cd'}\sigma^\mu{}_{ab'} = (B_{cd'})_a{}^f\sigma^\mu{}_{fb'} + \sigma^\mu{}_{af'}(B^\dagger{}_{d'c})^{f'}{}_{b'} \tag{9-70}$$

Here $(B_{cd'})_a{}^f$ and $(B^\dagger{}_{d'c})^{f'}{}_{b'}$ are the af and $f'b'$ elements of the matrices $B_{cd'}$ and $B^\dagger{}_{d'c}$, respectively.

Using Eqs. (9-70) and (9-48) in Eq. (9-67) one obtains:

$$\partial_{cd'} B_{ab'} - \partial_{ab'} B_{cd'} - (B_{cd'})_a{}^f B_{fb'} - (B^\dagger{}_{d'c})^{f'}{}_{b'} B_{af'} + (B_{ab'})_c{}^f B_{fd'} + (B^\dagger{}_{b'a})^{f'}{}_{d'} B_{cf'} + [B_{ab'}, B_{cd'}] = F_{ab'cd'} \tag{9-71}$$

Similarly, one can rewrite Eq. (9-65) in terms of the corresponding quantities with dyad indices to express relations between the matrices $B_{ab'}$ and $F_{ab'cd'}$ by using Eqs. (9-47) and (9-49). Multiplying Eq. (9-65) by $\sigma^\alpha{}_{ab'}\sigma^\beta{}_{cd'}\sigma^\gamma{}_{ef'}$, and using Eq. (9-70), one obtains:

$$\begin{aligned}\partial_{ab'} F_{cd'ef'} &+ \partial_{cd'} F_{ef'ab'} + \partial_{ef'} F_{ab'cd'} - (B_{ab'})_c{}^g F_{gd'ef'} \\ &- (B^\dagger{}_{b'a})^{g'}{}_{d'} F_{cg'ef'} - (B_{ab'})_e{}^g F_{cd'gf'} - (B^\dagger{}_{b'a})^{g'}{}_{f'} F_{cd'eg'} \\ &- (B_{cd'})_e{}^g F_{gf'ab'} - (B^\dagger{}_{d'c})^{g'}{}_{f'} F_{eg'ab'} - (B_{cd'})_a{}^g F_{ef'gb'} \\ &- (B^\dagger{}_{d'c})^{g'}{}_{b'} F_{ef'ag'} - (B_{ef'})_a{}^g F_{gb'cd'} - (B^\dagger{}_{f'e})^{g'}{}_{b'} F_{ag'cd'} \\ &- (B_{ef'})_c{}^g F_{ab'gd'} - (B^\dagger{}_{f'e})^{g'}{}_{d'} F_{ab'cg'} \\ &= [B_{ab'}, F_{cd'ef'}] + [B_{cd'}, F_{ef'ab'}] + [B_{ef'}, F_{ab'cd'}]\end{aligned} \tag{9-72}$$

One can easily verify that Eqs. (9-71) and (9-72) are the usual field equations obtained using the formalism of Newman and Penrose.[21]

We finally obtain the metric equation which connects $\sigma^\mu{}_{ab'}$ with $B_{ab'}$. This equation can easily be obtained from Eq. (9-70) and is given by:

$$\partial_{ab'}\sigma^\mu{}_{cd'} - \partial_{cd'}\sigma^\mu{}_{ab'} = (B_{ab'}\sigma^\mu)_{cd'} + (\sigma^\mu B^\dagger{}_{b'a})_{cd'} - (B_{cd'}\sigma^\mu)_{ab'} - (\sigma^\mu B^\dagger{}_{d'c})_{ab'} \tag{9-73}$$

Gravitational Lagrangian

The gravitational field equations, connecting the dynamical variables of general relativity, were obtained from Eqs. (9-39), (9-47), and (9-65) by rewriting them in terms of the corresponding quantities with dyad indices, and substituting the desired expression in terms of the energy–momentum tensor for the Ricci tensor components in the curvature matrices (9-63). However, one can also obtain two sets of these field equations from an action principle that is based on the analogy of the present theory to that of the Yang–Mills theory.

The simplest Lagrangian density which is invariant under both general coordinate transformation and spin frame transformation was shown by Carmeli and Yang to be given by[22]

$$\mathfrak{L}_0 = -\tfrac{1}{4}(-g)^{1/2}\,\mathrm{Tr}\,(\mathrm{F}_{\mu\nu}F^{\mu\nu}) \tag{9-74}$$

It is also a most natural generalization to the free-field Yang–Mills Lagrangian density (9-9). It follows, however, that the equation of motion obtained from such

[21] These equations are given in F. A. Pirani, *Lectures on General Relativity*, Prentice-Hall, Englewood Cliffs, N.J., 1964, p. 350.

[22] M. Carmeli, *Nucl. Phys.* **B38,** 621 (1972); C. N. Yang, *Phys. Rev. Letters* **33,** 445 (1974).

a Lagrangian density does not give the complete set of gravitational field equations but yields only the empty space ones.

Another Lagrangian density can be constructed, however, which gives the complete set of field equations (9-71) and (9-72). This was shown by Carmeli and Fickler[23] to be given by $-\frac{1}{4}(-g)^{1/2}\,\mathrm{Tr}\,(H^{\mu\nu}F_{\mu\nu})$, where

$$H^{\mu\nu} = \tilde{\sigma}^{\mu CB'}\tilde{\sigma}^{\nu AD'}\tilde{\sigma}^{\alpha}{}_{AB'}\tilde{\sigma}^{\beta}{}_{CD'}F_{\alpha\beta}$$

The first-order form of this Lagrangian density is:

$$\mathfrak{L} = -\tfrac{1}{2}(-g)^{1/2}\,\mathrm{Tr}\,\{H^{\mu\nu}(-\tfrac{1}{2}F_{\mu\nu} + \partial_\nu B_\mu - \partial_\mu B_\nu + [B_\mu, B_\nu])\} \qquad (9\text{-}75)$$

A complex conjugate expression can be added to (9-75) so that the Lagrangian density becomes real. The matrix elements of B_μ and $F_{\mu\nu}$ are considered to be the independent field variables, and Eq. (9-39) is assumed to be unknown. The matrices $\tilde{\sigma}^\mu$ are introduced in (9-75) as auxiliary quantities in order to accomplish invariance but they are not considered as part of the dynamical variables.

Application of the usual procedure of variational calculus then leads to the field equation (9-39) and to the following equation of motion:

$$\partial_\nu((-g)^{1/2}H^{\mu\nu}) - [B_\nu, (-g)^{1/2}H^{\mu\nu}] = 0 \qquad (9\text{-}76)$$

Equation (9-76) gives the dynamical equation of motion which the Riemann tensor has to satisfy, and accordingly we have a full description of the dynamical system. Equation (9-39) gives the Riemann tensor in terms of the spin coefficients, whereas Eq. (9-47) gives the spin coefficients in terms of the tetrad of null vectors.

To recover the gravitational field equations (9-71) to (9-73) one has merely to rewrite these equations in terms of the dynamical variables using Eqs. (9-48), (9-49), (9-53), and (9-63). One obtains the field equations (9-71), (9-73), and the following:

$$\partial^{cd'}F_{cb'ad'} - \{(B^{pd'})^c{}_p + (B^{\dagger q'c})_{q'}{}^{d'}\}F_{cb'ad'}$$
$$- \{\delta^{f'}{}_{b'}(B^{cd'})^e{}_a + \delta^e{}_a(B^{\dagger d'c})_{b'}{}^{f'}\}F_{cf'ed'} - [B^{cd'}, F_{cb'ad'}] = 0 \qquad (9\text{-}77)$$

When written in details, Eqs. (9-72) and (9-77) follow to be identical.

The Lagrangian density (9-75) is a natural generalization of the free-field Lagrangian density (9-74), and reduces to the latter in the free-field case. This can easily be seen since the expression in braces in $\mathfrak{L}$ can be written as $\frac{1}{2}H^{\mu\nu}F_{\mu\nu}$, and by Eq. (9-49) this is equal to $\frac{1}{2}F^{cb'ad'}F_{ab'cd'}$. In empty space (i.e., when all ϕ's and Λ are assumed to be zeros) this last expression can be seen, by Eq. (9-63), to be equal to $\frac{1}{2}F^{ab'cd'}F_{ab'cd'}$, or equal to $\frac{1}{2}F^{\mu\nu}F_{\mu\nu}$, thus giving the expression

$$-\tfrac{1}{2}(-g)^{1/2}\,\mathrm{Tr}\,\{F^{\mu\nu}(-\tfrac{1}{2}F_{\mu\nu} + \partial_\nu B_\mu - \partial_\mu B_\nu + [B_\mu, B_\nu])\} \qquad (9\text{-}78)$$

for the Lagrangian density (9-75) in free space, which is the Lagrangian density (9-74).

We finally remark that the two sets of equations (9-71), (9-72) [or (9-77)] and (9-73) are written in detailed form in Appendix D. The transformation laws of the field variables will be given explicitly in Appendix E.

[23] M. Carmeli and S. I. Fickler, *Phys. Rev.* **D5**, 290 (1972).

Conservation Laws

The gravitational Lagrangian density (9-75) is not satisfactory since: (1) it involves the Hermitian matrices $\tilde{\sigma}^{\mu}$ and the metric tensor $g_{\mu\nu}$ as auxiliary quantities in order to accomplish invariance but are not part of the variational procedure; (2) it leads to the Bianchi identities and to the definition of the fields $F_{\mu\nu}$ in terms of the gauge potentials B_{μ}; and (3) the Einstein condition is not incorporated in it and has to be added afterward in the field equations.

A modified Lagrangian density can be constructed in the form[24]

$$\mathfrak{L} = -\tfrac{1}{2}\,\mathrm{Tr}\,\{\varepsilon^{\alpha\beta\mu\nu}F^{J}_{\alpha\beta}(-\tfrac{1}{2}F^{J}_{\mu\nu} + \partial_{\nu}B_{\mu} - \partial_{\mu}B_{\nu} + [B_{\mu}, B_{\nu}])\} \tag{9-79}$$

to overcome these difficulties. In Eq. (9-79) $\varepsilon^{\alpha\beta\mu\nu}$ is the totally skew-symmetric tensor density of weight $+1$ with values $+1$ and -1, depending upon whether $\alpha\beta\mu\nu$ is an even or an odd permutation of 0123, and zero otherwise. The matrices B_{μ} and $F^{J}_{\mu\nu}$ are defined by $B_{\mu} = \sigma_{\mu}{}^{ab'}B_{ab'}$ and $F^{J}_{\mu\nu} = \sigma_{\mu}{}^{ab'}\sigma_{\nu}{}^{cd'}F^{J}_{ab'cd'}$, where $B_{ab'}$ are given by Eqs. (9-53) and $F^{J}_{ab'cd'}$ describe the Weyl tensor plus matter,

$$F^{J}_{ab'cd'} = F^{W}_{ab'cd'} + \kappa J_{ab'cd'} \tag{9-80}$$

Here κ is the Einstein gravitational constant, $F^{W}_{ab'cd'}$ are the Weyl matrices given by

$$F^{W}_{01'00'} = \begin{pmatrix} \psi_1 & -\psi_0 \\ \psi_2 & -\psi_1 \end{pmatrix}$$
$$F^{W}_{11'00'} = \begin{pmatrix} \psi_2 & -\psi_1 \\ \psi_3 & -\psi_2 \end{pmatrix} \tag{9-81}$$
$$F^{W}_{11'10'} = \begin{pmatrix} \psi_3 & -\psi_2 \\ \psi_4 & -\psi_3 \end{pmatrix}$$

along with $F^{W}_{10'00'} = F^{W}_{11'01'} = 0$ and $F^{W}_{10'01'} = -F^{W}_{11'00'}$, whereas the six matter matrices $J_{ab'cd'}$ are given by

$$J_{01'00'} = \frac{1}{2}\begin{pmatrix} 0 & 0 \\ T/6 & 0 \end{pmatrix}, \qquad J_{11'10'} = \frac{1}{2}\begin{pmatrix} 0 & -T/6 \\ 0 & 0 \end{pmatrix}$$
$$J_{10'00'} = \frac{1}{2}\begin{pmatrix} T_{10'00'} & -T_{00'00'} \\ T_{10'10'} & -T_{10'00'} \end{pmatrix}$$
$$J_{11'01'} = \frac{1}{2}\begin{pmatrix} T_{01'11'} & -T_{01'01'} \\ T_{11'11'} & -T_{01'11'} \end{pmatrix} \tag{9-82}$$
$$J_{11'00'} = \frac{1}{2}\begin{pmatrix} T_{11'00'} & -T_{01'00'} \\ T_{10'11'} & -T_{11'00'} \end{pmatrix} - \begin{pmatrix} T/6 & 0 \\ 0 & -T/6 \end{pmatrix}$$
$$J_{10'01'} = \frac{1}{2}\begin{pmatrix} T_{11'00'} & -T_{01'00'} \\ T_{10'11'} & -T_{11'00'} \end{pmatrix} - \frac{1}{2}\begin{pmatrix} T/6 & 0 \\ 0 & -T/6 \end{pmatrix}$$

[24] M. Carmeli, *Phys. Rev. D.*, **14**, 1727 (1976); A. Salam, in Proceedings of the Coral Gables Conference, January 1973.

Here $T_{ab'cd'} = \sigma^{\mu}_{ab'}\sigma^{\nu}_{cd'}T_{\mu\nu}$, where $T_{\mu\nu}$ is the usual energy-momentum tensor, and $T = g^{\mu\nu}T_{\mu\nu}$.

Carrying out variations with respect to the elements of the matrices $F^{J}_{\mu\nu}$ and B_{μ} then lead to the following field equations

$$F^{J}_{\mu\nu} = \partial_{\nu}B_{\mu} - \partial_{\mu}B_{\nu} + [B_{\mu}, B_{\nu}] \tag{9-83}$$

$$\varepsilon^{\mu\nu\alpha\beta}\{\partial_{\nu}F^{J}_{\alpha\beta} - [B_{\nu}, F^{J}_{\alpha\beta}]\} = 0 \tag{9-84}$$

The field equation (9-84) can now be written in the form

$$\varepsilon^{\mu\nu\alpha\beta}\{\partial_{\nu}F^{W}_{\alpha\beta} - [B_{\nu}, F^{W}_{\alpha\beta}]\} = \kappa J^{\mu}, \tag{9-85}$$

where J^{μ} represents the sources,

$$\begin{aligned} J^{\alpha} &= -\varepsilon^{\alpha\beta\gamma\delta}\{\partial_{\beta}J_{\gamma\delta} - [B_{\beta}, J_{\gamma\delta}]\} \\ J_{\alpha\beta} &= \sigma_{\alpha}{}^{ab'}\sigma_{\beta}{}^{cd'}J_{ab'cd'} \end{aligned} \tag{9-86}$$

Equations (9-85) and (9-83), consequently, give the Newman-Penrose field equations (see Appendix D) but with the Ricci tensor already being substituted for by the energy-momentum tensor through the Einstein field equations.

Choosing now the basis (9-42) in the inner complex 3-dimensional space of the group *SL*(2, *C*) one can then decompose all the matrices appearing in the field equations (9-85) in the space, $B_{\mu} = \mathbf{b}_{\mu}\cdot\mathbf{g}$, $F^{W}_{\mu\nu} = \mathbf{f}_{\mu\nu}\cdot\mathbf{g}$, and $J^{\mu} = \mathbf{J}^{\mu}\cdot\mathbf{g}$, where $\mathbf{g} = (g_1, g_2, g_3)$. Equation (9-85) may then have the form

$$\varepsilon^{\mu\nu\alpha\beta}\{\partial_{\nu}\mathbf{f}_{\alpha\beta} - \mathbf{b}_{\nu}\times\mathbf{f}_{\alpha\beta}\} = \kappa\mathbf{J}^{\mu} \tag{9-87}$$

Since the divergence of $\mathbf{J}^{\mu}$ does not vanish, one may define

$$\mathscr{J}^{\mu} = \mathbf{J}^{\mu} + \kappa^{-1}\varepsilon^{\mu\nu\alpha\beta}\mathbf{b}_{\nu}\times\mathbf{f}_{\alpha\beta}, \tag{9-88}$$

which leads to the equation of continuity

$$\partial\mathscr{J}^{\mu}/\partial x^{\mu} = 0 \tag{9-89}$$

The similarity to the Yang–Mills case is remarkable but should not be surprising. In terms of components, and because of the commutation relations that g_1, g_2, g_3 satisfy,

$$[g_1, g_2] = 2g_1, \qquad [g_3, g_1] = g_2, \qquad [g_2, g_3] = 2g_3 \tag{9-90}$$

one obtains

$$\begin{aligned} \mathscr{J}^{\mu} = \mathbf{J}^{\mu} + \kappa^{-1}\varepsilon^{\mu\nu\alpha\beta}\{&2(b_{\nu}{}^{1}f_{\alpha\beta}{}^{2} - b_{\nu}{}^{2}f_{\alpha\beta}{}^{1}), (b_{\nu}{}^{3}f_{\alpha\beta}{}^{1} - b_{\nu}{}^{1}f_{\alpha\beta}{}^{3}), \\ &2(b_{\nu}{}^{2}f_{\alpha\beta}{}^{3} - b_{\nu}{}^{3}f_{\alpha\beta}{}^{2})\} \end{aligned} \tag{9-91}$$

Here $b_{\mu}{}^{k}$ and $f_{\mu\nu}{}^{k}$ are given by $b_{\mu}{}^{k} = \sigma_{\mu}{}^{ab'}b_{ab'}{}^{k}$ and $f_{\mu\nu}{}^{k} = \sigma_{\mu}{}^{ab'}\sigma_{\nu}{}^{cd'}f_{ab'cd'}{}^{k}$, with $b_{ab'}{}^{k}$ and $f_{ab'cd'}{}^{k}$ given by $\mathbf{b}_{00'} = (\pi, \varepsilon, -\kappa)$, $\mathbf{b}_{01'} = (\mu, \beta, -\sigma)$, $\mathbf{b}_{10'} = (\lambda, \alpha, -\rho)$, $\mathbf{b}_{11'} = (\nu,$

$\gamma, -\tau)$, and $\mathbf{f}_{01'00'} = (\psi_2, \psi_1, -\psi_0)$, $\mathbf{f}_{11'00'} = -\mathbf{f}_{10'01'} = (\psi_3, \psi_2, -\psi_1)$, $f_{11'10'} = (\psi_4, \psi_3, -\psi_2)$, and $\mathbf{f}_{10'00'} = \mathbf{f}_{11'01'} = 0$. In the same way one easily finds the $SL(2, C)$ components of the "current" $\mathbf{J}^\mu$.

PROBLEMS

9.1 Show that the Hamiltonian derived from the Lagrangian density given in Eq. (9-11) is positive definite in the absence of the field of isotopic spin $\frac{1}{2}$.

9.2 To quantize the Yang–Mills field it is sometimes convenient to start with the Lagrangian density which is not obviously gauge-invariant:

$$\mathcal{L} = -\frac{1}{2}\frac{\partial \mathbf{b}_\mu}{\partial x^\nu}\cdot\frac{\partial \mathbf{b}_\mu}{\partial x^\nu} + (\mathbf{b}_\mu \times \mathbf{b}_\nu)\cdot\frac{\partial \mathbf{b}_\mu}{\partial x^\nu}$$
$$- (\mathbf{b}_\mu \times \mathbf{b}_\nu)^2 + \mathbf{J}_\mu \cdot \mathbf{b}_\mu - \bar{\psi}(\gamma_\mu\,\partial_\mu + m)\psi$$

Show that the equations of motion obtained from this Lagrangian density yield the equation

$$\frac{\partial^2 \mathbf{a}}{(\partial x^\nu)^2} = 0$$

where $\mathbf{a} = \partial \mathbf{b}_\mu/\partial x^\mu$.

9.3 Show that the Hamiltonian density derived from the Lagrangian density of Problem 9.2 is given by $H = H_0 + H_{\text{int}}$, where

$$H_0 = -\frac{1}{2}\boldsymbol{\pi}_\mu\cdot\boldsymbol{\pi}_\mu + \frac{1}{2}\frac{\partial \mathbf{b}_\mu}{\partial x^j}\cdot\frac{\partial \mathbf{b}_\mu}{\partial x^j} + \bar{\psi}(\gamma^\mu\partial_\mu + m)\psi$$

$$H_{\text{int}} = 2(\mathbf{b}_i \times \mathbf{b}_0)\cdot\boldsymbol{\pi}_i - 2(\mathbf{b}_\mu \times \mathbf{b}_j)\cdot(\partial \mathbf{b}_\mu/\partial x^j)$$
$$+ (\mathbf{b}_i \times \mathbf{b}_j)^2 - \mathbf{J}_\mu \cdot \mathbf{b}_\mu .$$

Here $\boldsymbol{\pi}_\mu$ is defined by

$$\boldsymbol{\pi}_\mu = -\partial \mathbf{b}_\mu/\partial x^0 + 2(\mathbf{b}_\mu \times \mathbf{b}_0)$$

Show also that the equal-time commutation rule between $\mathbf{b}_\mu$ and $\boldsymbol{\pi}_\mu$ is given by

$$[b_\mu{}^i(x), \pi_\nu{}^j(x')]_{t=t'} = -\delta_{ij}\,\delta_{\mu\nu}\,\delta^3(x - x')$$

[C. N. Yang and R. L. Mills, *Phys. Rev.* **96**, 191 (1954).]

9.4 Discuss the properties of the b quanta.

9.5 Show that if the field $F_{\mu\nu}$ defined by Eq. (9-3) vanishes everywhere, then the potential B_μ can be transformed away by a gauge transformation.

9.6 Show that the internal holonomy groups at different points are isomorphic.

9.7 Prove that compatibility of the internal holonomy group $\mathcal{H}$ and the gauge group $\mathcal{G}$ requires that either (1) $\mathcal{H}$ be a subgroup of $\mathcal{G}$, or (2) $\mathcal{G}$ be an invariant subgroup of $\mathcal{H}$. [H. G. Loos, *J. Math. Phys.* **8**, 2114 (1967).]

9.8 Use Eq. (9-29) to generalize the covariant derivative law (9-30) for the mixed tensor $\psi^{k\nu}$ to a tensor $\psi^{kl\cdots\rho\sigma\cdots}{}_{ab\cdots\alpha\beta\cdots}$.

9.9 Assuming $\Gamma'{}_{\mu\nu}{}^\rho = \Gamma'{}_{\nu\mu}{}^\rho$, prove Eqs. (9-32) and (9-33).

9.10 Prove Eq. (9-35).

9.11 Show that $\mathcal{T}^{\rho\sigma}$ of Eq. (9-36) is symmetric.

9.12 Prove Eq. (9-60).

9.13 Prove Eq. (9-61).

9.14 Prove Eq. (9-72).

9.15 Show that the Lagrangian density (9-74) yields the following field equations of gravitation:

$$\partial^{cd'} F_{ef'cd'} = \{(B^{pd'})^{c}{}_{p} + (B^{\dagger q'c})_{q'}{}^{d'}\} F_{ef'cd'} + \{\delta^{b'}_{f'}(B^{cd'})^{a}{}_{e} + \delta^{a}_{e}(B^{\dagger d'c})_{f'}{}^{b'}\} F_{ab'cd'} + [B^{cd'}, F_{ef'cd'}]$$

[See M. Carmeli, *Nucl. Phys.* B**38**, 621 (1972).]

9.16 Show that the Lagrangian density (9-75) can also be written in the form

$$-\tfrac{1}{4}\varepsilon^{\alpha\beta\mu\nu}\,\mathrm{Tr}\,(F_{\alpha\beta}F_{\mu\nu})$$

where $\varepsilon^{\alpha\beta\mu\nu}$ is the totally skew-symmetric tensor density, whose values are 1, 0, —1. [See M. Carmeli, *Phys. Rev. Letters* **36**, 59 (1976).]

CHAPTER

TEN

ANALYSIS OF THE GRAVITATIONAL FIELD

After having developed the gravitational field equations that connect the Riemann tensor, the spin coefficients, and the tetrad of null vectors in the previous chapter we are now in a position to apply these equations to specific problems and cases. The geometrical meaning of the spin coefficients and the components of the Weyl tensor are first discussed. This we follow by proving the theorem of Goldberg and Sachs, which states the conditions under which the Riemann tensor becomes algebraically special. The problem of choice of the coordinate system is then discussed. Subsequently, a specific coordinate system, in which we write down the free field equations of empty space, is chosen. For completeness, the Maxwell–Einstein equations and the neutrino equations are also written down for any coordinate system. We conclude the chapter by discussing the asymptotic behavior of the Riemann tensor, the spin coefficients, and the metric for a general type of radiative empty space.

10-1 GEOMETRICAL INTERPRETATION

We now proceed to examine the geometrical meaning of the spin coefficients and the components of the Weyl spinor. The components of the Weyl spinor were already introduced in Chapter 8. All of these dynamical variables will be frequently used in the sequel.[1]

Geometrical Meaning of the Spin Coefficients

Let ξ^μ be a congruence and let the vector l^μ, i.e. $\sigma^\mu{}_{00'}$ (see Sec. 9-3) be the tangent to this congruence, $l^\mu = d\xi^\mu/ds$. Then the spin coefficient κ is related to the first curvature of the congruence ξ^μ. To see this, one calculates the covariant derivative

[1] E. T. Newman and R. Penrose, *J. Math. Phys.* **3**, 566 (1962); **4**, 998 (1963).

of l^μ along the congruence, $\delta l^\mu/\delta s = l^\alpha \nabla_\alpha l^\mu = \sigma^\alpha{}_{00'} \nabla_\alpha \sigma^\mu{}_{00'} = \nabla_{00'} \sigma^\mu{}_{00'}$. Using Eq. (9-70), we obtain

$$\nabla_{00'} \sigma^\mu{}_{00'} = (\varepsilon + \bar{\varepsilon})\sigma^\mu{}_{00'} - \bar{\kappa}\sigma^\mu{}_{01'} - \kappa\sigma^\mu{}_{10'}$$

or

$$Dl^\mu = (\varepsilon + \bar{\varepsilon})l^\mu - \bar{\kappa}m^\mu - \kappa\bar{m}^\mu \tag{10-1}$$

In Eq. (10-1), and in the following, we use the notation according to which $D = \nabla_{00'}$, $\delta = \nabla_{01'}$, $\bar{\delta} = \nabla_{10'}$, $\Delta = \nabla_{11'}$ for the components of the covariant derivative operator $\nabla_{ab'}$, and $l^\mu = \sigma^\mu{}_{00'}$, $m^\mu = \sigma^\mu{}_{01'}$, $\bar{m}^\mu = \sigma^\mu{}_{10'}$, $n^\mu = \sigma^\mu{}_{11'}$, for the tetrad of the null vectors $\sigma^\mu{}_{ab'}$ (compare Sec. 9-3).

By a change in scale $l^\mu \to \phi l^\mu$, where ϕ is a real scalar function, one can get rid of the first term on the right-hand side of Eq. (10-1). This can be seen if one calculates Dl^μ in the new scale. One obtains

$$Dl^\mu = \phi^{-1}(\varepsilon + \bar{\varepsilon} - l^\alpha \partial_\alpha \phi)l^\mu - \phi^{-2}(\bar{\kappa}m^\mu + \kappa\bar{m}^\mu)$$

Hence if one chooses the function ϕ so that $\varepsilon + \bar{\varepsilon} = l^\alpha \partial_\alpha \phi$, Dl^μ will be proportional to m^μ and $\bar{m}^\mu$ alone. Consequently, if $\kappa = 0$ in the above choice of scaling for l^μ, it follows that $Dl^\mu = 0$ and the congruence ξ^μ describes geodesics. In this case, one finds that the spin coefficient ρ is given by

$$\rho = \{-\tfrac{1}{2}l^\mu{}_{;\mu} + i(\tfrac{1}{2}l_{[\mu;\nu]}\, l^{\mu;\nu})^{1/2}\} \tag{10-2}$$

and the spin coefficient σ is the complex *shear* of the vector l_μ satisfying

$$\sigma\bar{\sigma} = \tfrac{1}{2}\{l_{(\mu;\nu)}\, l^{\mu;\nu} - \tfrac{1}{2}(l^\mu{}_{;\mu})^2\} \tag{10-3}$$

In Eqs. (10-2) and (10-3), and in the following, a semicolon denotes a covariant derivative, $l_{\mu;\nu} = \nabla_\nu l_\mu$.

The spin coefficient τ describes how the direction of l_μ changes as we move in the direction n_μ according to the relation

$$\Delta l_\mu = l_{\mu;\nu}n^\nu = (\gamma + \bar{\gamma})l_\mu - \bar{\tau}m_\mu - \tau\bar{m}_\mu \tag{10-4}$$

Again the term with l_μ can be made to vanish by a change of scale $l_\mu \to \phi l_\mu$.

Equations (10-1), (10-2), (10-3), and (10-4) thus give a geometrical description for the spin coefficients κ, ρ, σ, and τ [compare Eq. (9-53)]. The spin coefficients ν, μ, λ, and π have similar meanings, the difference being that the congruence used is given by n_μ instead of l_μ.

If l_μ is taken to be tangent to a geodesic congruence and we propagate the remainder of the tetrad system parallelly along the congruence, then it follows that the three-vector $\mathbf{b}_{00'}$ of Eq. (9-52) vanishes,

$$\mathbf{b}_{00'} = (\pi, \varepsilon, -\kappa) = 0 \tag{10-5}$$

If, in addition to being tangent to geodesics, the vector l_μ is a hypersurface orthogonal, i.e., proportional to a gradient field, then one finds

$$\rho = \bar{\rho} \tag{10-6}$$

In the particular case where the vector l_μ is *equal* to a gradient field, one has

$$\begin{aligned} \rho &= \bar{\rho} \\ \tau &= \bar{\alpha} + \beta \end{aligned} \tag{10-7}$$

Geometrical Meaning of the Weyl Spinor Components

The components of the Weyl spinor $\psi_0, \ldots, \psi_4$ have already been discussed in Chapter 9. One may briefly understand their meaning by the following:

Consider the five cases: (1) $\psi_0 \neq 0$, all other four components vanish; (2) $\psi_1 \neq 0$, other components vanish; (3) $\psi_2 \neq 0$, other components vanish; (4) $\psi_3 \neq 0$, other components vanish; and (5) $\psi_4 \neq 0$, other components vanish.

The fields $F_{\mu\nu}$ (see Chapter 9) can easily be found for each k case. One finds for the five different cases the following:

$$\begin{aligned} F^{(1)}_{\mu\nu} &= 2\psi_0 \bar{m}_{[\mu} n_{\nu]} g_3 \\ F^{(2)}_{\mu\nu} &= 2\psi_1\{-\bar{m}_{[\mu} n_{\nu]} g_2 + (-l_{[\mu} n_{\nu]} + m_{[\mu} \bar{m}_{\nu]}) g_3\} \\ F^{(3)}_{\mu\nu} &= 2\psi_2\{-\bar{m}_{[\mu} n_{\nu]} g_1 + (l_{[\mu} n_{\nu]} - m_{[\mu} \bar{m}_{\nu]}) g_2 + l_{[\mu} m_{\nu]} g_3\} \\ F^{(4)}_{\mu\nu} &= 2\psi_3\{(l_{[\mu} n_{\nu]} - m_{[\mu} \bar{m}_{\nu]}) g_1 - l_{[\mu} m_{\nu]} g_2\} \\ F^{(5)}_{\mu\nu} &= 2\psi_4(-l_{[\mu} m_{\nu]}) g_1 \end{aligned} \tag{10-8}$$

In these equations the three matrices g_1, g_2, and g_3 are the infinitesimal generators of the group $SL(2, C)$ given by Eq. (9-42), and the index in parenthesis denotes the type number.

The Petrov type of the Weyl spinor corresponding to each one of these five cases is as follows: (1) Petrov type N with propagation vector n_μ; (2) Petrov type III with propagation vector n_μ; (3) Petrov type D with propagation vectors n_μ and l_μ; (4) Petrov type III with propagation vector l_μ; and (5) Petrov type N with propagation vector l_μ. A propagation vector in the above means a repeated principal null vector (compare Sec. 8-4).

If in empty space the vector field l_μ satisfies the equation [compare Eq. (8-86)]

$$l_{[\mu} R_{\alpha]\beta\gamma[\delta} l_{\nu]} l^\beta l^\gamma = 0 \tag{10-9a}$$

then the vector l_μ corresponds to one of the four principal null directions of the Riemann tensor, and $\psi_0 = 0$.

If two or more of the principal null directions coincide and are represented by l_μ, they must satisfy

$$R_{\alpha\beta\gamma[\delta}l_{\mu]}l^\beta l^\gamma = 0 \tag{10-9b}$$

or $\psi_0 = \psi_1 = 0$. A space which satisfies Eq. (10-9b) (i.e., $\psi_0 = \psi_1 = 0$) is algebraically special (see Sec. 8-4).

Goldberg–Sachs Theorem

The theorem of Goldberg and Sachs[2] states that, in empty space (i.e., $R_{\mu\nu} = 0$), a necessary and sufficient condition for a Riemann tensor to be algebraically special, with $\psi_0 = \psi_1 = 0$, is that $\sigma = \kappa = 0$ (see also Problem 10.5).

Let us first assume that $\psi_0 = \psi_1 = 0$. The field equations (D.2), that are relevant, then give

$$\begin{aligned} 3\sigma\psi_2 &= 0 \\ \delta\psi_2 &= 3\tau\psi_2 - 2\sigma\psi_3 \\ \Delta\psi_2 - \delta\psi_3 &= -3\mu\psi_2 + 2(\beta - \tau)\psi_3 + \sigma\psi_4 \\ 3\kappa\psi_2 &= 0 \\ D\psi_2 &= 3\rho\psi_2 - 2\kappa\psi_3 \\ \bar{\delta}\psi_2 - D\psi_3 &= -3\pi\psi_2 + 2(\varepsilon - \rho)\psi_3 + \kappa\psi_4 \end{aligned} \tag{10-10}$$

It is seen from these equations that $\sigma = \kappa = 0$, otherwise all the components of the Weyl spinor ψ vanish, i.e., the space is flat.

Let us assume now the contrary, i.e., $\sigma = \kappa = 0$ and we wish to prove that $\psi_0 = \psi_1 = 0$.

By a transformation of the form $m_\mu \to e^{i\theta}m_\mu$ and a suitable choice of scaling of the null vector l_μ, we can set $\varepsilon = 0$. The pertinent equations of the field equations (D-1) are consequently

$$\begin{aligned} D\rho &= \rho^2 \\ \psi_0 &= 0 \\ D\tau &= (\tau + \bar{\pi})\rho + \psi_1 \\ D\beta &= \beta\bar{\rho} + \psi_1 \\ \delta\rho &= \rho(\bar{\alpha} + \beta) + (\rho - \bar{\rho})\tau - \psi_1 \end{aligned} \tag{10-11}$$

Consequently, $\psi_0 = 0$, and as a result Eqs. (D-2) now give

$$\begin{aligned} \delta\psi_1 &= (4\tau + 2\beta)\psi_1 \\ D\psi_1 &= 4\rho\psi_1 \end{aligned} \tag{10-12}$$

[2] J. N. Goldberg and R. Sachs, *Acta Phys. Polonica* **22,** *Suppl.* 13 (1962).

There is still a freedom in the choice of the vector n_μ which enables one an additional transformation of the form of a one-parameter null rotation around the vector l_μ. This rotation is given by the transformation Eqs. (8-34) and (8-35), where the transformation matrix g of the group $SL(2, C)$ is given by $g_1(z)$ of Eq. (8-41). One obtains, using Eq. (8-40),

$$l_\mu \to l_\mu$$

$$m_\mu \to \bar{z}l_\mu + m_\mu$$

$$n_\mu \to z\bar{z}l_\mu + zm_\mu + \bar{z}\bar{m}_\mu + n_\mu$$

This rotation of the tetrad does not affect l_μ or the condition $\varepsilon = 0$. One can choose the parameter z in such a way that $\tau = 0$.[3] Accordingly, Eqs. (10-12) yield

$$\begin{aligned} \delta \ln \psi_1 &= 2\beta \\ D \ln \psi_1 &= 4\rho \end{aligned} \tag{10-13}$$

Hence, using Eqs. (10-11), one obtains

$$(D\delta - \delta D) \ln \psi_1 = 6\psi_1 + 2\beta\bar{\rho} - 4\rho(\bar{\alpha} + \beta) \tag{10-14}$$

On the other hand, the commutator $(D\delta - \delta D)$ when applied on a scalar function, is given in Appendix D by Eq. (D-3b) (see Problem 10.4). One has

$$(\delta D - D\delta) \ln \psi_1 = (\bar{\alpha} + \beta - \bar{\pi})D \ln \psi_1 - \bar{\rho}\delta \ln \psi_1$$

which, together with Eq. (10-13), gives

$$(D\delta - \delta D) \ln \psi_1 = 2\beta\bar{\rho} - 4\rho(\bar{\alpha} + \beta) + 4\rho\bar{\pi} \tag{10-15}$$

Equations (10-14) and (10-15) then lead to the result $\psi_1 = (2/3)\rho\bar{\pi}$. But Eqs. (10-11) show that $\psi_1 = -\bar{\pi}\rho$. Hence since ρ is supposed to be different from zero, one concludes that both ψ_1 and $\bar{\pi}$ are zeros, and the proof is completed.

10-2 CHOICE OF COORDINATE SYSTEM

We now choose a particular coordinate system and a tetrad of vectors.

Coordinate System

In the hyperbolic Riemannian space of general relativity it is always possible to introduce a family of null hypersurfaces, designated by a parameter $u =$ const., i.e.,

$$g^{\mu\nu}u_{,\mu}u_{,\nu} = 0 \tag{10-16}$$

where a comma denotes a partial differentiation, $u_{,\mu} = \partial_\mu u$. The vectors $l^\mu = g^{\mu\nu}u_{,\nu}$

[3] This is possible only if $\rho \neq 0$. However, if $\rho = 0$ one obtains $\psi_1 = 0$ as can be seen from the last of Eqs. (10-11) and therefore the theorem would be proven.

are tangent to the family of null geodesics lying in the hypersurfaces $u = \text{const.}$, and satisfy

$$Dl^\mu = l^\mu{}_{;\nu} l^\nu = 0 \tag{10-17}$$

It has been pointed out by Robinson and Trautman[4] that if one chooses as coordinates $u = x^0$, an *affine parameter*[5] along the geodesic $r = x^1$, and two coordinates x^2 and x^3 that label the geodesics on each surface $u = \text{const.}$, then the metric tensor will have the form

$$g^{\mu\nu} = \begin{pmatrix} 0 & 1 & 0 & 0 \\ 1 & g^{11} & g^{12} & g^{13} \\ 0 & g^{12} & & \\ 0 & g^{13} & & g^{ij} \end{pmatrix} \tag{10-18}$$

where $i, j = 2, 3$.

Tetrad

Having determined a coordinate system, the vector l_μ then has the form

$$l_\mu = \delta^0{}_\mu, \quad l^\mu = \delta_1{}^\mu \tag{10-19}$$

Since $l_\mu n^\mu = 1$ and $l_\mu m^\mu = 0$ (see footnote 9 of Chapter 8), we have

$$m^\mu = \omega\, \delta_1{}^\mu + \xi^i\, \delta_i{}^\mu \tag{10-20}$$

$$n^\mu = \delta_0{}^\mu + U\, \delta_1{}^\mu + X^i\, \delta_i{}^\mu \tag{10-21}$$

where $i = 2, 3$. The unknown components of the metric (10-18) are then given by[6]

$$\begin{aligned} g^{11} &= 2(U - \omega\bar{\omega}) \\ g^{1i} &= X^i - (\bar{\omega}\xi^i + \omega\bar{\xi}^i) \\ g^{ij} &= -(\xi^i\bar{\xi}^j + \bar{\xi}^i\xi^j) \end{aligned} \tag{10-22}$$

where $i, j = 2, 3$.

There is still a complete freedom for rotating the tetrad vectors m^μ and n^μ but leaving l^μ unchanged. This freedom is eliminated by demanding that the vectors m^μ and n^μ be parallelly propagated along the vector l^μ. This requirement, in addition to the fact that the vector l_μ has been chosen as a gradient field, is stated in terms of the spin coefficients by Eqs. (10-5) and (10-7) (i.e., $\kappa = \varepsilon = \pi = 0$; $\rho = \bar{\rho}$; $\tau = \bar{\alpha} + \beta$).

[4] I. Robinson and A. Trautman, *Phys. Rev. Letters* **4**, 431 (1960).

[5] An affine parameter is a parameter along the geodesic, such that the equation of the geodesic takes the standard form (see Sec. 7-1). See, for example, E. M. Schrödinger, *Expanding Universe*, Cambridge University Press, New York, 1956. Some authors use different parameters from the affine parameter, such as a *luminosity parameter*, which then yields $g^{01} \neq 1$. We here will use only an affine parameter.

[6] This follows from the fact that

$$g^{\mu\nu} = \sigma^\mu{}_{ab'}\sigma^{\nu ab'} = l^\mu n^\nu + n^\mu l^\nu - m^\mu \bar{m}^\nu - \bar{m}^\mu m^\nu$$

Operators

With these simplifications, the covariant differentiation operators D, δ, $\bar{\delta}$, and Δ, when applied on a scalar function, using Eqs. (10-20) and (10-21), are given by

$$\begin{aligned} D &= \partial/\partial r \\ \delta &= \omega\,\partial/\partial r + \xi^i\,\partial/\partial x^i \\ \Delta &= \partial/\partial u + U\,\partial/\partial r + X^i\,\partial/\partial x^i \end{aligned} \tag{10-23}$$

with $\bar{\delta}$ being the complex conjugate of δ, and $i = 2, 3$.

Free Field Equations

The free space field equations (i.e., when $R^{\mu\nu} = 0$) under the above conditions have a simplified form. Equations (D-2) become:

$$\bar{\delta}\psi_0 - D\psi_1 = 4\alpha\psi_0 - 4\rho\psi_1 \tag{10-24a}$$

$$\bar{\delta}\psi_1 - D\psi_2 = \lambda\psi_0 + 2\alpha\psi_1 - 3\rho\psi_2 \tag{10-24b}$$

$$\bar{\delta}\psi_2 - D\psi_3 = 2\lambda\psi_1 - 2\rho\psi_3 \tag{10-24c}$$

$$\bar{\delta}\psi_3 - D\psi_4 = 3\lambda\psi_2 - 2\alpha\psi_3 - \rho\psi_4 \tag{10-24d}$$

$$\Delta\psi_0 - \delta\psi_1 = (4\gamma - \mu)\psi_0 - (4\tau + 2\beta)\psi_1 + 3\sigma\psi_2 \tag{10-24e}$$

$$\Delta\psi_1 - \delta\psi_2 = \nu\psi_0 + 2(\gamma - \mu)\psi_1 - 3\tau\psi_2 + 2\sigma\psi_3 \tag{10-24f}$$

$$\Delta\psi_2 - \delta\psi_3 = 2\nu\psi_1 - 3\mu\psi_2 + 2(\beta - \tau)\psi_3 + \sigma\psi_4 \tag{10-24g}$$

$$\Delta\psi_3 - \delta\psi_4 = 3\nu\psi_2 - 2(\gamma + 2\mu)\psi_3 + (4\beta - \tau)\psi_4 \tag{10-24h}$$

whereas Eqs. (D-1) become:

$$D\rho = \rho^2 + \sigma\bar{\sigma} \tag{10-25a}$$

$$D\sigma = 2\rho\sigma + \psi_0 \tag{10-25b}$$

$$D\tau = \tau\rho + \bar{\tau}\sigma + \psi_1 \tag{10-25c}$$

$$D\alpha = \alpha\rho + \beta\bar{\sigma} \tag{10-25d}$$

$$D\beta = \beta\rho + \alpha\sigma + \psi_1 \tag{10-25e}$$

$$D\gamma = \tau\alpha + \bar{\tau}\beta + \psi_2 \tag{10-25f}$$

$$D\lambda = \lambda\rho + \mu\bar{\sigma} \tag{10-25g}$$

$$D\mu = \mu\rho + \lambda\sigma + \psi_2 \tag{10-25h}$$

$$D\nu = \tau\lambda + \bar{\tau}\mu + \psi_3 \tag{10-25i}$$

$$\Delta\lambda - \bar{\delta}\nu = 2\alpha\nu + (\bar{\gamma} - 3\gamma - \mu - \bar{\mu})\lambda - \psi_4 \tag{10-25j}$$

$$\delta\rho - \bar{\delta}\sigma = (\beta + \bar{\alpha})\rho + (\bar{\beta} - 3\alpha)\sigma - \psi_1 \tag{10-25k}$$

$$\delta\alpha - \bar{\delta}\beta = \mu\rho - \lambda\sigma - 2\alpha\beta + \alpha\bar{\alpha} + \beta\bar{\beta} - \psi_2 \tag{10-25l}$$

$$\delta\lambda - \bar{\delta}\mu = (\alpha + \bar{\beta})\mu + (\bar{\alpha} - 3\beta)\lambda - \psi_3 \tag{10-25m}$$

$$\delta\nu - \Delta\mu = \gamma\mu - 2\nu\beta + \bar{\gamma}\mu + \mu^2 + \lambda\bar{\lambda} \tag{10-25n}$$

$$\delta\gamma - \Delta\beta = \tau\mu - \sigma\nu + (\mu - \gamma + \bar{\gamma})\beta + \bar{\lambda}\alpha \tag{10-25o}$$

$$\delta\tau - \Delta\sigma = 2\tau\beta + (\bar{\gamma} + \mu - 3\gamma)\sigma + \bar{\lambda}\rho \tag{10-25p}$$

$$\Delta\rho - \bar{\delta}\tau = (\gamma + \bar{\gamma} - \bar{\mu})\rho - 2\alpha\tau - \lambda\sigma - \psi_2 \tag{10-25q}$$

$$\Delta\alpha - \bar{\delta}\gamma = \rho\nu - \tau\lambda - \lambda\beta + (\bar{\gamma} - \gamma - \bar{\mu})\alpha - \psi_3 \tag{10-25r}$$

The metric equation (9-73) gives

$$D\xi^i = \rho\xi^i + \sigma\bar{\xi}^i \tag{10-26a}$$

$$D\omega = \bar{\rho}\omega + \sigma\bar{\omega} - (\bar{\alpha} + \beta) \tag{10-26b}$$

$$DX^i = \tau\bar{\xi}^i + \bar{\tau}\xi^i \tag{10-26c}$$

$$DU = \tau\bar{\omega} + \bar{\tau}\omega - (\gamma + \bar{\gamma}) \tag{10-26d}$$

$$\delta X^i - \Delta\xi^i = (\mu + \bar{\gamma} - \gamma)\xi^i + \bar{\lambda}\bar{\xi}^i \tag{10-26e}$$

$$\delta\bar{\xi}^i - \bar{\delta}\xi^i = (\bar{\beta} - \alpha)\xi^i + (\bar{\alpha} - \beta)\bar{\xi}^i \tag{10-26f}$$

$$\delta\bar{\omega} - \bar{\delta}\omega = (\bar{\beta} - \alpha)\omega + (\bar{\alpha} - \beta)\bar{\omega} + (\mu - \bar{\mu}) \tag{10-26g}$$

$$\delta U - \Delta\omega = (\mu + \bar{\gamma} - \gamma)\omega + \bar{\lambda}\bar{\omega} - \bar{\nu} \tag{10-26h}$$

In the above equations $i = 2, 3$.

Maxwell–Einstein Equations

To conclude this section we give below the coupled Maxwell–Einstein equations and the neutrino equations for any coordinate system. Maxwell's equations in vacuum are then given by (Problem 10.13; for equations with sources see Problem 10.14; for the Dirac equation see Problem 10.15):

$$\delta\phi_0 - D\phi_1 = (2\alpha - \pi)\phi_0 - 2\rho\phi_1 + \kappa\phi_2 \tag{10-27a}$$

$$\bar{\delta}\phi_1 - D\phi_2 = \lambda\phi_0 - 2\pi\phi_1 + (2\varepsilon - \rho)\phi_2 \tag{10-27b}$$

$$\Delta\phi_0 - \delta\phi_1 = (2\gamma - \mu)\phi_0 - 2\tau\phi_1 + \sigma\phi_2 \tag{10-27c}$$

$$\Delta\phi_1 - \delta\phi_2 = \nu\phi_0 - 2\mu\phi_1 + (2\beta - \tau)\phi_2 \tag{10-27d}$$

where ϕ_0, ϕ_1, and ϕ_2 are the dyad components of the Maxwell spinor (see Chapter 8) which are given by

$$\begin{aligned} \phi_0 &= F_{\mu\nu} l^\mu m^\nu \\ \phi_1 &= \tfrac{1}{2} F_{\mu\nu}(l^\mu n^\nu + \bar{m}^\mu m^\nu) \\ \phi_2 &= F_{\mu\nu} \bar{m}^\mu n^\nu \end{aligned} \tag{10-28a}$$

In terms of the electromagnetic potentials A_μ, defined by $F_{\mu\nu} = \nabla_\nu A_\mu - \nabla_\mu A_\nu$, the three dyad components of the Maxwell spinor can be written as

$$\begin{aligned}
\phi_0 &= (\delta - \beta + \bar{\pi} - \bar{\alpha})A_{00'} - (D - \varepsilon + \bar{\varepsilon} - \bar{\rho})A_{01'} + \sigma A_{10'} - \kappa A_{11'} \\
\phi_1 &= \tfrac{1}{2}\{(\Delta - \gamma - \mu + \bar{\mu} - \bar{\gamma})A_{00'} - (\bar{\delta} + \bar{\beta} - \bar{\tau} - \alpha - \pi)A_{01'} \\
&\quad + (\delta + \tau + \beta + \bar{\pi} - \bar{\alpha})A_{10'} - (D + \bar{\varepsilon} - \bar{\rho} + \rho + \varepsilon)A_{11'}\} \\
\phi_2 &= -\nu A_{00'} + \lambda A_{01'} + (\Delta + \gamma + \bar{\mu} - \bar{\gamma})A_{10'} - (\bar{\delta} + \bar{\beta} - \bar{\tau} + \alpha)A_{11'}
\end{aligned} \tag{10-28b}$$

Here $A_{00'}$, $A_{01'}$, $A_{10'}$, and $A_{11'}$ are the dyad components of the electromagnetic potentials, $A_{cd'} = A_\mu \sigma^\mu{}_{cd'}$, thus $A_{00'}$ and $A_{11'}$ are real, whereas $A_{01'}$ and $A_{10'}$ are complex and conjugate to each other.

The Ricci tensor is proportional to the electromagnetic energy–momentum tensor, $R_{\mu\nu} = \kappa T_{\mu\nu}$. Hence one can write for the Ricci spinor ϕ_{mn}:

$$\phi_{mn} = \frac{2G}{c^4}\phi_m \bar{\phi}_n \tag{10-29}$$

where $m, n = 0, 1, 2$. The field equations (D-2) will accordingly have the form:

$$\begin{aligned}
&3\nu\psi_2 - (\Delta + 2\gamma + 4\mu)\psi_3 + (\delta - \tau + 4\beta)\psi_4 \\
&\quad = -\bar{\phi}_1 \Delta\phi_2 + \bar{\phi}_2 \bar{\delta}\phi_2 + 2(\nu\bar{\phi}_1\phi_1 - \lambda\bar{\phi}_2\phi_1 - \gamma\bar{\phi}_1\phi_2 + \alpha\bar{\phi}_2\phi_2) && \text{(10-30a)} \\
&2\nu\psi_1 - (\Delta + 3\mu)\psi_2 + (\delta - 2\tau + 2\beta)\psi_3 + \sigma\psi_4 \\
&\quad = -\bar{\phi}_1 \delta\phi_2 + \phi_2 D\phi_2 + 2(\mu\bar{\phi}_1\phi_1 - \pi\bar{\phi}_2\phi_1 - \beta\bar{\phi}_1\phi_2 + \varepsilon\bar{\phi}_2\phi_2) && \text{(10-30b)} \\
&\nu\psi_0 - (\Delta - 2\gamma + 2\mu)\psi_1 + (\delta - 3\tau)\psi_2 + 2\sigma\phi_3 \\
&\quad = -\bar{\phi}_1 \Delta\phi_0 + \bar{\phi}_2 \bar{\delta}\phi_0 + 2(\gamma\bar{\phi}_1\phi_0 - \alpha\bar{\phi}_2\phi_0 - \tau\bar{\phi}_1\phi_1 + \rho\bar{\phi}_2\phi_1) && \text{(10-30c)} \\
&(-\Delta + 4\gamma - \mu)\psi_0 + (\delta - 4\tau - 2\beta)\psi_1 + 3\sigma\psi_2 \\
&\quad = -\bar{\phi}_1 \delta\phi_0 + \bar{\phi}_2 D\phi_0 + 2(\beta\bar{\phi}_1\phi_0 - \varepsilon\bar{\phi}_2\phi_0 - \sigma\bar{\phi}_1\phi_1 + \kappa\bar{\phi}_2\phi_1) && \text{(10-30d)} \\
&3\lambda\psi_2 - (\bar{\delta} + 4\pi + 2\alpha)\psi_3 + (D + 4\varepsilon - \rho)\psi_4 \\
&\quad = -\bar{\phi}_0 \Delta\phi_2 + \bar{\phi}_1 \bar{\delta}\phi_2 + 2(\nu\bar{\phi}_0\phi_1 - \lambda\bar{\phi}_1\phi_1 - \gamma\bar{\phi}_0\phi_2 + \alpha\bar{\phi}_1\phi_2) && \text{(10-30e)} \\
&2\lambda\psi_1 - (\bar{\delta} + 3\pi)\psi_2 + (D + 2\varepsilon - 2\rho)\psi_3 + \kappa\psi_4 \\
&\quad = -\bar{\phi}_2 \delta\phi_2 + \phi_1 D\phi_2 + 2(\mu\bar{\phi}_0\phi_1 - \pi\bar{\phi}_1\phi_1 - \beta\bar{\phi}_0\phi_2 + \varepsilon\bar{\phi}_1\phi_2) && \text{(10-30f)} \\
&\lambda\psi_0 - (\bar{\delta} + 2\pi - 2\alpha)\psi_1 + (D - 3\rho)\psi_2 + 2\kappa\psi_3 \\
&\quad = -\bar{\phi}_0 \Delta\phi_0 + \bar{\phi}_1 \bar{\delta}\phi_0 + 2(\gamma\bar{\phi}_0\phi_0 - \alpha\bar{\phi}_1\phi_0 - \tau\bar{\phi}_0\phi_1 + \rho\bar{\phi}_1\phi_1) && \text{(10-30g)} \\
&(-\bar{\delta} - \pi + 4\alpha)\psi_0 + (D - 2\varepsilon - 4\rho)\psi_1 + 3\kappa\psi_2 \\
&\quad = -\bar{\phi}_0 \delta\phi_0 + \bar{\phi}_1 D\phi_0 + 2(\beta\bar{\phi}_0\phi_0 - \varepsilon\bar{\phi}_1\phi_0 - \sigma\bar{\phi}_0\phi_1 + \kappa\bar{\phi}_1\phi_1) && \text{(10-30h)}
\end{aligned}$$

where in the above equations the constant $2G/c^4$ was taken as equal to 1.

The other two sets of field equations, (D-1) and (9-73), are valid, of course, in general and need not be changed in the presence of an electromagnetic field.

Neutrino Equations

The field variables of the neutrino are the two components of a spinor, η_A. They satisfy Weyl's neutrino equation in curved space, namely

$$\nabla^{AB'}\eta_A = 0$$

The energy–momentum tensor for the neutrino field is given by

$$T_{\mu\nu} = i\{\tilde{\sigma}_\mu{}^{AX'}\eta_A\nabla_\nu\bar{\eta}_{X'} + \tilde{\sigma}_\nu{}^{AX'}\eta_A\nabla_\mu\bar{\eta}_{X'} - \tilde{\sigma}_\mu{}^{AX'}\bar{\eta}_{X'}\nabla_\nu\eta_A - \tilde{\sigma}_\nu{}^{AX'}\bar{\eta}_{X'}\nabla_\mu\eta_A\}$$

One can now obtain the equivalent equations in the tetrad formalism in a straightforward way. The Weyl equation then becomes

$$\nabla^{ab'}\eta_a - (B^{ab'})_a{}^d\eta_d = 0$$

or, in details:

$$\Delta\eta_0 - \delta\eta_1 = (\gamma - \mu)\eta_0 + (\beta - \tau)\eta_1 \tag{10-31a}$$

$$\bar{\delta}\eta_0 - D\eta_1 = (\alpha - \pi)\eta_0 + (\varepsilon - \rho)\eta_1 \tag{10-31b}$$

Here η_0 and η_1 are the dyad components of the Weyl spinor η_A (see Chapter 8), $\eta_a = \zeta_a{}^A\eta_A$, thus $\eta_0 = l^A\eta_A$ and $\eta_1 = n^A\eta_A$. The functions γ, μ, β, etc., are, of course, spin coefficients. Since $T_\mu{}^\mu = 0$, one has $R = 0$ and $\lambda = 0$, and hence the Einstein equations become $R_{\mu\nu} = \kappa T_{\mu\nu}$. Using Eq. (8-23), one therefore obtains for the spinorial form of the Einstein equations the following:

$$2\phi_{ACB'D'} = \kappa T_{AB'CD'}$$

where $\phi_{ACB'D'}$ is the trace-free part of the Ricci tensor, and $T_{AB'CD'}$ is the energy–momentum spinor, $T_{AB'CD'} = \tilde{\sigma}^\mu{}_{AB'}\tilde{\sigma}^\nu{}_{CD'}T_{\mu\nu}$. Hence the Einstein equations are given by

$$\phi_{ACB'D'} = \frac{\kappa i}{2}\{\eta_A\nabla_{CD'}\bar{\eta}_{B'} + \eta_C\nabla_{AB'}\bar{\eta}_{D'} - \bar{\eta}_{B'}\nabla_{CD'}\eta_A - \bar{\eta}_{D'}\nabla_{AB'}\eta_C\}$$

To find out the dyad components of the Ricci spinor, $\phi_{acb'd'}$, we rewrite the last equation in dyad notation. This can easily be done if one notices that

$$\zeta_b{}^B\nabla_{cd'}\eta_B = \nabla_{cd'}\eta_b - (B_{cd'})_b{}^e\eta_e$$

$$\zeta_{b'}{}^{B'}\nabla_{cd'}\bar{\eta}_{B'} = \nabla_{cd'}\bar{\eta}_{b'} - (\bar{B}_{d'c})_{b'}{}^{e'}\bar{\eta}_{e'}$$

where $(\bar{B}_{d'c})_{b'}{}^{e'}$ is the complex conjugate of the matrix element $(B_{dc'})_b{}^e$ (see Sec. 9.3). One obtains:

$$\begin{aligned}\phi_{acb'd'} = (i\kappa/2)\{&\eta_a[\nabla_{cd'}\bar{\eta}_{b'} - (\bar{B}_{d'c})_{b'}{}^{e'}\bar{\eta}_{e'}] \\ &+ \eta_c[\nabla_{ab'}\bar{\eta}_{d'} - (\bar{B}_{b'a})_{d'}{}^{e'}\bar{\eta}_{e'}] \\ &- \bar{\eta}_{b'}[\nabla_{cd'}\eta_a - (B_{cd'})_a{}^e\eta_e] \\ &- \bar{\eta}_{d'}[\nabla_{ab'}\eta_c - (B_{ab'})_c{}^e\eta_e]\}\end{aligned}$$

In terms of ϕ_{mn}, where m, $n = 0$, 1, 2 and $\phi_{a+c,\,b+d} \equiv \phi_{acb'd'}$ (see Sec. 9-3), one finally obtains:

$$\phi_{00} = -i\bar{\eta}_0(D\eta_0 - \varepsilon\eta_0 + \kappa\eta_1) + \text{c.c.} \tag{10-32a}$$

$$\begin{aligned}\phi_{11} = -(i/2)\{&\bar{\eta}_0(\Delta\eta_0 - \gamma\eta_0 + \tau\eta_1)\\ &+ \bar{\eta}_1(D\eta_1 - \pi\eta_0 + \varepsilon\eta_1)\} + \text{c.c.}\end{aligned} \tag{10-32b}$$

$$\phi_{22} = -i\bar{\eta}_1(\Delta\eta_1 - \nu\eta_0 + \gamma\eta_1) + \text{c.c.} \tag{10-32c}$$

$$\begin{aligned}\phi_{01} = (i/2)\{&\eta_0(D\bar{\eta}_1 + \delta\bar{\eta}_0 - \bar{\pi}\bar{\eta}_0 + \bar{\varepsilon}\bar{\eta}_1 - \bar{\alpha}\bar{\eta}_0 + \bar{\rho}\bar{\eta}_1)\\ &- \bar{\eta}_1(D\eta_0 - \varepsilon\eta_0 + \kappa\eta_1)\\ &- \bar{\eta}_0(\delta\eta_0 - \beta\eta_0 + \sigma\eta_1)\}\end{aligned} \tag{10-32d}$$

$$\begin{aligned}\phi_{02} = i\{&\eta_0(\delta\bar{\eta}_1 - \bar{\lambda}\bar{\eta}_0 + \bar{\alpha}\bar{\eta}_1)\\ &- \bar{\eta}_1(\delta\eta_0 - \beta\eta_0 + \sigma\eta_1)\}\end{aligned} \tag{10-32e}$$

$$\begin{aligned}\phi_{12} = (i/2)\{&\eta_0(\Delta\bar{\eta}_1 - \bar{\nu}\bar{\eta}_0 + \bar{\gamma}\bar{\eta}_1) + \eta_1(\delta\bar{\eta}_1 - \bar{\lambda}\bar{\eta}_0 + \bar{\alpha}\bar{\eta}_1)\\ &- \bar{\eta}_1(\Delta\eta_0 + \delta\eta_1 - \gamma\eta_0 + \tau\eta_1 - \mu\eta_0 + \beta\eta_1)\}\end{aligned} \tag{10-32f}$$

$$\phi_{10} = \bar{\phi}_{01},\ \phi_{20} = \bar{\phi}_{02},\ \phi_{21} = \bar{\phi}_{12} \tag{10-32g}$$

In Eqs. (10-32) the gravitational constant κ was taken to be 1, and primes were omitted from the dyad indices of complex components for the sake of clarity.

10-3 ASYMPTOTIC BEHAVIOUR

We now discuss the asymptotic behavior of the components of the Weyl spinor, the spin coefficients, and the metric tensor. This is done for a general type of a *radiative* field for the free space case.

Asymptotic Behaviour of Weyl Spinor

In order to discuss asymptotic behaviour in general, it is usually necessary to impose conditions on the field functions when they approach flatness at infinity of the space. These conditions are usually imposed on the metric tensor. However, as we saw in Chapter 9, the Riemann spinor and the spin coefficients can also be considered as field variables. Hence we will impose conditions on the components of the Weyl spinor.

Let us assume that the asymptotic behaviour of ψ_0 is given by[7]

$$\psi_0 = 0(r^{-5}) \tag{10-33a}$$

[7] A relation of the form $f(u, r, x^i) = 0(g(r))$ means that there exists a function F of the variables u and x^i such that $|f(u, r, x^i)| < |g(r)||F(u, x^i)|$ for large r. For more details see F. Erdyle, *Asymptotic Expansions*, Dover, New York, 1955.

and that the asymptotic behaviour of $D\psi_0$ is given by

$$D\psi_0 = 0(r^{-6})$$

It will be assumed, furthermore, that derivatives of the left-hand sides, with respect to the variables x^i, of these equations do not change their asymptotic behaviour. Thus we assume that $\partial_i \psi_0 = 0(r^{-5}), \ldots, \partial_i \partial_j D\psi_0 = 0(r^{-6})$, for $i, j = 2, 3$. It will also be assumed that the hypersurfaces $u =$ const. are not chosen to be asymptotically cylindrical or asymptotically plane.[8]

The main result to be shown, under the above conditions, is that the other four components $\psi_1, \psi_2, \psi_3, \psi_4$ of the Weyl spinor will have the asymptotic forms

$$\begin{aligned} \psi_1 &= 0(r^{-4}), & \psi_2 &= 0(r^{-3}) \\ \psi_3 &= 0(r^{-2}), & \psi_4 &= 0(r^{-1}) \end{aligned} \tag{10-33b}$$

To prove Eqs. (10-33b) we proceed as follows. Let us denote by P and Q the two matrices

$$P = \begin{pmatrix} \rho & \sigma \\ \bar{\sigma} & \rho \end{pmatrix}$$

$$Q = \begin{pmatrix} 0 & \psi_0 \\ \bar{\psi}_0 & 0 \end{pmatrix}$$

The first two equations of Eqs. (10-25) can then be written as

$$DP = p^2 + Q \tag{10-34}$$

Equation (10-34) has the solution

$$P = -(DY)Y^{-1} \tag{10-35}$$

Here the matrix Y, where

$$Y = \begin{pmatrix} y_1 & y_2 \\ \bar{y}_1 & \bar{y}_2 \end{pmatrix} \tag{10-36}$$

is a nonsingular solution of the matrix equation

$$DY = -PY \tag{10-37}$$

and therefore satisfies

$$D^2 Y = -QY \tag{10-38}$$

The asymptotic behaviour of the solution of Eq. (10-38) is[9]

$$DY = F + o(1) \tag{10-39}$$

$$Y = rF + o(r) \tag{10-40}$$

[8] This assumption means that certain very special choices of coordinate systems are to be ruled out. In the following both of these cases will be discussed.

[9] A relation of the form $f(u, r, x^i) = o(g(r))$ means $\lim_{r\to\infty} \{f(u, r, x^i)/g(r)\} = 0$ for any value of u and x^i, $i = 2, 3$. See F. Erdlye, *op. cit.*

where F is a constant matrix.[10] But $Q = 0(r^{-5})$. Hence, using Eqs. (10-38) and (10-40), we obtain

$$D^2 Y = -rQF + o(r^{-4}) = 0(r^{-4}) \tag{10-41}$$

By integrating this equation twice and comparing it with Eq. (10-39), we obtain

$$DY = F + 0(r^{-3}) \tag{10-42}$$

$$Y = rF + E + 0(r^{-2}) \tag{10-43}$$

where E is another constant matrix. Using Eqs. (10-42) and (10-43) in Eq. (10-35) then gives

$$P = -r^{-1}I + r^{-2}EF^{-1} + 0(r^{-3}) \tag{10-44}$$

where I is the 2×2 unit matrix and provided F is not singular.

If F is singular, i.e., its determinant vanishes, then the asymptotic behavior of the matrix P is quite different. When F is a nonzero matrix, then this is the asymptotically cylindrical case in which the matrix P becomes asymptotically proportional to a singular matrix, and ρ behaves as $\rho = -(1/2r) + 0(r^{-2})$. If $F = 0$, on the other hand, we obtain the asymptotically plane case for which $P = 0(r^{-3})$. In both cases the matrix E must be such that there are two linearly independent columns among those of the matrices E and F, since otherwise the matrix Y will have a vanishing determinant for all r and the solution for P of Eq. (10-34) does not exist anymore. We will exclude these asymptotically cylindrical and plane cases.

Equation (10-44) shows that the two spin coefficients ρ and σ are given by

$$\rho = -r^{-1} + 0(r^{-2}) \tag{10-45}$$

$$\sigma = 0(r^{-2}) \tag{10-46}$$

For further discussion the following lemma will be needed.

Lemma. *Let the $n \times n$ matrix A be independent of the affine parameter r and has no eigenvalue with a positive real part. Let also any pure imaginary eigenvalue of A be regular, i.e., its multiplicity is equal to the number of linearly independent eigenvectors corresponding to it. Then all the solutions of the equation*

$$Dy = (Ar^{-1} + B)y + b \tag{10-47}$$

are bounded as $r \to \infty$. Here y is a complex column matrix which is a function of r, B is an $n \times n$ complex matrix, and b is a column matrix with n elements, both of which are given as functions of r, where

$$B = 0(r^{-2}), \qquad b = 0(r^{-2}) \tag{10-48}$$

[10] E. Coddington and V. Levinson, *Theory of Ordinary Differential Equations*, McGraw-Hill, New York, 1955, p. 103.

The proof of this lemma is as follows. Let us denote r by exp l, then Eq. (10-47) can be written as

$$\frac{\mathrm{d}}{\mathrm{d}l}\begin{pmatrix} y \\ 1 \end{pmatrix} = \left\{ \left(\begin{array}{ccc|c} & A & & \begin{matrix} 0 \\ \vdots \\ 0 \end{matrix} \\ \hline 0 & \cdots & 0 & 0 \end{array} \right) + \left(\begin{array}{ccc|c} & C & & c \\ \hline 0 & \cdots & 0 & 0 \end{array} \right) \right\} \begin{pmatrix} y \\ 1 \end{pmatrix} \tag{10-49}$$

where the $n \times n$ matrix C is given by

$$C(l) = \mathrm{e}^{l} B(\mathrm{e}^{l}), \tag{10-50}$$

and the n-column matrix c is given by

$$c(l) = \mathrm{e}^{l} b(\mathrm{e}^{l}) \tag{10-51}$$

Now the equation

$$\frac{\mathrm{d}}{\mathrm{d}l}\begin{pmatrix} z \\ \zeta \end{pmatrix} = \left(\begin{array}{ccc|c} & A & & \begin{matrix} 0 \\ \vdots \\ 0 \end{matrix} \\ \hline 0 & \cdots & 0 & 0 \end{array} \right) \begin{pmatrix} z \\ \zeta \end{pmatrix} \tag{10-52}$$

has the solution

$$\begin{aligned} z &= \{\exp Al\} z_0 \\ \zeta &= \zeta_0 \end{aligned} \tag{10-53}$$

which is bounded. In Eqs. (10-53) z_0 and ζ_0 are constants. By Eqs. (10-48), (10-50), and (10-51) it follows that the integrals with respect to l of the moduli of the elements of the matrices C and c are bounded as $l \to \infty$. Hence, by Levinson's theorem,[11] the solutions of Eq. (10-49) are also all bounded and are given by

$$y = \{\exp Al\} y_0 + 0(1) \tag{10-54}$$

where y_0 is constant. This proves the lemma.

Suppose now the matrices B, b, and y are functions also of the coordinates x^2 and x^3. Then, differentiating Eq. (10-47) with respect to x^i, where $i = 2, 3$, yields

$$D(\partial_i y) = (Ar^{-1} + B)(\partial_i y) + \{(\partial_i B) y + (\partial_i b)\} \tag{10-55}$$

Equation (10-55) is also of the form of Eq. (10-47) provided that

$$\partial_i B = 0(r^{-2}), \qquad \partial_i b = 0(r^{-2}) \tag{10-56}$$

[11] N. Levinson, *Am. J. Math.* **68**, 1 (1946).

Assuming Eqs. (10-56) to be valid, it follows from the lemma that $\partial_i y$ should be bounded. By repeating the argument we obtain the corresponding results for higher derivatives.

Consider now the x^i derivatives, where $i = 2, 3$, of the first two equations of Eqs. (10-25). The equations obtained can be written as

$$D\left\{r^2\,\partial_i\begin{pmatrix}\rho\\ \sigma\\ \bar\sigma\end{pmatrix}\right\} = \begin{pmatrix}2\rho + 2r^{-1} & \bar\sigma & \sigma\\ 2\sigma & 2\rho + 2r^{-1} & 0\\ 2\bar\sigma & 0 & 2\rho + 2r^{-1}\end{pmatrix} \times r^2\,\partial_i\begin{pmatrix}\rho\\ \sigma\\ \bar\sigma\end{pmatrix} + r^2\,\partial_i\begin{pmatrix}0\\ \psi_0\\ \bar\psi_0\end{pmatrix} \tag{10-57}$$

Applying the lemma with $A = 0$, and using Eqs. (10-45) and (10-46), we obtain

$$\partial_i\rho = 0(r^{-2}) \tag{10-58}$$

$$\partial_i\sigma = 0(r^{-2}) \tag{10-59}$$

Applying the lemma again to the next two x^i derivatives of Eqs. (10-57) yields

$$\partial_i\,\partial_j\rho,\quad \partial_i\,\partial_j\sigma,\quad \partial_i\,\partial_j\,\partial_k\rho,\quad \partial_i\,\partial_j\,\partial_k\sigma = 0(r^{-2}) \tag{10-60}$$

Next, using Eqs. (10-23), one can apply the lemma to the first of Eqs. (10-24), the third and fourth of Eqs. (10-25), the first and second of Eqs. (10-26), and their complex conjugates, with the matrix y being the column built from the twelve elements

$$r^4\psi_1,\ r^4\bar\psi_1,\ r\alpha,\ r\bar\alpha,\ r\beta,\ r\bar\beta,\ r\xi^2,\ r\bar\xi^2,\ r\xi^3,\ r\bar\xi^3,\ \omega,\ \bar\omega$$

and the matrices b, A, are given by

$$b = 0 \tag{10-61}$$

$$A = \left(\begin{array}{ccc|cc} 0 & \cdots & 0 & 0 & 0\\ \vdots & & \vdots & \vdots & \vdots\\ 0 & \cdots & 0 & 0 & 0\\ \hline & & & -1 & 0\\ & -a & & & \\ & & & 0 & -1\end{array}\right) \tag{10-62}$$

$$a = \begin{pmatrix}0\,0\,0\,1\,1\,0\,0\,0\,0\,0\\ 0\,0\,1\,0\,0\,1\,0\,0\,0\,0\end{pmatrix} \tag{10-63}$$

whereas the elements of the matrix B are functions of order $0(r^{-2})$ of the variables r, ρ, σ, ψ_0, $D\psi_0$, and $\partial_i\psi_0$. It therefore follows that

$$\psi_1 = 0(r^{-4}) \tag{10-64}$$

and

$$\alpha, \beta, \xi^i = 0(r^{-1})$$

$$\omega = 0(1) \tag{10-65}$$

Furthermore, by the first of Eqs. (10-24) we obtain

$$D\psi_1 = 0(r^{-5}) \tag{10-66}$$

Taking now successive x^i derivatives, and using Eqs. (10-58), (10-59), and (10-60), the lemma then also gives

$$\partial_i \psi_1, \quad \partial_i \partial_j \psi_1, \quad \partial_i \partial_j \partial_k \psi_1 = 0(r^{-4}) \tag{10-67}$$

$$\partial_i \alpha, \quad \partial_i \beta, \quad \partial_i \xi^j, \quad \partial_i \partial_j \alpha, \quad \partial_i \partial_j \beta, \quad \partial_i \partial_j \xi^k,$$

$$\partial_i \partial_j \partial_k \alpha, \quad \partial_i \partial_j \partial_k \beta, \quad \partial_i \partial_j \partial_k \xi^l = 0(r^{-1}) \tag{10-68}$$

$$\partial_i \omega, \quad \partial_i \partial_j \omega, \quad \partial_i \partial_j \partial_k \omega = 0(1) \tag{10-69}$$

Hence, using the first of Eqs. (10-24), one obtains

$$\partial_i D\psi_1, \quad \partial_i \partial_j D\psi_1 = 0(r^{-5}) \tag{10-70}$$

We next apply the second of Eqs. (10-24) and the seventh and eighth of Eqs. (10-25). The lemma is now applied, with the matrix y being the column built from the elements $r^3\psi_2$, $r\lambda$, $r\mu$, and the matrix A is taken as zero, whereas the matrices B and b are built of elements of order $0(r^{-2})$ which are functions of r, ρ, σ, ψ_0, α, ψ_1, ω, $D\psi_1$, ξ^i, $\partial_i \psi_1$. It therefore follows that

$$\psi_2 = 0(r^{-3}) \tag{10-71}$$

$$\lambda, \mu = 0(r^{-1}) \tag{10-72}$$

Consequently

$$D\psi_2 = 0(r^{-4}) \tag{10-73}$$

$$\partial_i \psi_2, \quad \partial_i \partial_j \psi_2 = 0(r^{-3}) \tag{10-74}$$

$$\partial_i \lambda, \quad \partial_i \mu, \quad \partial_i \partial_j \lambda, \quad \partial_i \partial_j \mu = 0(r^{-1}) \tag{10-75}$$

$$\partial_i D\psi_2 = 0(r^{-4}) \tag{10-76}$$

Finally, the lemma may be applied to the third of Eqs. (10-24) with $y = r^2\psi_3$, and to the fourth of Eqs. (10-24) with $y = r\psi_4$ one obtains

$$\psi_3 = 0(r^{-2}) \tag{10-77}$$

$$\left.\begin{aligned} \partial_i \psi_3 &= 0(r^{-2}) \\ D\psi_3 &= 0(r^{-2}) \end{aligned}\right. \tag{10-78}$$

and

$$\psi_4 = 0(r^{-1}) \tag{10-79}$$

One furthermore obtains

$$\begin{aligned} \tau &= 0(r^{-1}) \\ \gamma,\ \nu,\ X^i &= 0(1) \\ U &= 0(r) \end{aligned} \tag{10-80}$$

which, by Eqs. (10-22), leads to

$$\begin{aligned} g^{11} &= 0(r) \\ g^{1j} &= 0(1) \\ g^{ij} &= 0(r^{-2}) \end{aligned} \tag{10-81}$$

with $i, j = 2, 3$.[12]

Sachs Peeling-off Theorem

We have seen that under certain general assumptions of approach to flatness at infinity, which one expects to find in radiative empty spaces, the Riemann tensor exhibits a characteristic asymptotic behaviour of the form

$$\psi_n = 0(r^{n-5}) \tag{10-82}$$

The space may thus be divided into five regions: a near zone, where all terms are important; three transition zones, where ψ_0, ψ_1, and ψ_2 become negligible in turn; and finally, the radiation zone in which one can neglect ψ_0, ψ_1, ψ_2, ψ_3 and only ψ_4 remains important. In the last zone the Riemann tensor is essentially null. The fourth zone is essentially type III whereas the third zone is essentially algebraically special or usually type II. The second zone is essentially a region in which there are "geodesic rays" in the terminology of Sachs.[13] Finally, the first zone is of "general" type. Consequently, as one moves backwards from infinity along a suitable null geodesic, the principal null directions "peel off" one by one from the outgoing radial direction. This behaviour was first pointed out by Sachs for the linearized general relativity.

Comparison with Electrodynamics

The analogy between the above property of the gravitational field and the electromagnetic field is striking. In the electromagnetic case there are three regions: the near zone, where r^{-3} terms are important; the transition zone, where r^{-2} terms are important; and the radiation zone where the field behaves like r^{-1} and is null.

[12] For a further discussion of the behaviour of asymptotically flat empty spaces see E. T. Newman and T. W. J. Unti, *J. Math. Phys.* **3,** 891 (1962).

[13] R. K. Sachs, *Proc. R. Soc. Lond.* (A)**270,** 103 (1962).

In terms of the electromagnetic field components ϕ_0, ϕ_1, and ϕ_2 of Eqs. (10-28), one has

$$\begin{aligned} \phi_0 &= 0(r^{-3}) \\ \phi_1 &= 0(r^{-2}) \\ \phi_2 &= 0(r^{-1}) \end{aligned} \tag{10-83}$$

The two electromagnetic principal null directions exhibit, in general, the same property of "peeling off" as in the gravitational case.[14]

PROBLEMS

10.1 Prove Eqs. (10-2) and (10-3).

10.2 Prove that the spin coefficients π, λ, μ, and ν are analogous to κ, ρ, σ and τ when the congruence used is n_μ instead of l_μ, by deriving the analogous equations to Eqs. (10-1) to (10-4).

10.3 Prove Eqs. (10-5) to (10-7).

10.4 Use Eq. (9-70) to show that if ϕ is a scalar function, then

$$\begin{aligned} &(\nabla_{ab'}\nabla_{cd'} - \nabla_{cd'}\nabla_{ab'})\phi \\ &= \{(B_{ab'})_c{}^f\nabla_{fd'} + (B^\dagger{}_{b'a})^{f'}{}_{d'}\nabla_{cf'} \\ &\quad - (B_{cd'})_a{}^f\nabla_{fb'} - (B^\dagger{}_{d'c})^{f'}{}_{b'}\nabla_{af'}\}\phi \end{aligned}$$

[This equation is written explicitly in Appendix D, Eqs. (D-3)]. Show, furthermore, that if one takes for the function ϕ in the above equation the coordinates x^μ, then the resulting equation is Eq. (9-73).

10.5 Prove that a necessary and sufficient condition for $\nu = \lambda = 0$ in empty space is $\psi_3 = \psi_4 = 0$.

10.6 Prove Eqs. (10-24), (10-25) and (10-26).

10.7 Prove Eqs. (10-30).

10.8 Show that the spin coefficients are related to the Ricci rotation coefficients $\gamma^m{}_{np}$ [see L. P. Eisenhart, *Riemannian Geometry*, Princeton University Press, 1949] by

$$\begin{aligned} &\kappa = \gamma_{131}, && \varepsilon = \tfrac{1}{2}(\gamma_{121} - \gamma_{341}), && \rho = \gamma_{134} \\ &\lambda = -\gamma_{244}, && \alpha = \tfrac{1}{2}(\gamma_{124} - \gamma_{344}), && \sigma = \gamma_{133} \\ &\mu = -\gamma_{243}, && \beta = \tfrac{1}{2}(\gamma_{123} - \gamma_{343}), && \nu = -\gamma_{242} \\ &\pi = -\gamma_{241}, && \gamma = \tfrac{1}{2}(\gamma_{122} - \gamma_{342}), && \tau = \gamma_{132} \end{aligned}$$

10.9 Show that the spin coefficients can be expressed as

$$\begin{aligned} &\pi = -n_{\mu;\nu}\bar{m}^\mu l^\nu, && \alpha = \tfrac{1}{2}(l_{\mu;\nu}n^\mu - m_{\mu;\nu}\bar{m}^\mu)\bar{m}^\nu \\ &\rho = l_{\mu;\nu}m^\mu\bar{m}^\nu, && \kappa = l_{\mu;\nu}m^\mu l^\nu \\ &\lambda = -n_{\mu;\nu}\bar{m}^\mu\bar{m}^\nu, && \varepsilon = \tfrac{1}{2}(l_{\mu;\nu}n^\mu - m_{\mu;\nu}\bar{m}^\mu)l^\nu \\ &\mu = -n_{\mu;\nu}\bar{m}^\mu m^\nu, && \gamma = \tfrac{1}{2}(l_{\mu;\nu}n^\mu - m_{\mu;\nu}\bar{m}^\mu)n^\nu \\ &\tau = l_{\mu;\nu}m^\mu n^\nu, && \sigma = l_{\mu;\nu}m^\mu m^\nu \\ &\nu = -n_{\mu;\nu}\bar{m}^\mu n^\nu, && \beta = \tfrac{1}{2}(l_{\mu;\nu}n^\mu - m_{\mu;\nu}\bar{m}^\mu)m^\nu \end{aligned}$$

[14] See also J. M. Blatt and V. F. Weisskopf, *Theoretical Nuclear Physics*, John Wiley, New York, 1952, footnote on page 803.

10.10 Show that the coordinate transformation $u = t - r$ transforms the Minkowski line elements $ds^2 = dt^2 - dr^2 - r^2(d\theta^2 + \sin^2\theta\, d\phi^2)$ into the null-coordinate line element $ds^2 = du^2 + 2\,du\,dr - r^2(d\theta^2 + \sin^2\theta\, d\phi^2)$. The surfaces $u =$ const. are then just the light cones emanating from the origin $r = 0$. Choose the tetrad so that l^μ is the outward vector tangent to the cone, n^μ is the inward null vector pointing toward the origin, and m^μ and $\bar{m}^\mu$ are vectors tangent to the two-dimensional sphere defined by $r =$ const. and $u =$ const. Show that these vectors in the null coordinate system $x^0 = u$, $x^1 = r$, $x^2 = \theta$, $x^3 = \phi$ will have the form

$$l^\mu = \delta_1{}^\mu$$

$$m^\mu = \frac{1}{\sqrt{2}}\frac{1}{r}\left(\delta_2{}^\mu + \frac{i}{\sin\theta}\delta_3{}^\mu\right)$$

$$n^\mu = \delta_0{}^\mu - \tfrac{1}{2}\delta_1{}^\mu$$

and $\bar{m}^\mu$ is the complex conjugate to m^μ. [See A. I. Janis and E. T. Newman, *J. Math. Phys.* **6**, 902 (1965).]

10.11 Use the results of Problem 10.10 to show that the spin coefficients in the null coordinate system in flat space are given by

$$\pi = \kappa = \varepsilon = \lambda = \gamma = \nu = \tau = \sigma = 0,$$

$$\rho = -\frac{1}{r}, \qquad \alpha = -\frac{1}{2\sqrt{2}r}\cot\theta$$

$$\beta = \frac{1}{2\sqrt{2}r}\cot\theta, \qquad \mu = -\frac{1}{2r}$$

Show that the Maxwell equations (10-27) now become

$$\partial\phi_1/\partial r + 2\phi_1/r = 2^{-1/2}r^{-1}(\mathscr{D} + \cot\theta)\phi_0$$

$$\partial\phi_2/\partial r + \phi_2/r = 2^{-1/2}r^{-1}\bar{\mathscr{D}}\phi_1$$

$$\partial\phi_0/\partial u - \tfrac{1}{2}\partial\phi_0/\partial r - \phi_0/2r = 2^{-1/2}r^{-1}\mathscr{D}\phi_1$$

$$\partial\phi_1/\partial u - \tfrac{1}{2}\partial\phi_1/\partial r - \phi_1/r = 2^{-1/2}r^{-1}(\mathscr{D} + \cot\theta)\phi_2$$

where $\mathscr{D} = \partial/\partial\theta + (i/\sin\theta)\,\partial/\partial\phi$, and $\bar{\mathscr{D}}$ is its complex conjugate. Solve these equations and find the monopole, dipole, and quadrupole solutions. In particular, show that the Coulomb solution is given by

$$\phi_0 = 0, \quad \phi_1 = e/2r^2, \quad \phi_2 = 0$$

where e is the charge. [See A. I. Janis and E. T. Newman, *J. Math. Phys.* **6**, 902 (1965).]

10.12 Show that if one identifies the neutrino spinor η_A with the spinor $\zeta_0{}^A$ of the spin frame, then the neutrino equations (10-31) reduce to the conditions $\rho = \varepsilon$ and $\tau = \beta$. Show that if one assumes that both spinors of the spin frame to satisfy the neutrino equation, then Eqs. (10-31) gives $\rho = \varepsilon$, $\tau = \beta$, $\alpha = \pi$ and $\gamma = \mu$. Discuss the physical meaning of this choice.

10.13 Show that Maxwell's equations (10-27) in free space can be obtained from combinations of Eqs. (7-48) and (7-50) of the form

$$\nabla^\nu(F_{\mu\nu} + H_{\mu\nu}) = 0$$

where $H_{\mu\nu}$ is defined by $\sigma_{\mu ab'}\sigma_{\nu cd'}\sigma^{\alpha cb'}\sigma^{\beta ad'}F_{\alpha\beta}$, and $F_{\mu\nu}$ is the Maxwell skew-symmetric tensor. Show that the above equation can be written in dyad notation as

$$\partial^{cd'}(F_{ab'cd'} + F_{cb'ad'}) = \{(B^{pd'})^c{}_p + (B^{\dagger q'c})_{q'}{}^{d'}\}(F_{ab'cd'} + F_{cb'ad'})$$
$$+ \{\delta_{b'}^{f'}(B^{cd'})^e{}_a + \delta_a^e(B^{\dagger d'c})_{b'}{}^{f'}\}(F_{ef'cd'}$$
$$+ F_{cf'ed'})$$

In the above equation $F_{ab'cd'} = \sigma^{\mu}{}_{ab'}\sigma^{\nu}{}_{cd'}F_{\mu\nu}$ are scalars (not matrices as in the gravitational case), and are related to ϕ_0, ϕ_1, ϕ_2 by $\phi_0 = F_{00'01'}$, $\phi_1 = \frac{1}{2}(F_{00'11'} + F_{10'01'})$, $\phi_2 = F_{10'11'}$. Show that the above dyad equation easily leads to Eqs. (10-27).

10.14 Show that Maxwell's equations *with sources* can be obtained from the equation

$$\nabla^{\nu}(F_{\mu\nu} + H_{\mu\nu}) = 4\pi J_{\mu}$$

where $F_{\mu\nu}$ and $H_{\mu\nu}$ are defined in Problem 10.13, and J_{μ} is the electromagnetic current. Show that this equation leads to the following equations for the electromagnetic field:

$$\delta\phi_0 - D\phi_1 = (2\alpha - \pi)\phi_0 - 2\rho\phi_1 + \kappa\phi_2 - 2\pi J_{00'}$$

$$\delta\phi_1 - D\phi_2 = \lambda\phi_0 - 2\pi\phi_1 + (2\varepsilon - \rho)\phi_2 - 2\pi J_{01'}$$

$$\Delta\phi_0 - \delta\phi_1 = (2\gamma - \mu)\phi_0 - 2\tau\phi_1 + \sigma\phi_2 - 2\pi J_{10'}$$

$$\Delta\phi_1 - \delta\phi_2 = \nu\phi_0 - 2\mu\phi_1 + (2\beta - \tau)\phi_2 - 2\pi J_{11'}$$

Here $J_{00'}$, $J_{01'}$, $J_{10'}$ and $J_{11'}$ are the dyad components of the current, $J_{ab'} = J_{\mu}\sigma^{\mu}{}_{ab'}$, thus $J_{00'}$ and $J_{11'}$ are real whereas $J_{01'}$ and $J_{10'}$ are complex and conjugate to each other. (Note that the π's appearing in the last four terms of the above equations are not spin coefficients but are the usual constant).

10.15 Find the Dirac equation for the electron written in terms of the 2-component spinors α^A and β^A instead of the usual 4-dimensional spinor. Show that when an electromagnetic potential A^{μ} is presented that the Dirac equation will have the form

$$(\nabla^{AB'} + ieA^{AB'})\alpha_A = \frac{m}{\sqrt{2}}\bar{\beta}^{B'}$$

$$(\nabla^{AB'} - ieA^{AB'})\beta_A = \frac{m}{\sqrt{2}}\bar{\alpha}^{B'}$$

where $A^{AB'} = \tilde{\sigma}^{AB'}_{\mu}A^{\mu}$ and m is the mass of the electron. Show that in dyad notation these equations yield:

$$(\Delta + ieA_{11'} - \gamma + \mu)\alpha_0 + (-\delta - ieA_{01'} - \beta + \tau)\alpha_1 = \frac{m}{\sqrt{2}}\bar{\beta}_1$$

$$(\bar{\delta} + ieA_{10'} - \alpha + \pi)\alpha_0 + (-D - ieA_{00'} - \varepsilon + \rho)\alpha_1 = \frac{m}{\sqrt{2}}\bar{\beta}_0$$

$$(\Delta - ieA_{11'} - \gamma + \mu)\beta_0 + (-\delta + ieA_{01'} - \beta + \tau)\beta_1 = \frac{m}{\sqrt{2}}\bar{\alpha}_1$$

$$(\bar{\delta} - ieA_{10'} - \alpha + \pi)\beta_0 + (-D + ieA_{00'} - \varepsilon + \rho)\beta_1 = \frac{m}{\sqrt{2}}\bar{\alpha}_0$$

where $\alpha_0, \alpha_1, \beta_0, \beta_1, A_{00}, A_{01'}, A_{10'}$, and $A_{11'}$ are the dyad components of the spinors α_A, β_A and $A_{AB'}$.

CHAPTER

ELEVEN

SOME EXACT SOLUTIONS OF THE GRAVITATIONAL FIELD EQUATIONS

After writing the gravitational field equations that connect the Riemann tensor, the spin coefficients, and the tetrad of null vectors in Chapter 9, and after analysing the field functions and their geometrical meaning in Chapter 10, we are now in a position to solve these equations. Exact solutions to the free-field Einstein equations are obtained. We first find exact solutions that contain hypersurface orthogonal geodesic rays. These include the familiar Robinson–Trautman solution, and the spherical and cylindrical classes of the Newman–Tamburino solutions. We consequently solve the field equations for another kind of solution, the Newman–Tamburino–Unti solution, which is a generalization of the Schwarzschild metric. Finally, we solve the field equations to find out all type D vacuum metrics. The eleven metrics obtained are due to Kinnersley, and include the familiar Kerr metric. This chapter is not a survey of exact solutions in general relativity but rather a demonstration of the power of the methods developed so far for obtaining such solutions. Other exact solutions are given in Section 7-5.

11-1 SOLUTIONS CONTAINING HYPERSURFACE ORTHOGONAL GEODESIC RAYS

We now solve the gravitational field equations, obtained in the previous two chapters, to find empty space solutions with hypersurface orthogonal null geodesic rays. These solutions include the Robinson–Trautman and the Newman–Tamburino solutions.

Divergence, Curl, and Shear

A congruence of null geodesics l_μ in a hyperbolic Riemannian space can be characterized by three scalar functions. These functions are the *divergence*, *curl*, and (complex) *shear*, denoted by θ, ω, and σ, respectively, of the congruence. Such a congruence is associated to any empty space solution of the Einstein field equations possessing a family of geodesic rays, i.e., principal null directions of the

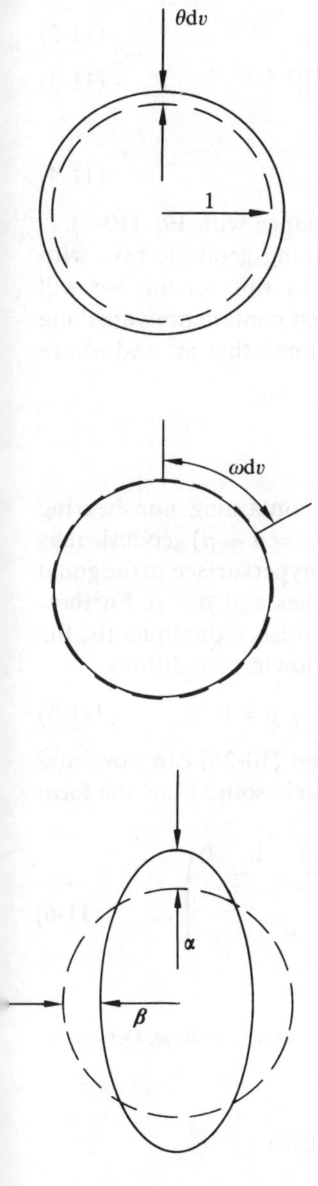

Figure 11.1 The shadow (solid lines) cast by an object will differ from the parallel displace of the object (dotted lines) by an expansion, a rotation, and shear. If the screen is an infinitesimal distance dv from the object then the shadow is expanded, rotated, and sheared by the respective amounts

$$-\theta \, dv = \tfrac{1}{2} l^\mu{}_{;\mu} \, dv$$

$$\omega \, dv = \{\tfrac{1}{2} l_{[\mu;\nu]} l^{\mu;\nu}\}^{1/2} \, dv$$

$$|\sigma| \, dv = \{\tfrac{1}{2}[l_{(\mu;\nu)} l^{\mu;\nu} - \tfrac{1}{2}(l^\mu{}_{;\mu})^2]\}^{1/2} \, dv = \tfrac{1}{2}(\alpha + \beta)$$

Riemann tensor.[1] The geometrical significance of the three scalar functions θ, ω, and σ was discussed by Sachs.[2] Figure 11.1 describes their geometrical meaning. It is then convenient to classify the vacuum solutions of the Einstein field equations according to the vanishing or nonvanishing of the divergence θ, the curl ω, and the shear σ of the geodesic rays.

The three scalar functions θ, ω, and σ are related to two of the spin coefficients ρ and σ. When the null congruence is affinely parametrized, one has:

$$\theta = -\tfrac{1}{2}\,\text{div}\, l^\mu = -\tfrac{1}{2}l^\mu{}_{;\mu} = \text{Re}\,\rho \tag{11-1}$$

$$\omega = \tfrac{1}{2}\,\text{curl}\, l^\mu = \{\tfrac{1}{2}l_{[\mu;\nu]}\,l^{\mu;\nu}\}^{1/2} = \text{Im}\,\rho \tag{11-2}$$

$$|\sigma| = \text{shear}\, l^\mu = \{\tfrac{1}{2}[l_{(\mu;\nu)}\,l^{\mu;\nu} - \tfrac{1}{2}(l^\mu{}_{;\mu})^2]\}^{1/2} \tag{11-3}$$

Hence one has

$$\rho = \theta + i\omega \tag{11-4}$$

as can be seen from Eqs. (11-1) and (11-2) and in accordance with Eq. (10-2).

Kundt[3] has considered the class of metrics containing geodesic rays with vanishing divergence, curl, and shear, $\theta = \omega = \sigma = 0$. In this section we will obtain solutions to the vacuum Einstein field equations that contain nonvanishing divergence, $\theta \neq 0$. Throughout the section it will be assumed that m^μ and n^μ are parallelly propagated along the geodesic rays.

Robinson–Trautman Solution

Robinson and Trautman[4] have presented all metrics containing nonshearing ($\sigma = 0$) and noncurling ($\omega = 0$) hypersurface orthogonal ($\tau = \bar{\alpha} + \beta$) geodesic rays l^μ, but with *nonvanishing* divergence, $\theta \neq 0$. Since l^μ is a hypersurface orthogonal geodesic ray, then $\kappa = 0$, and as a consequence ψ_0 vanishes and $\bar{\rho} = \rho$. Furthermore, since the geodesic ray is shear-free, ψ_1 must also vanish. Consequently, the Robinson–Trautman solution is characterized by the following conditions:

$$\psi_0 = \psi_1 = \kappa = \pi = \varepsilon = \sigma = \rho - \bar{\rho} = 0; \qquad \rho \neq 0 \tag{11-5}$$

The gravitational field equations (10-24), (10-25), and (10-26) can now, and throughout this section, be used and integrated. The metric sought has the form given by Eq. (10-18), namely

$$g^{\mu\nu} = l^\mu n^\nu + n^\mu l^\nu - m^\mu \bar{m}^\nu - \bar{m}^\mu m^\nu = \left(\begin{array}{c|ccc} 0 & 1 & 0 & 0 \\ \hline 1 & & & \\ 0 & & g^{ab} & \\ 0 & & & \end{array}\right) \tag{11-6}$$

[1] Principal null directions of the Riemann tensor are also known, sometimes, as Ruse, Debever, or Penrose vectors.

[2] R. Sachs, *Proc. R. Soc. Lond.* (A) **264,** 309 (1961).

[3] W. Kundt, *Z. Physik.* **163,** 77 (1961).

[4] I. Robinson and A. Trautman, *Proc. R. Soc. Lond.* (A) **265,** 463 (1962).

where the null tetrad is given by

$$\begin{aligned} l^\mu &= \delta_1{}^\mu \\ m^\mu &= \omega\, \delta_1{}^\mu + \xi^i\, \delta_i{}^\mu \\ \bar{m}^\mu &= \bar{\omega}\, \delta_1{}^\mu + \bar{\xi}^i\, \delta_i{}^\mu \\ n^\mu &= \delta_0{}^\mu + U\, \delta_1{}^\mu + X^i\, \delta_i{}^\mu \end{aligned} \tag{11-6a}$$

and $i = 2, 3$.

One can see that the form of this metric is invariant under the following coordinate transformations:

$$\begin{aligned} u' &= u \\ r' &= r + R(x^0, x^2, x^3) \\ x'^i &= x^i \end{aligned} \tag{11-7a}$$

which represents a *shift of the origin* of the coordinate r,

$$\begin{aligned} u' &= s(u) \\ r' &= r/\dot{s} \\ x'^i &= x^i \end{aligned} \tag{11-7b}$$

which *relabels hypersurfaces*, and

$$\begin{aligned} u' &= u \\ r' &= r \\ x'^i &= x'^i(x^0, x^2, x^3) \end{aligned} \tag{11-7c}$$

which *relabels geodesics*,[5] where $i = 2, 3$. Here a dot denotes differentiation with respect to the parameter u.

In addition to these transformations, there is a freedom in choosing tetrad transformations that do not affect the vectors l^μ or the orthogonality conditions (see Appendix E). These transformations are the *null* rotation and the spatial rotation. They are given by the transformations (8-34) and (8-35), where the matrices g of the group $SL(2, C)$ are given by $g_1(z)$ and $g_2(z)$ of Eq. (8-41). Using Eq. (8-40) one obtains (Problem 11.1):

$$\begin{aligned} l'^\mu &= l^\mu \\ m'^\mu &= \bar{z} l^\mu + m^\mu \\ n'^\mu &= z\bar{z} l^\mu + z m^\mu + \bar{z}\bar{m}^\mu + n^\mu \end{aligned} \tag{11-8a}$$

[5] It should be noted that where a hypersurface orthogonal geodesic ray exists, the null hypersurfaces $u =$ const. are fixed by the curvature tensor by Eq. (10-8) and the fact that $l_\mu = u_{,\mu}$.

for the null rotation, and

$$l'^{\mu} = l^{\mu}$$
$$m'^{\mu} = e^{iC} m^{\mu} \tag{11-8b}$$
$$n'^{\mu} = n^{\mu}$$

for the spatial rotation, where C is real and independent of the coordinate r.

From Eqs. (10-25) one obtains, under the conditions outlined above, the field equation[6]

$$D\rho = \rho^2$$

whose solution is easily found to be

$$\rho = -(r + \rho_0)^{-1}$$

where ρ_0 is a function that does not depend on r. One can choose ρ_0 as zero by means of the coordinate transformation (11-7a). Hence one obtains for the spin coefficient ρ the following:

$$\rho = -1/r \tag{11-9}$$

This value for ρ can now be used to integrate the rest of the field equations. Equations (10-25c), (10-25d), and (10-25e) then give:

$$D\tau = -\tau/r$$
$$D\alpha = -\alpha/r$$
$$D\beta = -\beta/r$$

The solutions of these equations are easily found to be given by

$$\tau = \tau_0/r \tag{11-10}$$
$$\alpha = \alpha_0/r \tag{11-11}$$
$$\beta = \beta_0/r \tag{11-12}$$

where τ_0, α_0, and β_0 are functions independent of r. However, by Eq. (10-7), one has the algebraic condition

$$\tau_0 = \bar{\alpha}_0 + \beta_0$$

Using the transformation (11-8) one can set τ_0 as zero, and consequently one obtains

$$\beta_0 = -\bar{\alpha}_0 \tag{11-13}$$
$$\tau = 0 \tag{11-14}$$

[6] E. T. Newman and L. A. Tamburino, *J. Math. Phys.* **3**, 902 (1962).

From Eq. (10-25k) one obtains

$$\delta\rho = \omega\,\partial(-1/r)/\partial r = 0$$

which yields[7]

$$\omega = 0 \tag{11-15}$$

Also, one finds that Eq. (10-25p) leads to

$$\lambda = 0 \tag{11-16}$$

Using the above results in Eqs. (10-24b), (10-25f), and (10-25h) then leads to the following:

$$\psi_2 = \psi_2{}^0/r^3 \tag{11-17}$$

$$\gamma = \gamma_0 - \psi_2{}^0/2r^2 \tag{11-18}$$

$$\mu = \mu_0/r - \psi_2{}^0/r^2 \tag{11-19}$$

where $\psi_2{}^0$, γ_0, and μ_0 are functions independent of r. But Eq. (10-26g) shows that $\mu = \bar{\mu}$. Hence both μ_0 and $\psi_2{}^0$ are real functions. Furthermore, Eqs. (10-26a), (10-26c) and (10-26d) give

$$\xi^i = \xi_0{}^i/r \tag{11-20}$$

$$X^i = X_0{}^i \tag{11-21}$$

$$U = U_0 - (\gamma_0 + \bar{\gamma}_0)r - \psi_2{}^0/r \tag{11-22}$$

where $\xi_0{}^i$ and $X_0{}^i$ are functions independent of r and $i = 2, 3$.

By using the coordinate transformation (11-7c) we can set

$$\xi_0{}^2 = -i\xi_0{}^3 = P(x^0, x^2, x^3) \tag{11-23}$$

The remaining coordinate freedom is

$$u' = s(u)$$

$$r' = r/\dot{s} \tag{11-24a}$$

$$\zeta' = f(u, \zeta)$$

where ζ and ζ' are complex variables defined by

$$\zeta = x^2 + ix^3 \tag{11-24b}$$

$$\zeta' = x'^2 + ix'^3 \tag{11-24c}$$

Substituting the above results into Eq. (10-26e) then leads to (Problem 11.2):

$$\xi_0{}^k X_0{}^i{}_{,k} - \partial\xi_0{}^i/\partial u - X_0{}^k \xi_0{}^i{}_{,k} = 2\bar{\gamma}_0\xi_0{}^i \tag{11-25a}$$

$$\mu_0 = U_0 \tag{11-25b}$$

from which one obtains

$$P\nabla X_0 = 0 \tag{11-26}$$

[7] The tetrad function ω here [see Eq. (10-20)] should not be confused with the curl function ω [see Eq. (11-2)].

In Eq. (11-26) the function P is defined by Eq. (11-23), the differentiation operator ∇ is defined by

$$\nabla = \partial/\partial x^2 + i\,\partial/\partial x^3 = 2\,\partial/\partial\bar{\zeta} \tag{11-27}$$

where ζ is given by Eq. (11-24b), and the function X_0 is defined by

$$X_0 = X_0{}^2 + iX_0{}^3 \tag{11-28}$$

Assume now the function $X_0 = X_0(u, \zeta)$ to be analytic. Using the coordinate transformation (11-24a), Eq. (11-26) then shows that one can set

$$X_0 = 0 \tag{11-29}$$

The remaining coordinate freedom is now

$$\begin{aligned} u' &= s(u) \\ r' &= r/\dot{s} \\ \zeta' &= f(\zeta) \end{aligned} \tag{11-30}$$

Equations (11-25a) and (11-29) then give

$$\bar{\gamma}^0 = -\tfrac{1}{2}\,\partial \ln P/\partial u \tag{11-31}$$

Furthermore, from Eq. (10-24f) one obtains

$$\delta\psi_2 = P\nabla\psi_2{}^0/r^4 = 0$$

Hence one obtains

$$\psi_2{}^0 = \psi_2{}^0(u) \tag{11-32}$$

From Eqs. (10-24c), (10-24d), and (10-25i) one obtains

$$\psi_3 = \psi_3{}^0/r \tag{11-33}$$

$$\psi_4 = \psi_4{}^0/r - (\bar{P}\bar{\nabla}\psi_3{}^0 + 2\alpha_0\psi_3{}^0)/r^2 \tag{11-34}$$

$$v = v_0 - \psi_3{}^0/r^2 \tag{11-35}$$

where $\psi_3{}^0$, $\psi_4{}^0$, and v_0 are functions independent of the variable r. The remaining equations (10-26f), (10-26h), (10-25j), (10-25l), and (10-25m) may be worked out in the same manner. They yield (Problem 11.3):

$$\alpha_0 = \tfrac{1}{2}\bar{P}\bar{\nabla} \ln P \tag{11-36}$$

$$\bar{v}_0 = -\tfrac{1}{2}P\nabla\,\partial(\ln P\bar{P})/\partial u \tag{11-37}$$

$$\psi_4{}^0 = \bar{P}\bar{\nabla}v_0 + 2\alpha_0 v_0 \tag{11-38}$$

$$\mu_0 = U_0 = -\tfrac{1}{2}P\bar{P}\nabla\bar{\nabla} \ln (P\bar{P}) \tag{11-39}$$

$$\psi_3{}^0 = \bar{P}\bar{\nabla}\mu_0 \tag{11-40}$$

Consequently, every function is now expressed in terms of the two functions P and $\psi_2{}^0$. Equation (10-24g) then relates the functions P and $\psi_2{}^0$ by

$$\{\partial/\partial u - 3(\partial P/\partial u)/P\}\psi_2{}^0 = P\nabla\psi_3{}^0 - 2\bar{\alpha}_0\psi_3{}^0 \tag{11-41}$$

All other equations are now identically safisfied. Furthermore, using Eq. (11-8b) one can finally set

$$P = \bar{P} \tag{11-42}$$

a fact that has already been incorporated in Eq. (11-41).

The metric tensor (11-6) now becomes

$$\begin{aligned} &g^{00} = 0,\ g^{01} = 1,\ g^{0i} = 0 \\ &g^{11} = 2U_0 - 4\gamma_0 r - 2\psi_2{}^0/r \\ &g^{1i} = 0 \\ &g^{ij} = -2(P^2/r^2)\,\delta^{ij} \end{aligned} \tag{11-43}$$

where U_0 is given by Eq. (11-39) and γ_0 by Eq. (11-31) and $i, j = 2, 3$. If $\psi_2 = 0$, then Eq. (11-41) becomes

$$\nabla^2(P^2\nabla^2 \ln P) = 0 \tag{11-44a}$$

where $\nabla^2 = \nabla\bar{\nabla}$. If $\psi_2 \neq 0$, on the other hand, then by using Eqs. (11-7b) one can set $\psi_2{}^0 = 1$ and Eq. (11-41) gives

$$\partial P/\partial u = (P^3/3)\nabla^2(P^2\nabla^2 \ln P) \tag{11-44b}$$

The only remaining coordinate freedom given by Eqs. (11-30) is

$$P' = P|\mathrm{d}f(\zeta)/\mathrm{d}\zeta|$$

This completes the Robinson–Trautman solution.

In the following we summarize the results that characterize the Robinson–Trautman metric:

(a) Spin Coefficients:

$$\begin{aligned} &\kappa = \varepsilon = \pi = \sigma = \tau = \lambda = 0 \\ &\rho = -1/r \\ &\alpha = -\beta = P\bar{\nabla} \ln P/2r \\ &\gamma = -\tfrac{1}{2}\,\partial \ln P/\partial u - \psi_2{}^0/2r^2 \\ &\mu = -\tfrac{1}{2}(P^2\nabla\bar{\nabla} \ln P^2)/r - \psi_2{}^0/r^2 \\ &\nu = -\tfrac{1}{2}P\nabla(\partial \ln P^2/\partial u) - \psi_3{}^0/r \end{aligned}$$

where $\psi_2{}^0 = 0, 1$, and $\psi_3{}^0 = -\tfrac{1}{2}P\bar{\nabla}(P^2\nabla\bar{\nabla} \ln P^2)$.

(b) Weyl Spinor:

$$\begin{aligned} &\psi_0 = \psi_1 = 0; \qquad \psi_2 = \psi_2{}^0/r^3 \\ &\psi_3 = -\tfrac{1}{2}r^{-2}\,P\bar{\nabla}(P^2\nabla\bar{\nabla} \ln P^2) \\ &\psi_4 = \psi_4{}^0/r - [P\bar{\nabla}\psi_3{}^0 + P(\bar{\nabla} \ln P)\psi_3{}^0]/r^2 \end{aligned}$$

where

$$\psi_4{}^0 = -\tfrac{1}{2}[P\bar{\nabla} + P(\bar{\nabla} \ln P)]P\nabla(\partial \ln P^2/\partial u)$$

(c) Tetrad: Using Eq. (11-6a) one obtains:

$$l^\mu = \delta_1{}^\mu$$

$$m^\mu = \xi^i\, \delta_i{}^\mu$$

$$\bar{m}^\mu = \bar{\xi}^i\, \delta_i{}^\mu$$

$$n^\mu = \delta_0{}^\mu + U\, \delta_1{}^\mu$$

where $i = 2, 3$, and $\xi^i = \xi_0{}^i/r$, $\xi_0{}^2 = -i\xi_0{}^3 = P(x^0, x^2, x^3)$, and

$$U = -P^2\nabla^2 \ln P + r\, \partial \ln P/\partial u - \psi_2{}^0/r$$

(d) The metric: The components of the metric tensor have already been given in Eq. (11-43).

Newman–Tamburino Solutions: Spherical Class

We now obtain the generalized solution for the shearing class ($\sigma \neq 0$) of noncurling ($\omega = 0$) hypersurface orthogonal geodesic rays with nonvanishing divergence ($\theta \neq 0$). There are two general subcases which are called *spherical* and *cylindrical.* The spherical class is characterized by

$$\begin{gathered} \psi_0 = \rho - \bar{\rho} = 0, \qquad \rho^2 \neq \sigma\bar{\sigma} \\ \rho \neq 0, \qquad \sigma \neq 0, \qquad \kappa = 0, \qquad \varepsilon = 0, \qquad \pi = 0 \end{gathered} \tag{11-45}$$

To integrate the gravitational field equations we proceed as follows. The field equations (10-25a) and (10-25b) may be expressed now as a matrix equation,

$$DM = M^2$$

or

$$DM^{-1} = -I$$

where I is the 2×2 unit matrix and M is a nonsingular matrix which, by Eq. (11-45), is given by

$$M = \begin{pmatrix} \rho & \sigma \\ \bar{\sigma} & \rho \end{pmatrix}$$

The solution of this matrix equation is given by

$$M^{-1} = \begin{pmatrix} \rho_0 - r & -\sigma_0 \\ -\bar{\sigma}_0 & \rho_0 - r \end{pmatrix}$$

and therefore the matrix M is given by

$$M = R^{-2}\begin{pmatrix} \rho_0 - r & \sigma_0 \\ \bar{\sigma}_0 & \rho_0 - r \end{pmatrix}$$

Here $R^2 = \det(M^{-1}) = (\rho_0 - r)^2 - \bar{\sigma}_0 \sigma_0 \neq 0$ by Eq. (11-45), and ρ_0 and σ_0 are functions independent of the coordinate r.

Using the coordinate transformation (11-7a) we can set $\rho_0 = 0$. Hence one obtains

$$\rho = -r/R^2 \tag{11-46}$$

$$\sigma = \sigma_0/R^2 \tag{11-47}$$

where $R^2 = r^2 - \bar{\sigma}_0 \sigma_0$.

Equation (10-24a), using Eq. (11-46), gives

$$\psi_1 = \psi_1{}^0/R^4 \tag{11-48}$$

where $\psi_1{}^0$ is a function independent of r. Equation (10-26a) and its complex conjugate are easily seen to be equivalent to the matrix equation

$$D\xi = M\xi$$

where ξ denotes the column matrix

$$\xi = \begin{pmatrix} \xi^i \\ \bar{\xi}^i \end{pmatrix}$$

If we now denote ξ by Mv, then v satisfies the equation

$$Dv = 0$$

whose solution can easily be found to be

$$v = -\begin{pmatrix} \xi_0{}^i \\ \bar{\xi}_0{}^i \end{pmatrix}$$

where $\xi_0{}^i$ are functions independent of the coordinate r. Hence one obtains from the relation $\xi = Mv$ the expression

$$\xi^i = R^{-2}(r\xi_0{}^i - \sigma_0 \bar{\xi}_0{}^i) \tag{11-49}$$

Using the coordinate transformation (11-7c), one can now set

$$\xi_0{}^2 = -i\xi_0{}^3 = P(x^0, x^2, x^3) \tag{11-50}$$

and the remaining coordinate freedom is now

$$\begin{aligned} u' &= s(u) \\ r' &= r/\dot{s} \\ \zeta' &= f(u, \zeta) \end{aligned} \tag{11-51}$$

where ζ and ζ' are given by Eqs. (11-24b) and (11-24c).

Next, Eqs. (10-25d) and (10-25e) are integrated to yield the r dependence of the functions α and β; the null transformation (11-8a) is used to eliminate a constant of integration. One obtains[8]

$$\alpha = \frac{\psi_1{}^0 \bar{\sigma}_0 L}{2a^3 R^2} + \frac{1}{R^2}\left[r\alpha_0 + \bar{\sigma}_0\left(\bar{\alpha}_0 - \frac{\psi_1{}^0}{2a^2}\right)\right] \tag{11-52}$$

$$\beta = -\frac{\psi_1{}^0 L}{2aR^2} - \frac{1}{R^2}(r\bar{\alpha}_0 + \sigma_0\alpha_0) \tag{11-53}$$

Here $a = |\sigma_0|$, α_0 is a function independent of r, whereas

$$L = \frac{1}{2}\ln\left(\frac{r+a}{r-a}\right)$$

We next turn to Eq. (10-26b) which yields

$$\omega = -\frac{L\bar{\psi}_1{}^0\sigma_0}{2a^3} + \frac{r\omega_0 - \sigma_0\bar{\omega}_0}{R^2} \tag{11-54}$$

where ω_0 is a function independent of r and whose value can be obtained from Eq. (10-25k). One finds

$$\omega_0 = \bar{\xi}_0{}^i\,\partial_i\sigma_0 + \bar{\psi}_1{}^0\sigma_0/2a^2 - 4\sigma_0\alpha_0 \tag{11-55}$$

Equations (10-26f) and (10-26c) give

$$\alpha_0 = \tfrac{1}{2}P\bar{\nabla}\ln P \tag{11-56}$$

$$X^i = X_0{}^i + \frac{z^i L}{2a^3} - \frac{rz^i - y^i}{2a^3R^2} - \frac{L(y^i r - a^2 z^i)}{2a^3R^2} \tag{11-57}$$

where the differentiation operator ∇ is given by Eq. (11-27), $X_0{}^i$ is a function independent of r, and y^i and z^i are defined by

$$y^i = \xi_0{}^i\psi_1{}^0\bar{\sigma}^0 + \bar{\xi}_0{}^i\bar{\psi}_1{}^0\sigma_0$$

$$z^i = \bar{\xi}_0{}^i\psi_1{}^0 + \xi_0{}^i\bar{\psi}_1{}^0$$

Equation (10-24b) then gives

$$\psi_2 = \frac{\psi_2{}^0}{R^3} - \frac{(\psi_2{}^0)^2\sigma_0 L}{a^3R^4} + \frac{r\bar{\sigma}_0 A_1}{a^2R^4} + \frac{B_1}{R^4} + \frac{\psi_1{}^0\bar{\sigma}_0}{a^3R^6}(rC - \sigma_0\bar{C}) \tag{11-58}$$

where $\psi_2{}^0$ is a function independent of the coordinate r, and A_1, B_1, and C are functions given by

$$A_1 = \xi_0{}^k\,\partial_k\psi_1{}^0 + 2\psi_1{}^0\bar{\alpha}_0 + 4\psi_1{}^0(\sigma_0\bar{\omega}_0 - \xi_0{}^k\,\partial_k a^2)/3a^2 \tag{11-59a}$$

$$B_1 = 2\psi_1{}^0\alpha_0 - \bar{\xi}_0{}^k\,\partial_k\psi_1{}^0 \tag{11-59b}$$

$$C = 2\sigma_0\bar{\omega}_0 + \xi_0{}^k\,\partial_k a^2 \tag{11-59c}$$

[8] L. A. Tamburino, Exact Empty Space Metrics Containing Geodesic Rays, Ph.D. Dissertation, University of Pittsburgh, 1962.

and Eq. (10-25f) gives

$$\gamma = \gamma_0 - \frac{r\psi_2{}^0}{a^2R} - \frac{|\psi_1{}^0|^2L^2r}{4a^4R^2} + \frac{(\psi_1{}^0)^2\bar{\sigma}_0L^2}{4a^4R^2} + \frac{rL}{2a^3R^2}(\psi_1{}^0\bar{\sigma}_0\bar{\alpha}_0 - \bar{\psi}_1{}^0\sigma_0\alpha_0)$$
$$+ \frac{L}{2aR^2}\left[\frac{|\psi_1{}^0|^2}{a^2} + (\psi_1{}^0\sigma_0 - \bar{\psi}_1{}^0\bar{\alpha}_0)\right] + \left(\frac{L}{2a^3} - \frac{r}{2a^2R^2}\right)\left[B_1 + \frac{3|\psi_1{}^0|^2}{4a^2}\right.$$
$$\left. + (\psi_1{}^0\alpha_0 - \bar{\psi}_1{}^0\bar{\alpha}_0) + \frac{\psi_1{}^0\bar{C}}{2a^2}\right] - \frac{1}{2a^2R^2}(\bar{\sigma}_0A_1 + \psi_1{}^0\bar{\sigma}_0\bar{\alpha}_0 - \bar{\psi}_1{}^0\sigma_0\alpha_0)$$
$$- \frac{\psi_1{}^0}{6a^2R^4}(\bar{\sigma}_0C - \bar{C}r) \tag{11-60}$$

where γ_0 is a function independent of r. Equations (10-25g) and (10-25h) now give

$$\lambda = -\frac{\bar{\sigma}_0\psi_2{}^0}{a^2R} - \frac{(\psi_1{}^0)^2\bar{\sigma}_0{}^2L^2r}{4a^6R^2} + \frac{rL}{2a^3R^2}\left[B_1\bar{\sigma}_0 + \frac{\psi_1{}^0\bar{\sigma}_0C}{2a^2}\right]$$
$$+ \frac{\bar{\sigma}_0{}^2}{2a^3}\left[\frac{(\psi_1{}^0)^2}{2a^2} + A_1 - \frac{\psi_1{}^0C}{6a^2}\right]\frac{L}{R^2} + \frac{r\lambda_0}{R^2}$$
$$- \frac{\bar{\sigma}_0}{R^2}\left[\mu_0 + \frac{B_1}{2a^2} + \frac{\psi_1{}^0\bar{C}}{4a^4}\right] + \frac{\psi_1{}^0\bar{\sigma}_0(Cr\bar{\sigma}_0 - \bar{C})}{12a^4R^4} \tag{11-61}$$

$$\mu = \frac{(\psi_1{}^0)^2\sigma_0L^2}{4a^4R^2} + \frac{rL}{a^3R^2}\left[\frac{(\psi_1{}^0)^2\bar{\sigma}_0}{4a^2} - \frac{\bar{\sigma}_0A_1}{2} + \frac{\psi_1{}^0\bar{\sigma}_0C}{12a^2}\right]$$
$$- \frac{L}{2aR^2}\left[B_1 + \frac{\psi_1{}^0\bar{C}}{2a^2}\right] + \frac{\mu_0r}{R^2} - \frac{1}{R^2}\left[\sigma_0\lambda_0 + \frac{\bar{\sigma}_0A_1}{2a^2}\right.$$
$$\left. + \frac{(\psi_1{}^0)^2\bar{\sigma}_0}{4a^4} + \frac{\psi_1{}^0\bar{\sigma}_0C}{12a^4}\right] + \frac{\psi_1{}^0(\bar{C}r - C\bar{\sigma}_0)}{4a^2R^4} \tag{11-62}$$

where λ_0 and μ_0 are functions which are independent of the coordinate r.

The next equation to be solved is Eq. (10-25l). Its solution leads to

$$\mu_0 = -\tfrac{1}{2}P\bar{P}\nabla\bar{\nabla}(\ln P\bar{P}) \tag{11-63}$$

This followed by Eq. (10-26d) which gives

$$U = U_0 - (\gamma_0 + \bar{\gamma}_0)r + \frac{R}{a^2}(\psi_2{}^0 + \bar{\psi}_2{}^0) + \frac{|\psi_1{}^0|^2L^2}{4a^4} + \frac{WL}{2aR^2}$$
$$- \frac{VLr}{2a^3R^2} + \frac{1}{12a^2R^2}(6V + \psi_1{}^0\bar{C} + \bar{\psi}_1{}^0C)$$
$$- \frac{r}{12a^4R^2}(6a^2W + \psi_1{}^0\bar{\sigma}_0C + \bar{\psi}_1{}^0\sigma_0\bar{C}) - \frac{SLr}{2a^3} + \frac{TL}{2a^3} \tag{11-64}$$

where U_0 is independent of r, and the functions S, T, V, W are given by

$$S = B_1 + \bar{B}_1 + 3|\psi_1{}^0|^2/2a^2 + (\psi_1{}^0\bar{C} + \bar{\psi}_1{}^0 C)/2a^2 \tag{11-65a}$$

$$T = W - (\bar{\sigma}_0 A_1 + \sigma_0 \bar{A}_1) + (\psi_1{}^0\bar{\sigma}_0 C + \bar{\psi}_1{}^0\sigma_0\bar{C})/6a^2 \tag{11-65b}$$

$$V = \psi_1{}^0\omega_0\bar{\sigma}_0 + \bar{\psi}_1{}^0\bar{\omega}_0\sigma_0 \tag{11-65c}$$

$$W = \psi_1{}^0\bar{\omega}_0 + \bar{\psi}_1{}^0\omega_0 \tag{11-65d}$$

and Eq. (10-26g) which gives

$$\xi_0{}^i\,\partial_i\bar{\omega}_0 = \bar{\xi}_0{}^i\,\partial_i\omega_0 = 2(\bar{\alpha}_0\bar{\omega}_0 - \alpha_0\omega_0) + (\bar{\sigma}_0\bar{\lambda}_0 - \sigma_0\lambda_0) + (\bar{\psi}_1{}^0\omega_0 - \psi_1{}^0\bar{\omega}_0)/a^2 - [(\psi_1{}^0)^2\bar{\sigma}_0 - (\bar{\psi}_1{}^0)^2\sigma_0]/4a^2 \tag{11-66}$$

So far the results obtained above contain the Robinson–Trautman solution when $\psi_1{}^0$ and σ tend to zero.

The next equation to be considered is Eq. (10-26e). This equation is by far the longest and most tedious equation that is examined in closed form. However, the results obtained from it both simplify the above results and, in addition, give extra differential equations that eventually *rule out* the possibility of obtaining the Robinson–Trautman solution. The following is a summary of these results:

$$\psi_2{}^0 = 0 \tag{11-67}$$

$$\xi_0{}^j\,\partial_j X_0{}^i - \partial_j\xi_0{}^i X_0{}^j - \partial\xi_0{}^i/\partial u = 2\bar{\gamma}_0\xi_0{}^i \tag{11-68}$$

As in the Robinson–Trautman case, Eq. (11-68) and the coordinate transformation (11-51) yield

$$X_0{}^i = 0 \tag{11-69}$$

and

$$\bar{\gamma}_0 = -\tfrac{1}{2}(\partial \ln P/\partial u) \tag{11-70}$$

and the remaining coordinate freedom is now

$$u' = s(u)$$

$$r' = r/\dot{s} \tag{11-71}$$

$$\zeta' = f(\zeta)$$

Furthermore, one obtains from Eq. (10-26e):

$$U_0 = \mu_0 + (\bar{\psi}_1{}^0\sigma_0\bar{\omega}_0 + \psi_1{}^0\bar{\sigma}_0\omega_0 + |\psi_1{}^0|^2/2)/2a^4 \tag{11-72}$$

$$\bar{\lambda}_0 = \partial\sigma_0/\partial u + \sigma_0(3\bar{\gamma}_0 - \gamma_0) \tag{11-73}$$

$$\xi_0{}^k\,\partial_k a^2 = -2\sigma_0\bar{\omega}_0 \tag{11-74}$$

$$\bar{\xi}_0{}^k\,\partial_k\psi_1{}^0 = \psi_1{}^0(\bar{\psi}_1{}^0 + 4\alpha_0 a^3 + \bar{\xi}_0{}^k\,\partial_k a^2)/2a^2 \tag{11-75}$$

$$\xi_0{}^k\,\partial_k\psi_1{}^0 = -\psi_1{}^0(2\bar{\alpha}_0 + 3\sigma_0\bar{\omega}_0/a^2) \tag{11-76}$$

Using these results one obtains for the functions A_1, B_1, and C the following expressions:

$$A_1 = \psi_1{}^0\sigma_0\bar{\omega}_0/a^2 \tag{11-77a}$$

$$B_1 = \psi_1{}^0\bar{\sigma}_0\omega_0/a^2 - |\psi_1{}^0|^2/2a^2 \tag{11-77b}$$

$$C = 0 \tag{11-77c}$$

The last equation to be worked out in a closed form is Eq. (10-24e). One simply finds that it is identically satisfied.

It has been shown by Tamburino that the main features of the above calculations are analogous to those of the Robinson–Trautman case discussed above. However, some of the functions become increasingly complicated and their manipulation is extremely tedious, though straightforward. Some equations are complicated by having r dependence appear in various products of the functions R and L, such as L/R^2, Lr/R^2, L/R^4, etc. These products were shown to be linearly independent, and therefore the matching of coefficients of like products becomes equivalent to matching powers of r. All field functions have been obtained in closed form except for the spin coefficient ν and the Weyl components ψ_3 and ψ_4. In order to avoid lengthy expressions in the final phases of the calculations, a power series expansion in r for all equations containing the functions ν, ψ_3, and ψ_4 can be resorted to. It is conceivable to assume that this use of power series yields all obtainable information in the lowest powers, and that a more exhaustive treatment would be fruitless.

The following system of equations are then easily derived from Eqs. (11-55), (11-74), (11-75), and (11-76):

$$\nabla \ln (\psi_1{}^0\bar{P}/a^3) = 0 \tag{11-78a}$$

$$\bar{\nabla} \ln (\psi_1{}^0\sigma_0/P^3) = 0 \tag{11-78b}$$

$$\psi_1{}^0 = a^2 P\nabla \ln (\bar{P}^4/a^2\bar{\sigma}_0{}^2) \tag{11-78c}$$

In the Robinson–Trautman case the functions $\psi_1{}^0$ and σ_0 are zeros and hence Eqs. (11-78) are trivial identities. In the present case $\psi_1{}^0 \neq 0$ and $\sigma_0 \neq 0$, and therefore Eqs. (11-78) yield the following results:

$$P = \tfrac{1}{8}C_1^{-1} \exp [4(G + \bar{G}) + C_1(T_1 + \bar{T}_1)] \tag{11-79a}$$

$$a = \tfrac{1}{8}C_1^{-1} \exp [2C_1(T_1 + \bar{T}_1)] \tag{11-79b}$$

$$\exp (i\phi) = \exp [8(\bar{G} - G) + 4C_1(T_1 - \bar{T}_1)] \tag{11-79c}$$

where

$$P = p \exp (i\phi p) \tag{11-79d}$$

$$\sigma = a \exp (i\phi a) \tag{11-79e}$$

$$\phi = 2\phi p - \phi a \tag{11-79f}$$

Here C_1 is a real function of u only, G is an arbitrary analytic function of the variables u and ζ, and

$$T_1 = \int \exp(-8G)\, d\zeta. \tag{11-79g}$$

Equations (11-79) are enough to demonstrate that the Robinson–Trautman solution is *no* longer contained in this class of metrics. From Eqs. (11-66), (10-25i), and (10-26h) it can be shown that Eqs. (11-79) reduce to

$$P = \tfrac{1}{8} \exp[4(g + \bar{g}) + (t + \bar{t})] \tag{11-80a}$$

$$a = A \exp[2(t + \bar{t})] \tag{11-80b}$$

$$\exp(i\phi) = \exp[4(t - 2g) - 4(\bar{t} - 2\bar{g})] \tag{11-80c}$$

$$\psi_1{}^0 = P\nabla a^2 \tag{11-80d}$$

where g is an analytic function of the variable ζ,

$$t = \int \exp(-8g)\, d\zeta \tag{11-80e}$$

and A is a real function of u only.

We may set $\sigma_0 = a$, by using the spatial rotation (11-8b), and therefore

$$P = \tfrac{1}{8} \exp(8\bar{g} + 3t - \bar{t}) \tag{11-81}$$

From the coordinate transformation $\zeta' = f(\zeta)$ of Eq. (11-71) one obtains

$$P' = P(\partial\bar{f}/\partial\bar{\zeta}) \tag{11-82}$$

If we choose

$$\partial\bar{f}/\partial\bar{\zeta} = \exp(-8\bar{g} + 4\bar{t}) = \tfrac{1}{4}(\partial|e^{4\bar{t}}/\partial\bar{\zeta})$$

or

$$4\bar{f} = 4\bar{\zeta}' = e^{4\bar{t}}$$

then Eqs. (11-80) reduce to the following:

$$P = \tfrac{1}{8} \exp[3(t + \bar{t})] = (\zeta\bar{\zeta})^{3/4} \tag{11-83a}$$

$$\sigma_0 = a = A(\zeta\bar{\zeta})^{1/2} \tag{11-83b}$$

$$\psi_1{}^0 = 2A^2(\zeta\bar{\zeta})^{3/4}\zeta \tag{11-83c}$$

From Eq. (10-25n) the function $A(u)$ is then found to satisfy the differential equation

$$\frac{\partial^2 A}{\partial u^2} = 0 \tag{11-83d}$$

Hence one can choose A as equals to either Bu or B, where B is an arbitrary constant.

The results from the remaining equations, when only considered up to several orders in r, are given below. All constants of integrations are determined uniquely up to the arbitrary constant B. In these results the coordinates transformation

$u' = s(u)$, $r' = r/\dot{s}$, of Eq. (11-71) have been tacitly used to eliminate the integration constant in the indefinite integral (11-80e). All other available transformations have been, by now, completely used up.

Summarizing these results, one obtains for the spherical class of the Newman–Tamburino solution the following:

(a) The Metric Tensor:

$$\begin{aligned}
g^{00} &= g^{02} = g^{03} = g^{23} = 0, \qquad g^{01} = 1 \\
g^{11} &= -\frac{2r^2(\zeta\bar{\zeta})^{1/2}}{R^2} + \frac{2rL}{A} + \frac{2r^3A(\zeta^2+\bar{\zeta}^2)}{R^4} - \frac{4r^2A^2(\zeta\bar{\zeta})^{3/2}}{R^4} \\
g^{12} &= 4A^2(\zeta\bar{\zeta})^{3/2}x^2\left\{\frac{L}{2a^3} - \frac{(r-2a)}{2a^2R^2} - \frac{(r-a)}{R^4}\right\} \\
g^{13} &= 4A^2(\zeta\bar{\zeta})^{3/2}x^3\left\{\frac{L}{2a^3} - \frac{(r+2a)}{2a^2R^2} - \frac{(r+a)}{R^4}\right\} \\
g^{22} &= -\frac{2(\zeta\bar{\zeta})^{3/2}}{(r+a)^2}, \qquad g^{33} = -\frac{2(\zeta\bar{\zeta})^{3/2}}{(r-a)^2}
\end{aligned} \tag{11-84a}$$

where $x^0 = u$, $x^1 = r$, $\zeta = x^2 + ix^3$, $a = A(\zeta\bar{\zeta})^{1/2}$, $R^2 = r^2 - a^2$,

$$L = \tfrac{1}{2}\ln\left[(r+a)/(r-a)\right],$$

and A is a function of u satisfying $\mathrm{d}^2A/\mathrm{d}u^2 = 0$, and therefore A can be taken as either equal to Bu or B, where B is a real constant.

(b) The Components of the Weyl Spinor:

$$\begin{aligned}
\psi_0 &= 0 \\
\psi_1 &= 2A^2(\zeta\bar{\zeta})^{3/4}\zeta/R^4 \\
\psi_2 &= -[4A^2(\zeta\bar{\zeta})^{1/2}\zeta^2L + 2A\zeta^2r + 4A^2(\zeta\bar{\zeta})^{3/2}]/R^4 \\
\psi_3 &= (1/r^2)[2(\mathrm{d}A/\mathrm{d}u)(\zeta\bar{\zeta})^{1/4}\zeta] + (1/r^3)8A\zeta + \cdots \\
\psi_4 &= -(1/r^2)[8(\mathrm{d}A/\mathrm{d}u)(\zeta\bar{\zeta})] + \cdots
\end{aligned} \tag{11-84b}$$

In the above solutions one can verify that the limit $A \to 0$ yields the flat space case.

Cylindrical Class

The cylindrical class of metrics of Newman and Tamburino containing hypersurface orthogonal geodesic rays with nonvanishing shear and divergence is characterized by the conditions:

$$\psi_0 = \rho - \bar{\rho} = 0, \qquad \rho^2 = \sigma\bar{\sigma} \neq 0, \qquad \kappa = 0, \qquad \varepsilon = 0, \qquad \pi = 0 \tag{11-85}$$

In the following we give the only empty space metric that satisfies Eqs. (11-85). It will be noted that although the metric obtained is complicated, it is still given in a closed form.

Equations (10-25a) and (10-25b) yield

$$\rho = 1/(R_0 - 2r)$$

$$\sigma = \sigma_0/(R_0 - 2r)$$

where R_0 and σ_0 are functions independent of the coordinate r. One can, using the transformations (11-7a) and (11-7b), set

$$\rho = \sigma = -1/2r$$

All the remaining scalar field functions can be obtained in a closed form in a manner analogous to that used in the spherical case discussed above.

The solution obtained does not depend on the coordinate x^3 and contains two arbitrary constants denoted by a and e. We obtain (Problem 11.5):

(a) The Metric Tensor:

$$\begin{aligned}
g_{00} &= -4a^2(\mathrm{cn}^2 ay)(\ln r)^2 - [e + a^2 \ln(r^2 \mathrm{cn}^4 ay)]/\mathrm{cn}^2 ay \\
g_{01} &= 1, \qquad g_{11} = g_{12} = g_{13} = 0 \\
g_{02} &= -4Y[r + 4a^2 u(\mathrm{cn}^2 ay) \ln r] \\
g_{03} &= -2(\mathrm{cn}^2 ay) \ln r \\
g_{22} &= -r^2/2 - 4a^2u^2Y^2(\mathrm{cn}^2 ay) \ln r \\
g_{23} &= -8uY\mathrm{cn}^2(ay) \\
g_{33} &= -\mathrm{cn}^2(ay)/a^2
\end{aligned} \tag{11-86a}$$

where $x^0 = u$, $x^1 = r$, $x^2 = y$, $\mathrm{cn}(ay)$ is an elliptic function having modules $k = 1/\sqrt{2}$, and Y is given by

$$Y = \pm \frac{a(1 - \mathrm{cn}^4 ay)^{1/2}}{2\sqrt{2}\mathrm{cn}(ay)} \tag{11-86b}$$

where the + or − sign are taken according to whether $y > 0$ or $y < 0$, respectively.

(b) The Weyl Spinor:

$$\begin{aligned}
\psi_0 &= 0 \\
\psi_1 &= 2b/r^2 \\
\psi_2 &= 8b(ib_2 - bL)/r^2 \\
\psi_3 &= 4b\{\lambda_0 - 8ib_1 b_2 + 4b(\bar{b} - ib_2) + 4Lb(b_1 - 4ib_2) + 6b^2L^2\}/r^2 \\
\psi_4 &= 16b\{-b_1 U_0 + 4b_1{}^2(3ib_2 - b_1) \\
&\quad + L[-bU_0 + 8bb_1(3ib_2 - \bar{b})] + 4b^2L^2(7ib_2 - 2b - bL)\}/r^2
\end{aligned} \tag{11-86c}$$

where b_1 and b_2 are defined by

$$b = b_1 + ib_2 = Y + i\,\frac{a\,\mathrm{cn}(ay)}{2\sqrt{2}} \tag{11-86d}$$

the function L is defined by $L = \ln r$, and

$$\lambda_0 = \frac{e + a^2 \ln(\mathrm{cn}(ay))}{\mathrm{cn}^2(ay)} + \frac{a^2\mathrm{cn}^2(ay)}{4} \tag{11-86e}$$

$$U_0 = \frac{2e + a^2 + a^2 \ln(\mathrm{cn}^2(ay))}{\mathrm{cn}^2(ay)} + \frac{a^2\mathrm{cn}(ay)}{2} \tag{11-86f}$$

(c) The Spin Coefficients:

$$\begin{aligned}
&\rho = \sigma = -1/2r, \qquad \kappa = \varepsilon = \pi = 0\\
&\alpha = b(L+1)/r, \qquad \beta = b(L-1)/r\\
&\tau = 2(b_1 L - ib_2)/r\\
&\gamma = 4b(-b_1 L^2/r + 2ib_2 L + ib_2)/r\\
&\mu = \{\lambda_0 + 4\bar{b}b_1 + 4b(b + ib_2)L - 2b^2L^2\}/r\\
&\lambda = (\lambda_0 - 8ib_1 b_2 - 4bb_1 - 4bb_1 L - 2b^2L^2)/r\\
&\nu = 4\{-\lambda_0(b + b_1) - 8b^2\bar{b} + 4i\bar{b}b_1 b_2\\
&\qquad - L(\lambda_0 b_1 + 4b^2 b_1 - 8b_1 {b_2}^2) - 2ibb_2 L^2(3b + 2b_1) + 2L^3b^2b_1\}/r
\end{aligned} \tag{11-86g}$$

(d) Finally, the components of the tetrad are given by:

$$\begin{aligned}
&U = U_0 + 8{b_2}^2 L + 4b\bar{b}L^2\\
&\omega = 2(b_1 - \bar{b}L)\\
&\xi^2 = 1/r, \qquad \xi^3 = {A_0}^4/r + i{B_0}^4\\
&X^2 = -4b_1(L+1)/r\\
&X^3 = -4b_1 {A_0}^4(L+1)/r - 4{B_0}^4 b_2 L
\end{aligned} \tag{11-86h}$$

where

$${A_0}^4 = -8a^2 b_1 u, \qquad {B_0}^4 = a^2/4b_2 \tag{11-86i}$$

Using Eq. (11-6a) one can obtain the tetrad of vectors l^μ, m^μ, $\bar{m}^\mu$, and n^μ.

Final Remarks on the Newman–Tamburino Solutions

The metrics of Newman and Tamburino discussed above are of Petrov type I nondegenerate. Together with the metrics of Kundt and Robinson–Trautman, which are degenerate, they exhaust all metrics that contain hypersurface orthogonal geodesic rays.

Since the Robinson–Trautman solution contains the Schwarzschild metric as a special case, one expects that the solutions of the generalized class containing shearing geodesic rays would be interesting in discussing gravitational radiation problems. The generalized solutions, however, do not depend on any arbitrary function and therefore seem not suitable to describe a gravitational radiation field. The spherical class does not yield the Robinson–Trautman metric in the limit of vanishing shear, but instead as $A \to 0$ the metric degenerates into flat space.

In the next section the class of metrics containing nonvanishing geodesic rays with nonvanishing curl will be given.

11-2 THE NUT-TAUB METRIC

In the previous section we presented the Robinson–Trautman metric which contains nonshearing and noncurling geodesic rays with nonvanishing divergence. We also presented the Newman–Tamburino metrics containing noncurling geodesic rays but with shear and divergence. As has been pointed out at the beginning of Section 11-1, Kundt has considered the class of metrics containing geodesic rays with vanishing divergence, curl, and shear.

In this section we present a subset of the class of metrics possessing nonshearing but curling geodesic rays with nonvanishing divergence. A total of three stationary metrics and their groups of motion are obtained. All the solutions, which were obtained by Newman, Tamburino, and Unti, are determined up to several constants. One of the solutions is of particular interest for it is a generalization of the Schwarzschild metric.[9]

Tetrad System and Coordinate Conditions

Once again the gravitational field equations are greatly simplified by suitable choices of the tetrad system and coordinate conditions. Specifically, the vector l^μ will be taken as a normalized geodesic ray, and the remainder of the tetrad system is propagated in a parallel manner along this congruence. Since l^μ is a geodesic ray, then ψ_0 vanishes. Furthermore, since the geodesic ray is shear free, ψ_1 must also vanish.

The coordinate conditions are chosen such that the components of the vector l^μ are equal to $\delta_1{}^\mu$. The coordinate $x^1 = r$ then becomes the affine parameter along the null geodesics. With this choice, the pertinent nonvanishing spin coefficients are: ρ, τ, λ, μ, ν, α, β, and γ. The tetrad components will be denoted by

$$
\begin{aligned}
l^\mu &= \delta_1{}^\mu \\
m^\mu &= \omega\, \delta_1{}^\mu + \xi^a\, \delta_a{}^\mu \\
\bar{m}^\mu &= \bar{\omega}\, \delta_1{}^\mu + \bar{\xi}^a\, \delta_a{}^\mu \\
n^\mu &= U\, \delta_1{}^\mu + X^a\, \delta_a{}^\mu
\end{aligned}
\tag{11-87a}
$$

[9] E. T. Newman, L. Tamburino, and T. Unti, *J. Math. Phys.* **4**, 915 (1963).

where $a = 0, 2, 3$, and the intrinsic derivatives by

$$\begin{aligned} D\phi &= l^\mu \nabla_\mu \phi = \partial\phi/\partial r \\ \delta\phi &= m^\mu \nabla_\mu \phi = \omega\, \partial\phi/\partial r + \xi^a\, \partial\phi/\partial x^a \\ \bar{\delta}\phi &= \bar{m}^\mu \nabla_\mu \phi = \bar{\omega}\, \partial\phi/\partial r + \bar{\xi}^a\, \partial\phi/\partial x^a \\ \Delta\phi &= n^\mu \nabla_\mu \phi = U\, \partial\phi/\partial r + X^a\, \partial\phi/\partial x^a \end{aligned} \tag{11-87b}$$

Field Equations

With the above conditions, the empty space field equations obtained from Eqs. (D-1), (D-2) and (D-4) of Appendix D that describe a space containing shear-free, but curling, geodesic rays are given by:[10]

$$D\rho = \rho^2 \tag{11-88a}$$

$$D\tau = \rho\tau \tag{11-88b}$$

$$D\alpha = \rho\alpha \tag{11-88c}$$

$$D\beta = \bar{\rho}\beta \tag{11-88d}$$

$$D\gamma = \tau\alpha + \bar{\tau}\beta + \psi_2 \tag{11-88e}$$

$$D\lambda = \rho\lambda \tag{11-88f}$$

$$D\mu = \bar{\rho}\mu + \psi_2 \tag{11-88g}$$

$$D\nu = \bar{\tau}\mu + \tau\lambda + \psi_3 \tag{11-88h}$$

$$D\psi_2 - 3\rho\psi_2 = 0 \tag{11-88i}$$

$$D\psi_3 - 2\rho\psi_3 = \bar{\delta}\psi_2 \tag{11-88j}$$

$$D\psi_4 - \rho\psi_4 = \bar{\delta}\psi_3 + 2\alpha\psi_3 - 3\lambda\psi_2 \tag{11-88k}$$

$$DU = \tau\bar{\omega} + \bar{\tau}\omega - (\gamma + \bar{\gamma}) \tag{11-88l}$$

$$DX^a = \tau\bar{\xi}^a + \bar{\tau}\xi^a \tag{11-88m}$$

$$D\omega = \bar{\rho}\omega - (\bar{\alpha} + \beta) \tag{11-88n}$$

$$D\xi^a = \bar{\rho}\xi^a \tag{11-88o}$$

and

$$\delta\rho = \rho(\bar{\alpha} + \beta) + (\rho - \bar{\rho})\tau \tag{11-89a}$$

$$\delta\alpha - \bar{\delta}\beta = \rho\mu + \alpha\bar{\alpha} + \beta\bar{\beta} - 2\alpha\beta + \gamma(\rho - \bar{\rho}) - \psi_2 \tag{11-89b}$$

$$\delta\lambda - \bar{\delta}\mu = (\rho - \bar{\rho})\nu + \mu(\alpha + \bar{\beta}) + \lambda(\bar{\alpha} - 3\beta) - \psi_3 \tag{11-89c}$$

$$\Delta\lambda - \bar{\delta}\nu = \lambda(\bar{\gamma} - 3\gamma - \mu - \bar{\mu}) + \nu(3\alpha + \bar{\beta} - \bar{\tau}) - \psi_4 \tag{11-89d}$$

$$\delta\nu - \Delta\mu = \mu^2 + \lambda\bar{\lambda} + \mu(\gamma + \bar{\gamma}) + \nu(\tau - \bar{\alpha} - 3\beta) \tag{11-89e}$$

[10] Note that now $\rho \neq \bar{\rho}$.

$$\delta\tau = \rho\bar{\lambda} + \tau(\tau + \beta - \bar{\alpha}) \tag{11-89f}$$

$$\Delta\rho - \bar{\delta}\tau = \rho(\gamma + \bar{\gamma} - \bar{\mu}) + \tau(\bar{\beta} - \alpha - \bar{\tau}) - \psi_2 \tag{11-89g}$$

$$\Delta\alpha - \bar{\delta}\gamma = \rho\nu - \lambda(\tau + \beta) + \alpha(\bar{\gamma} - \bar{\mu}) + \gamma(\bar{\beta} - \bar{\tau}) - \psi_3 \tag{11-89h}$$

$$\delta\gamma - \Delta\beta = \alpha\bar{\lambda} + \mu\tau + \gamma(\tau - \bar{\alpha} - \beta) + \beta(\mu + \bar{\gamma} - \gamma) \tag{11-89i}$$

$$\delta\psi_2 = 3\tau\psi_2 \tag{11-89j}$$

$$\Delta\psi_2 - \delta\psi_3 = -3\mu\psi_2 + 2(\beta - \tau)\psi_3 \tag{11-89k}$$

$$\Delta\psi_3 - \delta\psi_4 = 3\nu\psi_2 - (2\gamma + 4\mu)\psi_3 + (4\beta - \tau)\psi_4 \tag{11-89l}$$

$$\delta U - \Delta\omega = -\bar{\nu} + U(\tau - \bar{\alpha} - \beta) + \bar{\omega}\bar{\lambda} + \omega(\mu - \gamma + \bar{\gamma}) \tag{11-89m}$$

$$\delta X^a - \Delta\xi^a = X^a(\tau - \bar{\alpha} - \beta) + \bar{\xi}^a\bar{\lambda} + \xi^a(\mu - \gamma + \bar{\gamma}) \tag{11-89n}$$

$$\bar{\delta}\omega - \delta\bar{\omega} = (\bar{\mu} - \mu) + U(\bar{\rho} - \rho) + \bar{\omega}(\beta - \bar{\alpha}) + \omega(\alpha - \bar{\beta}) \tag{11-89o}$$

$$\bar{\delta}\xi^a - \delta\bar{\xi}^a = (\bar{\rho} - \rho)X^a + (\beta - \bar{\alpha})\bar{\xi}^a + (\alpha - \bar{\beta})\xi^a \tag{11-89p}$$

where $a = 0, 2, 3$.

Coordinate and Tetrad Transformations

There are a number of coordinate and tetrad transformations that preserve the coordinate conditions and the orthogonality conditions as well as the parallel propagation of the tetrad system.

The coordinate transformations are

$$\begin{aligned} x'^a &= x'^a(x^0, x^2, x^3) \\ r' &= r + R(x^a) \end{aligned} \tag{11-90a}$$

where $a = 0, 2, 3$.

The allowed tetrad transformations are

$$\begin{aligned} l'^\mu &= l^\mu \\ m'^\mu &= \bar{z}l^\mu + m^\mu \\ n'^\mu &= z\bar{z}l^\mu + zm^\mu + \bar{z}\bar{m}^\mu + n^\mu \end{aligned} \tag{11-90b}$$

where z is a complex scalar independent of r, and

$$\begin{aligned} l'^\mu &= l^\mu \\ m'^\mu &= e^{iC}m^\mu \\ n'^\mu &= n^\mu \end{aligned} \tag{11-90c}$$

where C is real and independent of r.

A combination of tetrad and coordinate transformation is

$$
\begin{aligned}
l'^{\mu} &= A^{-1}\, l^{\mu} \\
m'^{\mu} &= m^{\mu} \\
n'^{\mu} &= A n^{\mu} \\
x'^{a} &= x^{a} \\
r' &= A^{-1} r
\end{aligned}
\tag{11-90d}
$$

where $A = A(x^a)$, and $a = 0, 2, 3$.

Integration of Field Equations

We are now in a position to integrate the field equations. The following simplifying assumptions will be made:

$$\psi_3 = \psi_4 = 0 \tag{11-91a}$$

$$\tau = 0 \tag{11-91b}$$

Hence the only nonzero component of the Weyl spinor is ψ_2. Since $\psi_3 = \psi_4 = 0$, it follows (Problem 10.5) that

$$\nu = \lambda = 0 \tag{11-91c}$$

In the remainder of this section the component of the Weyl spinor ψ_2 will be denoted simply by ψ. Furthermore, a subscript zero will indicate that a function is independent of the coordinate r.

The solution of Eq. (11-88a) is

$$\rho = -1/(r + \rho_0)$$

Using the coordinate transformation (11-90a) one can set the real part of the constant of integration ρ_0 equal to zero. Furthermore, using the combined coordinate and tetrad transformation $r' = A^{-1}r$ of Eqs. (11-90d) gives $\rho'(r') = -1/(Ar' + \rho_0' A)$ and therefore $\rho_0' = A^{-1}\rho_0$. The scalar function A can now be chosen to make ρ_0' a constant. Dropping the primes, one has

$$\rho_0 = i\,|\rho_0| = \text{const.} \tag{11-92}$$

The constant could be given a definite value, but it is preferable to let ρ_0 be a parameter.

From Eqs. (11-88c) to (11-88o) one can easily obtain the r dependence of the remaining variables. The solutions to these equations are given below:

$$\alpha = \rho\alpha_0 \tag{11-93a}$$

$$\beta = \bar{\rho}\beta_0 \tag{11-93b}$$

$$\psi = \rho^3\psi_0 \tag{11-93c}$$

$$\gamma = \gamma_0 + \tfrac{1}{2}\rho^2\psi_0 \tag{11-93d}$$

$$\mu = \bar{\rho}\mu_0 + \tfrac{1}{2}(\rho^2 + \rho\bar{\rho})\psi_0 \tag{11-93e}$$

$$U = U_0 - (\gamma_0 + \bar{\gamma}_0)r - \tfrac{1}{2}(\rho\psi_0 + \bar{\rho}\bar{\psi}_0) \tag{11-93f}$$

$$\omega = \bar{\rho}\omega_0 + (\bar{\alpha}_0 + \beta_0) \tag{11-93g}$$

$$X^a = X_0{}^a \tag{11-93h}$$

$$\xi^a = \bar{\rho}\xi_0{}^a \tag{11-93i}$$

where $a = 0, 2, 3$. So far the radial dependence of all field variables has been determined.

From Eq. (11-88j), using Eqs. (11-87b), one obtains by comparing the coefficients of equal powers of r,

$$\omega_0 = 0 \tag{11-94a}$$

$$\bar{\xi}_0{}^a\,\partial_a\psi_0 = -3(\alpha_0 + \bar{\beta}_0)\psi_0 \tag{11-94b}$$

Equation (11-89j) gives

$$(\bar{\rho}/\rho)\xi_0{}^a\,\partial_a\psi_0 = -3(\bar{\alpha}_0 + \beta_0)\psi_0 \tag{11-94c}$$

Expanding the expression $(\bar{\rho}/\rho)$ into a power series in $1/r$ gives

$$\bar{\rho}/\rho = 1 + (\rho_0 - \bar{\rho}_0)/r + (\bar{\rho}_0{}^2 - \rho_0\bar{\rho}_0)/r^2 + \cdots$$

Using this expansion in Eq. (11-94c), and comparing coefficients of equal powers of r, gives

$$\xi_0{}^a\,\partial_a\psi_0 = -3(\bar{\alpha}_0 + \beta_0)\psi_0$$

$$(\rho_0 - \bar{\rho}_0)\xi_0{}^a\,\partial_a\psi_0 = 0, \text{ etc.}$$

Hence, unless $\rho_0 = \bar{\rho}_0$, one obtains

$$\xi_0{}^a\,\partial_a\psi_0 = 0 \tag{11-94d}$$

$$\bar{\alpha}_0 + \beta_0 = 0 \tag{11-94e}$$

Using Eqs. (11-94a) and (11-94e) in Eq. (11-93g) then shows that

$$\omega = 0 \tag{11-94f}$$

Equation (11-89k) yields, again by comparing the coefficients of equal powers of r, the following relations:

$$3(\gamma_0 + \bar{\gamma}_0)\psi_0 + X_0{}^a\,\partial_a\psi_0 = 0 \tag{11-95a}$$

$$-U_0 - (\gamma_0 + \bar{\gamma}_0)\rho_0 = \mu_0 \tag{11-95b}$$

$$U_0\rho_0 + (\gamma_0 + \bar{\gamma}_0)\rho_0{}^2 + (\psi_0 - \bar{\psi}_0)/2 = \mu_0\rho_0 \tag{11-95c}$$

Hence from Eqs. (11-95b) and (11-95c) one obtains

$$\psi_0 - \bar{\psi}_0 = 4\mu_0\rho_0 \tag{11-95d}$$

which shows that μ_0 is real. Therefore Eq. (11-95b) yields

$$\gamma_0 + \bar{\gamma}_0 = 0 \tag{11-95e}$$

$$U_0 = -\mu_0 = -\bar{\mu}_0 \tag{11-95f}$$

and consequently Eq. (11-95a) leads to

$$X_0{}^a\, \partial_a \psi_0 = 0 \tag{11-95g}$$

where $a = 0, 2, 3$.

From Eqs. (11-89c) and (11-89e) we obtain

$$\xi_0{}^a\, \partial_a \mu_0 = 0 \tag{11-96a}$$

$$X_0{}^a\, \partial_a \mu_0 = 0 \tag{11-96b}$$

Also, Eq. (11-89n) yields

$$\xi_0{}^a\, \partial_a X_0{}^b - X_0{}^a\, \partial_a \xi_0{}^b = 2\bar{\gamma}_0\, \xi_0{}^b \tag{11-96c}$$

At this stage it is convenient to use the available coordinate and tetrad transformations to simplify the above results. The transformation $m'^\mu = \exp(iC_0)m^\mu$ of Eqs. (11-90c), along with Eq. (11-95e), are used to set

$$\gamma_0 = 0 \tag{11-97}$$

Under the coordinate transformation $x'^a = \theta^a(x^b)$ of Eqs. (11-90a), the variables $X_0{}^a$ and $\xi_0{}^a$ transform as follows:

$$X_0{}'^a = X_0{}^b\, \partial_b \theta^a \tag{11-98a}$$

$$\xi_0{}'^a = \xi_0{}^b\, \partial_b \theta^a \tag{11-98b}$$

The first of these equations can be used to set

$$X_0{}^0 = 1 \tag{11-99a}$$

$$X_0{}^2 = X_0{}^3 = 0 \tag{11-99b}$$

The remaining coordinate freedom is now

$$\begin{aligned} x'^0 &= x^0 + \theta^0(x^2, x^3) \\ x'^2 &= \theta^2(x^2, x^3) \\ x'^3 &= \theta^3(x^2, x^3) \end{aligned} \tag{11-100}$$

and consequently one obtains

$$\begin{aligned} \xi'_0{}^0 &= \xi_0{}^0 + \xi_0{}^i\, \partial_i \theta^0 \\ \xi'_0{}^2 &= \xi_0{}^i\, \partial_i \theta^2 \\ \xi'_0{}^3 &= \xi_0{}^i\, \partial_i \theta^3 \end{aligned} \tag{11-101}$$

where $i = 2, 3$.

Equation (11-96c) now becomes, using Eqs. (11-99),

$$\partial \xi_0{}^b/\partial x^0 = \partial \xi_0{}^b/\partial u = 0$$

Hence it is possible to use Eqs. (11-100) and (11-101) to set

$$\xi_0{}^2 = -i\xi_0{}^3 = P(x^2, x^3) \tag{11-102}$$

and to use the tetrad transformation $m'^\mu = \exp[iC(x^2, x^3)]m^\mu$ to set

$$P = \bar{P} = p(x^2, x^3) \tag{11-103}$$

The coordinate freedom now available is

$$x'^0 = x^0 + f(x^2, x^3) \tag{11-104a}$$

$$\xi'_0{}^0 = \xi_0{}^0 + p\nabla f \tag{11-104b}$$

and

$$\zeta' = g(\zeta) \tag{11-104c}$$

where the complex coordinate ζ is defined by $\zeta = x^2 + ix^3$, and the differentiation operator ∇ is defined by $\nabla = 2\,\partial/\partial\bar{\zeta}$.

Putting Eqs. (11-99) and (11-102) into Eqs. (11-94), (11-95), and (11-96) one finds μ_0 and ψ_0 to be constants which satisfy Eq. (11-95d).

From Eqs. (11-89p) and (11-89b) one obtains the following three formulae:

$$2\alpha_0 = p\bar{\nabla} \ln p \tag{11-105a}$$

$$2\rho_0 = p^2[\nabla(\bar{\xi}_0{}^0/p) - \bar{\nabla}(\xi_0{}^0/p)] \tag{11-105b}$$

$$2\mu_0 = (\sqrt{2}p)^2\nabla\bar{\nabla} \ln(\sqrt{2}p) \tag{11-105c}$$

With these three equations, we have obtained all the available information from the gravitational field equations. Equation (11-105a) is taken as defining the function α_0, and it remains to integrate Eqs. (11-105b) and (11-105c).

It can be shown that the constant $2\mu_0$ in Eq. (11-105c) is the Gaussian curvature of a two-dimensional space with the metric $(\sqrt{2}p)^2\,\delta^{ij}$ (Problem 11.6). This metric is equal to the limit, at $r \to \infty$, of $(r^2 g^{ij})$, where g^{ij} $(i, j = 2, 3)$ are the components of the metric obtained [Eq. (11-109d) below]. With the coordinate transformation (11-104c), it is possible to choose coordinates such that

$$\sqrt{2}p = 1 + \mu_0 \zeta\bar{\zeta}/2 \tag{11-106}$$

The curvature $2\mu_0$ may be reduced, by means of a scale change in the coordinate r, to one of the three values, 1, 0, or -1. It is for this reason that the constant ρ_0 has been chosen as a parameter.

Using the expression (11-106) for p into the inhomogeneous equation (11-105b), one obtains the following solution for $\xi_0{}^0$:

$$\xi_0{}^0 = -\rho_0\zeta/\sqrt{2} \tag{11-107}$$

Let η_0 be the general solution to the homogeneous equation (obtained from the inhomogeneous equation by putting $\rho_0 = 0$). The most general solution to Eq. (11-105b) is then

$$\xi_0{}^0 = \eta_0 - \rho_0 \zeta/\sqrt{2} \tag{11-108}$$

The coordinate transformations (11-104a) and (11-104b) then yield

$$\xi_0{}'^0 = \eta_0 + p\nabla f(x^2, x^3) - \rho_0 \zeta/\sqrt{2}$$

The function f can be chosen such that $\xi_0{}'^0 = -\rho_0 \zeta/\sqrt{2}$, which means that (11-107) becomes the most general solution to Eq. (11-105b). This is possible since one can always satisfy $\nabla f = -\eta_0/p$, because the integrability condition $\nabla\bar{\nabla} f = \bar{\nabla}\nabla f$ is equivalent to the homogeneous equation itself, and hence is satisfied.

This completes the formal calculations. The metric tensor is obtained, of course, from $g^{\mu\nu} = l^\mu n^\nu + n^\mu l^\nu - m^\mu \bar{m}^\nu - \bar{m}^\mu m^\nu$, where the null tetrad l^μ, m^μ, $\bar{m}^\mu$ and n^μ is given by Eq. (11-87a).

The metric so obtained, is usually referred to as the NUT–Taub metric.

Summary of Calculations

In the following we give a summary of the calculations presented above.

(a) Spin Coefficients:

$$\begin{aligned}
&\sigma = \tau = \lambda = \nu = \kappa = \varepsilon = \pi = 0\\
&\rho = -1/(r + \rho_0); \qquad \rho_0 = i|\rho_0| = \text{const.},\\
&\mu = \mu_0 \bar{\rho} + \psi_0(\rho^2 + \rho\bar{\rho})/2; \qquad 2\mu_0 = -1, 0, 1\\
&\alpha = -\beta = \rho\mu_0 \zeta/2\sqrt{2}; \qquad \zeta = x^2 + ix^3\\
&\gamma = \rho^2 \psi_0/2; \qquad \psi_0 = \tilde{\psi}_0 + 2i\mu_0|\rho_0|
\end{aligned} \tag{11-109a}$$

where $\tilde{\psi}_0$ is a real constant.

(b) Tetrad Components: Using Eq. (11-87a) one obtains:

$$\begin{aligned}
l^\mu &= \delta_1{}^\mu\\
m^\mu &= \xi^a\, \delta_a{}^\mu\\
\bar{m}^\mu &= \bar{\xi}^a\, \delta_a{}^\mu\\
n^\mu &= \delta_0{}^\mu + U\, \delta_1{}^\mu
\end{aligned} \tag{11-109b}$$

where $a = 0, 2, 3$ and the functions ξ^a and U are given by

$$\begin{aligned}
\xi^0 &= -\bar{\rho}\rho_0 \zeta/\sqrt{2}\\
\xi^2 &= -i\xi^3 = \bar{\rho}p\\
U &= -\mu_0 + \rho\bar{\rho}(r\tilde{\psi}_0 + 2\mu_0|\rho_0|^2)
\end{aligned}$$

and $\sqrt{2}p = 1 + \mu_0 \zeta\bar{\zeta}/2$.

(c) Weyl Spinor:

$$\begin{aligned} \psi_0 = \psi_1 = \psi_3 = \psi_4 = 0 \\ \psi_2 = \psi = \rho^3(\tilde{\psi}_0 + 2i\mu_0|\rho_0|) \end{aligned} \tag{11-109c}$$

(d) The Metric:

$$g^{\mu\nu} = \begin{pmatrix} -(A^2+B^2)/R^2 & 1 & -A/R^2 & -B/R^2 \\ 1 & 2U & 0 & 0 \\ -A/R^2 & 0 & -1/R^2 & 0 \\ -B/R^2 & 0 & 0 & -1/R^2 \end{pmatrix} \tag{11.109d}$$

Here $A = |\rho_0|x^3/\sqrt{2}p$, $B = -|\rho_0|x^2/\sqrt{2}p$, and $1/R^2 = 2\rho\bar{\rho}p^2 = (-\det g^{\mu\nu})^{1/2}$.

(e) Killing Fields: There is a four-parameter group of motions which is described by the following solutions to the Killing equation $\xi_{\mu;\nu} + \xi_{\nu;\mu} = 0$:

$$\begin{aligned} \xi_t{}^\mu &= [1, 0, 0, 0] \\ \xi_a{}^\mu &= [0, 0, -y, x] \\ \xi_b{}^\mu &= [-2|\rho_0|y, 0, 2+(x^2-y^2)\mu_0, 2\mu_0 xy] \\ \xi_c{}^\mu &= [2|\rho_0|x, 0, 2\mu_0 xy, 2-\mu_0(x^2-y^2)] \end{aligned} \tag{11-109e}$$

where $\mu = 0, 1, 2, 3$, $x = x^2$, and $y = x^3$.

Generalized Schwarzschild Metric

The metric given by Eq. (11-109d) is an extension of a metric first obtained by Taub.[11] The solution with the choice $2\mu_0 = 1$ yields the Schwarzschild metric in the limit $\rho_0 = 0$.

To see this one writes the metric (11-109d) in terms of spherical coordinates t, r', θ, and ϕ by use of the following coordinate transformation:

$$\begin{aligned} t &= u - \int (2U)^{-1}\,dr \\ r' &= r \\ \sin\theta &= \zeta\bar{\zeta}/(1+\zeta\bar{\zeta}/4) \\ \tan\phi &= x^3/x^2 \end{aligned} \tag{11-110}$$

The transformed covariant metric tensor is consequently given by

$$ds^2 = f(r)[dt + 4l\sin^2(\theta/2)\,d\phi]^2 - (1/f)\,dr^2 - (r^2+l^2)(d\theta^2 + \sin^2\theta\,d\phi^2) \tag{11-111}$$

where

$$f(r) = 1 - 2(mr + l^2)/(r^2 + l^2) \tag{11-112}$$

and m and l are given by $m = \tilde{\psi}_0$ and $l = |\rho_0|$, and the prime has been dropped from r.

[11] A. H. Taub, *Ann. Math.* **53**, 472 (1951).

The contravariant metric tensor is given by

$$g^{\mu\nu} = \begin{pmatrix} g^{00} & 0 & 0 & g^{03} \\ 0 & g^{11} & 0 & 0 \\ 0 & 0 & g^{22} & 0 \\ g^{03} & 0 & 0 & g^{33} \end{pmatrix} \tag{11-113}$$

where

$$\begin{aligned} g^{00} &= f^{-1} - 16l^2 \sin^4 (\theta/2) \operatorname{cosec}^2 \theta/(r^2 + l^2) \\ g^{03} &= 4l \sin^2 (\theta/2) \operatorname{cosec}^2 \theta/(r^2 + l^2) \\ g^{11} &= -f \\ g^{22} &= -1/(r^2 + l^2) \\ g^{33} &= -\operatorname{cosec}^2 \theta/(r^2 + l^2) \end{aligned} \tag{11-114}$$

and $x^0 = t$, $x^1 = r$, $x^2 = \theta$, and $x^3 = \phi$. Furthermore, one finds that

$$\det g_{\mu\nu} = -(r^2 + l^2)^2 \sin^2 \theta. \tag{11-115}$$

It can easily be seen that one obtains the Schwarzschild solution from the metric tensor given by Eqs. (11-111) and (11-113) if one sets $l = |\rho_0| = 0$, where m is the mass in units for which the gravitational constant and the speed of light are taken to be equal to unity.

Expressed in terms of the new coordinate system, the Killing vectors (11-109e) are now given by:

$$\begin{aligned} \xi_t{}^\mu &= [1, 0, 0, 0] \\ \xi_a{}^\mu &= [-2l, 0, 0, 1] \\ \xi_b{}^\mu &= [-2l \tan (\theta/2) \sin \phi, 0, \cos \phi, -\cot \theta \sin \phi] \\ \xi_c{}^\mu &= [2l \tan (\theta/2) \cos \phi, 0, \sin \phi, \cot \theta \cos \phi] \end{aligned} \tag{11-116}$$

The Killing fields associated with the Schwarzschild solution are obtained by setting $l = 0$.

The Groups of Motions

In order to gain physical insight into the NUT–Taub metric, it is convenient to consider a test particle whose Lagrangian, four momentum, and Hamiltonian are given by:

$$L = \tfrac{1}{2} g_{\mu\nu} \dot{x}^\mu \dot{x}^\nu \tag{11-117a}$$

$$P_\mu = \partial L/\partial \dot{x}^\mu = g_{\mu\nu} \dot{x}^\nu \tag{11-117b}$$

$$H = \tfrac{1}{2} g_{\mu\nu} P^\mu P^\nu \tag{11-117c}$$

where $\dot{x}^\mu = dx^\mu/ds$. Killing generators are defined by

$$G_A = P_\mu \xi_A{}^\mu \tag{11-118}$$

where $\xi_A{}^\mu$ is a Killing vector. Using the above definition one can show (Problem 11.7) that Killing's equation is equivalent to the Poisson bracket relation

$$[G_A, H] = 0 \tag{11-119}$$

Consequently, the Killing generators are constants of the motion.

The Killing generators obtained from Eqs. (11-116) are:

$$\begin{aligned}
G_t &= P_0 \\
G_a &= -2lP_0 + P_3 \\
G_b &= -2lP_0 \tan(\theta/2) \sin\phi + P_2 \cos\phi - P_3 \cot\theta \sin\phi \\
G_c &= 2lP_0 \tan(\theta/2) \cos\phi + P_2 \sin\phi + P_3 \cot\theta \cos\phi
\end{aligned} \tag{11-120}$$

These four Killing generators are constants of the motion. In particular, the energy of the test particles, P_0, and the ϕ component of the four-momentum, P_3, are constants of the motion.

The following Poisson bracket relations are readily obtained:

$$[G_A, G_B] = \varepsilon_{ABC} G_C \tag{11-121a}$$

$$[G_t, G_A] = 0 \tag{11-121b}$$

In these equations A, B, C each takes the values a, b, c, and ε_{ABC} is the completely skew-symmetric symbol with $\varepsilon_{abc} = +1$. The group characterized by Eq. (11-121a) is the covering group of the rotation group O_3, namely the group SU_2, discussed in Chapter 1.

A natural definition for the angular momenta are the generators of rotations about the three axes, which are the Killing generators of the Schwarzschild metric:

$$\begin{aligned}
M_a &= P_3 \\
M_b &= P_2 \cos\phi - P_3 \cot\theta \sin\phi \\
M_c &= P_2 \sin\phi - P_3 \cot\theta \cos\phi
\end{aligned} \tag{11-122}$$

Finally, from the relations

$$\begin{aligned}
G_a - M_a &= -2lP_0 \\
G_b - M_b &= -2lP_0 \tan(\theta/2) \sin\phi \\
G_c - M_c &= 2lP_0 \tan(\theta/2) \cos\phi
\end{aligned} \tag{11-123}$$

one sees that the components of angular momentum differ from the constants of motion by terms proportional to lP_0.

Discussion

We have seen that the constant $m = \tilde{\psi}_0$ becomes the mass in the Schwarzschild limit of $l = |\rho_0| = 0$. Hence it is reasonable to assume that m describes the mass of the source even when $l \neq 0$.

The parameter l, which is a measure of the curl of the geodesic ray l_μ, describes the departure of the stationary NUT–Taub metric from the static Schwarzschild metric. It is also a measure of the deviation between two components of a test particle's angular momentum and its constants of the motion. The ϕ component is singled out as the only constant, which implies that the field has axial symmetry. Hence, one can interpret l as the angular momentum of the source, the latter being considered as a spinning particle. The field of a rotating body, in the linear approximation, was given many years ago by Lense and Thirring.[12] The NUT–Taub solution resembles that solution in the sense that the same metric components are zero, the angular momentum appears linearly in the mixed components, and both pass to the Schwarzschild metric when the angular momentum vanishes.

It has been pointed out by Misner[13] that the orbit of a point under the group of time translations is a circle, rather than a line as in the Schwarzschild case. The time-like hypersurfaces $r = \text{const.}$ which are left invariant by the group of motions are topologically three-spheres S^3, in contrast to the topology $S^2 \times R$ (or $S^2 \times S^1$) for the $r = \text{const.}$ surfaces in the Schwarzschild case. In the Schwarzschild case, the intersection of a spacelike surface $t = \text{const.}$ and an $r = \text{const.}$ surface is a sphere S^2. If σ is any spacelike hypersurface in the NUT–Taub metric, then its (two-dimensional) intersection with an $r = \text{const.}$ surface is not any closed two-dimensional manifold, that is, the metric admits no reasonable spacelike surfaces. Thus, even though the only nonvanishing curvature invariant $\psi = \psi_2$ vanishes as $r \to \infty$, in fact $\psi = 0(1/r^3)$, as in the Schwarzschild case, the metric is not asymptotically flat in the sense that coordinates can be introduced for which $g_{\mu\nu} - \eta_{\mu\nu} = 0(1/r)$, where $\eta_{\mu\nu}$ is the Minkowskian metric.

To conclude this section we emphasize that the NUT–Taub solution is of Petrov type D (type I degenerate), or in Penrose' notation (in which the Weyl spinor is decomposed into a product of four spinors)—see Chapter 8—the four spinors coincide in pairs. In the next section we find all type D vacuum metrics.

11-3 TYPE D VACUUM METRICS

We are now in a position to solve the vacuum field equations for Petrov type D solutions. An exhaustive set of eleven metrics were obtained by Kinnersley. These metrics present all type D vacuum metrics, including the Kerr solution[14] and the NUT–Taub solution, discussed in Sec. 11-2.[15] Figure 11.2 describes these metrics.

Petrov type D fields are of interest to study for several reasons. The Schwarzschild and Kerr metrics are familiar cases of this class. Besides being free of the

[12] J. Lense and H. Thirring, *Physik. Z.* **19,** 156 (1918).

[13] C. W. Misner, *J. Math. Phys.* **4,** 924 (1963).

[14] R. Kerr, *Phys. Rev. Letters* **11,** 237 (1963).

[15] W. Kinnersley, *J. Math. Phys.* **10,** 1195 (1969). See also W. Kinnersley, Recent progress in exact solutions, Proceedings of the International Conference on Gravitation and Relativity (held in Tel Aviv, 24–28 June 1974).

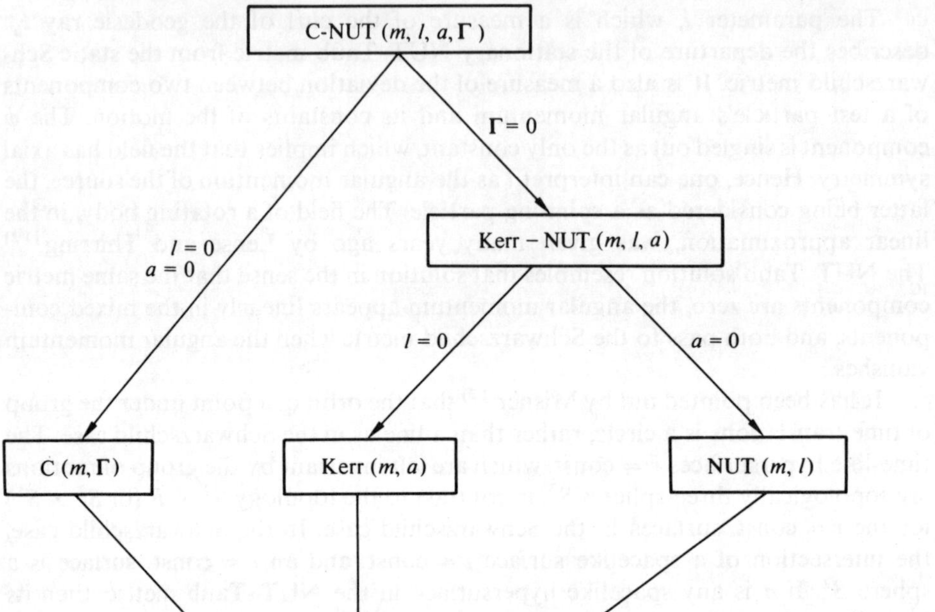

Figure 11.2 Kinnersley's diagram describing type D vacuum metrics.

line singularities which plague some algebraically special metrics such as the Robinson–Trautman metric (see Sec. 11-1), these two metrics are also stationary in time. The relation between the Petrov type and the existence of Killing vectors is not known. One can prove a weak converse, namely that an algebraically special field which is stationary, time-reversible, and with a finite source can only be type D.

Field Equations

Let the space under consideration be vacuum and of type D, and let l^μ, m^μ, $\bar{m}^\mu$, n^μ be the tetrad, and choose l^μ and n^μ to lie along the principal null directions. This implies $\psi_0 = \psi_1 = \psi_3 = \psi_4 = 0$. As in the last section we will refer to ψ_2 simply as ψ. The Goldberg–Sachs theorem implies $\kappa = \sigma = \nu = \lambda = 0$. As before, we choose coordinates such that $l^\mu = \delta_1{}^\mu$, making $x^1 = r$ an affine parameter along l^μ, and we set $\varepsilon = 0$. The tetrad components then take the form for the vector components.

$$l^\mu = (0, 1, 0, 0) \tag{11-124a}$$

$$m^\mu = (\xi^0, \omega, \xi^2, \xi^3) \tag{11-124b}$$

$$\bar{m}^\mu = (\bar{\xi}^0, \bar{\omega}, \bar{\xi}^2, \bar{\xi}^3) \tag{11-124c}$$

$$n^\mu = (X^0, U, X^2, X^3) \tag{11-124d}$$

identical to those given by Eq. (11-87a) of the last section.

The metric equation (9-73) gives, under the above conditions,

$$DU = (\bar{\tau} + \pi)\omega + (\tau + \bar{\pi})\bar{\omega} - (\gamma + \bar{\gamma}) \tag{11-125a}$$

$$DX^i = (\bar{\tau} + \pi)\xi^i + (\tau + \bar{\pi})\bar{\xi}^i \tag{11-125b}$$

$$D\omega = \bar{\rho}\omega + (\bar{\pi} - \bar{\alpha} - \beta) \tag{11-125c}$$

$$D\xi^i = \bar{\rho}\xi^i \tag{11-125d}$$

where $i = 0, 2, 3$, and

$$\delta U - \Delta\omega = (\tau - \bar{\alpha} - \beta)U + (\mu - \gamma + \bar{\gamma})\omega \tag{11-126a}$$

$$\delta X^i - \Delta\xi^i = (\tau - \bar{\alpha} - \beta)X^i + (\mu - \gamma + \bar{\gamma})\xi^i \tag{11-126b}$$

$$\bar{\delta}\omega - \delta\bar{\omega} = (\bar{\mu} - \mu) + (\bar{\rho} - \rho)U - (\bar{\beta} - \alpha)\omega - (\bar{\alpha} - \beta)\bar{\omega} \tag{11-126c}$$

$$\bar{\delta}\xi^i - \delta\bar{\xi}^i = (\bar{\rho} - \rho)X^i - (\bar{\beta} - \alpha)\xi^i - (\bar{\alpha} - \beta)\bar{\xi}^i \tag{11-126d}$$

Equations (D-1) of Appendix D give

$$D\rho = \rho^2 \tag{11-127a}$$

$$D\beta = \bar{\rho}\beta \tag{11-127b}$$

$$D\alpha = \rho(\alpha + \pi) \tag{11-127c}$$

$$D\tau = \rho(\tau + \bar{\pi}) \tag{11-127d}$$

$$D\gamma = \alpha(\tau + \bar{\pi}) + \beta(\bar{\tau} + \pi) + \tau\pi + \psi \tag{11-127e}$$

$$D\mu - \delta\pi = \bar{\rho}\mu + \pi(\bar{\pi} - \bar{\alpha} + \beta) + \psi \tag{11-127f}$$

$$\delta\rho = \rho(\bar{\alpha} + \beta) + (\rho - \bar{\rho})\tau \tag{11-128a}$$

$$\delta\tau = \tau(\tau - \bar{\alpha} + \beta) \tag{11-128b}$$

$$\delta\alpha - \bar{\delta}\beta = \rho\mu + \alpha\bar{\alpha} + \beta\bar{\beta} - 2\alpha\beta + (\rho - \bar{\rho})\gamma - \psi \tag{11-128c}$$

$$\bar{\delta}\pi = -\pi(\pi + \alpha - \bar{\beta}) \tag{11-128d}$$

$$\bar{\delta}\mu = -\mu(\alpha + \bar{\beta}) - (\mu - \bar{\mu})\pi \tag{11-128e}$$

$$\Delta\rho - \bar{\delta}\tau = -\rho\bar{\mu} - \tau(\bar{\tau} + \alpha - \bar{\beta}) + \rho(\gamma + \bar{\gamma}) - \psi \qquad (11\text{-}129a)$$

$$\Delta\pi = -\mu(\bar{\tau} + \pi) - \pi(\gamma - \bar{\gamma}) \qquad (11\text{-}129b)$$

$$\Delta\alpha - \bar{\delta}\gamma = -\bar{\mu}\alpha + \gamma(\bar{\beta} - \bar{\tau}) + \bar{\gamma}\alpha \qquad (11\text{-}129c)$$

$$\Delta\beta - \delta\gamma = -\mu(\beta + \tau) + \gamma(2\beta + \bar{\alpha} - \tau) - \bar{\gamma}\beta \qquad (11\text{-}129d)$$

$$\Delta\mu = -\mu(\mu + \gamma + \bar{\gamma}) \qquad (11\text{-}129e)$$

Equations (D-2) give

$$D\psi = 3\rho\psi \qquad (11\text{-}130a)$$

$$\Delta\psi = -3\mu\psi \qquad (11\text{-}130b)$$

$$\delta\psi = 3\tau\psi \qquad (11\text{-}130c)$$

$$\bar{\delta}\psi = -3\pi\psi \qquad (11\text{-}130d)$$

The commutation relations (D-3) of Appendix D give

$$\Delta D - D\Delta = (\gamma + \bar{\gamma})D - (\bar{\tau} + \pi)\,\delta - (\tau + \bar{\pi})\,\bar{\delta} \qquad (11\text{-}131a)$$

$$\delta D - D\delta = (\bar{\alpha} + \beta - \bar{\pi})D - \bar{\rho}\delta \qquad (11\text{-}131b)$$

$$\delta\Delta - \Delta\delta = (\tau - \bar{\alpha} - \beta)D + (\mu - \gamma + \bar{\gamma})\,\delta \qquad (11\text{-}131c)$$

$$\bar{\delta}\delta - \delta\bar{\delta} = (\bar{\mu} - \mu)D + (\bar{\rho} - \rho)\,\Delta - (\bar{\beta} - \alpha)\,\delta - (\bar{\alpha} - \beta)\,\bar{\delta} \qquad (11\text{-}131d)$$

In addition to the above equations we obtain four more important relations among the spin coefficients variables by applying the commutators to the Weyl spinor component ψ:

$$\begin{aligned}(\Delta D - D\Delta)\psi &= \Delta(3\rho\psi) - D(-3\mu\psi)\\ &= 3\psi(\Delta\rho + D\mu)\\ &= 3\psi[\rho(\gamma + \bar{\gamma}) - \tau(\bar{\tau} + \pi) + \pi(\tau + \bar{\pi})\end{aligned}$$

Hence we obtain

$$\Delta\rho + D\mu = \rho(\gamma + \bar{\gamma}) + \pi\bar{\pi} - \tau\bar{\tau} \qquad (11\text{-}132a)$$

Likewise, from $(\bar{\delta}D - D\bar{\delta})\psi$, $(\delta\Delta - \Delta\delta)\psi$, and $(\bar{\delta}\delta - \delta\bar{\delta})\psi$ we obtain

$$\bar{\delta}\rho + D\pi = \rho(\alpha + \bar{\beta}) \qquad (11\text{-}132b)$$

$$\delta\mu + \Delta\tau = -\mu(\bar{\alpha} + \beta) + \tau(\gamma - \bar{\gamma}) \qquad (11\text{-}132c)$$

$$\bar{\delta}\tau + \delta\pi = \bar{\mu}\rho - \mu\bar{\rho} - (\bar{\beta} - \alpha)\tau + (\bar{\alpha} - \beta)\pi \qquad (11\text{-}132d)$$

The integration of the above equations falls naturally into two cases, $\rho \neq 0$ and $\rho = 0$.

Solution for $\rho \neq 0$. Radial Integration

The solution to the equation $D\rho = \rho^2$, Eq. (11-127a), is

$$\rho = -1/(r + i\rho_0) \qquad (11\text{-}133)$$

where ρ_0 is real, and the subscript indicates it is independent of the coordinate r. The choice of the integration "constant" as $i\rho_0$ is made possible by means of a coordinate transformation (see Sec. 11-2). The case $\rho = 0$, resulting from $\rho_0 \to \infty$, will be considered separately later.

An equation for $\bar{\delta}\rho$ may be derived from the commutator

$$\begin{aligned}(\bar{\delta}D - D\bar{\delta})\rho &= 2\rho\bar{\delta}\rho - D\bar{\delta}\rho \\ &= \rho^2(\alpha + \bar{\beta} - \pi) - \rho\bar{\delta}\rho\end{aligned}$$

giving

$$D\bar{\delta}\rho - 3\rho\,\bar{\delta}\rho = \rho^2(\pi - \alpha - \bar{\beta}) \tag{11-134}$$

Using now Eqs. (11-127a) to (11-127c), the general solution of the above equation is found to be

$$\bar{\delta}\rho = \rho(\alpha + \bar{\beta}) - 2\bar{\tau}_0\rho^3 \tag{11-135}$$

where $\bar{\tau}_0$ is a "constant" of integration, i.e., a function independent of the coordinate r. This result is now substituted in Eq. (11-132b) to yield

$$D\pi = 2\bar{\tau}_0\rho^3 \tag{11-136}$$

which has the solution

$$\pi = \pi^0 + \bar{\tau}_0\rho^2 \tag{11-137}$$

Equations (11-125), (11-127b) to (11-127e), and (11-130a) can now be integrated easily one at a time, each one yielding a new integration "constant." In this way the radial dependence of every tetrad field function, spin coefficient function, except μ, and the Weyl spinor component ψ, is determined; the solutions are as follows:

$$\xi^i = \bar{\rho}\xi_0{}^i \tag{11-138a}$$

$$\omega = \bar{\rho}\omega_0 + \bar{\alpha}_0 + \beta_0 - \bar{\pi}_0/\bar{\rho} \tag{11-138b}$$

$$X^i = X_0{}^i + \rho\bar{\rho}(\bar{\tau}_0\xi_0{}^i + \tau_0\bar{\xi}_0{}^i) + \rho\eta_0\bar{\xi}_0{}^i + \bar{\rho}\bar{\eta}_0\xi_0{}^i \tag{11-138c}$$

$$\begin{aligned}U = U_0 &- r(\gamma_0 + \bar{\gamma}_0 + \eta_0\pi_0 + \bar{\eta}_0\bar{\pi}_0) + r^2\pi_0\bar{\pi}_0 \\ &+ \rho[\bar{\tau}_0(\bar{\alpha}_0 + \beta_0) - \bar{\tau}_0\eta_0 + \eta_0\bar{\omega}_0 - \psi_0/2] \\ &+ \bar{\rho}[\tau_0(\alpha_0 + \bar{\beta}_0) - \tau_0\bar{\eta}_0 + \bar{\eta}_0\omega_0 - \bar{\psi}_0/2] \\ &+ \rho\bar{\rho}(\bar{\tau}_0\omega_0 + \tau_0\bar{\omega}_0 - \tau_0\bar{\tau}_0) - (\rho/\rho_0)\bar{\tau}_0\bar{\pi}_0 \\ &- (\bar{\rho}/\rho)\tau_0\pi_0\end{aligned} \tag{11-138d}$$

$$\beta = \bar{\rho}\beta_0 \tag{11-138e}$$

$$\alpha = \rho\alpha_0 - \pi_0 + \rho^2\bar{\tau}_0 \tag{11-138f}$$

$$\tau = \rho\eta_0 + \rho\bar{\rho}\tau_0 - \bar{\pi}_0 \tag{11-138g}$$

$$\begin{aligned}\gamma = \gamma_0 &+ \rho(\eta_0\alpha_0 - \bar{\tau}_0\bar{\pi}_0) + \bar{\rho}(\bar{\eta}_0\beta_0 - \tau_0\pi_0) \\ &+ \rho^2(\psi_0/2 + \bar{\tau}_0\eta_0) + \rho\bar{\rho}(\tau_0\alpha_0 + \bar{\tau}_0\beta_0) \\ &+ \rho^2\bar{\rho}\tau_0\bar{\tau}_0 - r\pi_0\bar{\pi}_0\end{aligned} \tag{11-138h}$$

$$\psi = \rho^3\psi_0 \tag{11-138i}$$

In order to obtain an expression for $\Delta\rho$ and μ we again use a commutator:

$$\begin{aligned}(\Delta D - D\Delta)\rho &= 2\rho\,\Delta\rho - D\,\Delta\rho\\ &= \rho^2(\gamma+\bar\gamma) - (\tau+\bar\pi)\,\bar\delta\rho - (\bar\tau+\pi)\,\delta\rho\end{aligned}$$

The last term of this equation can be written out in full using Eqs. (11-128a), (11-135), (11-137), and (11-138). The result is an equation for $\Delta\rho$ which can be integrated to give

$$\begin{aligned}\Delta\rho = &-\rho^2 M_0 + \rho^2\eta_0(\alpha_0+\bar\beta_0) + \rho(\gamma_0+\bar\gamma_0+\eta_0\pi_0)\\ &+ \bar\rho_0\bar\eta_0\bar\pi_0 + \rho\bar\rho[\bar\eta_0(\bar\alpha_0+\beta_0) - \tau_0\pi_0 + \bar\tau_0\bar\pi_0 - \eta_0\bar\eta_0]\\ &- \rho^3(\psi_0/2+\bar\tau_0\eta_0) - \rho^2\bar\rho[\bar\psi_0/2+\bar\tau_0\eta_0\\ &- \tau_0(\alpha_0+\bar\beta_0) - \bar\tau_0(\bar\alpha_0+\beta_0)]\\ &- \rho^3\bar\rho\tau_0\bar\tau_0 + r^2\rho^2\pi_0\bar\pi_0 \end{aligned} \tag{11-139}$$

where M_0 is a constant of integration.

Finally, we substitute Eq. (11-139) into Eq. (11-132a) and perform the radial integration for the spin coefficient μ. We obtain:

$$\begin{aligned}\mu = \mu_0 &+ \rho(M_0 - \bar\tau_0\bar\pi_0) + \bar\rho\tau_0\pi_0\\ &+ \rho^2(\psi_0/2+\bar\tau_0\eta_0) + \rho\bar\rho\bar\psi_0/2\\ &+ \rho^2\bar\rho\tau_0\bar\tau_0 - r^2\rho\pi_0\bar\pi_0\end{aligned} \tag{11-140}$$

Transverse Equations

In the second stage of the solution of the field equations we complete the elimination of the coordinate r by substituting the above results into the remaining equations and equating the coefficients of similar powers of ρ. In this manner we obtain differential equations involving $\xi_0{}^i$ and $X_0{}^i$, and also some purely algebraic constraints between the integration constants.

To find out the derivatives of ρ_0 we differentiate Eq. (11-133). The results are:

$$\begin{aligned}\delta\rho &= \rho^2(\omega + i\,\delta\rho_0)\\ \bar\delta\rho &= \rho^2(\bar\omega + i\,\bar\delta\rho_0)\\ \Delta\rho &= \rho^2(U + i\,\Delta\rho_0)\end{aligned}$$

Expanding now these equations and comparing them with Eqs. (11-128a), (11-135), and (11-139), we obtain the following relations:

$$\xi_0{}^i\rho_{0,i} = -\rho_0(\bar\alpha_0+\beta_0-\eta_0) - i\tau_0 + 2i\rho_0{}^2\bar\pi_0 \tag{11-141a}$$

$$\begin{aligned}X_0{}^i\rho_{0,i} = &-\rho_0(\gamma_0+\bar\gamma_0+2\pi_0\eta_0+2\bar\pi_0\bar\eta_0)\\ &+ i(M_0-\bar M_0)/2 + i(\tau_0\pi_0 - \bar\tau_0\bar\pi_0)\end{aligned} \tag{11-141b}$$

$$\omega_0 = -i\rho_0(\bar{\alpha}_0 + \beta_0 - \eta_0) - 2\rho_0{}^2\bar{\pi}_0 \tag{11-141c}$$

$$\begin{aligned} U_0 = {} & \eta_0(\alpha_0 + \bar{\beta}_0) + \bar{\eta}_0(\bar{\alpha}_0 + \beta_0) - \eta_0\bar{\eta}_0 \\ & + i\rho_0(\pi_0\eta_0 - \bar{\pi}_0\bar{\eta}_0) - (M_0 + \bar{M}_0)/2 \\ & + \tau_0\pi_0 + \bar{\tau}_0\bar{\pi}_0 \end{aligned} \tag{11-141d}$$

where a comma, as usual, denotes a partial differentiation, $f_{,i} = \partial f/\partial x^i$.

The three other field equations, Eqs. (11-130b) to (11-130d), provide the derivatives of ψ_0:

$$\xi_0{}^i\psi_{0,i} = -3\psi_0(\bar{\alpha}_0 + \beta_0 - \eta_0 - 2i\rho_0\bar{\pi}_0) \tag{11-142a}$$

$$\bar{\xi}_0{}^i\psi_{0,i} = -3\psi_0(\alpha_0 + \bar{\beta}_0) \tag{11-142b}$$

$$X_0{}^i\psi_{0,i} = -3\psi_0(\gamma_0 + \bar{\gamma}_0 + \mu_0 + \pi_0\eta_0 + \bar{\pi}_0\bar{\eta}_0) \tag{11-142c}$$

Substitution in Eqs. (11-128b) and (11-129a) then gives the transverse derivatives of π_0, τ_0, and η_0:

$$\xi_0{}^i\pi_{0,i} = -\mu_0 + \pi_0(\bar{\alpha}_0 - \beta_0) \tag{11-143a}$$

$$\bar{\xi}_0{}^i\pi_{0,i} = \pi_0(\bar{\beta}_0 - \alpha_0) \tag{11-143b}$$

$$\xi_0{}^i\tau_{0,i} = -\tau_0(3\bar{\alpha}_0 + \beta_0) \tag{11-143c}$$

$$\begin{aligned} \bar{\xi}_0{}^i\tau_{0,i} = {} & -\tau_0(\alpha_0 + 3\bar{\beta}_0 - \bar{\eta}_0 + 2i\rho_0\pi_0) - 2i\rho_0\bar{M}_0 \\ & - 2i(\rho_0)^3\pi_0\bar{\pi}_0 + (\psi_0 - \bar{\psi}_0)/2 \end{aligned} \tag{11-143d}$$

$$\xi_0{}^i\eta_{0,i} = -\eta_0(2\bar{\alpha}_0 - \eta_0 - 2i\rho_0\bar{\pi}_0) + 2\tau_0\bar{\pi}_0 \tag{11-143e}$$

$$\bar{\xi}_0{}^i\eta_{0,i} = -2\bar{\beta}_0\eta_0 - M_0 + \bar{M}_0 + 2\bar{\tau}_0\bar{\pi}_0 \tag{11-143f}$$

Equations (11-127f) and (11-128d) confirm these results. Equation (11-128c) yields

$$\begin{aligned} \xi_0{}^i\alpha_{0,i} - \bar{\xi}_0{}^i\beta_{0,i} = {} & 2\beta_0(\bar{\beta}_0 - \alpha_0) \\ & + 2i\rho_0(\gamma_0 + \mu_0 + \alpha_0\bar{\pi}_0 + \beta_0\pi_0) \\ & + M_0 + 3\rho_0{}^2\pi_0\bar{\pi}_0 \end{aligned} \tag{11-144}$$

Equation (11-128e) yields the relations

$$\begin{aligned} \bar{\xi}_0{}^i\mu_{0,i} = {} & -\mu_0(\alpha_0 + \bar{\beta}_0) - 2i\rho_0\bar{\mu}_0\pi_0 + 2i\rho_0\pi_0\bar{\pi}_0\bar{\eta}_0 \\ & + (M_0 + \bar{M}_0)\pi_0 + 6\rho_0{}^2\pi_0{}^2\bar{\pi}_0 \end{aligned} \tag{11-145a}$$

$$\begin{aligned} \bar{\xi}_0{}^iM_{0,i} = {} & -2M_0(\alpha_0 + \bar{\beta}_0) + (\psi_0 + 2\bar{\psi}_0)\pi_0 \\ & + \rho_0{}^2\bar{\mu}_0\pi_0 - 2\mu_0\bar{\tau}_0 + 2\bar{\tau}_0\pi_0\eta_0 \\ & - 2\rho_0{}^2\pi_0\bar{\pi}_0\bar{\eta}_0 + 2i\rho_0\bar{\tau}_0\bar{\pi}_0\pi_0 + 4i\rho_0{}^3\pi_0{}^2\bar{\pi}_0 \end{aligned} \tag{11-145b}$$

and the relation

$$\eta_0 = 2i\rho_0\bar{\pi}_0 \tag{11-145c}$$

When the last equation (11-145c) is introduced, many of the above equations become simplified. If we differentiate it and compare it with Eq. (11-143f) we obtain a further condition:

$$M_0 - \bar{M}_0 = 2i\rho_0\bar{\mu}_0 + 4\bar{\tau}_0\bar{\pi}_0 - 8\rho_0{}^2\pi_0\bar{\pi}_0 \tag{11-146a}$$

Since the real part of the right-hand side of this equation must vanish we also obtain the relation

$$2\tau_0\pi_0 + 2\bar{\tau}_0\bar{\pi}_0 = 8\rho_0{}^2\pi_0\bar{\pi}_0 + i\rho_0(\mu_0 - \bar{\mu}_0) \tag{11-146b}$$

Continuing in the same manner, Eqs. (11-129) and (11-132c) then gives the following set of equations:

$$\begin{aligned}
\xi_0{}^i\mu_{0,i} &= -\mu_0(\bar{\alpha}_0 + \beta_0) + \bar{\pi}_0(M_0 + \bar{M}_0) \\
&\quad + 2i\rho_0\bar{\pi}_0\mu_0 + 10\rho_0{}^2\pi_0\bar{\pi}_0{}^2
\end{aligned} \tag{11-147a}$$

$$\begin{aligned}
\xi_0{}^iM_{0,i} &= -2M_0(\bar{\alpha}_0 + \beta_0) + 3\bar{\pi}_0\bar{\psi}_0 + 2i\rho_0\bar{\pi}_0(5M_0 + \bar{M}_0) \\
&\quad - 3\rho_0{}^2\mu_0\bar{\pi}_0 - 6i\rho_0\tau_0\pi_0\bar{\pi}_0 + 20i\rho_0{}^3\pi_0\bar{\pi}_0{}^2
\end{aligned} \tag{11-147b}$$

$$X_0{}^i\pi_{0,i} = -\pi_0(\gamma_0 - \bar{\gamma}_0 + 2i\rho_0\pi_0\bar{\pi}_0) \tag{11-147c}$$

$$\begin{aligned}
X_0{}^i\tau_{0,i} &= -\tau_0(\mu_0 + \gamma_0 + 3\bar{\gamma}_0) + \bar{\pi}_0(\psi_0 - \bar{\psi}_0)/2 \\
&\quad - 2i\rho_0\bar{\pi}_0M_0 + 6i\rho_0\tau_0\pi_0{}^2 - 2i\rho_0{}^3\pi_0\bar{\pi}_0{}^2
\end{aligned} \tag{11-147d}$$

$$\begin{aligned}
X_0{}^i\alpha_{0,i} - \xi_0{}^i\gamma_{0,i} &= -\alpha_0(\bar{\mu}_0 + \gamma_0) + \bar{\beta}_0\gamma_0 \\
&\quad + i\rho_0\pi_0(2\gamma_0 + 2\mu_0 - \bar{\mu}_0) + \bar{M}_0\pi_0 \\
&\quad + 2i\rho_0\pi_0{}^2\beta_0 - 4i\rho_0\pi_0\bar{\pi}_0\alpha_0 \\
&\quad + \bar{\tau}_0\pi_0\bar{\pi}_0 + 7\rho_0{}^2\pi_0{}^2\bar{\pi}_0
\end{aligned} \tag{11-147e}$$

$$\begin{aligned}
X_0{}^i\beta_{0,i} - \xi_0{}^i\gamma_{0,i} &= -\beta_0(\mu_0 - \gamma_0 + 2\bar{\gamma}_0) + \bar{\alpha}_0\gamma_0 \\
&\quad - i\rho_0\bar{\pi}_0(2\gamma_0 + \mu_0) + M_0\bar{\pi}_0 + 4i\rho_0\pi_0\bar{\pi}_0\beta_0 \\
&\quad - 2i\rho_0\bar{\pi}_0{}^2\alpha_0 + 3\tau_0\pi_0\bar{\pi}_0 + 3\rho_0{}^2\pi_0\bar{\pi}_0{}^2
\end{aligned} \tag{11-147f}$$

$$X_0{}^i\mu_{0,i} = -\mu_0(\mu_0 + \gamma_0 + \bar{\gamma}_0 - 2i\rho_0\pi_0\bar{\pi}_0) - 2i\rho_0\pi_0\bar{\pi}_0\bar{\mu}_0 \tag{11-147g}$$

$$\begin{aligned}
X_0{}^iM_{0,i} &= -2M_0(\mu_0 + \gamma_0 + \bar{\gamma}_0) - \pi_0\bar{\pi}_0(\psi_0 - \bar{\psi}_0) + 4i\rho_0M_0\pi_0\bar{\pi}_0 \\
&\quad + 2\mu_0\bar{\tau}_0\bar{\pi}_0 + \rho_0{}^2\pi_0\bar{\pi}_0(\mu_0 - \bar{\mu}_0) + 4i\rho_0\bar{\pi}_0\pi_0{}^2\tau_0 \\
&\quad - 4i\rho_0\bar{\pi}_0{}^2\pi_0\tau_0 - 12i\rho_0{}^3\pi_0{}^2\bar{\pi}_0{}^2
\end{aligned} \tag{11-147h}$$

and also an algebraic constraint

$$\mu_0 = \bar{\mu}_0 \tag{11-147i}$$

It is interesting to note that Eq. (11-147i) is implied only by the very last equation, Eq. (11-129e). Differentiation of Eq. (11-146b) leads to

$$\begin{aligned}
\pi_0(\psi_0 - \bar{\psi}_0)/2 &= -\mu_0\tau_0 + 4\rho_0{}^2\mu_0\bar{\pi}_0 + 2i\rho_0M_0\bar{\pi}_0 \\
&\quad + 2i\rho_0\tau_0\pi_0\bar{\pi}_0 - 2i\rho_0\bar{\tau}_0\bar{\pi}_0\pi_0 - 6i\rho_0{}^3\pi_0\bar{\pi}_0{}^2
\end{aligned} \tag{11-148}$$

Finally, the metric equations (11-126) give

$$\xi_0{}^i X_0{}^j{}_{,i} - X_0{}^i \xi_0{}^j{}_{,i} = (\mu_0 + 2\bar{\gamma}_0 - 4i\rho_0 \bar{\pi}_0 \pi_0)\xi_0{}^j + 2i\rho_0 \pi_0 \bar{\pi}_0 \bar{\xi}_0{}^j + (2i\rho_0 \bar{\pi}_0 - \bar{\alpha}_0 - \beta_0) X_0{}^j \quad (11\text{-}149a)$$

$$\bar{\xi}_0{}^i \xi_0{}^j{}_{,i} - \xi_0{}^i \bar{\xi}_0{}^j{}_{,i} = -2(\beta_0 + i\rho_0 \pi_0)\xi_0{}^j + 2(\bar{\beta}_0 - i\rho_0 \bar{\pi}_0)\bar{\xi}_0{}^j - 2i\rho_0 X_0{}^j \quad (11\text{-}149b)$$

Choice of Tetrad and Coordinates

So far the tetrad has not been completely specified; it still has the freedom of the two-parameter group of rotations which leaves the directions of the vectors l^μ and n^μ unchanged:

$$\begin{aligned} l^\mu &\to A_0{}^{-1} l^\mu \\ m^\mu &\to \exp(i\theta_0) m^\mu \\ \bar{m}^\mu &\to \exp(-i\theta_0)\bar{m}^\mu \\ n^\mu &\to A_0 n^\mu \end{aligned} \quad (11\text{-}150)$$

where A_0 and θ_0 are arbitrary real functions independent of the coordinate r.[16] This transformation induces a transformation on all the field variables. For instance, one obtains

$$\begin{aligned} U_0 &\to A_0{}^2 U_0 \\ \alpha_0 + \bar{\beta}_0 &\to [A_0(\alpha_0 + \bar{\beta}_0) - \bar{\xi}_0{}^j A_{0,j}] \exp(i\theta_0) \\ \pi_0 &\to \pi_0 \exp(-i\theta_0) \\ \tau_0 &\to A_0{}^2 \tau_0 \exp(i\theta_0) \\ \psi_0 &\to A_0{}^3 \psi_0 \end{aligned}$$

The fact that some variables transform inhomogeneously permits a partial check on the results obtained so far, since all the equations obtained above must be covariant under the transformation (11-150). It also suggests the possiblity of setting some of the variables as zeros.

We now perform a rotation with A_0 chosen in such a way that $\psi_0 \bar{\psi}_0$ becomes a constant. Then Eqs. (11-142a) to (11-142c) yield the following equations:

$$\bar{\alpha}_0 + \beta_0 = 2i\rho_0 \bar{\pi}_0 \quad (11\text{-}151a)$$

$$\gamma^0 + \bar{\gamma}^0 + \mu_0 = 0 \quad (11\text{-}151b)$$

[16] If the functions A_0 and θ_0 were allowed to depend on the coordinate r, the condition $\varepsilon = 0$ would be violated.

Classification of Solutions

At this stage of the calculation it is worthwhile to resolve several cases.

Case I: Let us assume that $\pi_0 = \tau_0 = 0$. This is the case already treated by Newman, Tamburino, and Unti which leads to the three NUT–Taub metrics discussed in the last section.

Case II: $\pi_0 = 0$, $\tau_0 \neq 0$. We select θ_0 such that $i\tau_0$ is everywhere real and positive. Equations (11-143a), (11-147d), and (11-151b) give us $\mu_0 = \gamma_0 = 0$. Now $M_0 = -U_0$ and $\psi_0 \equiv m + il$ become constants. From a comparison of Eqs. (11-143c), (11-143d), and (11-144) we obtain

$$\begin{aligned} \beta_0 &= \bar{\beta}_0 \\ 2\rho_0 U_0 &= -l - 4i\beta_0\tau_0 \\ \xi_0{}^i\beta_{0,i} &= U_0/2 - 2\beta_0{}^2 \\ \xi_0{}^i\tau_{0,i} &= 2\beta_0\tau_0 \end{aligned} \tag{11-152}$$

Let us now choose coordinates such that $X_0{}^i = \delta_0{}^i$. Use ρ_0 to define a coordinate x^2 by $\rho_0 = \rho_0(x^2)$. Since $\xi_0{}^i\rho_{0,i}$ is real, it follows that $\xi_0{}^2$ must also be real. By the coordinate transformations

$$\begin{aligned} x^0 &\to x^0 + f(x^2, x^3) \\ x^2 &\to g(x^2) \\ x^3 &\to x^3 + h(x^2, x^3) \end{aligned}$$

we can make $\xi_0{}^2 = -1/\sqrt{2}$ and $\xi_0{}^0$ and $\xi_0{}^3$ imaginary. Then the relations

$$\begin{aligned} \xi_0{}^i\xi_0{}^j{}_{,i} &= \bar{\xi}_0{}^i\xi_0{}^j{}_{,i} \\ \xi_0{}^i\tau_{0,i} &= \bar{\xi}_0{}^i\tau_{0,i} \\ \xi_0{}^i\beta_{0,i} &= \bar{\xi}_0{}^i\beta_{0,i} \\ X_0{}^i\xi_0{}^j{}_{,i} = X_0{}^i\tau_{0,i} &= X_0{}^i\beta_{0,i} = 0 \end{aligned}$$

show that $\xi_0{}^i$, τ_0, and β_0 are functions of only the coordinate x^2. Integration is now straightforward and leads to the following metrics:

Case II.A: $U_0 < 0$. By a further rotation by Eq. (11-150), with A_0 constant, one can set $U_0 = -\frac{1}{2}$. The solution is known as the Kerr–NUT space,[17] with a being the Kerr angular momentum per unit mass parameter, and $x^0 = u$, $x^1 = r$, $x^2 = x$, $x^3 = y$:

$$\begin{aligned} \beta_0 &= -\tfrac{1}{4}2^{1/2} \cot x \\ \tau_0 &= -\tfrac{1}{2}ia2^{1/2} \sin x \\ \rho &= -(r + il - ia \cos x)^{-1} \\ l^\mu &= (0, 1, 0, 0) \\ m^\mu &= -\tfrac{1}{2}2^{1/2}\bar{\rho}[ia \sin x + 2il \cot x, 0, 1, i \operatorname{cosec} x] \\ n^\mu &= \rho\bar{\rho}[r^2 + l^2 + a^2, -(r^2 - 2mr - l^2 + a^2)/2, 0, a] \end{aligned} \tag{11-153a}$$

[17] M. Demianski and E. Newman, *Bull. Acad. Polon. Sci.* **14**, 653 (1966).

The only nonzero components of the metric are

$$\begin{aligned} g_{uu} &= \rho\bar{\rho}(r^2 - 2mr - l^2 + a^2 \cos^2 x) \\ g_{ur} &= 1 \\ g_{uy} &= -2\rho\bar{\rho}l \cos x(r^2 - 2mr - l^2 + a^2) + 2\rho\bar{\rho}a \sin^2 x(mr + l^2) \\ g_{ry} &= -a \sin^2 x - 2l \cos x \\ g_{xx} &= -r^2 - (l - a \cos x)^2 \\ g_{yy} &= \rho\bar{\rho}(r^2 - 2mr - l^2 + a^2)(a \sin^2 x + 2l \cos x)^2 \\ &\quad - \rho\bar{\rho} \sin^2 x(r^2 + l^2 + a^2)^2 \end{aligned} \tag{11-153b}$$

Case II.B: $U_0 > 0$, $\beta_0 < -2^{1/2}/4$. Now we use the rotation to set $U_0 = \frac{1}{2}$ and obtain

$$\begin{aligned} \beta_0 &= -\tfrac{1}{4}2^{1/2} \coth x \\ \tau_0 &= -\tfrac{1}{2}ia2^{1/2} \sinh x \\ \rho &= -(r - il + ia \cosh x)^{-1} \\ l^\mu &= (0, 1, 0, 0) \\ m^\mu &= -\tfrac{1}{2}2^{1/2}\bar{\rho}(-ia \sinh x + 2il \coth x, 0, 1, i \operatorname{cosech} x) \\ n^\mu &= \rho\bar{\rho}[r^2 + l^2 + a^2, (r^2 + 2mr - l^2 + a^2)/2, 0, a] \end{aligned} \tag{11-154a}$$

whereas the nonzero components of the metric are given by

$$\begin{aligned} g_{uu} &= -\rho\bar{\rho}(r^2 + 2mr - l^2 + a^2 \cosh^2 x) \\ g_{ur} &= 1 \\ g_{uy} &= 2\rho\bar{\rho}l \cosh x(r^2 + 2mr - l^2 + a^2) - 2\rho\bar{\rho}a \sinh^2 x(mr - l^2) \\ g_{ry} &= a \sinh^2 x - 2l \cosh x \\ g_{xx} &= -r^2 - (-l + a \cosh x)^2 \\ g_{yy} &= -\rho\bar{\rho}(r^2 + 2mr - l^2 + a^2)(-a \sinh^2 x + 2l \cosh x)^2 \\ &\quad - \rho\bar{\rho} \sinh^2 x(r^2 + l^2 + a^2)^2 \end{aligned} \tag{11-154b}$$

Case II.C: $U_0 = +\frac{1}{2}$, $\beta_0 > -2^{1/2}/4$. We obtain

$$\begin{aligned} \beta_0 &= -\tfrac{1}{4}2^{1/2} \tanh x \\ \tau_0 &= -\tfrac{1}{2}ia2^{1/2} \cosh x \\ \rho &= -(r - il + ia \sinh x)^{-1} \\ l^\mu &= (0, 1, 0, 0) \\ m^\mu &= -\tfrac{1}{2}2^{1/2}\bar{\rho}(-ia \cosh x + 2il \tanh x, 0, 1, i \operatorname{sech} x) \\ n^\mu &= \rho\bar{\rho}[r^2 + l^2 - a^2, (r^2 + 2mr - l^2 - a^2)/2, 0, a] \end{aligned} \tag{11-155a}$$

whereas the nonzero components of the metric are given by

$$\begin{aligned}
g_{uu} &= -\rho\bar{\rho}(r^2 + 2mr - l^2 + a^2 \sinh^2 x) \\
g_{ur} &= 1 \\
g_{uy} &= 2\rho\bar{\rho}l \sinh x(r^2 + 2mr - l^2 - a^2) - 2\rho\bar{\rho}a \cosh^2 x(mr - l^2) \\
g_{ry} &= a \cosh^2 x - 2l \sinh x \\
g_{xx} &= -r^2 - (-l + a \sinh x)^2 \\
g_{yy} &= -\rho\bar{\rho}(r^2 + 2mr - l^2 - a^2)(-a \cosh^2 x + 2l \sinh x)^2 \\
&\quad - \rho\bar{\rho} \cosh^2 x(r^2 + l^2 - a^2)^2
\end{aligned} \tag{11-155b}$$

Case II.D: $U_0 = +\frac{1}{2}$, $\beta_0 = -2^{1/2}/4$. We obtain

$$\begin{aligned}
\beta_0 &= -2^{1/2}/4 \\
\tau_0 &= -\tfrac{1}{2}ia2^{1/2}\, e^x \\
\rho &= -(r - il + ia\, e^x)^{-1} \\
l^\mu &= (0, 1, 0, 0) \\
m^\mu &= -\tfrac{1}{2}2^{1/2}\bar{\rho}(-ia\, e^x + 2il, 0, 1, i\, e^{-x}) \\
n^\mu &= \rho\bar{\rho}[r^2 + l^2, (r^2 + 2mr - l^2)/2, 0, a]
\end{aligned} \tag{11-156a}$$

whereas the nonzero components of the metric are given by

$$\begin{aligned}
g_{uu} &= -\rho\bar{\rho}(r^2 + 2mr - l^2 + a^2\, e^{2x}) \\
g_{ur} &= 1 \\
g_{uy} &= 2\rho\bar{\rho}l\, e^x(r^2 + 2mr - l^2) - 2\rho\bar{\rho}a\, e^{2x}(mr - l^2) \\
g_{ry} &= a\, e^{2x} - 2l\, e^x \\
g_{xx} &= -r^2 - (-l + a\, e^x)^2 \\
g_{yy} &= -\rho\bar{\rho}(r^2 + 2mr - l^2)(-a\, e^{2x} + 2l\, e^x)^2 - \rho\bar{\rho}(r^2 + l^2)^2\, e^{2x}
\end{aligned} \tag{11-156b}$$

Case II.E: $U_0 = 0, l \neq 0$. We may now use a tetrad rotation and a rescaling of coordinates in order to set $l = +1$. Then the solution is given by

$$\begin{aligned}
\beta_0 &= -\tfrac{1}{4}2^{1/2}/x \\
\tau_0 &= -\tfrac{1}{2}i2^{1/2}x \\
\rho &= -(r + ib + ix^2/2)^{-1} \\
l^\mu &= (0, 1, 0, 0) \\
m^\mu &= -\tfrac{1}{2}2^{1/2}\bar{\rho}(-ix^3/4 - ibx, 0, 1, i/x) \\
n^\mu &= \rho\bar{\rho}(r^2 + b^2, mr + b, 0, 1)
\end{aligned} \tag{11-157a}$$

whereas the metric is given by

$$
\begin{aligned}
g_{uu} &= -\rho\bar{\rho}(2mr + 2b + x^2) \\
g_{ur} &= 1 \\
g_{uy} &= \rho\bar{\rho}x^2(r^2 - 2mbr - mrx^2/2 - b^2 - bx^2/2) \\
g_{ry} &= bx^2 + x^4/4 \\
g_{xx} &= -r^2 - (b + x^2/2)^2 \\
g_{yy} &= -\rho\bar{\rho}(2mr + 2b)(bx^2 + x^4/4)^2 - \rho\bar{\rho}x^2(r^2 + b^2)^2
\end{aligned}
\tag{11-157b}
$$

Case II.F: $U_0 = 0$, $l = 0$. We obtain

$$
\begin{aligned}
\beta_0 &= 0 \\
\tau_0 &= -\tfrac{1}{2}i2^{1/2} \\
\rho &= -(r + ix)^{-1} \\
l^\mu &= (0, 1, 0, 0) \\
m^\mu &= -\tfrac{1}{2}2^{1/2}\bar{\rho}(-ix^2, 0, 1, i) \\
n^\mu &= \rho\bar{\rho}(r^2, mr - \tfrac{1}{2}, 0, 1)
\end{aligned}
\tag{11-158a}
$$

whereas the metric is given by

$$
\begin{aligned}
g_{uu} &= -2\rho\bar{\rho}mr \\
g_{ur} &= 1 \\
g_{uy} &= \rho\bar{\rho}(r^2 - 2mrx^2 + x^2) \\
g_{ry} &= x^2 \\
g_{xx} &= -r^2 - x^2 \\
g_{yy} &= -\rho\bar{\rho}(r^4 + 2mx^4r - x^4)
\end{aligned}
\tag{11-158b}
$$

Case III: Here we assume $\pi_0 \neq 0$. Now we choose θ_0 in such a way that π_0 is everywhere real. Equations (11-143a), (11-143b), (11-147c), and (11-151a) then imply the relations:

$$
\begin{aligned}
\mu_0 &= -2\pi_0(\beta_0 + \bar{\beta}_0) \\
\beta_0 - \bar{\beta}_0 &= 2i\rho_0\pi_0 \\
\gamma_0 - \bar{\gamma}_0 &= -2i\rho_0\pi_0{}^2 \\
\alpha_0 &= -\beta_0
\end{aligned}
$$

Likewise, Eqs. (11-146a), (11-146b), and (11-148) reduce to the following set of equations:

$$M_0 - \bar{M}_0 = 4i\rho_0\pi_0(\beta_0 + \bar{\beta}_0) - 2\pi_0(\tau_0 - \bar{\tau}_0)$$

$$\tau_0 + \bar{\tau}_0 = 4\rho_0{}^2\pi_0$$

$$\psi_0 - \bar{\psi}_0 = 2i\rho_0(M_0 + \bar{M}_0) + 2(\tau_0 - \bar{\tau}_0)(\beta_0 + \bar{\beta}_0) - 6i\rho_0{}^3\pi_0{}^2$$

It is advantageous to abandon the complex notation at this point of calculation and work with real and imaginary parts. Accordingly we write:

$$\begin{aligned}
\beta_0 &= b_0 + i\rho_0\pi_0 \\
\tau_0 &= 2\rho_0{}^2\pi_0 + i\pi_0 t_0 \\
\mu_0 &= -4\pi_0 b_0 \\
\gamma_0 &= 2\pi_0 b_0 - i\rho_0\pi_0{}^2 \\
M_0 &= -U_0 + 4\rho_0{}^2\pi_0{}^2 - 2i\pi_0{}^2 t_0 - 4i\rho_0\pi_0 b_0 \\
\psi_0 &\equiv m_0 + il_0 \\
&= m_0 + 2i(-\rho_0 U_0 + 2b_0\pi_0 t_0 + \rho_0{}^3\pi_0{}^2)
\end{aligned} \tag{11-159}$$

Then the equations to be solved are the following:

$$\xi_0{}^i\rho_{0,i} = \bar{\xi}_0{}^i\rho_{0,i} = \pi_0 t_0 \tag{11-160a}$$

$$\xi_0{}^i t_{0,i} = \bar{\xi}_0{}^i t_{0,i} = -8\rho_0{}^3\pi_0 \tag{11-160b}$$

$$\xi_0{}^i\pi_{0,i} = \bar{\xi}_0{}^i\pi_{0,i} = 2b_0\pi_0 \tag{11-160c}$$

$$\xi_0{}^i b_{0,i} = \bar{\xi}_0{}^i b_{0,i} = U_0/2 - 2b_0{}^2 - \tfrac{9}{2}\rho_0{}^2\pi_0{}^2 \tag{11-160d}$$

$$\begin{aligned}
\xi_0{}^i U_{0,i} &= \bar{\xi}_0{}^i U_{0,i} \\
&= -3m_0\pi_0 - 6\rho_0\pi_0{}^3 t_0 - 12\rho_0{}^2\pi_0{}^2 b_0
\end{aligned} \tag{11-160e}$$

$$\xi_0{}^i\psi_{0,i} = \bar{\xi}_0{}^i\psi_{0,i} = 6i\rho_0\pi_0\psi_0 \tag{11-160f}$$

$$\bar{\xi}_0{}^i\xi_0{}^j{}_{,i} - \xi_0{}^i\bar{\xi}_0{}^j{}_{,i} = -2i\rho_0 X_0{}^j - 2b_0(\xi_0{}^j - \bar{\xi}_0{}^j) \tag{11-160g}$$

$$\xi_0{}^i X_0{}^j{}_{,i} - X_0{}^i\xi_0{}^j{}_{,i} = -2i\rho_0\pi_0{}^2(\xi_0{}^j - \bar{\xi}_0{}^j) \tag{11-160h}$$

$$\begin{aligned}
X_0{}^i\rho_{0,i} &= X_0{}^i t_{0,i} = X_0{}^i\pi_{0,i} = X_0{}^i b_{0,i} \\
&= X_0{}^i U_{0,i} = X_0{}^i\psi_{0,i} = 0
\end{aligned} \tag{11-160i}$$

Again by Eq. (11-152) we can make $\xi_0{}^0$ and $\xi_0{}^3$ to be imaginary and $\xi_0{}^2 = d\pi_0$, where d is a real constant to be chosen in the sequel. Equations (11-160a) and (11-160b) then imply

$$\rho_{0,xx} = -8\rho_0{}^3/d^2 \tag{11-161}$$

The solution of this equation is easily seen to be given by

$$\rho_0 = a \operatorname{cn}[(2a2^{1/2}/d)x]$$

where $\operatorname{cn} x$ is an elliptic function of modulus $k = 1/\sqrt{2}$, and a is a constant of integration.

One consequently has the following subcases:

Case III.A: $a = 0$. Here we choose $d = 1$. The remaining integrations are simple and lead to the following vacuum solution:

$$\begin{aligned}
\tau_0 &= 0 \\
m_0 &= m \\
l_0 &= 0 \\
\pi_0 &= (\tfrac{1}{2})^{1/2} f(x) \qquad\qquad (11\text{-}162a) \\
l^\mu &= (0, 1, 0, 0) \\
m^\mu &= \{0, (\tfrac{1}{2})^{1/2} r f(x), -(\tfrac{1}{2})^{1/2} r^{-1} f(x), \\
&\quad - i(\tfrac{1}{2})^{1/2} r^{-1} [f(x)]^{-1}\} \\
n^\mu &= \{1, \tfrac{1}{2} r^2 [f(x - 1/r)]^2, 0, 0\}
\end{aligned}$$

and the metric becomes:

$$\begin{aligned}
g_{uu} &= -r^2 [f(x - 1/r)]^2 \\
g_{ur} &= 1 \\
g_{ux} &= -r^2 \qquad\qquad (11\text{-}162b) \\
g_{xx} &= -r^2 [f(x)]^{-2} \\
g_{yy} &= -r^2 [f(x)]^2
\end{aligned}$$

where

$$f(x) = (-2mx^3 + \tilde{a}x + \tilde{b})^{1/2}$$

and m, $\tilde{a}$, and $\tilde{b}$ are constants.

This solution is the static "C" metric discussed by Ehlers and Kundt.[18]

Case III.B: $a \neq 0$. Choose $d = -2^{3/2} a$. Then one obtains

$$\rho_0 = a \operatorname{cn} x \qquad\qquad (11\text{-}163)$$

$$t_0 = 2^{3/2} a^2 \operatorname{sn} x \operatorname{dn} x$$

where $\operatorname{sn} x$ and $\operatorname{dn} x$ are elliptic functions.

Equation (11-160f) is next integrated to yield

$$\psi_0 = (m + il)(\operatorname{dn} x - i2^{-1/2} \operatorname{sn} x)^3 \qquad\qquad (11\text{-}164)$$

[18] J. Ehlers and W. Kundt, in *Gravitation: An Introduction to Current Research*, (L. Witten, Ed.), John Wiley, New York, 1962.

where m and l are constants. Equation (11-160e) can then be written as

$$\xi_0{}^i(U_0 + 3\rho_0{}^2\pi_0{}^2)_{,i} = -3m_0\pi_0$$

whose integration yields

$$U_0 + 3\rho_0{}^2\pi_0{}^2 = b + \tfrac{3}{4}2^{1/2}a^{-1}\text{cn } x(m \text{ sn } x - 2^{1/2}l \text{ dn } x) \qquad (11\text{-}165)$$

where b is a constant. When Eqs. (11-160c), (11-160d), and (11-165) are combined, they give the following equation for $\pi_0{}^2$:

$$\pi_0{}^2{}_{,x} \text{ sn } x \text{ dn } x - \pi_0{}^2\text{cn}^3 x = -\tfrac{1}{4}ba^{-2}\text{cn } x + \tfrac{1}{4}la^{-3}\text{dn}^3 x + \tfrac{1}{16}2^{1/2}ma^{-3}\text{sn}^3 x \qquad (11\text{-}166)$$

whose solution is given by

$$\pi_0{}^2 = c \text{ sn } x \text{ dn } x + \tfrac{1}{4}ba^{-2}\text{cn}^2 x - \tfrac{1}{8}2^{1/2}a^{-3}\text{cn } x(m \text{ sn } x + 2^{1/2}l \text{ dn } x) \qquad (11\text{-}167)$$

Finally, the solution of Eqs. (11-160g) and (11-160h), for $j = 0, 3$, is found to be

$$X_0{}^j = D^j \text{ sn } x + E^j \text{ dn } x \qquad (11\text{-}168a)$$

$$\xi_0{}^j = \tfrac{1}{2}2^{1/2}i\pi_0{}^{-1}(-D^j \text{ dn } x + \tfrac{1}{2}E^j \text{ sn } x) \qquad (11\text{-}168b)$$

We choose $D^j = \delta_3{}^j$ and $E^j = \delta_0{}^j$, and the resulting vacuum solution is, consequently, given by

$$\begin{aligned} \rho &= -(r + i\,a \text{ cn } x)^{-1} \\ l^\mu &= (0, 1, 0, 0) \\ m^\mu &= -2^{-1/2}\bar\rho[-\tfrac{1}{2}i\pi_0{}^{-1}\text{sn } x,\ 2^{1/2}\pi_0(r^2 + 3a^2\text{cn}^2 x), \\ &\qquad 4a\pi_0,\ i\pi_0{}^{-1}\text{dn } x] \\ n^\mu &= (X^0, U, 0, X^3) \end{aligned} \qquad (11\text{-}169a)$$

where

$$\begin{aligned} X^0 &= \text{dn } x + 2^{1/2}a\rho\bar\rho(r \text{ cn } x + 2^{1/2}a \text{ sn } x \text{ dn } x) \text{ sn } x \\ X^3 &= \text{sn } x - 2^{3/2}a\rho\bar\rho(r \text{ cn } x + 2^{1/2}a \text{ sn } x \text{ dn } x) \text{ dn } x \\ U &= b + \tfrac{3}{4}2^{1/2}a^{-1}\text{cn } x(m \text{ sn } x - 2^{1/2}l \text{ dn } x) \\ &\quad + 2^{3/2}acr \text{ cn}^3 x - 2^{1/2}a^{-1}br \text{ sn } x \text{ cn } x \text{ dn } x \\ &\quad - \tfrac{1}{2}a^{-2}mr(\text{dn}^3 x - \tfrac{3}{2}\text{sn}^2 x \text{ dn } x) + \pi_0{}^2(r^2 - 3a^2\text{cn}^2 x) \\ &\quad + \tfrac{1}{2}a^{-2}lr(\tfrac{3}{2}\text{dn}^2 x \text{ sn } x - \tfrac{1}{4}\text{sn}^3 x) + \rho\bar\rho(rm_0 + al_0 \text{ cn } x) \\ &\quad - 4\rho\bar\rho a^2\pi_0{}^2(r \text{ cn } x - 2^{1/2}a \text{ sn } x \text{ dn } x)^2 \\ m_0 &= m(\text{dn}^3 x - \tfrac{3}{2}\text{dn } x \text{ sn}^2 x) + 2^{1/2}l(\tfrac{3}{2}\text{dn}^2 x \text{ sn } x - \tfrac{1}{4}\text{sn}^3 x) \\ l_0 &= -2^{1/2}m(\tfrac{3}{2}\text{dn}^2 x \text{ sn } x - \tfrac{1}{4}\text{sn}^3 x) + l(\text{dn}^3 x - \tfrac{3}{2}\text{dn } x \text{ sn}^2 x) \\ \pi_0{}^2 &= c \text{ sn } x \text{ dn } x + \tfrac{1}{4}a^{-2}b \text{ cn}^2 x - \tfrac{1}{8}2^{1/2}a^{-3}\text{cn } x(m \text{ sn } x + 2^{1/2}l \text{ dn } x) \end{aligned} \qquad (11\text{-}169b)$$

and a, b, c, l, and m are arbitrary constants.

The nonzero components of the metric, when written in these terms, are given by:

$$
\begin{aligned}
g_{uu} &= -2U\,\mathrm{dn}^2 x - (X^3\pi_0)^2(r^2 + a^2\mathrm{cn}^2 x) \\
g_{ur} &= \mathrm{dn}\, x \\
g_{ux} &= -\tfrac{1}{4}2^{1/2}a^{-1}\mathrm{dn}\, x(r^2 + 3a^2\mathrm{cn}^2 x) \\
g_{uy} &= -U\,\mathrm{sn}\, x\,\mathrm{dn}\, x + X^0X^3\pi_0{}^2(r^2 + a^2\mathrm{cn}^2 x) \qquad \text{(11-169c)} \\
g_{ry} &= \tfrac{1}{2}\,\mathrm{sn}\, x \\
g_{xx} &= -\tfrac{1}{16}(a\pi_0)^{-2}(r^2 + a^2\mathrm{cn}^2 x) \\
g_{xy} &= -\tfrac{1}{8}2^{1/2}a^{-1}\mathrm{sn}\, x(r^2 + 3a^2\mathrm{cn}^2 x) \\
g_{yy} &= -\tfrac{1}{2}U\,\mathrm{sn}^2 x - (X^0\pi_0)^2(r^2 + a^2\mathrm{cn}^2 x)
\end{aligned}
$$

Generalized Kerr Metric

The solutions of Case II, Eqs. (11-153) to (11-159), can be interpreted physically as follows.

If we examine the two-dimensional positive-definite metric of the wavefronts on the equipotentials $u = \text{const.}$, $r = \text{const.}$, we see that asymptotically, as $r \to \infty$, they become spheres in Case II.A, pseudospheres in cases II.B to II.D, and planes in Cases II.E and II.F.

Case II.A is the Kerr–NUT space. It has been pointed out by Demianski and Newman[19] that this solution describes the field of a particle possessing a mass m, an angular momentum ma, and a "magnetic monopole of mass" l. When the parameter l is taken to be zero, the resulting metric is reduced into the following:

$$
\begin{aligned}
g_{uu} &= (r^2 + a^2\cos^2 x - 2mr)/(r^2 + a^2\cos^2 x) \\
g_{ur} &= 1 \\
g_{uy} &= (2amr\sin^2 x)/(r^2 + a^2\cos^2 x) \qquad \text{(11-170)} \\
g_{ry} &= -a\sin^2 x \\
g_{xx} &= -(r^2 + a^2\cos^2 x) \\
g_{yy} &= \sin^2 x[a^2(r^2 + a^2 - 2mr)\sin^2 x - (r^2 + a^2)^2]/(r^2 + a^2\cos^2 x)
\end{aligned}
$$

The solution (11-170) is known as the Kerr metric.[20] If we expand the metric (11-170) as a power series in m and a in the sense of Einstein–Infeld–Hoffmann (see Sec. 7-7), assuming m to be of order two and a of order one, and compare it with the Einstein–Infeld–Hoffmann approximation for a spinning particle,

[19] M. Demianski and E. Newman, *op. cit.*

[20] R. Kerr, *op. cit.*

one finds that m is the Schwarzschild mass (i.e. Gm/c^2) and ma the angular momentum about the z axis. It has no higher order multipole moments in this approximation (see Problem 11.9; for generalizations of the Kerr metric see Problem 11.14, 11.16 and 11.17).

The metrics of Case II have been interpreted by Kinnersley to represent spinning particles and correspond to the six different ways we can pick a velocity four-vector and an angular momentum vector orthogonal to it. Cases II.B and II.D are attributed to a particle with spacelike velocity, and Cases II.E and II.F to one with lightlike velocity. The angular momentum vector is supposed to be spacelike for Cases II.A, II.C, and II.F, null for Cases II.D and II.E, and timelike for Case II.B. These assertions are reinforced by an examination of the geometry of the principal congruences in the flat space limit $m = l = 0$, and also from the fact that the metrics may be obtained from one another by infinite Lorentz transformations.

As has been mentioned before, case III.A, Eqs. (11-162), is the static degenerate "C" metric given by Ehlers and Kundt.[21] Case III.B seems to be closely related to it, but with rotation added. Both are asymptotically flat at $r \to \infty$.

Solution for $\rho = 0$

In what follows we can assume the spin coefficient $\mu = 0$ since otherwise after interchange of l^μ and n^μ the previous calculation would apply. Since l^μ is now proportional to a gradient, we can follow Chapter 10 and set $\tau = \bar{\alpha} + \beta$, $X^0 = 1$, and $\xi^0 = 1$. Furthermore, Eqs. (11-127d), (11-132a), and (11-132b) imply $D\tau = D\pi = 0$, and $\pi\bar{\pi} = \tau\bar{\tau}$, so we may use the tetrad rotation, Eq. (11-150), to set $\tau = -\pi$. The field equations (11-127f), (11-128b), (11-128d), and (11-129a) now yield the following expression for ψ:

$$\psi = 2\tau(\beta - \alpha) \tag{11-171}$$

We must insist on $\psi \neq 0$ and hence $\tau \neq 0$ since otherwise the space will be flat. Some of the other immediate consequences of the field equations are

$$D\psi = \Delta\psi = 0 \tag{11-172a}$$

$$D\tau = \Delta\tau = 0 \tag{11-172b}$$

$$\delta\psi = \bar{\delta}\psi = 3\tau\psi \tag{11-172c}$$

$$\delta\tau = \bar{\delta}\tau = 2\beta\tau \tag{11-172d}$$

$$\alpha + \beta = \bar{\alpha} + \bar{\beta} \tag{11-172e}$$

$$\gamma = \bar{\gamma} \tag{11-172f}$$

$$\delta\gamma = 0 \tag{11-172g}$$

[21] See J. Ehlers and W. Kundt, *op. cit.*

The radial integration can easily be performed. One finds that all the variables are independent of the coordinate r except for

$$\gamma = \gamma_0 + r[(\tau + \bar{\tau})(\beta - \alpha) - \tau^2] \quad (11\text{-}173a)$$

$$X^i = X_0{}^i - (\tau - \bar{\tau})(\xi^i - \bar{\xi}^i) \quad (11\text{-}173b)$$

$$\omega = \omega_0 - r(\tau + \bar{\tau}) \quad (11\text{-}173c)$$

$$U = U_0 - r[2\gamma_0 + (\tau - \bar{\tau})\omega_0] - r^2[(\tau + \bar{\tau})(\beta - \alpha) - \tau^2] \quad (11\text{-}173d)$$

When these equations are substituted in the remaining field equations we obtain the following equations:

$$\delta\gamma_0 = -\omega_0[(\tau + \bar{\tau})(\beta - \alpha) - \tau^2] \quad (11\text{-}174a)$$

$$\delta U_0 - X_0{}^i\omega_{0,i} = \omega_0[2\gamma_0 + (\tau - \bar{\tau})(\omega_0 - \bar{\omega}_0)] - U_0(\tau + \bar{\tau}) \quad (11\text{-}174b)$$

$$\bar{\delta}\omega_0 - \delta\bar{\omega}_0 = -(\omega_0 - \bar{\omega}_0)(\alpha + 2\beta + \bar{\beta}) \quad (11\text{-}174c)$$

$$\bar{\delta}\xi^i - \delta\bar{\xi}^i = (\xi^i - \bar{\xi}^i)(\alpha - \bar{\beta}) \quad (11\text{-}174d)$$

$$\delta X_0{}^i - X_0{}^k\xi^i{}_{,k} = \omega_0(\tau - \bar{\tau})(\xi^i - \bar{\xi}^i) \quad (11\text{-}174e)$$

First, we will show how the function ω_0 may be eliminated. The coordinate transformation $r \to r + f(x^0, x^2, x^3)$ leaves previous conditions unchanged. Under this transformation one finds that

$$\gamma_0 \to \gamma_0 - f[(\tau + \bar{\tau})(\beta - \alpha) - \tau^2] \quad (11\text{-}175a)$$

$$\omega_0 \to \omega_0 + \delta f + f(\tau + \bar{\tau}) \quad (11\text{-}175b)$$

The quantity in square brackets in Eq. (11-175a) can easily be shown to be equal $D\gamma$. If $D\gamma$ is nonzero, $f = \gamma_0/D\gamma$ will make $\gamma_0 \to 0$ and Eq. (11-174a) will then imply $\omega_0 \to 0$. If $D\gamma = 0$ we have $\delta\gamma_0 = 0$. Then Eq. (11-175b) may be used to make $\omega_0 \to 0$, provided the integrability conditions are satisfied. In other words we must specify Df and Δf such that all the commutators applied to f are given correctly. Choose $Df = 0$, $X_0{}^i f_{,i} = U_0 - 2\gamma_0 f$. The only nontrivial commutators are $(\Delta\delta - \delta\Delta)f$ and $(\bar{\delta}\delta - \delta\bar{\delta})f$; and these are automatically satisfied by virtue of Eqs. (11-174b) to (11-174e).

Next we show how to eliminate the functions U_0, γ_0 by means of a combined coordinate transformation and tetrad rotation. The tetrad rotation is given by

$$\begin{aligned} l'^\mu &= A^{-1}(x^0)l^\mu \\ n'^\mu &= A(x^0)n^\mu \end{aligned} \quad (11\text{-}176a)$$

whereas the change of coordinates is given by

$$\begin{aligned} x'^0 &= \int_0^{x_0} A^{-1}(u)\,\mathrm{d}u \\ r' &= rA(x^0) + U_0 R(x^0) \end{aligned} \quad (11\text{-}176b)$$

which together preserve all previous conditions, but transform $U = U_0 - 2r\gamma_0(x^0) - r^2 D\gamma$ into something new. We want to pick A and R such that the new U_0, γ_0 are zeros. The observation that $\delta(U_0 D\gamma) = 0$ is sufficient to reduce the problem to the solution of two total differential equations for $A(x^0)$ and $R(x^0)$, given γ_0 and $U_0 D\gamma$ as arbitrary functions of x^0. Under sufficient assumptions of continuity, such equations always have solutions, which is all we need to know.

Finally, from $\delta\psi = \bar{\delta}\psi = 0$, we can choose a coordinate $x^2 = x$ such that ξ^2 is real, and then the usual transformation $x^3 \to x^3 + f(x^2, x^3)$ makes ξ^3 imaginary. Equations (11-172a) and (11-172b) show that the remaining variables depend only on x.

From Eqs. (11-172d) and (11-172e) we find that

$$\begin{aligned} \tau - \bar{\tau} &= \alpha - \bar{\alpha} + \beta + \bar{\beta} \\ &= 2(\beta - \bar{\beta}) \\ &= \tau^{-1}\,\delta\tau - \bar{\tau}^{-1}\,\delta\bar{\tau} \end{aligned} \tag{11-177}$$

Solve now Eq. (11-172c) for τ and substitute it in Eq. (11-177). The result may be written in the form

$$\delta(\psi^{4/3}\,\delta\bar{\psi}/\bar{\psi}^{4/3}\,\delta\psi) = 0 \tag{11-178}$$

The integration of this equation yields

$$C\bar{\psi}^{-1/3} = \psi^{-1/3} + D \tag{11-179}$$

The coordinate freedom $x' = f(x)$ would let us set Re ψ equal to an arbitrary function of x. But due to Eq. (11-179) we can choose x such that

$$\psi = (m + il)(x + ia)^{-3} \tag{11-180}$$

where m, l, and a are again real constants. Next, τ, α, and β are all expressed in terms of ξ^2 and substituted into Eq. (11-171) which consequently becomes

$$\partial_x[(\xi^2)^2] + 2ia(\xi^2)^2/(x^2 + a^2) = -(m + il)/(x + ia)^2 \tag{11-181}$$

The case for which $a \neq 0$, Eq. (11-181) yields the real solution

$$(\xi^2)^2 = \frac{2amx + l(a^2 - x^2)}{2a(x^2 + a^2)} \tag{11-182}$$

whereas if $a = 0$, the solution is simplified into

$$(\xi^2)^2 = C + m/x \tag{11-183}$$

where C is an arbitrary constant, and necessarily $l = 0$. Then by a coordinate transformation we can set $C = \pm\frac{1}{2}, 0$.

Resulting Metrics

The resulting subcases are:

Case IV.A:

$$l^\mu = (0, 1, 0, 0)$$
$$m^\mu = (0, 2rx\xi/(x^2 + a^2), \xi, i/\xi) \quad (11\text{-}184a)$$
$$n^\mu = (1, -r^2 l/2a(x^2 + a^2), 0, 4ar/(x^2 + a^2))$$

whereas the metric is given by

$$g_{uu} = r^2 l a^{-1}(x^2 + a^2)^{-1}$$
$$g_{ur} = 1$$
$$g_{ux} = -2rx(x^2 + a^2)^{-1} \quad (11\text{-}184b)$$
$$g_{xx} = -\tfrac{1}{2}\xi^{-2}$$
$$g_{yy} = -2\xi^2$$

where

$$\xi = \left[\frac{2amx + l(a^2 - x^2)}{2a(x^2 + a^2)}\right]^{1/2} \quad (11\text{-}184c)$$

Case IV.B:

$$l^\mu = (0, 1, 0, 0)$$
$$m^\mu = (0, 2r\xi/x, \xi, i/\xi) \quad (11\text{-}185a)$$
$$n^\mu = (1, Cr^2/x^2, 0, 0)$$

and the metric is given by

$$g_{uu} = -2Cr^2/x^2$$
$$g_{ur} = 1$$
$$g_{ux} = -2r/x \quad (11\text{-}185b)$$
$$g_{xx} = -\tfrac{1}{2}\xi^{-2}$$
$$g_{yy} = -2\xi^2$$

where

$$C = \pm\tfrac{1}{2}, 0$$
$$\xi = (C + m/x)^{1/2}. \quad (11\text{-}185c)$$

Table 11.1: Vacuum metrics

Optical parameters σ	ω	θ	Metric	Petrov type	Metric parameters	Comments
0	✓	✓	General case (C–NUT)	D	m, l, a, Γ	$\pi_0 \neq 0$, $\tau_0 \neq 0$.
0	✓	✓	Case I (NUT–Taub)	D	m, l	$\pi_0 = \tau_0 = 0$, $2\mu_0 = \pm 1, 0$, $\rho_0 \neq 0$. ($2\mu_0 = 1$ is Generalized Schwarzschild).
0	0	✓	Case I (Schwarzschild)	D	m	$\pi_0 = \tau_0 = 0$, $2\mu_0 = 1$, $\rho_0 = 0$.
0	✓	✓	Case IIA (Kerr–NUT)	D	m, l, a	$\pi_0 = 0$, $\tau_0 \neq 0$, $U_0 = -\frac{1}{2}$. ($l = 0$ yields Kerr).
0	✓	✓	Case IIB	D	m, l, a	$\pi_0 = 0$, $\tau_0 \neq 0$, $U_0 = +\frac{1}{2}$, $\beta_0 < -2^{1/2}/4$.
0	✓	✓	Case IIC	D	m, l, a	$\pi_0 = 0$, $\tau_0 \neq 0$, $U_0 = +\frac{1}{2}$, $\beta_0 > -2^{1/2}/4$.
0	✓	✓	Case IID	D	m, l, a	$\pi_0 = 0$, $\tau_0 \neq 0$, $U_0 = +\frac{1}{2}$, $\beta_0 = -2^{1/2}/2$.
0	✓	✓	Case IIE	D	m, b	$\pi_0 = 0$, $\tau_0 \neq 0$, $U_0 = 0$, $l = 1$.
0	✓	✓	Case IIF	D	m	$\pi_0 = 0$, $\tau_0 \neq 0$, $U_0 = 0$, $l = 0$.
0	✓	✓	Case IIIA (C metric)	D	m, Γ	$\pi_0 \neq 0$, $\tau_0 = 0$, $a = 0$.
0	✓	✓	Case IIIB	D	m, Γ, a	$\pi_0 \neq 0$, $\tau_0 = 0$, $a \neq 0$.
0	0	0	Case IVA	D	m, l, a	$\pi = -\tau$, $a \neq 0$, $\rho = 0$. (Case IVA → IVB as $l \to 0$, $a \to 0$, $l/a \to C$).
0	0	0	Case IVB (B metric)	D	m, l	$\pi = -\tau$, $a = 0$, $\rho = 0$, $C = 0, \pm\frac{1}{2}$.
0	0	✓	Robinson–Trautman	Algebraically Special	m, P	$\rho^2 \neq \sigma\bar{\sigma}$. (Spherical Rays. Includes Schwarzschild).
✓	0	✓	Newman–Tamburino	I	A	$\rho^2 \neq \sigma\bar{\sigma} \neq 0$. (Spherical Rays. Does not include RT as $\sigma \to 0$. $A \to 0$ yields flat space).
✓	0	✓	Newman–Tamburino	I	a, e	$\rho^2 = \sigma\bar{\sigma} \neq 0$. (Cylindrical rays).

Discussion

Case IV.B, Eqs. (11-185), was also given by Ehlers and Kundt,[22] referred to by them as the "B" metrics. Case IV.A is a rotating generalization which tends smoothly to Case IV.B in the limit $l \to 0$, $a \to 0$, $l/a \to C$.

To summarize this chapter we list in the accompanying table all metrics discussed in it. In this table the parameter m describes the mass, l describes "magnetic monopole of mass," a describes angular momentum per unit mass, and Γ describes other parameters that the metric might have.

PROBLEMS

11.1 Prove Eqs. (11-8).

11.2 Prove Eqs. (11-25a), (11-25b), and (11-26).

11.3 Prove Eqs. (11-36), (11-37), (11-38), (11-39), and (11-40).

11.4 Show that the Robinson–Trautman solution contains the Schwarzschild metric.

11.5 Prove Eqs. (11-86).

11.6 Show that the constant $2\mu_0$ of Eq. (11-105c) is the Gaussian curvature of a two-space with the metric $(\sqrt{2}p)^2\,\delta^{ij}$. [See E. T. Newman, L. Tamburino, and T. Unti, *J. Math. Phys.* **4**, 915 (1963).]

11.7 Prove Eq. (11-119). [See E. T. Newman, L. Tamburino, and T. Unti, *J. Math. Phys.* **4**, 915 (1963).]

11.8 Prove Eqs. (11-138).

11.9 Show that the parameter a in the Kerr metric (11-170) has the meaning that ma is the angular momentum. [See R. Kerr, *Phys. Rev. Letters* **11**, 237 (1963).]

11.10 Use the coordinate transformation $u = t - r - 2m\,|\ln(r - 2m)$, $r' = r$, $\theta' = \theta$, $\phi' = \phi$ to write the Schwarzschild metric in the form $g^{\mu\nu} = l^\mu n^\nu + n^\mu l^\nu - m^\mu \bar{m}^\nu - \bar{m}^\mu m^\nu$, where

$$l^\mu = \delta_1{}^\mu$$

$$m^\mu = \frac{1}{\sqrt{2}r}\left(\delta_2{}^\mu + \frac{i}{\sin\theta}\,\delta_3{}^\mu\right)$$

$$\bar{m}^\mu = \frac{1}{\sqrt{2}r}\left(\delta_2{}^\mu - \frac{i}{\sin\theta}\,\delta_3{}^\mu\right)$$

$$n^\mu = \delta_0{}^\mu - \tfrac{1}{2}(1 - 2m/r)\,\delta_1{}^\mu$$

with $x^0 = u$, $x^1 = r$, $x^2 = \theta$, $x^3 = \phi$.

11.11 Allow the coordinate r in Problem 11.10 to take complex values and rewrite the tetrad in the form

$$l^\mu = \delta_1{}^\mu$$

$$m^\mu = \frac{1}{\sqrt{2}\bar{r}}\left(\delta_2{}^\mu + \frac{i}{\sin\theta}\,\delta_3{}^\mu\right)$$

$$\bar{m}^\mu = \frac{1}{\sqrt{2}r}\left(\delta_2{}^\mu - \frac{i}{\sin\theta}\,\delta_3{}^\mu\right)$$

$$n^\mu = \delta_0{}^\mu - \tfrac{1}{2}[1 - m(1/r + 1/\bar{r})]\,\delta_1{}^\mu$$

[22] See J. Ehlers and W. Kundt, *op. cit.*

where $\bar{r}$ is the complex conjugate of r. Perform now the complex coordinate transformation $u' = u - ia \cos\theta$, $r' = r + ia \cos\theta$, $\theta' = \theta$, $\phi' = \phi$, on the vectors of the tetrad.

Show that if one allows r', u' to be real, one obtains the following tetrad in terms of the new coordinate system:

$$l^\mu = \delta_1{}^\mu$$
$$m^\mu = [\sqrt{2}(r + ia\cos\theta)]^{-1}[ia\sin\theta(\delta_0{}^\mu - \delta_1{}^\mu) + \delta_2{}^\mu + (i/\sin\theta)\,\delta_3{}^\mu]$$
$$n^\mu = \delta_0{}^\mu - \tfrac{1}{2}\{1 - 2m[r/(r^2 + a^2\cos^2\theta)]\}\,\delta_1{}^\mu$$

with $\bar{m}^\mu$ being the complex conjugates of m^μ. Show that $g^{\mu\nu}$ obtained from the above tetrad by $g^{\mu\nu} = l^\mu n^\nu + l^\nu n^\mu - m^\mu \bar{m}^\nu - \bar{m}^\mu m^\nu$ to be equivalent to the Kerr metric (up to a coordinate transformation). [See E. T. Newman and A. I. Janis, *J. Math. Phys.* **6,** 915 (1965).]

11.12 Show that one cannot treat the coordinates r, θ, and ϕ of Problem 11.11 as the usual polar coordinates, for even in the flat space limit ($m = 0$), the metric $g^{\mu\nu}$ is not the polar coordinate version of $\eta_{\mu\nu}$, the Minkowski metric. Show that the following coordinate transformation does lead to polar coordinates r', θ', ϕ', and $u' \equiv t - r'$:

$$u' = u + r - (r^2 + a^2\sin^2\theta)^{1/2}$$
$$(r')^2 = r^2 + a^2\sin^2\theta$$
$$\cos\theta' = r\cos\theta/(r^2 + a^2\sin^2\theta)^{1/2}$$
$$\tan\phi' = (\tan\phi - a/r)/(1 + \tan\phi a/r)$$

[See E. T. Newman and A. I. Janis, *J. Math. Phys.* **6,** 915 (1965).]

11.13 Show that the Reissner metric, Eq. (7-69), can be written in null coordinates as $g^{\mu\nu} = l^\mu n^\nu + l^\nu n^\mu - m^\mu \bar{m}^\nu - \bar{m}^\mu m^\nu$, with

$$l^\mu = \delta_1{}^\mu$$
$$m^\mu = (1/\sqrt{2}r)[\delta_2{}^\mu + (i/\sin\theta)\,\delta_3{}^\mu]$$
$$n^\mu = \delta_0{}^\mu - \left(\frac{1}{2} - \frac{m}{r} + \frac{e^2}{2r^2}\right)\delta_1{}^\mu$$

and $\bar{m}^\mu$ is the complex conjugate of m^μ.

11.14 Allow the coordinate r in Problem 11.13 to take complex values and rewrite the tetrad in the form

$$l^\mu = \delta_1{}^\mu$$
$$m^\mu = (1/\sqrt{2}\bar{r})[\delta_2{}^\mu + (i/\sin\theta)\,\delta_3{}^\mu]$$
$$n^\mu = \delta_0{}^\mu - \tfrac{1}{2}\{1 - m(1/r + 1/\bar{r}) + e^2/r\bar{r}\}\,\delta_1{}^\mu$$

with $\bar{r}$ being the complex conjugate of r. Perform now the same complex coordinate transformation used in Problem 11.11 to obtain the following tetrad

$$l^\mu = \delta_1{}^\mu$$
$$m^\mu = [\sqrt{2}(r + ia\cos\theta)]^{-1}[ia\sin\theta(\delta_0{}^\mu - \delta_1{}^\mu) + \delta_2{}^\mu + (i/\sin\theta)\,\delta_3{}^\mu]$$
$$n^\mu = \delta_0{}^\mu - [\tfrac{1}{2} - (mr - e^2/2)(r^2 + a^2\cos^2\theta)^{-1}]\,\delta_1{}^\mu$$

Calculate the metric $g^{\mu\nu} = l^\mu n^\nu + n^\mu l^\nu - m^\mu \bar{m}^\nu - \bar{m}^\mu m^\nu$, and show that it is given by

$$\begin{aligned}
g^{00} &= -a^2 \sin^2 \theta/(r^2 + a^2 \cos^2 \theta) \\
g^{01} &= (r^2 + a^2)/(r^2 + a^2 \cos^2 \theta) \\
g^{03} &= -a/(r^2 + a^2 \cos^2 \theta) \\
g^{11} &= (2mr - r^2 - a^2 - e^2)/(r^2 + a^2 \cos^2 \theta) \\
g^{13} &= a/(r^2 + a^2 \cos^2 \theta) \\
g^{22} &= -1/(r^2 + a^2 \cos^2 \theta) \\
g^{33} &= -1/(r^2 + a^2 \cos^2 \theta) \sin^2 \theta
\end{aligned}$$

Also show that the electromagnetic field components ϕ_0, ϕ_1, ϕ_2 of Eq. (10-28) are given by

$$\begin{aligned}
\phi_0 &= 0 \\
\phi_1 &= e/2(r - ia \cos \theta)^2 \\
\phi_2 &= iea \sin \theta/(r - ia \cos \theta)^3 \sqrt{2}.
\end{aligned}$$

Show that the above metric and electromagnetic field constitutes a solution of the Einstein–Maxwell equations. The above solution is known as the Kerr-Newman solution, and represents the metric of a rotating, charged mass. [See E. T. Newman *et al.*, *J. Math. Phys.* **6**, 918 (1965); M. M. Schiffer, R. J. Adler, J. Mark, and C. Sheffield, *J. Math. Phys.* **14**, 52 (1973); E. T. Newman, *J. Math. Phys.* **14**, 774 (1973).]

11.15 Transform the null tetrad that generates the Kerr-Newman metric, given in Problem 11.14, by means of a null rotation about the vector l_μ,

$$g_1(z) = \begin{pmatrix} 1 & 0 \\ z & 1 \end{pmatrix}$$

(see Appendix E). Show that if one chooses the complex variable z as

$$z = (ia\rho \sin \theta)/\sqrt{2}$$

where

$$\rho = -1/(r - ia \cos \theta)$$

then one obtains for the components of the null tetrad the following:

$$\begin{aligned}
l^\mu &= (0, 1, 0, 0) \\
m^\mu &= -2^{-\frac{1}{2}}\bar{\rho}(ia \sin \theta, 0, 1, i \operatorname{cosec} \theta) \\
n^\mu &= \rho\bar{\rho}(r^2 + a^2, -\Delta/2, 0, a)
\end{aligned}$$

where

$$\Delta = r^2 + a^2 + e^2 - 2mr$$

and for the covariant components

$$\begin{aligned}
l_\mu &= (1, 0, 0, -a \sin^2 \theta) \\
m_\mu &= -(\bar{\rho}/\sqrt{2})[ia \sin \theta, 0, -1/\rho\bar{\rho}, -i(a^2 + r^2) \sin \theta] \\
n_\mu &= [\Delta\rho\bar{\rho}/2, 1, 0, -(\Delta\rho\bar{\rho}a/2)\sin^2 \theta]
\end{aligned}$$

Show that with the above null tetrad one obtains for the components of the Weyl spinor the following:

$$\psi_0 = \psi_1 = \psi_3 = \psi_4 = 0$$
$$\psi_2 = \rho^3(m + e^2\bar{\rho})$$

The spin coefficients are given by

$$\kappa = \nu = \sigma = \lambda = \varepsilon = 0$$
$$\rho = -(r - ia\cos\theta)^{-1}$$
$$\tau = -ia\sin\theta\,\rho\bar{\rho}/\sqrt{2}$$
$$\pi = ia\sin\theta\,\rho^2/\sqrt{2}$$
$$\beta = -\cot\theta\,\bar{\rho}/2\sqrt{2}$$
$$\alpha = \pi - \bar{\beta}$$
$$\mu = \Delta\rho^2\bar{\rho}/2$$
$$\gamma = \mu + (r - m)\rho\bar{\rho}/2$$

Finally, one finds for the components of the Maxwell spinor the following:

$$\phi_0 = 0$$
$$\phi_1 = e\rho^2/2$$
$$\phi_2 = 0$$

11.16 Assume that the mass parameter appearing in the null tetrad of the Kerr metric (given in Problem 11.15 with $e = 0$) to depend on the retarded time $x^0 = u$, i.e., $m = m(u)$. Calculate the spin coefficients and show that they are identical with those of the Kerr metric except for the presence of the spin coefficient ν in this case which is given by

$$\nu = -i\dot{m}(u)ra\sin\theta\,\rho^2\bar{\rho}/\sqrt{2}$$

and otherwise vanishes when m is a constant. Show that in the present case the components of the Weyl spinor are given by

$$\psi_0 = \psi_1 = 0$$
$$\psi_2 = m(u)\rho^3$$
$$\psi_3 = -i\dot{m}(u)a\sin\theta\,\rho^2\bar{\rho}/2\sqrt{2} - 2i\dot{m}(u)ra\sin\theta\,\rho^3\bar{\rho}/\sqrt{2}$$
$$\psi_4 = \ddot{m}(u)ra^2\sin^2\theta\,\rho^3\bar{\rho}/2 + \dot{m}(u)ra^2\sin^2\theta\,\rho^4\bar{\rho}$$

Also show that the Ricci spinor is given by

$$\phi_{00} = \phi_{01} = \phi_{02} = \phi_{11} = \Lambda = 0$$
$$\phi_{12} = -i\dot{m}(u)a\sin\theta\,\rho^2\bar{\rho}/2\sqrt{2}$$
$$\phi_{22} = -\ddot{m}ra^2\sin^2\theta\rho^2\bar{\rho}^2/2 - \dot{m}(u)r^2\,\rho^2\bar{\rho}^2$$

In the above equations a dot denotes differentiation with respect to the retarded time u. [See M. Carmeli and M. Kaye, *Phys. Lett.* **53A**, 14 (1975)].

11.17 Show that the Kerr-Newman metric can be generalized to include a magnetic charge g. Show that the metric obtained has the form

$$\begin{aligned} ds^2 = {} & [1 - \rho\bar{\rho}(2mr - e^2 - g^2)]du^2 + 2dudr \\ & + 2a\rho\bar{\rho} \sin^2 \theta(2mr - e^2 - g^2)dud\phi \\ & - 2a \sin^2 \theta dr d\phi - d\theta^2/\rho\bar{\rho} \\ & - \sin^2 \theta[r^2 + a^2 + a^2\rho\bar{\rho} \sin^2 \theta(2mr - e^2 - g^2)]d\phi^2 \end{aligned}$$

Show that the tetrad components of the electromagnetic field are given by

$\phi_0 = 0$

$\phi_1 = (e + ig)\rho^2/2$

$\phi_2 = 0$

Here $\rho = -(r - ia \cos \theta)^{-1}$. [*Hint*: use the results of Problem 8.15.]

CHAPTER

TWELVE

THE BONDI–METZNER–SACHS GROUP

The group $SL(2, C)$, whose representations were found and applied in previous chapters, is a finite parameter group. In this chapter we introduce another kind of group that occurs in gravitational theory. The new group has an infinite number of parameters. It is known as the Bondi–Metzner–Sachs group, and was originated in the study of gravitational radiation problems.

After the Bondi–Metzner–Sachs group is defined, we introduce the conformal group, and the relationship between the two groups is discussed. Consequently, a certain class of functions is introduced; these are the spin-s spherical harmonics and the spin-weighted functions. Applications of these functions to Maxwell's theory and a further generalization are subsequently made. The structure of the Bondi–Metzner–Sachs group is then discussed, and its normal subgroups are found. This is followed by finding out the infinitesimal transformations. Finally, a brief account of the representations theory of the Bondi–Metzner–Sachs group is given.

12-1 THE BONDI–METZNER–SACHS GROUP

The group $SL(2, C)$ introduced and discussed in previous chapters is a finite parameter group. In this chapter we introduce another kind of a group which occurs in gravitational theory, a group having an infinite number of parameters. This is known as the Bondi–Metzner–Sachs group (BMS group).

The BMS group originated in the study of gravitational radiation problems,[1, 2] and was discussed in detail by Sachs.[3] It follows from this study that, even in space–times which are asymptotically Minkowskian, one apparently cannot extract the inhomogeneous Lorentz group (see Chapter 2) as an asymptotic symmetry group if finite, retarded, time intervals only are considered. Instead, one obtains the BMS group.

Definition of the BMS Group

Let u, r, θ, and ϕ be standard coordinates for asymptotically flat space–time. This means the coordinate u is a *retarded time* parameter such that $u =$ const. are null hypersurfaces opening into future, r is a radial coordinate, namely an affine parameter on each of the generators of the hypersurfaces, and θ and ϕ are spherical polar coordinates for the sphere at infinity on each hypersurface $u =$ const., with $u, \theta, \phi =$ const. giving the null geodesic generators of these hypersurfaces.

The BMS transformation is defined by the following transformation of the coordinates u, θ, and ϕ.[4]

$$\begin{aligned} u' &= K(\theta, \phi)\{u - \alpha(\theta, \phi)\} \\ \theta' &= \theta'(\theta, \phi) \\ \phi' &= \phi'(\theta, \phi) \end{aligned} \tag{12-1}$$

where the transformation of the coordinates θ and ϕ into θ' and ϕ' is a *conformal transformation* of the (θ, ϕ) sphere into itself, with K being the corresponding *conformal factor* defined by

$$d\theta'^2 + \sin^2\theta'\, d\phi'^2 = K^2(\theta, \phi)(d\theta^2 + \sin^2\theta\, d\phi^2) \tag{12-2}$$

and where $\alpha(\theta, \phi)$ is a suitably smooth, twice differentiable, real function defined on the sphere.[5] The expression $d\theta^2 + \sin^2\theta\, d\phi^2$ is, of course, the metric of the unit sphere. The aggregate of all BMS transformations forms a group called the BMS group.

The particular BMS transformations for which $\theta' = \theta$ and $\phi' = \phi$ are called, in general, *supertranslations*.

In order to understand the meaning of the BMS group we first discuss the conformal transformations.

[1] H. Bondi, M. G. J. van der Burg, and A. W. K. Metzner, *Proc. R. Soc. Lond.* (A) **269,** 21 (1962).
[2] R. K. Sachs, *Proc. R. Soc. Lond.* (A) **270,** 103 (1962).
[3] R. K. Sachs, *Phys. Rev.* **128,** 2851 (1962).
[4] E. T. Newman and R. Penrose, *J. Math. Phys.* **7,** 863 (1966).
[5] The r coordinate may also be involved in the BMS group of transformations but such a transformation is not relevant to the structure of the group.

The Conformal Group

The aggregate of all conformal transformations, given by Eq. (12-2), provides a group. It is called the *conformal group*. Obviously, the rigid rotations, described in Chapter 1, which form the rotation group O_3, form that subgroup of the conformal group for which the conformal factor K satisfies $K^2 = 1$.

The conformal group, which preserves the angle between two curves and its direction (see Problem 12.2), is a six-parameter Lie group which is isomorphic to the proper, orthochronous, homogeneous Lorentz group (see Chapter 2). The isomorphism between the two groups can be verified as follows.[6]

Introducing *stereographic* coordinates, namely a complex coordinate ζ related to the spherical coordinates θ and ϕ by

$$\zeta = \cot \frac{\theta}{2} e^{i\phi} \tag{12-3}$$

In terms of the new coordinate ζ, the metric on the unit sphere will then have the form

$$ds^2 = d\theta^2 + \sin^2 \theta \, d\phi^2 = 4(1 + \zeta\bar{\zeta})^{-2} \, d\zeta \, d\bar{\zeta} \tag{12-4}$$

Now the complex coordinate ζ defines a point in the complex plane. Therefore, the conformal transformations of the complex plane induces the conformal transformations of the unit sphere onto itself. The only transformations with a simple pole and a simple zero at the new north and south poles, respectively, are given by the Möbius transformation:

$$\zeta' = \frac{a\zeta + b}{c\zeta + d}; \qquad ad - bc = 1 \tag{12-5}$$

where a, b, c, and d are four complex constants which, together with the restriction indicated in Eq. (12-5), represent six real parameters.

Applying now the transformation of coordinates (12-5), one obtains for the metric on the unit sphere,

$$ds^2 = 4K^2(1 + \zeta'\bar{\zeta}')^{-2} \, d\zeta' \, d\bar{\zeta}' \tag{12-6}$$

with the conformal factor K given by

$$K = \frac{(a\zeta + b)(\bar{a}\bar{\zeta} + \bar{b}) + (c\zeta + d)(\bar{c}\bar{\zeta} + \bar{d})}{1 + \zeta\bar{\zeta}} \tag{12-7}$$

To show the isomorphism of the conformal group of transformations (12-2) to the proper, orthochronous, homogeneous Lorentz group (see Chapter 2) one notices that the transformation (12-5) is exactly as that obtained by transforma-

[6] See, for example, P. Roman, *Theory of Elementary Particles*, North-Holland Publishing Company, Amsterdam, 1960.

tions of the group $SL(2, C)$. To see this, one introduces a two-dimensional complex linear vector space. Let v_1 and v_2 be the components of a vector in this space. Then to each transformation (12-5) there corresponds a linear transformation of the group $SL(2, C)$ of the form

$$\begin{aligned} v_1' &= av_1 + bv_2 \\ v_2' &= cv_1 + dv_2 \end{aligned} \tag{12-8}$$

as can be seen by the identification $\zeta = v_1/v_2$. Accordingly, one can identify the group of transformations of the form (12-5) with the group $SL(2, C)$. Because of the invariance of the conformal factor given by Eq. (12-7) under the simultaneous change $a \to -a$, $b \to -b$, $c \to -c$, and $d \to -d$, it follows that the group $SL(2, C)$ furnishes a double covering of the conformal group of transformations (12-2) exactly as it furnishes a double covering of the proper orthochronous, homogeneous Lorentz group. Thus the required isomorphism between the conformal group and the proper, orthochronous, homogeneous Lorentz group is established.

The Irreducible Representation $D^{(j_1, j_2)}$

Let ξ and η be two independent basis vectors in the two-dimensional linear space of the vectors which satisfy the transformation law (12-8). Then a basis for the linear vector space defining an irreducible representation of the Lorentz group, which we denote by $D^{(j_1, j_2)}$, is given by[7]

$$(\xi^{2j_1 - m_1}\eta^{m_1})(\bar{\xi}^{2j_2 - m_2}\bar{\eta}^{m_2}) \tag{12-9}$$

where $0 \leqslant m_1 \leqslant 2j_1$ and $0 \leqslant m_2 \leqslant 2j_2$. The parentheses in Eq. (12-9) indicate complete symmetrization of the factors. This linear vector space is a $(2j_1 + 1) \times (2j_2 + 1)$-dimensional. Therefore, an arbitrary vector in this space is determined by $(2j_1 + 1)(2j_2 + 1)$ numbers $a_{m_1 m_2}$.

The transformation (12-5), which maps the components (v_1, v_2) into (v_1', v_2'), induces a corresponding mapping of the components $a_{m_1 m_2}$ into new components $a'_{m_1 m_2}$. By considering the transformation of the quantities

$$\begin{aligned} &(v_1')^{2j_1 - m_1}(v_2')^{m_1}(\bar{v}_1')^{2j_2 - m_2}(\bar{v}_2')^{m_2} \\ &\quad = (av_1 + bv_2)^{2j_1 - m_1}(cv_1 + dv_2)^{m_1}(\bar{a}\bar{v}_1 + \bar{b}\bar{v}_2)^{2j_2 - m_2}(\bar{c}\bar{v}_1 + \bar{d}\bar{v}_2)^{m_2} \\ &\quad = \sum_{n_1 = 0}^{2j_1} \sum_{n_2 = 0}^{2j_2} A^{(j_1, j_2)}_{m_1 m_2; n_1 n_2} v_1^{2j_1 - n_1} v_2^{n_1} \bar{v}_1^{2j_2 - n_2} \bar{v}_2^{n_2} \end{aligned} \tag{12-10}$$

one establishes the transformation

$$a'_{m_1 m_2} = \sum_{n_1 = 0}^{2j_1} \sum_{n_2 = 0}^{2j_2} A^{(j_1, j_2)}_{m_1 m_2; n_1 n_2} a_{n_1 n_2} \tag{12-11}$$

[7] J. N. Goldberg, A. J. Macfarlane, E. T. Newman, F. Rohrlich and E. C. G. Sudarshan, *J. Math. Phys.* **8**, 2155 (1967).

Spin-s Spherical Harmonics

Consider now the set of functions defined by

$$ {}_sZ^L_{m_1m_2} = (1+\zeta\bar{\zeta})^{-L}\zeta^{L-s-m_1}\bar{\zeta}^{L+s-m_2} \tag{12-12}$$

where $|s| \leqslant L$, $0 \leqslant m_1 \leqslant L-s$, and $0 \leqslant m_2 \leqslant L+s$. Applying the transformation (12-5) one obtains for the transformed set

$$ {}_sZ'^L{}_{m_1m_2} = K^{-L}(1+\zeta\bar{\zeta})^{-L}\,\mathrm{e}^{is\lambda}(a\zeta+b)^{L-m_1}(c\zeta+d)^{m_1}(\bar{a}\bar{\zeta}+\bar{b})^{L+s-m_2}(\bar{c}\bar{\zeta}+\bar{d})^{m_2} \tag{12-13}$$

where

$$\frac{1}{1+\zeta'\bar{\zeta}'} = \frac{(c\zeta+d)(\bar{c}\bar{\zeta}+\bar{d})}{K(1+\zeta\bar{\zeta})}$$

$$\mathrm{e}^{i\lambda} = \frac{c\zeta+d}{\bar{c}\bar{\zeta}+\bar{d}}$$

and K is given by Eq. (12-7). Comparing Eq. (12-13) with Eqs. (12-10) and (12-11) one finds that

$$ {}_sZ'^L_{m_1m_2} = K^{-L}\,\mathrm{e}^{is\lambda}\sum_{n_1=0}^{L-s}\sum_{n_2=0}^{L+s} A^{\left(\frac{L-s}{2},\frac{L+s}{2}\right)}_{m_1m_2n_1n_2}\;{}_sZ^L_{n_1n_2} \tag{12-14}$$

Therefore, up to the conformal factor $K^{-L}\,\mathrm{e}^{is\lambda}$, the functions ${}_sZ^L_{m_1m_2}$ transform under the irreducible representation $D^{\left(\frac{L-s}{2},\frac{L+s}{2}\right)}$ of the Lorentz group.

The set of functions ${}_sZ^L_{m_1m_2}$ do not form an orthogonal set of functions on the sphere for fixed s. Indeed, for all $L > |s|$ they form a redundant set of functions for definite *spin weight s*. However, the *spin-s spherical harmonics* ${}_sY_{lm}(\theta, \phi)$ do form an orthogonal set for fixed s. For $l \leqslant L$, the ${}_sY_{lm}$ are given uniquely in terms of the ${}_sZ^L_{m_1m_2}$ by:

$$ {}_sY_{lm} = \sum_{m_1=0}^{L-s}\sum_{m_2=0}^{L+s} {}_sB^{Lm_1m_2}_{lm}\;{}_sZ^L_{m_1m_2} \tag{12-15}$$

with $s \leqslant l \leqslant L$, $|m| \leqslant l$, and

$$ {}_sB^{Lm_1m_2}_{lm} = \frac{a_{lm}}{\sqrt{[(l-s)!\,(l+s)!]}}\sum_{\rho=0}^{\rho_m}(-1)^{\rho+s-m}\binom{l-s}{\rho}\binom{l+s}{\rho+s-m} \times\binom{L-l}{L-s-m_1-\rho}\delta_{m_2,\,m_1+s+m} \tag{12-16}$$

In Eq. (12-16) $\rho_m = \min\,(L-s-m_1,\, l-s,\, l+m)$, and the a_{lm} are the numbers defined by

$$ a_{lm} = (-1)^{l-m}[(l+m)!\,(l-m)!\,(2l+1)/4\pi]^{1/2} \tag{12-17}$$

For fixed s and L, the coefficients ${}_sB^{Lm_1m_2}_{lm}$ form a nonsingular $(L-s+1)(L+s+1) \times (L-s+1)(L+s+1)$ matrix $[(l, m),\ (m_1, m_2)]$ connecting the

${}_sZ^L_{m_1m_2}$ to the ${}_sY_{lm}$. Since the ${}_sZ^L_{m_1m_2}$ transform under the representation $D^{\left(\frac{L-s}{2},\frac{L+s}{2}\right)}$ of the Lorentz group up to the factor $K^{-L}\,e^{is\lambda}$, it follows that the ${}_sY_{lm}$, with $|s| \leqslant l \leqslant L$ and $|m| \leqslant l$, transform under an equivalent representation up to the same factor.

The above results hold both for L and s integral or half integral.

Spin-Weighted Functions on a Sphere

The spin-s spherical harmonics ${}_sY_{lm}$ introduced above can also be introduced in conjection with a certain differential operator, denoted by $\eth$, that appears as follows.

In three-dimensional Euclidean space with polar coordinates, r, θ, and ϕ we introduce an orthogonal triad **a**, **b**, and **c** of vector fields. The vectors **a** and **b** are tangent to the sphere of radius r at each of its points while **c** is in the direction of the radius vector r. Of course the two vectors **a** and **b** are only defined up to a rotation of angle ψ about the vector **c**. It is very convenient to introduce in place of the two vectors **a** and **b** the complex vector **m** and its complex conjugates $\bar{\mathbf{m}}$ by means of

$$\sqrt{2}\mathbf{m} = \mathbf{a} + i\mathbf{b} \tag{12-18}$$

Then the complex vector **m** is defined up to a phase factor, i.e.

$$\mathbf{m}' = e^{i\psi}\mathbf{m} \tag{12-19}$$

where ψ is a real function.

A quantity η is now said to be of *spin weight s* if, under the transformation (12-19), it transforms according to

$$\eta' = e^{is\psi}\eta \tag{12-20}$$

Here, s is in general integral, but half-integral values can also occur. Examples of quantities of spin weights $s = -1, 0, 1$, respectively, are obtained from the scaler products of an arbitrary vector function **V** with the complex vector **m**:

$$\mathbf{V}\cdot\bar{\mathbf{m}}, \qquad \mathbf{V}\cdot\mathbf{c}, \qquad \mathbf{V}\cdot\mathbf{m}$$

More generally, examples of quantities of spin-weight s are furnished by three-dimensional tensors of rank n contracted k_1, k_2, and k_3 times with $\bar{\mathbf{m}}$, **c**, and **m**, respectively, where $k_3 - k_1 = s$, and $k_1 + k_2 + k_3 = n$. The convention adopted is that one for which the real and imaginary parts of the vector **m** point along the coordinate lines and hence transform according to Eq. (12-20) under coordinate transformations.

The function η could have, in addition, a conformal weight. Generally, a quantity η defined on the (θ, ϕ)-sphere has *conformal weight w* if under conformal transformation of the sphere with conformal factor K as in Eq. (12-2) (and with fixed vectors m^μ) one has

$$\eta' = K^w\eta$$

K is, in effect, the relativistic Doppler factor $(c+v)^{1/2}(c-v)^{1/2}$. For consistency with the coordinate conditions, the conformal transformation should be accompanied by the transformation

$$l'^{\mu} = Kl^{\mu}, \qquad r' = K^{-1}r$$

with $m'^{\mu} = m^{\mu}$.

Effectively, the concepts of spin-weight and conformal weight refer to the behaviour of functions on the (θ, ϕ)-sphere at infinity only, and do not refer to the remainder of the space–time. Indeed, the concepts will apply to *any* two-dimensional abstract surface, with a Riemannian or conformal structure. Quantities with spin weights correspond to irreducible tensor quantities on the surface.

The differential operator $\eth$, acting on a quantity η of spin-weight s, is defined by the following equation:

$$\eth\eta = -(\sin\theta)^{s}\left(\frac{\partial}{\partial\theta} + \operatorname{cosec}\theta\,\frac{\partial}{\partial\phi}\right)(\sin\theta)^{-s}\eta \tag{12-21a}$$

Since, under the transformation (12-20), one has

$$(\eth\eta)' = e^{i(s+1)\psi}(\eth\eta) \tag{12-22}$$

It is seen that the operator $\eth$ has the important property of raising the spin weight of the function η by 1. Similarily, if one defines the operator $\bar{\eth}$ by

$$\bar{\eth}\eta = -(\sin\theta)^{-s}\left(\frac{\partial}{\partial\theta} - i\operatorname{cosec}\theta\,\frac{\partial}{\partial\phi}\right)(\sin\theta)^{s}\eta \tag{12-21b}$$

with η here also a quantity of spin-weight s, one can see that $\bar{\eth}$ lowers the spin weight by 1. Furthermore, one easily sees that

$$(\bar{\eth}\,\eth - \eth\,\bar{\eth})\eta = 2s\eta \tag{12-23}$$

Of importance, too, is the effect of the operator $\eth$ on ordinary spherical harmonics $Y_{lm}(\theta, \phi)$, with $-l \leqslant m \leqslant l$, $l = 0, 1, 2, \ldots$. Indeed one may define the spin-s spherical harmonics ${}_sY_{lm}$, for integral s, l, and m, by

$${}_sY_{lm}(\theta, \phi) = [(l-s)!/(l+s)!]^{1/2}\,\eth^{s}Y_{lm}(\theta, \phi) \tag{12-24a}$$

for $0 \leqslant s \leqslant l$, and by

$${}_sY_{lm}(\theta, \phi) = [(l+s)!/(l-s)!]^{1/2}(-1)^{s}\,\bar{\eth}^{s}Y_{lm}(\theta, \phi) \tag{12-24b}$$

for $-l \leqslant s \leqslant 0$. Hence ${}_sY_{lm}$ are not defined for $|s| > l$. They form a complete orthonormal set for each value of s, namely any spin-weight s function can be expanded in a series in terms of ${}_sY_{lm}$. The spin-s spherical harmonics have, furthermore, the properties:

$${}_s\bar{Y}_{lm} = (-1)^{m+s}\,{}_{-s}Y_{Lm} \tag{12-25a}$$

$$\eth\,{}_sY_{lm} = [(l-s)(l+s+1)]^{1/2}\,{}_{s+1}Y_{lm} \tag{12-25b}$$

$$\bar{\eth}\, {}_sY_{lm} = -[(l+s)(l-s+1)]^{1/2}\, {}_{s-1}Y_{lm} \tag{12-25c}$$

$$\bar{\eth}\,\eth\, {}_sY_{lm} = -(l-s)(l+s+1)\, {}_sY_{lm} \tag{12-25d}$$

$${}_s\bar{Y}_{lm} = (-1)^{m+s}\, {}_{-s}Y_{l,\,-m} \tag{12-25e}$$

Accordingly, the operators $\eth$ and $\bar{\eth}$ act on raising and lowering operators on the "quantum number" s and the ${}_sY_{lm}$ are eigenfunctions of the operator $\bar{\eth}\,\eth$.

In terms of the stereographic coordinater $\zeta = \cot(\theta/2)\, e^{i\phi}$ of Eq. (12-3), the operators $\eth$ and $\bar{\eth}$ become

$$\eth\eta = 2P^{1-s}\,\partial(P^s\eta)/\partial\zeta \tag{12-26a}$$

$$\bar{\eth}\eta = 2P^{1+s}\,\partial(P^{-s}\eta)/\partial\bar{\zeta} \tag{12-26b}$$

with $P = \frac{1}{2}(1+\zeta\bar{\zeta})$. In terms of the coordinates ζ and $\bar{\zeta}$, the spin-s spherical harmonics take the form

$${}_sY_{lm} = \frac{a_{lm}}{\sqrt{[(l-s)!\,(l+s)!]}}(1+\zeta\bar{\zeta})^{-l}\sum_p \binom{l-s}{p}\binom{l+s}{p+s-m}\zeta^p(-\bar{\zeta})^{p+s-m} \tag{12-27}$$

with a_{lm} given by Eq. (12-17). Expression (12-27) applies also to spin-s spherical harmonics for which l, m, and s are all half-odd integers.

Simple Example: Maxwell's Equations

As a simple example illustrating the use of the operator $\eth$ and the spin-s spherical harmonics ${}_sY_{lm}$, we consider the Maxwell equations in free space:

$$\begin{aligned} \nabla\cdot(\mathbf{E}+i\mathbf{B}) &= 0 \\ \nabla\times(\mathbf{E}+i\mathbf{B}) - i(\partial/\partial t)(\mathbf{E}+i\mathbf{B}) &= 0 \end{aligned} \tag{12-28}$$

The quantities

$$\begin{aligned} \eta_- &= (\mathbf{E}+i\mathbf{B})\cdot\bar{\mathbf{m}} \\ \eta_0 &= (\mathbf{E}+i\mathbf{B})\cdot\mathbf{c} \\ \eta_+ &= (\mathbf{E}+i\mathbf{B})\cdot\mathbf{m} \end{aligned} \tag{12-29}$$

of spin weight -1, 0, 1, respectively, then satisfy the equations

$$\begin{aligned} \left(\frac{\partial}{\partial t}-\frac{\partial}{\partial r}\right)r^2\left(\frac{\partial}{\partial t}+\frac{\partial}{\partial r}\right)r\eta_- - \bar{\eth}\,\eth r\eta_- &= 0 \\ \left(\frac{\partial^2}{\partial t^2}-\frac{\partial^2}{\partial r^2}\right)r^2\eta_0 - \bar{\eth}\,\eth r\eta_0 &= 0 \\ \left(\frac{\partial}{\partial t}+\frac{\partial}{\partial r}\right)r^2\left(\frac{\partial}{\partial t}-\frac{\partial}{\partial r}\right)r\eta_+ - \eth\,\bar{\eth} r\eta_+ &= 0 \end{aligned} \tag{12-30}$$

in which the quantities η_-, η_0, η_+ are already uncoupled. If we assume solutions of these equations of the form

$$\begin{aligned}\eta_\pm(t, r, \theta, \phi) &= \sum_{j=1}^{\infty} \sum_{m=-j}^{j} \alpha^j_{\pm 1, m}(t, r)_{\pm 1} Y_{jm}(\theta, \phi) \\ \eta_0(t, r, \theta, \phi) &= \sum_{j=0}^{\infty} \sum_{m=-j}^{j} \alpha^j_{0, m}(t, r)\,_0 Y_{jm}(\theta, \phi)\end{aligned} \tag{12-31}$$

it is seen that

$$\begin{aligned}&\left(\frac{\partial}{\partial t} \mp \frac{\partial}{\partial r}\right) r^2 \left(\frac{\partial}{\partial t} \pm \frac{\partial}{\partial r}\right) r\alpha^j_{\pm 1, m}(t, r) + j(j+1) r\alpha^j_{\pm 1, m}(t, r) = 0 \\ &\left(\frac{\partial^2}{\partial t^2} - \frac{\partial^2}{\partial r^2}\right) r^2 \alpha^j_{0, m}(t, r) + j(j+1) r\alpha^j_{0, m}(t, r) = 0\end{aligned} \tag{12-32}$$

and the dependence on the angular variables have been cancelled out. These latter equations can be solved easily.

Further Remarks on the Spin-*s* Spherical Harmonics

The spin-*s* spherical harmonics and spin-weighted functions introduced above have some group-theoretical interpretation. To see this one proceeds as follows.[8]

One starts as before, by introducing a triad of unit vectors ξ_1, ξ_2, ξ_3 on each point of a sphere of radius r. The two vectors ξ_1 and ξ_2 are taken to be in the tangent plane to the sphere at the spherical angles ϕ and θ, whereas ξ_3 is taken to be normal to the sphere there. The vectors ξ_1 and ξ_2 are defined up to a rotation with an angle, which we denote by ϕ_2, in the tangent plane about an axis in the direction of ξ_3. The rotation of ξ_1 and ξ_2 about ξ_3 is given a definite mathematical expression in the sequel.

By introducing the above rotation, we have actually added a new variable ϕ_2 upon which the two vectors ξ_1 and ξ_2 depend. Accordingly, these two vectors depend on the spherical angles ϕ, θ as well as the new angle ϕ_2:

$$\xi_1 = \xi_1(\phi, \theta, \phi_2), \qquad \xi_2 = \xi_2(\phi, \theta, \phi_2)$$

The vector ξ_3, on the other hand, depends only on the angle ϕ and θ:

$$\xi_3 = \xi_3(\phi, \theta).$$

Of particular interest to us is the behaviour of ξ_1 and ξ_2 under the rotation about ξ_3. Such a rotation can be presented by the orthogonal matrix

$$\Omega \equiv \begin{pmatrix} \cos\psi & \sin\psi & 0 \\ -\sin\psi & \cos\psi & 0 \\ 0 & 0 & 1 \end{pmatrix}$$

[8] M. Carmeli, *J. Math. Phys.* **10,** 569 and 1699 (1969); *Nuovo Cim.* **67**B, 103 (1970).

Denoting the column of the three vectors ξ_1, ξ_2, ξ_3 by ξ,

$$\xi = \begin{pmatrix} \xi_1 \\ \xi_2 \\ \xi_3 \end{pmatrix} \tag{12-33}$$

then under the rotation Ω we have

$$\xi' = \Omega\xi$$

It is convenient to introduce the complex vector ξ_+ and its complex conjugate ξ_- defined by

$$\xi_\pm = 2^{-1/2}(\xi_1 \mp i\xi_2)$$

Under the rotation Ω these complex vectors transform as

$$\xi'_\pm = e^{\pm i\psi}\xi_\pm \tag{12-34a}$$

similar to the vector **m** appearing in Eq. (12-19).

A quantity η is now said to be of spin weight s if under the rotation Ω it transforms as

$$\eta \to \eta' = e^{is\psi}\eta \tag{12-34b}$$

where s is an integer, as in Eq. (12-20).

Examples of quantities of spin weights 1, 0 and -1 are obtained by scalar multiplication of a vector field $\mathbf{v}$ with $\xi_\pm$, ξ_3:

$$\begin{aligned} \eta_{\pm 1}(\phi, \theta, \phi_2) &= \mathbf{v}(\phi, \theta) \cdot \xi_\pm(\phi, \theta, \phi_2) \\ \eta(\phi, \theta) &= \mathbf{v}(\phi, \theta) \cdot \xi_3(\phi, \theta) \end{aligned} \tag{12-35}$$

as before.

Further examples of quantities of spin weights of any order can be obtained in a similar way from tensor fields.

In the following we restrict ourselves to quantities η obtained by contraction of tensor fields with the triad $\xi_\pm$, ξ_3. In other words, the η functions are components of tensor fields along the complex triad. These components generally depend on the three angles ϕ, θ, ϕ_2. By relating these angles to the three parameters appearing in the three-dimensional rotation group (such as Euler's angles or direction and angle of rotation, see Chapter 1), the quantities η can be considered as functions of g:

$$\eta = \eta(g) \tag{12-36a}$$

where g is an element of O_3.

Using the homomorphism relationship between the rotation group and the special unitary group of order two, SU_2, we consider η as functions on the group SU_2 also:

$$\eta = \eta(u) \tag{12-36b}$$

where $u \in SU_2$.

In the following, we give the explicit dependence of g on the angles ϕ, θ, and ϕ_2. We also find out what should be substituted for g in order that η can be written as a linear combination of components of the original tensor field.

First we relate the triad field ξ, given by Eq. (12-33), to $g \in O_3$ in such a way that for each triad ξ there corresponds a rotation $g \in O_3$. This correspondence can be achieved by assigning to each triad ξ a rotation $g \in O_3$ which transforms a certain given triad on the sphere at $\phi = \theta = 0$ to ξ. To find the rotation g, we proceed as follows.

We introduce a new triad of unit vectors

$$\mathbf{e} = \begin{pmatrix} \mathbf{e}_\theta \\ \mathbf{e}_\phi \\ \mathbf{e}_r \end{pmatrix}$$

whose vectors are directed along the coordinates θ, ϕ, and r, respectively, and whose origin coincides with that of the triad ξ. The triad of vectors $\mathbf{e}$ is a function of the spherical coordinates, $\mathbf{e} = \mathbf{e}(\phi, \theta)$. The particular triad at $\phi = \theta = 0$ is denoted by

$$\mathbf{e}^0 = \begin{pmatrix} \mathbf{e}_\theta^0 \\ \mathbf{e}_\phi^0 \\ \mathbf{e}_r^0 \end{pmatrix}$$

Accordingly we have

$$\mathbf{e}^0 = [\mathbf{e}(\phi, \theta)]_{\phi=\theta=0}$$

It is easily seen that the three vectors $\mathbf{e}_\theta^0$, $\mathbf{e}_\phi^0$, and $\mathbf{e}_r^0$ are pointing in the same directions as the Cartesian coordinates x, y, and z of the fixed system. The transformation g is then defined as that one which transforms the triad $\mathbf{e}^0$ into the triad ξ:

$$\xi = g\mathbf{e}^0$$

Now the vectors $\mathbf{e}_\theta$, $\mathbf{e}_\phi$, and $\mathbf{e}_r$ can be decomposed along the Cartesian coordinates, hence along $\mathbf{e}_\theta^0$, $\mathbf{e}_\phi^0$, and $\mathbf{e}_r^0$. One easily finds that the matrix of rotation R which transforms the triad $\mathbf{e}^0$ into the triad $\mathbf{e}$,

$$\mathbf{e} = R\mathbf{e}^0 \tag{12-37}$$

is given by

$$R = \begin{pmatrix} \cos\theta\cos\phi & \cos\theta\sin\phi & -\sin\theta \\ -\sin\phi & \cos\phi & 0 \\ \sin\theta\cos\phi & \sin\theta\sin\phi & \cos\theta \end{pmatrix}$$

The above matrix R can be written as a product of three orthogonal matrices

$$R = DBA' \tag{12-38}$$

where

$$D = \begin{pmatrix} 0 & 1 & 0 \\ -1 & 0 & 0 \\ 0 & 0 & 1 \end{pmatrix}, \qquad B = \begin{pmatrix} 1 & 0 & 0 \\ 0 & \cos\theta & -\sin\theta \\ 0 & \sin\theta & \cos\theta \end{pmatrix} \tag{12-39}$$

and where

$$A' = \begin{pmatrix} \sin\phi & -\cos\phi & 0 \\ \cos\phi & \sin\phi & 0 \\ 0 & 0 & 1 \end{pmatrix}$$

By changing the variable ϕ, the matrix A' may be written as

$$A = \begin{pmatrix} \cos\phi_1 & -\sin\phi_1 & 0 \\ \sin\phi_1 & \cos\phi_1 & 0 \\ 0 & 0 & 1 \end{pmatrix} \tag{12-40}$$

where $\phi_1 = \pi/2 - \phi$.

It remains to find the transformation from the triad $\mathbf{e}$ to the triad $\boldsymbol{\xi}$. Since the angle of rotation ϕ_2 of the two vectors $\boldsymbol{\xi}_1$ and $\boldsymbol{\xi}_2$ was left undetermined, we define it by

$$\boldsymbol{\xi} = CD^{-1}\mathbf{e} \tag{12-41}$$

where the matrix C is given by

$$C = \begin{pmatrix} \cos\phi_2 & -\sin\phi_2 & 0 \\ \sin\phi_2 & \cos\phi_2 & 0 \\ 0 & 0 & 1 \end{pmatrix} \tag{12-42}$$

For later calculations we have to know the relation between the triad $\boldsymbol{\xi}$ when $\phi_2 = 0$ and the triad $\mathbf{e}$. From Eq. (12-41) we obtain

$$\begin{pmatrix} \boldsymbol{\xi}_1 \\ \boldsymbol{\xi}_2 \\ \boldsymbol{\xi}_3 \end{pmatrix}_{|\phi_2=0} = \begin{pmatrix} -\mathbf{e}_\phi \\ \mathbf{e}_\theta \\ \mathbf{e}_r \end{pmatrix}$$

Using Eqs. (12-41), (12-37), (12-38) and (12-40), we find the explicit form of the rotation $g \in O_3$ which transforms the fixed triad $\mathbf{e}^0$ into the triad $\boldsymbol{\xi}$:

$$g(\phi_1, \theta, \phi_2) = C(\phi_2)B(\theta)A(\phi_1) \tag{12-43}$$

The transformation $g(\phi_1, \theta, \phi_2)$ represents three rotations with Euler's angles ϕ_1, θ, and ϕ_2 around the z, x, and z axis, respectively, where $\phi_1 = \pi/2 - \phi$. Hence, for each value of the variables ϕ, θ, and ϕ_2 of the triad $\boldsymbol{\xi}$ there corresponds a rotation $g(\pi/2 - \phi, \theta, \phi_2) \in O_3$, and any function of these variables can be considered as a function of $g \in O_3$.

In particular, the vectors $\boldsymbol{\xi}_\pm$, $\boldsymbol{\xi}_3$ can be considered as functions over the group O_3:

$$\boldsymbol{\xi}_\pm = \boldsymbol{\xi}_\pm(g), \qquad \boldsymbol{\xi}_3 = \boldsymbol{\xi}_3(g)$$

A direct calculation shows that

$$\begin{aligned} \boldsymbol{\xi}_\pm(g) &= -2^{-1/2}[\mathbf{e}_\phi(\phi_1, \theta) \pm i\mathbf{e}_\theta(\phi_1, \theta)]e^{\mp\phi_2} \\ \boldsymbol{\xi}_3(g) &= \mathbf{e}_r(\phi_1, \theta) \end{aligned} \tag{12-44}$$

where $\phi_1 = \pi/2 - \phi$.

The above considerations show that all functions η obtained by contraction of tensor fields with the complex vectors, such as those given by Eqs. (12-35), are functions of $g \in O_3$:

$$\eta = [\eta(g)]_{\phi_1 = \pi/2 - \phi} \tag{12-45}$$

For example, the functions η_s, $s = 1, 0, -1$, given by Eqs. (12-35), are given by

$$\begin{aligned} \eta_{\pm 1}(g) &= -2^{-1/2}[v_\phi(\phi_1, \theta) \pm i v_\theta(\phi_1, \theta)]e^{\mp i\phi_2} \\ \eta_0(g) &= v_r(\phi_1, \theta) \end{aligned} \tag{12-46}$$

When we put $\phi_2 = 0$, the function η become functions of the spherical coordinates ϕ and θ only:

$$\eta(\phi, \theta) = [\eta(g)]_{\phi_1 = \pi/2 - \phi,\, \phi_2 = 0} \tag{12-47}$$

Using Eqs. (12-46), we obtain the result for η_s, $s = 1, 0, -1$, for example:

$$\begin{aligned} \eta_\pm(\phi, \theta) &= 2^{-1/2}(v_\phi \pm i v_\theta) \\ \eta_0(\phi, \theta) &= v_r \end{aligned} \tag{12-48}$$

In general, one can relate these functions to the matrix elements of the irreducible representations of O_3 and SU_2.

A tensor field **W** of rank two provides nine quantities W_s of weights $s = 2, 1, 0, -1, -2$. These are obtained by contraction of the tensor field **W** with $\xi_\mp$, ξ_3 and inserting $\phi_2 = 0$.

Table 12.1: Various spin-weight functions W_s obtained from the tensor field W and their expansion modes in D^j_{sn}.

Weight s	W_s	Tensor components	Expanded in D^j_{sn}
± 2	$\mathbf{W} \cdot \xi_\pm \xi_\pm$	$W_{\phi\phi} - W_{\theta\theta} \pm i(w_{\theta\phi} + W_{\phi\theta})$	$D^j_{\pm 2, n}(\pi/2 - \phi, \theta, 0)$
± 1	$\mathbf{W} \cdot \xi_\pm \xi_3$	$-(W_{\phi r} \pm i W_{\theta r})$	$D^j_{\pm 1, n}(\pi/2 - \phi, \theta, 0)$
	$\mathbf{W} \cdot \xi_3 \xi_\pm$	$-(W_{r\phi} \pm i W_{r\theta})$	
0	$\mathbf{W} \cdot \xi_+ \xi_-$	$W_{\phi\phi} + W_{\theta\theta} + i(W_{\theta\phi} - W_{\phi\theta})$	$D^j_{0, n}(\pi/2 - \phi, \theta, 0)$
	$\mathbf{W} \cdot \xi_3 \xi_3$	W_{rr}	
	$\mathbf{W} \cdot \xi_- \xi_+$	$W_{\phi\phi} + W_{\theta\theta} - i(W_{\theta\phi} - W_{\phi\theta})$	

We give in the table the various spin-weight functions W_s obtained from the tensor field **W**. We also give their expansion moded in D^j_{sn}; the latter are evaluated at $\phi_1 = (\pi/2) - \phi$ and $\phi_2 = 0$.

Using this formalism one can, for example, write down all the electromagnetic field variables as functions over the rotation group or the group SU_2 and then isolate completely the dynamical field variables (see Problems 12.5 and 12.6).

Isometries

In the next section we will give a detailed discussion of the BMS group. This preliminary section will be concluded by a discussion of the geometrical notion of symmetry in a curved space–time (see also Section 7-4).

Suppose one has the metric tensor $g_{\mu\nu}$ as a function of the coordinates x^α. Suppose, furthermore, that there exists a one-parametric set of transformations $\bar{x}^\alpha = \bar{x}^\alpha(W; x^\beta)$, $\bar{x}^\alpha(0; x^\beta) = x^\alpha$, which are such that the transformed metric is the same function of the new coordinates as $g_{\mu\nu}$ is of the coordinates x^α. Then the transformations are called symmetry or *isometry* transformations. Let $\xi^\alpha = [\partial \bar{x}^\alpha/\partial W]_{W=0}$; ξ^α is a contravariant vector. The basic way to look for isometries in a given space–time is to use the following theorem: *The vector ξ^α obeys Killing's equation*

$$\xi_{\alpha;\beta} + \xi_{\beta;\alpha} = 0$$

if, and only if, the corresponding transformations are isometries.[9] Here, as before, semicolons denote covariant derivatives.

By means of specific examples one can convince oneself that any reasonably general space–time permits either no solution of Killing's equation or at most one solution of Killing's equation. To look for general, physically realistic properties of gravitational radiation from bounded sources in a space–time that permits two linearly independent solutions of Killing's equation, is not a promising procedure. Minkowski space, of course, permits ten linearly independent solutions, corresponding to the ten parameters of the Poincaré group (see Chapter 2).

The Euclidean Group

To illustrate the above considerations we discuss a much simpler group than the BMS group that arises from the Euclidean group in the plane, $E(2)$.[10] Using polar coordinates $x^1 = r$ and $x^2 = \theta$, with $0 < r < \infty$ and $0 \leqslant \theta \leqslant 2\pi$, and choosing the metric to be positive-definite, then one has

$$g^{ij}(r, \theta) = \begin{pmatrix} 1 & 0 \\ 0 & r^{-2} \end{pmatrix}$$

where $i, j = 1, 2$. Solving the Killing equation $\mathscr{L}_\xi g^{ij} = 0$, which can be written in the form (see Chapter 7)

$$g^{ik}\xi^j{}_{,k} + g^{jk}\xi^i{}_{,k} - g^{ij}{}_{,k}\xi^k = 0$$

gives

$$\xi^1 = f(\theta)$$
$$\xi^2 = g(\theta) + (1/r)f'(\theta)$$

[9] L. P. Eisenhart, *Riemannian Geometry*, Princeton, New Jersey, 1949.
[10] P. J. McCarthy, *J. Math. Phys.* **13**, 1837 (1972).

with the additional condition that

$$f''(\theta) + f(\theta) + rg'(\theta) = 0$$

Here $f(\theta)$ and $g(\theta)$ are arbitrary, differentiable functions of the variable θ, subject to the condition stated above, and a prime denotes a differentiation, $f'(\theta) = \mathrm{d}f(\theta)/\mathrm{d}\theta$ and $g'(\theta) = \mathrm{d}g(\theta)/\mathrm{d}\theta$. Since the last condition is supposed to be satisfied for any r, one concludes that

$$f''(\theta) + f(\theta) = 0$$

$$g'(\theta) = 0$$

Consequently, the solution of the Killing equation is given by

$$\xi^1 = f^{(a,\, b)}(\theta) = a \cos \theta + b \sin \theta$$

$$\xi^2 = K + (1/r)f'^{(a,\, b)}(\theta) = K + (1/r)(-a \sin \theta + b \cos \theta)$$

where K, a, and b are constants.

The above solution shows that we have three degrees of freedom. These degrees of freedom describe the infinitesimal group of motions of the plane, a and b correspond to translations in the plane, whereas K corresponds to rotations around the point $r = 0$.

Asymptotic Isometries

Let us now consider the positive-definite, two-dimensional, differentiable Riemannian manifold with metric $g^{ij}(x)$ given in some coordinate patch x^i, $i = 1, 2$. Define in this coordinate patch two differentiable scalar functions $r(x)$ and $\theta(x)$ having the property of "polar" coordinates. This means that the metric in the new coordinate system r, θ should be such as $g^{rr} = 1$ and $g^{r\theta} = 0$. Hence the new coordinates should satisfy

$$g^{ij} \frac{\partial r}{\partial x^i} \frac{\partial r}{\partial x^j} = 1$$

$$g^{ij} \frac{\partial r}{\partial x^i} \frac{\partial \theta}{\partial x^j} = 0$$

Consequently, the metric tensor takes the form

$$g^{ij}(r, \theta) = \begin{pmatrix} 1 & 0 \\ 0 & f(r, \theta) \end{pmatrix}$$

where $f(r, \theta)$ is an arbitrary, differentiable, function.

We now impose conditions so as the manifold be "asymptotically flat." These conditions are:

(a) The patch of coordinates x^i should cover the whole manifold outside some bounded region. The coordinate ranges outside the region is given by $r_0 \leqslant r < \infty$ and $0 \leqslant \theta \leqslant 2\pi$ for some $r_0 > 0$;

(b) In these intervals of r and θ, the function $f(r, \theta)$ should behave as $f(r, \theta) = r^{-2} + 0(r^{-\lambda})$, for $2 < \lambda \leqslant 3$; and

(c) The function $f(r, \theta)$ should also satisfy $(\partial/\partial r)[f(r, \theta) - r^{-2}] = 0(r^{-q})$, where $q \geqslant \lambda$. It is also assumed that the derivatives of $f(r, \theta)$ with respect to the coordinate θ satisfy the same condition. By transforming to "Cartesian" coordinates one can show (Problem 12.7) that

$$g^{ij} \rightarrow \begin{pmatrix} 1 & 0 \\ 0 & 1 \end{pmatrix}$$

as $r \rightarrow \infty$, in "Cartesian" coordinates.

Hence the metric so obtained

$$g^{ij} = \begin{pmatrix} 1 & 0 \\ 0 & r^{-2} + 0(r^{-\lambda}) \end{pmatrix}$$

describes a wide class of "asymptotically flat" manifolds. The Killing equations $\mathscr{L}_{\xi} g^{ij} = 0$, unlike the previous case, now have no nonzero solutions. However, if we require only *asymptotic* isometries, i.e., transformations that preserve the metric asymptotically, there will be nonzero solutions. Hence we look for solutions of the approximate equations

$$\mathscr{L}_{\xi} g^{11} = 0$$

$$\mathscr{L}_{\xi} g^{12} = \mathscr{L}_{\xi} g^{21} = 0$$

$$\mathscr{L}_{\xi} g^{22} = 0(r^{-\lambda})$$

which are, for large r, independent of the $0(r^{-\lambda})$ term. Indeed, the solution of the first two of the above equations are

$$\xi^1 = f(\theta)$$

$$\xi^2 = g(\theta) + (1/r) f'(\theta) + 0(r^{-\mu})$$

for $1 < \mu \leqslant 2$, where $f(\theta)$ and $g(\theta)$ are arbitrary differentiable functions. The last Killing equation gives the condition

$$2r^{-2} g'(\theta) + 2r^{-3}[f''(\theta) + f(\theta)] + 0(r^{-s}) = 0(r^{-\lambda})$$

with $s \geqslant \lambda$. Multiplying this equation by r^2 and letting $r \rightarrow \infty$, one sees that $g'(\theta) = 0$. Hence $g(\theta) = K$, where K is a constant. This is similar to the flat space case. However, since $2 < \lambda \leqslant 3$, one obtains no condition on the function $f(\theta)$, and the solutions have, consequently, the form

$$\xi^1 = f(\theta)$$

$$\xi^2 = K + (1/r) f'(\theta) + 0(r^{-\mu})$$

with $1 < \mu \leqslant 2$.

As has been pointed out before, these solutions are, for larger r, independent the $0(r^{-\lambda})$ term. When compared with the solution of the flat space case, we see that the two-parameter family of functions $f^{(a,b)}(\theta)$ is now enlarged into an *infinite parameter* family of arbitrary functions $f(\theta)$. This is exactly what happens in the BMS group. If we would assume that $\lambda > 3$, the solutions would have been

$$\xi^1 = f^{(a,b)}(\theta) = a \cos \theta + b \sin \theta$$
$$\xi^2 = K + (1/r) f'^{(a,b)}(\theta) + 0(r^{-\mu})$$

with $\mu > 2$. This means if the metric approaches flatness *too rapidly* as $r \to \infty$, the two-parameter functions $f^{(a,b)}(\theta)$ will *not* be enlarged into an infinite parameter family of functions.

Finally, we notice that the finite coordinate transformation which *preserve the metric asymptotically* can be calculated here also just as for the BMS group case. One finds

$$r \to r + f(\theta) + \tfrac{1}{2} r^{-1}[f'(\theta)]^2 + 0(r^{-2})$$
$$\theta \to \theta + K + r^{-1} f'(\theta) + 0(r^{-2})$$

12-2 THE STRUCTURE OF THE BONDI–METZNER–SACHS GROUP

It will be useful to introduce at this stage some of the subgroups of the Bondi–Metzner–Sachs group.

Supertranslation, Translation, and Proper Subgroups of the BMS Group

As has been previously mentioned, the particular BMS transformations for which $\theta' = \theta$ and $\phi' = \phi$, are called supertranslations. In this case one obtains

$$u' = u - \alpha, \qquad \theta' = \theta, \qquad \phi' = \phi \tag{12-49}$$

for Eq. (12-1). The aggregate of all supertranslations forms the *supertranslation subgroup N*. Under a supertranslation, the system of null hypersurfaces $u =$ const. is transformed into a different system of null hypersurfaces $u' =$ const., but no Lorentz rotation is involved. We may write the function α in terms of spherical harmonics:

$$\alpha(\theta, \phi) = \sum_{l=0}^{\infty} \sum_{m=-l}^{l} \alpha_{lm} Y_{lm}(\theta, \phi)$$

where α_{lm} are constants. The infinite set of parameters α_{lm}, subjected to the condition $\alpha_{l,-m} = (-1)^m \bar{\alpha}_{lm}$, since α is real, then define the supertranslations.

If $\alpha_{lm} = 0$ for $l > 2$, so that α takes the form:

$$\alpha = \varepsilon_0 + \varepsilon_1 \sin \theta \cos \phi + \varepsilon_2 \sin \theta \sin \phi + \varepsilon_3 \cos \theta \tag{12-50}$$

then the supertranslations reduce to a special case, called the *translations*, with just four parameters $\varepsilon_0, \ldots, \varepsilon_3$, and one obtains the *translation* subgroup.

This terminology is, in fact, consistent with that for the Lorentz translations in Minkowski space. We may call a hypersurface $u = \text{const.}$ a "good" cone in Minkowski space if it is the null cone of some point, and a "bad" cone if, on the other hand, the generators of $u = \text{const.}$ do not all meet in a point. Thus, an actual translation in Minkowski space must send "good" cones into "good" cones; under a general supertranslation, the "good" cones will be warped into "bad" cones. It is precisely the condition that α be given by (12-50), which is required to preserve the "goodness" of the Minkowski null cones.

In curved asymptotically flat space–times the difficulty is to find an appropriate analog of the Minkowskian concept of "good" and "bad" cones. It is not, in fact, necessary to do this in order to single out the translations from the remaining supertranslations since the translations are already determined by (12-50). But if we wish, in addition, to isolate the "pure" Lorentz rotations from Lorentz rotations which have a "super-translations component", then some concept of a distinction between "good" and "bad" cones is necessary. For we might try to define Lorentz rotations (homogeneous Lorentz transformation) as given by (12-1) with $\alpha = 0$. The hypersurface $u = 0$ is then transformed into itself. In Minkowski space, if $u = 0$ is a "good" cone, the resulting transformation indeed represents a Lorentz rotation and it sends other "good" cones into "good" cones. If, on the other hand, $u = 0$ is a "bad" cone, then we do not get a Lorentz rotation in general. Thus, for asymptotically flat spaces, in order to know which of the BMS transformations are to be regarded as "supertranslation-free Lorentz rotations", we must have some definition of "goodness" of $u = \text{const.}$ hypersurfaces.

In Minkowski space, the "good" cones can be characterized locally by the fact that the null rays generating them possess no shear. In asymptotically Minkowskian spaces, it will only be the asymptotic behaviour of the (complex) shear σ of these null rays that will concern us. With a null vector l^μ tangent to the null rays and a complex null vector m^μ orthogonal to l^μ satisfying $m^\mu \overline{m}_\mu = -1$, l^μ and m^μ being parallelly propagated along each ray, we have

$$l_{\mu;\nu} m^\mu m^\nu = \sigma(u, \theta, \phi, r) = \sigma^0(u, \theta, \phi)/r^2 + 0(r^{-4}) \tag{12-51}$$

where r is scaled so that $r_{,\mu} l^\mu \to 1$ at infinity. Thus σ^0 defines the asymptotic shear of the hypersurface given, say, by $u = \text{const.}$ The complex quantity σ^0 is of special interest in gravitational radiation theory. It forms part of the initial data on $u = 0$ used to determine the space–time asymptotically. Furthermore, $\partial\sigma^0/\partial u$ and $\partial^2\sigma^0/\partial u^2$ both have physical significance. We may call $\partial^2\sigma^0/\partial u^2$ the gravitational radiation field since it represents the r^{-1} part of the Riemann curvature field. Bondi *et al.* and Sachs call $\partial\sigma^0/\partial u$ the *news function* since it can be used as asymptotic initial data for the gravitational radiation field and $|\partial\sigma^0/\partial u|^2$ represents the flux energy of the gravitational radiation in their analysis.

We cannot, however, attempt to define "good" cones, in general, simply by requiring $\sigma^0 = 0$. In many cases it is simply not possible to arrange $\sigma^0 = 0$ for all values of θ, ϕ, on one hypersurface, but even in cases where it is possible (e.g. in the axially symmetric, reflection symmetric cases), it is clear from the above remarks that, in the presence of gravitational radiation, if $\sigma^0 = 0$ for one value of u, we will generally have $\sigma^0 \neq 0$ for a later value of u (i.e., "goodness" would not be invariant under time translation). The idea presented here suggests that, if we make apparently reasonable physical assumptions as to how the gravitational radiation falls off at $u = -\infty$ (or alternately at $u = +\infty$), then we can effectively minimize σ^0 at $u = -\infty$ (or alternatively at $u = +\infty$). This will restrict our coordinates to such an extent that only a subgroup of the BMS group remains—which is isomorphic to the improper orthochronous inhomogeneous Lorentz group.

Finally, the BMS transformations (12-1) for which the Jacobian J, defined by

$$K^4 = \sin^2 \theta J^2(\theta, \phi; \theta', \phi') \sin^{-2} \theta' \tag{12-52}$$

where K is the conformal factor, is positive, form the subgroup of all *proper* BMS transformations.

Normal Subgroups

We now discuss the important question of what normal subgroups the BMS group contains.[11] A normal subgroup N of a group G is characterized by the property that if n is an element of N and g is an element of G, then $n' = g^{-1}ng$ is an element of N (see Appendix A).

Theorem 12.1 *The supertranslations form an Abelian normal subgroup N of the BMS group; the factor group is isomorphic to the orthochronous homogeneous Lorentz group.*

Proof. The supertranslations are characterized by the fact that they leave the angles θ and ϕ unchanged. Using this fact, one finds that the supertranslations form a normal subgroup. The factor group is isomorphic to the conformal subgroup defined above. We have seen that this conformal subgroup is isomorphic to the homogeneous orthochronous Lorentz group; that any two supertranslations commute follows from Eq. (12-1).

Lemma 12.1 *The translations form a normal four-dimensional subgroup of the proper BMS group.*

In fact, any translation commutes with any supertranslation. As can be seen from the Lorentz group the commutator of a translation with a conformal transformation is some translation. Therefore, the translations form a normal subgroup. This normal subgroup is four-dimensional since it requires, as one sees from Eq. (12-50), exactly four parameters to span the translation group.

[11] R. K. Sachs, *Phys. Rev.* **128,** 2851 (1962).

Lemma 12.2 *If N' is a four-dimensional normal subgroup of the proper BMS group then N' is contained in the supertranslation group N.*

In fact, let G be the proper BMS group and consider the image N'/N of N' under the homomorphism $G \to G/N$; since N' is a normal subgroup of G, N'/N is a normal subgroup of G/N. Therefore, according to Theorem 12.1, N'/N is a normal subgroup of the proper orthochronous homogeneous Lorentz group L. However, the only normal subgroups of L are L itself and the identity e of L. If $N'/N = L$ then N'/N must be six-dimensional; then N' is at least six-dimensional, contrary to hypothesis. Therefore, $N'/N = e$; N' is, therefore, contained in N.

Lie Transformation Group, Lie Commutator, and Lie Algebra

We now introduce the infinitesimal BMS group transformations.[12] Suppose one has an S-dimensional Lie transformation group of an R-dimensional space. Let the coordinates of the space be $y^\alpha(\alpha, \beta = 1, \ldots, R)$ and the parameters of the group be $z^\mu(\mu, \nu = 1, \ldots, S)$, where z^μ is the identity of the group. Then the transformations have the form

$$y'^\alpha = f^\alpha(y^\beta; z^\mu), \quad \text{where} \quad f(y^\beta; 0) = y^\alpha; \tag{12-53}$$

the functions f^α are assumed to be twice differentiable. Consider now the quantities

$$q_\mu{}^\alpha = (\partial f^\alpha/\partial z^\mu)_{z^\mu=0}, \qquad (\alpha = 1, \ldots, R; \mu = 1, \ldots, S) \tag{12-54}$$

$q_\mu{}^\alpha$ is a vector defined everywhere on the R-dimensional space; there are S such vectors, one for each of the group parameters. If the group is truly S-dimensional and the z^μ are chosen suitably, these vectors are linearly independent; that is, if one has S constants B^μ then

$$B^\mu q_\mu{}^\alpha = 0 \Rightarrow B^\mu = 0, \qquad (\alpha = 1, \ldots, R; \mu = 1, \ldots, S) \tag{12-55}$$

To investigate the group structure one introduces the differential operators P_μ:

$$P_\mu = q_\mu{}^\alpha(\partial/\partial y^\alpha), \qquad (\alpha = 1, \ldots, R; \mu = 1, \ldots, S) \tag{12-56}$$

these are again linearly independent in the above sense. Consider a sufficiently small, finite neighbourhood of the identity $z^\mu = 0$ in the space of different group elements. Every group element in such a neighbourhood lies in precisely one one-dimensional subgroup of the full group. To each one-dimensional subgroup corresponds precisely one set $\{Q'\}$ of multiples of a particular linear combination of the basic linear operators

$$Q = B^\mu P_\mu, \qquad Q' = BQ \tag{12-57}$$

[12] The BMS group is not locally compact (compare, for example, L. Pontrjagin, *Topological Groups*, Princeton University Press, 1946).

Here, the B^μ are again constants and B is an arbitrary constant. Conversely, to every set $\{Q'\}$, defined by Eq. (12-57) with fixed B^μ and arbitrary B corresponds precisely one-dimensional subgroup of the group.

To the commutator of two group elements g and h that lie in the relevant neighbourhood of the group identity corresponds the Lie commutator

$$\begin{aligned} [G, H] &= -[H, G] = G^\mu H^\nu [P_\mu, P_\nu] \\ &= [G^\mu H^\nu - H^\mu G^\nu](q_\mu{}^\alpha \, \partial q_\nu{}^\beta / \partial y^\alpha)(\partial/\partial y^\beta) \\ G &= G^\mu P_\mu, \; H = H^\mu P_\mu, \qquad (\alpha, \beta = 1, \ldots, R; \mu, \nu = 1, \ldots, S) \end{aligned} \tag{12-58}$$

of differential operators G and H that correspond to the subgroups in which g and h lie. The group axioms imply that the Lie commutator must be a linear combination of the basic differential operators:

$$[P_\mu, P_\nu] = A_{\mu\nu}{}^\rho P_\rho, \qquad (\mu, \nu, \rho = 1, \ldots, S) \tag{12-59}$$

Here the quantities $A_{\mu\nu}{}^\rho$ are constants, the *structure constants* (see Chapter 1). The differential operators, considered as abstract quantities whose only relevant properties are given by their commutator table, are a complete linearly independent set of basic elements for what is called the *Lie algebra* of the group. The Lie algebra itself consists of the linear combinations of the differential operators; commutators are imposed on these linear combinations in the obvious way and then obey all the usual abstract properties of Poisson or commutator brackets, such as antisymmetry and bilinearity.[13]

To an $S' < S$ dimensional subgroup correspond S' linearly independent operators Q_μ (the symbol B with indices will denote constants):

$$Q_\mu = B_\mu{}^\nu P_\nu, \qquad (\mu = 1, \ldots, S'; \nu = 1, \ldots, S; S' < S) \tag{12-60}$$

These operators have the property that their Lie commutators are linear combinations of themselves:

$$[Q_\mu, Q_\nu] = B_{\mu\nu}{}^\rho Q_\rho, \qquad (\mu, \nu, \rho = 1, \ldots, S') \tag{12-61}$$

If the S' dimensional subgroup is a normal subgroup then the Q_μ obey the stronger conditions

$$[Q_\mu, P_\gamma] = Q_\rho B'_{\mu\gamma}{}^\rho, \qquad (\mu, \rho = 1, \ldots, S'; \gamma = 1, \ldots, S) \tag{12-62}$$

The fundamental theorem on Lie groups states that these relations can be inverted. If one can find in the Lie algebra combinations Q_μ that obey Eqs. (12-61) or (12-62), then there exists, respectively, a subgroup or normal subgroup to which the Q_μ correspond. Thus the structure of the Lie algebra characterizes the structure of the Lie group up to those global properties that cannot be analysed by analysing a small finite neighbourhood of the group identity.

[13] L. Pontrjagin, *op. cit.*

Infinitesimal Transformations

To apply these ideas to the BMS group, we expand the function $\alpha(\theta, \phi)$ of Eq. (12-1) as before in terms of spherical harmonics[14]

$$\alpha(\theta, \phi) = \sum_{l=0}^{\infty} \sum_{m=-l}^{l} \alpha_{lm} Y_{lm}(\theta, \phi), \qquad \alpha_{l,-m} = (-1)^m \bar{\alpha}_{lm} \tag{12-63}$$

From Eq. (12-56) one finds for the supertranslations

$$P_{lm} = Y_{lm}(\theta, \phi)(\partial/\partial u), \qquad (\bar{P}_{lm} = (-1)^m P_{l,-m}, \; P_{lm} = 0 \text{ for } |m| > l) \tag{12-64}$$

The six differential operators corresponding to the conformal group will be denoted by L_{ab}, with $L_{ab} = -L_{ba}$. These are the infinitesimal generators that generate rotations within the (x^a, x^b) plane of the Minkowski space (see Chapter 2). For example, a rotation in the (x, y) plane is given by

$$u' = u, \qquad \theta' = \theta, \qquad \phi' = \phi + \text{const.} \tag{12-65}$$

a rotation in the (z, t) plane is given by

$$\begin{gathered} u' = K(\theta)u, \qquad \phi' = \phi \\ \cot(\theta'/2) = (1 - \text{const.}) \cot(\theta/2) \end{gathered} \tag{12-66}$$

Using Eqs. (12-54) and (12-56), one then obtains for the generators of the above rotations

$$L_{12} = \frac{\partial}{\partial \phi} \tag{12-67a}$$

and

$$L_{30} = u \cos\theta \frac{\partial}{\partial u} + \sin\theta \frac{\partial}{\partial \theta} \tag{12-67b}$$

It is often convenient to introduce the following linear combinations:

$$\begin{aligned} L_{\pm} &= \pm i L_{23} + L_{13}, \qquad L_3 = L_{12} \\ K_{\pm} &= \mp i L_{20} + L_{10}, \qquad K_3 = L_{30} \end{aligned} \tag{12-68a}$$

A simple calculation then leads to the following:

$$\begin{aligned} L_{\pm} &= e^{\pm i\phi}\left(\frac{\partial}{\partial \theta} \pm i \cot\theta \frac{\partial}{\partial \phi}\right) \\ L_3 &= \frac{\partial}{\partial \phi} \\ K_{\pm} &= e^{\pm i\phi}\left(u \sin\theta \frac{\partial}{\partial u} - \cos\theta \frac{\partial}{\partial \theta} \mp i \operatorname{cosec}\theta \frac{\partial}{\partial \phi}\right) \\ K_3 &= u \cos\theta \frac{\partial}{\partial u} + \sin\theta \frac{\partial}{\partial \theta} \end{aligned} \tag{12-68b}$$

[14] Such an expansion is always possible, since the function $\alpha(\theta, \phi)$ is twice differentiable. See R. Courant and D. Hilbert, *Methods of Mathematical Physics*, Interscience, New York, 1953.

The P_{lm} and L_{ab} together form a complete set of linearly independent differential operators for the infinitesimal operators of the BMS group. From Eqs. (12-68) one finds the basic commutators

$$[L_{ab}, L_{cd}] = \eta_{ad} L_{bc} + \eta_{bc} L_{ad} - \eta_{ac} L_{bd} - \eta_{bd} L_{ac}$$

$$\left[L_{ab}, \alpha \frac{\partial}{\partial u}\right] = (L_{ab}\alpha - \alpha W) \frac{\partial}{\partial u}$$

implying

$$\begin{aligned}
&[L_3, P_{lm}] = imP_{lm} \\
&[L_+, P_{lm}] = -[(l-m)(l+m+1)]^{1/2} P_{l,\, m+1} \\
&[K_3, P_{lm}] = (l-1)[(l-m+1)(l+m+1)]^{1/2}[(2l+1)(2l+3)]^{-1/2} P_{l+1,\, m} \\
&\qquad -(l+2)(l^2-m^2)^{1/2}(4l^2-1)^{-1/2} P_{l-1,\, m} \qquad (12\text{-}69) \\
&[K_+, P_{lm}] = (l-1)[(l+m+2)(l+m+1)]^{1/2}[(2l+1)(2l+3)]^{-1/2} P_{l+1,\, m+1} \\
&\qquad -(l+2m)[(l-m)(l-m-1)]^{1/2}(4l^2-1)^{-1/2} P_{l-1,\, m-1}
\end{aligned}$$

Here W is defined by the relation $\partial(L_{ab} f)/\partial u \equiv L_{ab}\, \partial f/\partial u + W\, \partial f/\partial u$ for arbitrary $f(u)$. All other commutators can be obtained from those given in Eq. (12-69) by taking linear combinations or complex conjugate. For example,

$$\begin{aligned}
&[L_+, L_-] = -[K_+, K_-] = 2iL_3 \\
&[L_+, K_-] = 2K_3 \\
&[L_3, K_3] = [L_+, K_+] = 0 \qquad (12\text{-}70) \\
&[L_3, L_+] = i[K_3, K_+] = iL_+ \\
&[L_3, K_+] = -i[L_+, K_3] = iK_+ \\
&[L_-, P_{lm}] = \overline{[L_+, \bar{P}_{lm}]} = (-1)^m \overline{[L_+, P_{l,\, -m}]}
\end{aligned}$$

From the last four relations in Eq. (12-69) one sees that the conformal transformations transform the translations only among themselves (because of the factor $l-1$), but completely mix up all the other supertranslations with each other. This fact can be used to prove the uniqueness of the translation group.

Theorem 12.2 *The only normal four-dimensional subgroup of the BMS group is the translation group.*

Proof. In fact, suppose there was a second four-dimensional normal subgroup. Label the four linearly independent differential operators that correspond to the supposed second group as $P_a (a = 0, \ldots, 3)$. Then from Lemma 12.2 one infers:

$$P_a = \sum_{l=0}^{\infty} \sum_{m=-l}^{l} B_a{}^{lm} P_{lm} \qquad (12\text{-}71)$$

Moreover, there must be at least one value of m, one value of "a", and one value of $l \geqslant 2$ for which $B_a{}^{lm} \neq 0$, since the four linearly independent operators P_{lm} with $l < 2$ are merely the four operators of the translation group. Now let $n \geqslant 2$ be the minimum value of $l \geqslant 2$ for which there is at least one $B_a{}^{nm} \neq 0$. Choose one value of "a" for which $B_a{}^{nm} \neq 0$. Let m_0 be the minimum value of m for which $B_a{}^{nm} \neq 0$ with this choice of "a". Now commute $P_a(n - m_0)$ times with the operator L_+. The resulting operator Q then has the form

$$Q = BP_{nn} + \sum_{l \neq n} \sum_{m} B^{lm} P_{lm} \qquad B \neq 0, \text{ no summation on "}n\text{"} \tag{12-72}$$

as one sees from the commutator formulae (12-69). Commute Q four times successively with the operator L_-; one obtains:

$$\begin{aligned} [L_-, Q] &\equiv Q' = B'P_{n,\,n-1} + \cdots, \; (n \geqslant 2;\, B' \neq 0) \\ [L_-, Q'] &\equiv Q'' = B''P_{n,\,n-2} + \cdots, \; (B'' \neq 0) \\ [L_-, Q''] &\equiv Q''' = B'''P_{n,\,n-3} + \cdots, \; (B''' \neq 0) \\ [L_-, Q'''] &= B^{iv}P_{n,\,n-4} + \cdots, \; (B^{iv} \neq 0) \end{aligned} \tag{12-73}$$

Now the five operators shown in Eqs. (12-73) are all linearly independent, since the P_{nm} appearing in them are linearly independent. On the other hand, by virtue of Eq. (12-62) and the hypothesis that we are dealing with a normal four-dimensional subgroup, these five operators must depend linearly on the original four P_a (with complex coefficients). This is a contradiction and establishes the theorem.

As desired, the theorem characterizes translations uniquely.

The homogeneous Lorentz transformations are not similarly unique. In fact, let L be the conformal subgroup and t any finite supertranslation. Then the group $M = tLt^{-1}$ is a subgroup of the BMS group distinct from L and isomorphic to the homogeneous orthochronous Lorentz group. If t is the infinitesimal supertranslation $\alpha(\theta, \phi)(\partial/\partial u)$, then the infinitesimal elements of M have the form

$$\begin{aligned} L_3' &= L_3 + L_3\alpha \frac{\partial}{\partial u} \\ L_+' &= L_+ + L_+\alpha \frac{\partial}{\partial u} \\ K_3' &= K_3 + (K_3\alpha - \alpha \cos\theta) \frac{\partial}{\partial u} \\ K_+' &= K_+ + (K_+\alpha + \alpha \sin\theta\, e^{i\phi}) \frac{\partial}{\partial u} \end{aligned} \tag{12-74}$$

Representations of the BMS Group

We now discuss some representations of the BMS group and prove two theorems on the BMS Lie algebra.

Theorem 12.3 *There is at least one irreducible Hermitian representation of the BMS Lie algebra; the induced representation of the orthochronous Lorentz group is equivalent to the rest-mass zero, spin zero representation.*

Proof. Consider the (indefinite) scalar product for any two functions $f(u, \theta, \phi)$ and $g(u, \theta, \phi)$:

$$\langle f, g \rangle = i \int_{-\infty}^{\infty} du \int_0^{\pi} d\theta \int_0^{2\pi} \sin\theta \, d\phi (\partial \bar{f}/\partial u) g \tag{12-75}$$

Consider also the set of all twice-differentiable functions $\{f\}$ which are, together with their first two derivatives, integrable in the sense that Eq. (12-75) remains finite when any pair of functions are integrated. Suppose that they and their first derivatives vanish at $u = \pm\infty$. With the scalar product (12-75) one obtains a Hilbert space.

Consider now the linear operators

$$\begin{aligned} P'_{lm} &= P_{lm}, & K'_3 &= K_3 + \cos\theta \\ L'_3 &= L_3, & K'_\pm &= K_\pm + e^{\pm i\phi} \sin\theta \\ L'_\pm &= L_\pm \end{aligned} \tag{12-76}$$

By a direct calculation one verifies that these linear operators again obey the commutation relation of the BMS Lie algebra. Moreover, let L'_{ab} be the linear operators that correspond to the operators $L'_\pm$, L'_3, $K'_\pm$, and K'_3 via Eqs. (12-68). For the scalar product given one finds

$$\overline{\langle f, iP_{lm} g \rangle} = \langle g, iP_{lm} f \rangle$$

$$\overline{\langle f, iL_{ab} g \rangle} = \langle g, iL_{ab} f \rangle$$

the last relations verify the existence of a Hermitian representation.

Let us leave the question of irreducibility aside for the moment and examine the relation to representations of the Lorentz group.

Consider the solutions of D'Alembert's equation

$$\eta^{\alpha\beta} \partial_\alpha \partial_\beta \mu$$

where $\eta^{\alpha\beta}$ is the Lorentz metric, which are nonsingular, vanish at spatial infinity, and obey the Sommerfeld outgoing radiation condition for fixed μ and large r.[15] Consider the quantity

$$\rho(u, \theta, \phi) = \lim r\mu(u, r, \theta, \phi) \tag{12-77}$$

Because of the outgoing radiation condition, this limit always exists. Moreover, it obeys the conditions placed on f and g above. We show below that the relation between ρ and μ is one-to-one.

[15] A. Sommerfeld, *Partial Differential Equations in Physics*, Academic Press, New York, N.Y. 1949.

In fact, expand the field μ in spherical harmonics Y_{lm}, spherical Bessel functions $j_l(kr)$ and a Fourier time integral:[16]

$$\mu(t, r, \theta, \phi) = \sum_{l=0}^{\infty} \sum_{m=-l}^{l} \int_0^{\infty} k^{1/2} \, \mathrm{d}k \, Y_{lm}(\theta, \phi) \times j_l(kr) \, \mathrm{e}^{ikt} A_{lm}(k) + \text{c.c.} \tag{12-78}$$

Here the $A_{lm}(k)$ are the expansion coefficients. From the Fourier–Bessel theorem[17]

$$\int_0^{\infty} r^2 j_l(kr) j_l(k'r) \, \mathrm{d}r = \pi \, \delta(k - k')(2kk')^{-1} \tag{12-79}$$

one obtains

$$A_{lm}(k) = (i\pi)^{-1} \int \mathrm{d}^3x \, \bar{Y}_{lm} j_l(kr)^{1/2} (\partial\mu/\partial t + ik\mu)_{t=0} \tag{12-80}$$

Inserting the asymptotic values

$$j_l(y) \underset{r\to\infty}{\rightarrow} y^{-1} \sin(y - l\pi/2) \tag{12-81}$$

into Eq. (12-78), one finds

$$\rho(u, \theta, \phi) = \sum_{l=0}^{\infty} (-i)^{l+1} \sum_{m=-l}^{l} \int_0^{\infty} k^{1/2} \, \mathrm{d}k \, Y_{lm} \, \mathrm{e}^{iku} A_{lm} + \text{c.c.} \tag{12-82}$$

From Eq. (12-82) one infers that the knowledge of ρ enables one to calculate all the coefficients A_{lm}; from these coefficients one in turn obtains μ. Thus, ρ and μ determine each other uniquely.

If one now writes down the standard rest-mass zero, spin zero, representation of the Lorentz group Lie algebra acting on μ, for example $P_0 \mu = i \, \partial\mu/\partial t$, etc., one is able to induce a corresponding representation of differential operators acting on ρ. The latter turns out to be just given by the relevant quantities in Eq. (12-76). Therefore, we have verified the equivalence of the two representations as far as the orthochronous inhomogeneous Lorentz group[18] algebra is concerned.

We now come to the problem of irreducibility of the given representation of the BMS Lie algebra. The irreducibility now also follows. In fact, suppose the above BMS Lie algebra representation contains an invariant subspace. By means of the above one-to-one correspondence between ρ and μ there would be an induced invariant subspace of the corresponding representation of the inhomogeneous Lorentz group algebra by operators acting on μ. But the latter representation is well known to be irreducible; consequently, there is also no

[16] The proof is carried out for *real* ρ and μ. The extension to the complex case can easily be made.

[17] I. N. Sneddon, in *Handbuch der Physik*, (Ed. S. Flügge), Springer-Verlag, Berlin, 1955, Vol. 2.

[18] Representations of the inhomogeneous Lorentz group (Poincaré group) were given by E. P. Wigner, *Ann. Math.* **40**, 149 (1939).

invariant subspace for the given BMS Lie algebra representation. This completes the proof of the theorem.

The next theorem concerns an operator that commutes with all the operators of the BMS Lie algebra. Consider the "rest mass" operator

$$m^2 = P_{00}{}^2 - P_{10}{}^2 - P_{11}\bar{P}_{11} \tag{12-83}$$

Using the differential operators (12-69), one finds[19]

$$m^2 = 0 \tag{12-84}$$

Equations (12-69) also imply the following:

Theorem 12.4 *In every representation of the BMS Lie algebra, m^2 commutes with all other operators.*[20]

In the case of the Lorentz group there is a second operator, the "spin" operator, which also commutes with all the other operators.[21] By a direct calculation one verifies that the analogous operator in the case of the BMS group fails to commute with those supertranslations that are not merely translations. Therefore, Sachs has conjectered that the general unitary irreducible representation of the BMS group contains some mixture of Lorentz group representations with different spins, and that perhaps one such representation may contain precisely the mixture of spins that is found in nature.

PROBLEMS

12.1 Prove that the aggregate of all BMS transformations forms a group

12.2 Prove that the conformal transformations preserve the angle between two curves and its direction.

12.3 Show that the spin-s spherical harmonics satisfy the orthogonality relation

$$\int {}_sY_{lm}\, {}_{s'}\bar{Y}_{l'm'}\, dS = \delta_{ss'}\, \delta_{ll'}\, \delta_{mm'}$$

where dS is the surface area element on the sphere.

12.4 Prove that the spin-s spherical harmonics ${}_sY_{lm}$ provide a complete set for spin weight s quantities on the sphere.

12.5 Write down the Maxwell equations with or without sources when the field variables are considered as functions over the group SU_2. [See M. Carmeli, *J. Math. Phys.* **10,** 1699 (1969).]

12.6 Use Problem 12.5 in order to isolate a single complex function to describe the dynamics of the Maxwell field. Apply the canonical quantization procedure to quantize the wave equation so obtained. [See M. Carmeli, *Nuovo Cim.* **67B**, 103, (1970).]

[19] Since it is only the Lie commutators that can be taken over from the Lie algebra to representations of the Lie algebra, Eq. (12-83) does not imply the (possibly correct) statement that the matrix corresponding to m^2 must vanish in every representation of the BMS Lie algebra.

[20] Sachs has conjectored that Theorem 12.4 implies that in any irreducible representation of the BMS group, m^2 is represented by a constant—a linear operator which has only one eigenvalue and has every function as eigenfunction. See, however, P. J. McCarthy, *Phys. Rev. Letters* **29,** 817 (1972).

[21] See, for example, S. Schweber, *Relativistic Field Theory*, Row–Peterson, Evanston, Ill., 1961.

12.7 Make a coordinate transformation to "Cartesian" coordinates of the metric obtained on a generalization of the Euclidean plane

$$g^{ij}(r, \theta) = \begin{pmatrix} 1 & 0 \\ 0 & f(r, \theta) \end{pmatrix}$$

discussed at the end of Section 12.1. Show that in "Cartesian" coordinates, as $r \to \infty$, the metric tensor satisfies

$$g^{ij} \to \begin{pmatrix} 1 & 0 \\ 0 & 1 \end{pmatrix}$$

12.8 Consider the group G of conformal mappings on the Riemann sphere S. Let χ be the vector space of smooth, real-valued functions on S. Define the action of G on χ as follows: If g belongs to G and f belongs to χ, then $gf(p) = Kf(g(p))$, where p belongs to S and K is the conformal factor associated with the action of g on S. Show that the semidirect product of G and χ, denoted by $G \circledS \chi$, is the BMS group. [See R. Geroeh and E. T. Newman, *J. Math. Phys.* **12**, 314 (1971); see, however, a correction in P. J. McCarthy, *J. Math. Phys.* **13**, 1837 (1972).]

12.9 Use the method of Problem 12.6 to quantize the linearized gravitational field equations. [See S. Malin, *Phys. Rev. D* **10**, 2338 (1974).]

APPENDIX

A

REVIEW OF GROUP THEORY

Group theory is well covered in textbooks, including those by Pontrjagin,[1] van der Waerden,[2] and Wigner.[3] This Appendix is devoted to an exposition of the fundamental concepts of the theory.

A-1 GROUP AND SUBGROUP

A set G of elements is called a *group* if the following *axioms* are satisfied:

(1) There exists an operation in G which associates with each two elements a, b of G a third element c of G. This operation is called *multiplication*, and the element c is called the *product* of a and b, $c = ab$;

(2) The multiplication is *associative*, i.e., if a, b, and c are elements of G, then $(ab)c = a(bc)$;

(3) G contains a *right identity*, i.e., an element e such that $ae = a$ for any element a of G; and

(4) For each element a of G there exists a *right inverse* element, a^{-1}, such that $aa^{-1} = e$.

If the set G is finite the group G is called *finite* and the number of elements of G is called its *order*. Otherwise the group G is called *infinite*. If the product of any two elements a and b of G is commutative, $ab = ba$, the group is called *abelian*. In abelian groups the multiplicative notation ab is replaced by additive notation $a + b$ and the group operation is called *addition*. The identity is called *zero* and

[1] L. Pontrjagin, *Topological Groups*, Princeton University Press, 1946.

[2] B. L. van der Waerden, *Modern Algebra*, Fredric Ungar, New York, 1953.

[3] E. P. Wigner, *Group Theory and its Application to the Quantum Mechanics of Atomic Spectra*, Academic Press, New York, 1959.

denoted by 0, and the inverse of a is called the negative of a and denoted by $-a$.

Since the product of group elements is associative, one writes for $(ab)c = a(bc)$ simply abc. The same holds for products of any number of elements. One can easily show that a right identity e is also a left identity, $ea = a$, for any element a of G, and that a right inverse a^{-1} of a is also a left inverse, $a^{-1}a = e$. Hence the inverse of a^{-1} is a. Moreover, it follows that both the identity and the inverse are unique. This allows the use of the notation of algebra such as $a^{m+1} = a^m a$, with $a^1 = a$, for any natural number m. Negative powers of a are introduced by $a^{-m} = (a^{-1})^m$, $a^0 = e$. Hence $a^p a^q = a^{p+q}$, and $(a^p)^q = a^{pq}$, where p and q are integers.

A set H of elements of a group G is called a *subgroup* of G if it is a group with the same law of multiplication which operates in G. A necessary and sufficient condition for a subset H of a group G to be a subgroup is that if H contains two elements a and b it must also contain the element ab^{-1}.

A-2 NORMAL SUBGROUP AND FACTOR GROUP

Let G be a group and H a subgroup, and let a and b be two elements of G. One calls a and b *equivalent*,[4] $a \sim b$, if ab^{-1} is an element of H. The group G is thus divided into classes of equivalent elements each called a *right coset* of H relative to G. It follows that if A is a right coset of H and a is an element of A then $A = Ha$.[5] Moreover, every subset of the form Hb is a right coset and the subgroup H itself is one of the cosets. One can also introduce *left cosets* of H, written in the form aH. They are obtained from an equivalence relation such that $a \sim b$ if $a^{-1}b$ belongs to H.

A subgroup N of a group G is called an *invariant* or *normal subgroup* of G if for every element n of N and a of G the element $a^{-1}na$ belongs to N. It follows that a necessary and sufficient condition for right and left cosets of a subgroup N to coincide is that N be a normal subgroup.[6]

If N is a normal subgroup of a group G and A and B are two cosets of N, $A = Na$, $B = Nb$, then AB is also a coset of N. The multiplication of cosets thus defined satisfies the group axioms, and the set of all cosets is called the *factor group* of G by the normal subgroup N and is denoted by G/N.

[4] A *relation of equivalence* is said to be established in a set M if every two elements a, b of M are either equivalent, $a \sim b$, or not equivalent, a not $\sim b$. A relation of equivalence should be (a) reflexive: $a \sim a$; (b) symmetric: If $a \sim b$ then $b \sim a$; and (c) transitive: If $a \sim b$ and $b \sim c$ then $a \sim c$. A relation of equivalence in M divides M into disjoint classes of equivalent elements.

[5] If A and B are two subsets of a group G, one denotes by AB the subset of all elements of the form ab, where $a \in A$ and $b \in B$. The subset A^{-1} denotes all elements a^{-1}, where $a \in A$. The subset A^{m+1} is defined by $A^{m+1} = A^m A$, where $A^1 = A$, the subset A^{-m} is defined by $A^{-m} = (A^{-1})^m$ for a natural number m, and the subset A° is the set containing the identity only.

[6] Every group has at least two normal subgroups, the subgroup which includes only the identity, and the subgroup which coincides with the group itself. A group which has no normal subgroup except for these two subgroups is called *simple*.

A-3 ISOMORPHISM AND HOMOMORPHISM

A mapping f of a group G on another group G' is called *isomorphism* if it (1) is one-to-one; and (2) preserves the multiplication. G and G' are then called *isomorphic*. The inverse f^{-1} of an isomorphism f is itself an isomorphism. An isomorphism of a group onto itself is called *automorphism*. The aggregate of all automorphisms of a group forms a group.

A mapping f of a group G on another group G' is called *homomorphism* if it preserves the operation of multiplication. The set N of all elements of G which go over into the identity of G' under the homomorphism is called the *kernel* of the homomorphism. If the kernel coincides with the identity of G then the homomorphism is an isomorphism. It follows that N is a normal subgroup of G, and G' is isomorphic to G/N. The isomorphism between G' and G/N is called the *natural isomorphism*. The mapping f of a group G on G/N defined by associating with each element a of G the element $f(a) = A$ of G/N containing a is a homomorphism, called the *natural homomorphism* of a group on its factor group. If f is a homomorphism of a group G on another group G' and H is a (normal) subgroup of G, then $f(H)$ is a (normal) subgroup of G'. If f is a homomorphism of a group G on another group G', and g is a homomorphism of G' on a third group G'', then the mapping gf is a homomorphism of G on G''.

One finally notes that if f is a homomorphism of a group G on part of another group G' then the set of all elements of G' which are images of elements of G forms a subgroup of G'. Also, if $f^{-1}(H')$ is the set of all elements of G which go into $H' \subset G'$ under the homomorphism f, and if H' is a (normal) subgroup of the group G', then $f^{-1}(H')$ is also a (normal) subgroup of the group G.

APPENDIX

B

BASIC CONCEPTS OF REPRESENTATIONS THEORY

In this Appendix the basic concepts of the theory of finite-dimensional representations are reviewed. For more details the reader is referred to the books of Wigner,[1] Naimark,[2] and of Gelfand, Graev, and Vilenkin.[3]

B-1 LINEAR OPERATORS

Let R be a linear space and x a vector in R. A function $D(x)$ is called an *operator* in R if for any vector x of R there corresponds a vector $y = D(x)$ of R. An operator D in R is called *linear* if $D(x + y) = D(x) + D(y)$ and $D(\alpha x) = \alpha D(x)$, for any x, y of R and any complex number α. Addition of two operators A and B is defined in R by $(A + B)x = Ax + Bx$ for all vectors x of R. Similarly, multiplication by a number α and multiplication of operators A and B in R are defined by $(\alpha A)x = \alpha(Ax)$ and $(AB)x = A(Bx)$. If, furthermore, A and B are linear operators then $A + B$, αA, and AB are also linear operators.

Linear operators in a finite-dimensional space R can be represented as matrices by introducing a basis $e_1, \ldots, e_n$ in R. Accordingly, if A is a linear operator in R, then Ae_k can be written as a linear combination of $e_1, \ldots, e_n$, or,

[1] E. P. Wigner, *Group Theory and its Application to the Quantum Mechanics of Atomic Spectra*, Academic Press, New York, 1959.

[2] M. A. Naimark, *Linear Representations of the Lorentz Group*, Pergamon Press, New York, 1964.

[3] I. M. Gelfand, M. I. Graev, and N. Ya. Vilenkin, *Generalized Functions, Vol. 5: Integral Geometry and Representation Theory*, Academic Press, New York, 1966.

$Ae_k = \sum_{j=1}^{n} A_{jk} e_j$, for $k = 1, \ldots, n$. A_{jk} are the elements of the matrix representing the operator A relative to the basis $e_1, \ldots, e_n$. One can show that the operator A is completely determined by its matrix A_{ij}. Furthermore, the operations of addition, multiplication by a number, and multiplication of operators correspond to the same operations of their matrices relative to a fixed basis.

B-2 FINITE-DIMENSIONAL REPRESENTATION OF A GROUP

Let G be a group and g an arbitrary element of G. A correspondence $g \to D(g)$ of each element g of G to a linear operator $D(g)$ in a finite-dimensional space R is called a *representation* if: (1) $D(g_1 g_2) = D(g_1)D(g_2)$; and (2) $D(e)$ is the unit operator in R where e is the identity element of G. R is called the *space of representation* and its dimension is called the *dimension of representation.*

Two finite-dimensional representations $g \to D(g)$ and $g \to D'(g)$ in two spaces R and R' having the same dimensions, respectively, are called *equivalent* if basis in R and R' can be chosen so that the matrices of the operators $D(g)$ and $D'(g)$ are identical. A subspace S of R is called *invariant* with respect to the representation $g \to D(g)$ if for every vector x of S one finds that $D(g)x$ is also a vector in S for all $g \in G$. If there is no invariant subspaces in R with respect to the representation $g \to D(g)$, except for the trivial cases of the null subspace and the whole space, the representation is called *irreducible*. A representation $g \to D(g)$ of a group G is called *continuous* if $D(g)$ is a continuous operator function on the group G.[4] We will consider here only continuous representations.

B-3 UNITARY REPRESENTATIONS

A linear space is called *Euclidean* if from each two vectors x and y of it one can define a function, called the *scalar product* of x and y, denoted by (x, y), which satisfies: (1) $(x, x) \geqslant 0$, $(x, x) = 0$ if and only if $x = 0$; (2) $(y, x) = \overline{(x, y)}$; (3) $(\alpha x, y) = \alpha(x, y)$; and (4) $(x_1 + x_2, y) = (x_1, y) + (x_2, y)$. A scalar product can be introduced in every finite-dimensional space.

An operator A in a finite-dimensional Euclidean space R is called *unitary* if it preserves the scalar product, namely, $(Ax, Ay) = (x, y)$ for all x, y of R. A representation $g \to D(g)$ is called *unitary* if all the operators $D(g)$ are unitary.

[4] An operator function $D(g)$ is called continuous on a group G if the elements of the matrix of $D(g)$, relative to a fixed basis, are continuous functions on G. This definition of continuity of $D(g)$ does not depend on the choice of the basis since the matrix elements relative to another basis are linear combinations, with constant coefficients, of the matrix elements relative to the original basis.

APPENDIX

C

INFINITE-DIMENSIONAL REPRESENTATIONS

In Appendix B the elementary concepts of finite-dimensional representations were discussed. In this Appendix the theory of infinite-dimensional representations is briefly outlined.[1]

C-1 BANACH SPACE

A linear space R is called *normed* if a function $|x|$ is defined in it and satisfies the conditions: (1) $|x| \geqslant 0$, $|x| = 0$ if and only if the vector $x = 0$; (2) $|\alpha x| = |\alpha|\,|x|$ for any number α and any vector x of R; and (3) $|x + y| \leqslant |x| + |y|$ for any vectors x and y of the space R. The function $|x|$ is called a *norm*.[2]

A sequence of elements x_n of a normed space R is called *convergent in norm* to the element x of R if $|x - x_n| \to 0$ as $n \to \infty$. A sequence x_n of R is called *fundamental* if it satisfies the Cauchy condition (i.e. if for every $\varepsilon > 0$ there exists a number $N = N(\varepsilon)$ such that $|x_n - x_m| < \varepsilon$ for $n, m > N$.) A space R is called

[1] For more details see M. A. Naimark, *Linear Representations of the Lorentz Group*, Pergamon Press, New York, 1964; I. M. Gelfand, M. I. Graev, and N. Ya. Vilenkin, *Generalized Functions, Vol. 5: Integral Geometry and Representation Theory*, Academic Press, New York, 1966.

[2] An example of a normed space is the aggregate C of all complex numbers x. The norm of a complex number is taken as its modulus. Another example is provided by the aggregate of all sequences $x = \{\xi_1, \xi_2, \ldots\}$ of complex numbers $\xi_1, \xi_2, \ldots$ for which the series $|\xi_1|^2 + |\xi_2|^2 + \cdots$ converges. The operations in the space are defined as $\alpha x = \{\alpha\xi_1, \alpha\xi_2, \ldots\}$ and $x + y = \{\xi_1 + \eta_1, \xi_2 + \eta_2, \ldots\}$ for $x = \{\xi_1, \xi_2, \ldots\}$ and $y = \{\eta_1, \eta_2, \ldots\}$. The norm is defined as $|x| = \{|\xi_1|^2 + |\xi_2|^2 + \cdots\}^{1/2}$. This space is sometimes denoted by l^2.

complete if every fundamental sequence in R converges in norm to some element x of R.[3] A complete normed space is called a *Banach space*.

Let S be an arbitrary set in a Banach space R. The set $\bar{S}$ obtained from S by adding to it all the limits in norm of sequences of elements x_n of S is called the *closure* of the set S. A set S is called *dense* in R if $\bar{S} = R$. A set S is called *closed* if $\bar{S} = S$. A closed subspace of a Banach space is itself a Banach space.

A series $x_1 + x_2 + \cdots$ of elements x_n of R is called *convergent* and the element x of R is called the sum of the series if $x_1 + x_2 + \cdots + x_n \to x$ as $n \to \infty$ in the sense of the norm in R. A series $x_1 + x_2 + \cdots$ is called *absolutely convergent* if the series $|x_1| + |x_2| + \cdots$ of real numbers is convergent. In a Banach space every absolutely convergent series converges. This follows from the inequality $|x_{n+1} + \cdots + x_{n+p}| \leqslant |x_{n+1}| + \cdots + |x_{n+p}|$ and the fact that the space is complete.

C-2 OPERATORS IN BANACH SPACE

A linear operator A in a Banach space R is called bounded if there exists a constant $c \geqslant 0$ such that $|Ax| \leqslant c|x|$ for all x of R. The smallest number c satisfying this condition is called the *norm* of the bounded operator A and is denoted by $|A|$. Hence $|Ax| \leqslant |A|\,|x|$. If A and B are bounded operators, then also the operators αA, $A + B$, and BA are bounded and satisfy

$$|\alpha A| = |\alpha|\,|A|; \qquad |A + B| \leqslant |A| + |B|; \qquad |AB| \leqslant |A|\,|B|$$

It then follows that every bounded linear operator A is continuous. Furthermore, if two bounded operators A and B coincide on a set S which is dense in a space R, then they coincide on the whole of R.

C-3 GENERAL DEFINITION OF A REPRESENTATION

A mapping $g \to D(g)$ of a group G on a Banach space R is called a *representation* if to every element g of G there corresponds a bounded linear operator $D(g)$ in R such that $D(e) = 1$ and $D(g_1 g_2) = D(g_1)D(g_2)$. A representation $g \to D(g)$ in a Banach space R is called *irreducible* if R contains no closed subspace (other than the null one and R itself) which is invariant with respect to all operators $D(g)$. This definition coincides with that of irreducibility for the finite-dimensional one. This is so since every finite-dimensional subspace is closed.

A *linear functional* $f(x)$ in a linear space R is a numerical function satisfying $f(\alpha x) = \alpha f(x)$ and $f(x + y) = f(x) + f(y)$ for any number α and x and y of R. A linear functional in a normed space R is called *bounded* if there exists a constant

[3] The space C of all complex numbers discussed in footnote 2 above, is complete. The space l^2 defined in footnote 2 is also complete. An example of a non-complete normed space is the set of all sequences $x = \{\xi_1, \xi_2, \ldots\}$ in which only a finite number of ξ_n is non-zero, all other operations of the space are the same as those of the space l^2.

$c \geqslant 0$ such that $|f(x)| \leqslant c|x|$ for all x of R. The smallest number $c \geqslant 0$ satisfying this condition is called the norm of the functional and is denoted by $|f|$. Thus $|f(x)| \leqslant |f|\,|x|$. The bound linear functionals in R form a normed linear space where the sum and product are defined by $(f_1 + f_2)(x) = f_1(x) + f_2(x)$ and $(\alpha f)(x) = \bar{\alpha} f(x)$. This space is called the conjugate to the space R. It is a complete space.

Let a linear space be denoted by R and let its conjugate space be denoted by R'. Then for every element x of R there exists a functional f of R' such that $f(x) = |x|$ and $|f| = 1$. Hence if $f(x) = 0$ for all f of R' then $x = 0$. Furthermore, if M is a closed subspace of a Banach space R and x_0 is a vector in R not belonging to M, then there exists a functional f of R' satisfying $f(x_0) \neq 0$, and $f(x) = 0$ for all x of M.

Since the conjugate space R' is a normed space, one can therefore consider the linear bounded functional $F(f)$ in it. Such functionals are obtained, for example, if we put $F_x(f) = \overline{f(x)}$ for a fixed element x of R since $F_x(f)$ is a bounded linear functional in R'. A space R is called reflexive if the functionals $F_x(f)$, for all x of R, exhaust all the bounded linear functionals in R'. In other words if every bounded linear functional $F(f)$ in R' is given by $F(f) = f(x)$ for some x of R.[4]

C-4 CONTINUOUS REPRESENTATIONS

Let $x(t) = x(t_1, t_2, \ldots, t_m)$ be a vector function of a point $t = (t_1, \ldots, t_m)$ in an m-dimensional space with values in R. A vector function $x(t)$ is called *continuous* in a set D in m-dimensional space if for every functional f of the conjugate space R' the numerical function $f[x(t)]$ is continuous in D. A bounded linear operator function $A(t)$ in R is called *continuous* in D if for every x of R and f of R' the numerical function $f(A(t)x)$ is continuous in D. For example if G is a group of matrices, then it may be regarded as a subset of m-dimensional space for a sufficiently large m. Hence one may speak of a vector-function $x(g)$ or an operator function $A(g)$ as continuous in the group G. A representation $g \to D(g)$ of a group of matrices is called continuous if $D(g)$ is a continuous operator function.[5] It then follows that if $x(t)$ and $A(t)$ are vector and operator functions, respectively, which are continuous in a closed bounded set D, then the numerical functions $|x(t)|$ and $|A(t)|$ are bounded in that set.

C-5 UNITARY REPRESENTATIONS

The concept of a unitary representation discussed in Appendix B can be generalized as follows.

Let R be a Euclidean space which might be infinite. One can easily introduce a norm in R by putting $|x| = (x, x)^{1/2}$, where (x, y) is a scalar product in R. A

[4] Throughout our discussion we consider only representations in reflexive Banach spaces.

[5] Throughout the text the term representation stands for continuous representation (unless otherwise stated).

Euclidean space R, complete with respect to the norm $|x| = (x, x)^{1/2}$, is called a *Hilbert space*.[6] In a Hilbert space R every bounded linear functional $f(x)$ is represented in the form $f(x) = (x, y)$, where y belongs to R and $|f| = |y|$. Hence the space R', conjugate to a Hilbert space R, may be identified with R itself, $R' = R$.

Two Hilbert spaces R_1 and R_2 are called *isometric* if there exists a linear operator U mapping R_1 onto R_2 and preserving the scalar product, $(Ux, Uy) = (x, y)$ for all x, y of R_1. The operator U itself is then called *isometric* and it satisfies $|Ux| = |x|$ for all x of R_1.

If U is an isometric operator mapping R onto itself, then U is called a *unitary operator* in R. A representation $g \to D(g)$ of a group G in a space R is called *unitary* if R is a Hilbert space and $D(g)$ is a unitary operator for all g of G. A representation $g \to D(g)$ in a Hilbert space R is unitary if $(D(g)x, D(g)y) = (x, y)$ for all g of G and x, y of R.

Let A be a bounded operator in a Hilbert space R. An operator $A^\dagger$ is called *adjoint* to A if $(Ax, y) = (x, A^\dagger y)$ for all x, y of R. One can show that $A^{\dagger\dagger} = A$, $(\alpha A)^\dagger = \alpha A^\dagger$, $(A + B)^\dagger = A^\dagger + B^\dagger$, $(AB)^\dagger = B^\dagger A^\dagger$, and $|A^\dagger| = |A|$. An operator U is unitary if and only if $U^\dagger U = UU^\dagger = 1$. The operator A^{-1} is called *inverse* to A if $AA^{-1} = A^{-1}A = 1$. Hence a unitary operator satisfies $U^\dagger = U^{-1}$. An operator A is called Hermitian if $A^\dagger = A$. A Hermitian operator P is called a *projection* operator if $P^2 = P$.

Finally, let $R_1, R_2, \ldots$ be closed, mutually orthogonal, subspaces of a Hilbert space R. The aggregate of all sums $x = x_1 + x_2 + \cdots$ of convergent series of elements $x_k \in R_k$ is called the *orthogonal sum* of the Hilbert spaces $R_1, R_2, \ldots$, and is denoted by $R_1 \oplus R_2 \oplus \cdots$. It follows that $R_1 \oplus R_2 \oplus \cdots$ is a closed subspace of R. If E_n is a projection operator in R onto $R_1 \oplus R_2 \oplus \cdots \oplus R_n$, then $E_n x = x_1 + \cdots + x_n$ for any vector $x = x_1 + x_2 + \cdots$ of R, where $x_k \in R_k$. The bounded linear operator A in a space R is called the *orthogonal sum* of the operators A_k in R_k, denoted by $A_1 \oplus A_2 \oplus \cdots$, if $R = R_1 \oplus R_2 \oplus \cdots$ and $Ax = A_1 x_1 + A_2 x_2 + \cdots$, where $x = x_1 + x_2 + \cdots$. A unitary representation $g \to D(g)$ of a group G in a Hilbert space R is called the orthogonal sum of the representations $g \to D^{(k)}(g)$ in the closed subspaces R_k if $D(g) = D^{(1)}(g) + D^{(2)}(g) + \cdots$ for all g of G.

[6] The space l^2 discussed in footnotes 2 and 3 above is a Hilbert space if the scalar product is defined by $(x, y) = \Sigma \xi_k \bar{\eta}_k$ for $x = \{\xi_1, \xi_2, \ldots\}$ and $y = \{\eta_1, \eta_2, \ldots\}$. Another example of a Hilbert space is the aggregate of all functions $f(x)$, measurable in a fixed interval (a, b) and satisfying the conditions $\int_a^b |f(x)|^2\, dx < \infty$, if the operations of addition and multiplication by a number are defined in the usual way, and the scalar product is defined by $(f_1, f_2) = \int_a^b f_1(x)\bar{f_2}(x)\, dx$. This Hilbert space is sometimes denoted by $L^2(a, b)$. In the same way the Hilbert space $L^2(SU_2)$ is defined as the aggregate of all functions $f(u)$ satisfying $\int |f(u)|^2\, du < \infty$, where the scalar product is defined by $(f_1, f_2) = \int f_1(u)\bar{f_2}(u)\, du$.

APPENDIX

D

GRAVITATIONAL FIELD EQUATIONS

In Chapter 9 we wrote down the gravitational field equations of general relativity that connect the three sets of dynamical variables: (1) the components of the Riemann tensor (the Weyl tensor, the trace-free parts of the Ricci tensor, and the Ricci scalar); (2) the spin coefficients; and (3) a tetrad of null vectors. These equations were given by Eq. (9-71), Eq. (9-72) (or equivalently Eq. (9-77)), and Eq. (9-73).

We here write down explicitly Eqs. (9-71), (9-72) and (9-73). We shall use the standard notation for the operators $\partial_{ab'}$ according to which $D = \partial_{00'}$, $\delta = \partial_{01'}$, $\bar{\delta} = \partial_{10'}$ and $\Delta = \partial_{11'}$. Also given in this appendix are the commutation relations that these operators satisfy.

D-1 GRAVITATIONAL FIELD EQUATIONS

For Eq. (9-71) one obtains:

$$D\begin{pmatrix} \beta & -\sigma \\ \mu & -\beta \end{pmatrix} - \delta\begin{pmatrix} \varepsilon & -\kappa \\ \pi & -\varepsilon \end{pmatrix} - \begin{pmatrix} \psi_1 & -\psi_0 \\ \psi_2 + 2\Lambda & -\psi_1 \end{pmatrix}$$
$$= \begin{pmatrix} -\kappa\gamma + \bar{\pi}\varepsilon - \bar{\varepsilon}\beta + \sigma\alpha & -3\varepsilon\sigma + \kappa\tau - \bar{\pi}\kappa + \bar{\varepsilon}\sigma \\ -\bar{\alpha}\varepsilon + \bar{\rho}\beta + \sigma\pi - \kappa\mu & -\sigma\rho + \bar{\alpha}\kappa - \bar{\rho}\sigma + 3\beta\kappa \\ & \\ -\kappa\nu + \bar{\pi}\pi - \bar{\varepsilon}\mu + \sigma\lambda & \kappa\gamma - \bar{\pi}\varepsilon + \bar{\varepsilon}\beta - \sigma\alpha \\ -\bar{\alpha}\pi + \bar{\rho}\mu + \pi\beta - \varepsilon\mu & +\bar{\alpha}\varepsilon - \bar{\rho}\beta + \mu\kappa - \pi\sigma \end{pmatrix} \tag{D-1a}$$

$$D\begin{pmatrix} \alpha & -\rho \\ \lambda & -\alpha \end{pmatrix} - \bar{\delta}\begin{pmatrix} \varepsilon & -\kappa \\ \pi & -\varepsilon \end{pmatrix} - \begin{pmatrix} \phi_{10} & -\phi_{00} \\ \phi_{20} & -\phi_{10} \end{pmatrix}$$
$$= \begin{pmatrix} \pi\varepsilon - 2\varepsilon\alpha + \bar{\varepsilon}\alpha - \bar{\kappa}\gamma & -\pi\kappa - \bar{\varepsilon}\rho + \bar{\kappa}\tau + 3\alpha\kappa \\ +\rho\alpha - \bar{\beta}\varepsilon + \bar{\sigma}\beta + \rho\pi - \kappa\lambda & -\rho^2 + \bar{\beta}\kappa - \bar{\sigma}\sigma - \rho\varepsilon \\ & \\ \pi^2 - 3\varepsilon\lambda + \bar{\varepsilon}\lambda - \bar{\kappa}\nu & -\pi\varepsilon + 2\varepsilon\alpha - \bar{\varepsilon}\alpha + \bar{\kappa}\gamma \\ +\rho\lambda - \bar{\beta}\pi + \bar{\sigma}\mu + \pi\alpha & -\rho\alpha + \bar{\beta}\varepsilon - \bar{\sigma}\beta + \lambda\kappa - \pi\rho \end{pmatrix} \tag{D-1b}$$

$$D\begin{pmatrix} \gamma & -\tau \\ \nu & -\gamma \end{pmatrix} - \Delta\begin{pmatrix} \varepsilon & -\kappa \\ \pi & -\varepsilon \end{pmatrix} - \begin{pmatrix} \psi_2 + \phi_{11} - \Lambda & -\psi_1 - \phi_{01} \\ \psi_3 + \phi_{21} & -\psi_2 - \phi_{11} + \Lambda \end{pmatrix}$$
$$= \begin{pmatrix} \pi\beta + \bar{\pi}\alpha - \bar{\varepsilon}\gamma - 2\gamma\varepsilon & -\pi\sigma - \bar{\pi}\rho + \bar{\varepsilon}\tau - \tau\rho \\ +\tau\alpha - \bar{\gamma}\varepsilon + \bar{\tau}\beta + \tau\pi - \kappa\nu & +\bar{\gamma}\kappa - \bar{\tau}\sigma - \tau\varepsilon + 3\kappa\gamma \\ & \\ \pi\mu - 3\varepsilon\nu + \bar{\pi}\lambda - \bar{\varepsilon}\nu & -\pi\beta + 2\varepsilon\gamma - \bar{\pi}\alpha + \bar{\varepsilon}\gamma \\ +\tau\lambda - \bar{\gamma}\pi + \bar{\tau}\mu + \gamma\pi & -\tau\alpha + \bar{\gamma}\varepsilon - \bar{\tau}\beta + \nu\kappa - \pi\tau \end{pmatrix} \tag{D-1c}$$

$$\delta\begin{pmatrix} \alpha & -\rho \\ \lambda & -\alpha \end{pmatrix} - \bar{\delta}\begin{pmatrix} \beta & -\sigma \\ \mu & -\beta \end{pmatrix} + \begin{pmatrix} \psi_2 - \phi_{11} - \Lambda & -\psi_1 + \phi_{01} \\ \psi_3 - \phi_{21} & -\psi_2 + \phi_{11} + \Lambda \end{pmatrix}$$
$$= \begin{pmatrix} \mu\varepsilon - 2\beta\alpha + \bar{\alpha}\alpha - \bar{\rho}\gamma & -\mu\kappa - \bar{\alpha}\rho + \bar{\rho}\tau - \rho\tau \\ +\rho\gamma - \bar{\mu}\varepsilon + \bar{\beta}\beta + \rho\mu - \sigma\lambda & +\bar{\mu}\kappa - \bar{\beta}\sigma - \beta\rho + 3\sigma\alpha \\ & \\ \mu\pi - 3\beta\lambda + \bar{\alpha}\lambda - \bar{\rho}\nu & -\mu\varepsilon + 2\beta\alpha - \bar{\alpha}\alpha + \bar{\rho}\gamma \\ +\rho\nu - \bar{\mu}\pi + \bar{\beta}\mu + \mu\alpha & -\mu\rho - \rho\gamma + \bar{\mu}\varepsilon - \bar{\beta}\beta + \lambda\sigma \end{pmatrix} \tag{D-1d}$$

$$\delta\begin{pmatrix} \gamma & -\tau \\ \nu & -\gamma \end{pmatrix} - \Delta\begin{pmatrix} \beta & -\sigma \\ \mu & -\beta \end{pmatrix} - \begin{pmatrix} \phi_{12} & -\phi_{02} \\ \phi_{22} & -\phi_{12} \end{pmatrix}$$
$$= \begin{pmatrix} \mu\beta - 2\beta\gamma + \bar{\lambda}\alpha - \bar{\alpha}\gamma & -\mu\sigma - \bar{\lambda}\rho + \bar{\alpha}\tau + 3\gamma\sigma \\ +\tau\gamma - \bar{\nu}\varepsilon + \bar{\gamma}\beta + \tau\mu - \nu\sigma & -\tau^2 + \bar{\nu}\kappa - \bar{\gamma}\sigma - \tau\beta \\ & \\ \mu^2 - 3\beta\nu + \bar{\lambda}\lambda - \bar{\alpha}\nu & -\mu\beta + 2\beta\gamma - \bar{\lambda}\alpha + \bar{\alpha}\gamma + \nu\sigma \\ +\tau\nu - \bar{\nu}\pi + \bar{\gamma}\mu + \gamma\mu & -\tau\gamma + \bar{\nu}\varepsilon - \bar{\gamma}\beta - \mu\tau \end{pmatrix} \tag{D-1e}$$

$$\bar{\delta}\begin{pmatrix} \gamma & -\tau \\ \nu & -\gamma \end{pmatrix} - \Delta\begin{pmatrix} \alpha & -\rho \\ \lambda & -\alpha \end{pmatrix} - \begin{pmatrix} \psi_3 & -\psi_2 - 2\Lambda \\ \psi_4 & -\psi_3 \end{pmatrix}$$
$$= \begin{pmatrix} \lambda\beta + \bar{\mu}\alpha - \bar{\beta}\gamma - \nu\varepsilon & -\lambda\sigma - \bar{\mu}\rho + \bar{\beta}\tau + \nu\kappa \\ -\bar{\gamma}\alpha + \bar{\tau}\gamma + \tau\lambda - \rho\nu & +\bar{\gamma}\rho - \bar{\tau}\tau + \gamma\rho - \tau\alpha \\ & \\ \lambda\mu - 3\alpha\nu + \bar{\mu}\lambda - \bar{\beta}\nu & -\lambda\beta - \bar{\mu}\alpha + \bar{\beta}\gamma + \nu\varepsilon \\ -\nu\pi + 3\gamma\lambda - \bar{\gamma}\lambda + \bar{\tau}\nu & +\bar{\gamma}\alpha - \bar{\tau}\gamma - \tau\lambda + \rho\nu \end{pmatrix} \tag{D-1f}$$

Equation (9-72) gives

$$
\bar{\delta}\begin{pmatrix} \psi_1 & -\psi_0 \\ \psi_2 + 2\Lambda & -\psi_1 \end{pmatrix} - D\begin{pmatrix} \psi_2 - \phi_{11} - \Lambda & -\psi_1 + \phi_{01} \\ \psi_3 - \phi_{21} & -\psi_2 + \phi_{11} + \Lambda \end{pmatrix} - \delta\begin{pmatrix} \phi_{10} & -\phi_{00} \\ \phi_{20} & -\phi_{10} \end{pmatrix}
$$

$$
= \begin{pmatrix} \lambda\psi_0 + 2(\alpha - \pi)\psi_1 - 3\rho\psi_2 + 2\kappa\psi_3 & (\pi - 4\alpha)\psi_0 + (2\varepsilon + 4\rho)\psi_1 - 3\kappa\psi_2 \\ 2\lambda\psi_1 - 3\pi\psi_2 + 2(\varepsilon - \rho)\psi_3 + \kappa\psi_4 & -\lambda\psi_0 - 2(\alpha - \pi)\psi_1 + 3\rho\psi_2 - 2\kappa\psi_3 \end{pmatrix}
$$

$$
+ \begin{pmatrix} -\mu\phi_{00} + \pi\phi_{01} + (\bar{\pi} - 2\bar{\alpha})\phi_{10} & (2\bar{\alpha} + 2\beta - \bar{\pi})\phi_{00} - 2(\varepsilon + \bar{\rho})\phi_{01} \\ \quad + 2\bar{\rho}\phi_{11} - \bar{\kappa}\phi_{12} + \sigma\phi_{20} - \kappa\phi_{21} & \quad + \bar{\kappa}\phi_{02} - 2\sigma\phi_{10} + 2\kappa\phi_{11} \\ -2\mu\phi_{10} + 2\pi\phi_{11} - (2\bar{\alpha} - \bar{\pi} - 2\beta)\phi_{20} & \mu\phi_{00} - \pi\phi_{01} + (2\bar{\alpha} - \bar{\pi})\phi_{10} \\ \quad + 2(\bar{\rho} - \varepsilon)\phi_{21} - \bar{\kappa}\phi_{22} & \quad - 2\bar{\rho}\phi_{11} + \bar{\kappa}\phi_{12} - \sigma\phi_{20} + \kappa\phi_{21} \end{pmatrix} \tag{D-2a}
$$

$$
\Delta\begin{pmatrix} \psi_1 & -\psi_0 \\ \psi_2 + 2\Lambda & -\psi_1 \end{pmatrix} - \delta\begin{pmatrix} \psi_2 + \phi_{11} - \Lambda & -\psi_1 - \phi_{01} \\ \psi_3 + \phi_{21} & -\psi_2 - \phi_{11} + \Lambda \end{pmatrix} + D\begin{pmatrix} \phi_{12} & -\phi_{02} \\ \phi_{22} & -\phi_{12} \end{pmatrix}
$$

$$
= \begin{pmatrix} \nu\psi_0 + 2(\gamma - \mu)\psi_1 - 3\tau\psi_2 + 2\sigma\psi_3 & (\mu - 4\gamma)\psi_0 + (4\tau + 2\beta)\psi_1 - 3\sigma\psi_2 \\ 2\nu\psi_1 - 3\mu\psi_2 + 2(\beta - \tau)\psi_3 + \sigma\psi_4 & -\nu\psi_0 - 2(\gamma\ \mu)\psi_1 + 3\tau\psi_2 - 2\sigma\psi_3 \end{pmatrix}
$$

$$
+ \begin{pmatrix} \mu\phi_{01} + \pi\phi_{02} - \bar{\lambda}\phi_{10} + 2\bar{\pi}\phi_{11} & \bar{\lambda}\phi_{00} - 2(\bar{\pi} - \beta)\phi_{01} \\ \quad + (\bar{\rho} - 2\bar{\varepsilon})\phi_{12} + \sigma\phi_{21} - \kappa\phi_{22} & \quad + (2\bar{\varepsilon} - 2\varepsilon - \bar{\rho})\phi_{02} - 2\sigma\phi_{11} + 2\kappa\phi_{12} \\ -2\mu\phi_{11} + 2\pi\phi_{12} - \bar{\lambda}\phi_{20} & \mu\phi_{01} - \pi\phi_{02} + \bar{\lambda}\phi_{10} - 2\bar{\pi}\phi_{11} \\ \quad + 2(\beta + \bar{\pi})\phi_{21} + (\bar{\rho} - 2\varepsilon - 2\bar{\varepsilon})\phi_{22} & \quad + (2\bar{\varepsilon} - \bar{\rho})\phi_{12} - \sigma\phi_{21} + \kappa\phi_{22} \end{pmatrix} \tag{D-2b}
$$

$$
\bar{\delta}\begin{pmatrix} \psi_2 + \phi_{11} - \Lambda & -\psi_1 - \phi_{01} \\ \psi_3 + \phi_{21} & -\psi_2 - \phi_{11} + \Lambda \end{pmatrix} - D\begin{pmatrix} \psi_3 & -\psi_2 - 2\Lambda \\ \psi_4 & -\psi_3 \end{pmatrix} - \Delta\begin{pmatrix} \phi_{10} & -\phi_{00} \\ \phi_{20} & -\phi_{10} \end{pmatrix}
$$

$$
= \begin{pmatrix} 2\lambda\psi_1 - 3\pi\psi_2 - 2(\rho - \varepsilon)\psi_3 + \kappa\psi_4 & -\lambda\psi_0 - 2(\alpha - \pi)\psi_1 + 3\rho\psi_2 - 2\kappa\psi_3 \\ 3\lambda\psi_2 - (4\pi + 2\alpha)\psi_3 - (\rho - 4\varepsilon)\psi_4 & -2\lambda\psi_1 + 3\pi\psi_2 - 2(\varepsilon - \rho)\psi_3 - \kappa\psi_4 \end{pmatrix}
$$

$$
+ \begin{pmatrix} -\nu\phi_{00} + \lambda\phi_{01} + (\bar{\mu} - 2\bar{\gamma})\phi_{10} & (2\gamma + 2\bar{\gamma} - \bar{\mu})\phi_{00} - 2(\alpha + \bar{\tau})\phi_{01} \\ \quad + 2\bar{\tau}\phi_{11} - \bar{\sigma}\phi_{12} + \tau\phi_{20} - \rho\phi_{21} & \quad + \bar{\sigma}\phi_{02} - 2\tau\phi_{10} + 2\rho\phi_{11} \\ -2\nu\phi_{10} + 2\lambda\phi_{11} + (2\gamma - 2\bar{\gamma} + \bar{\mu})\phi_{20} & \nu\phi_{00} - \lambda\phi_{01} + (2\bar{\gamma} - \bar{\mu})\phi_{10} \\ \quad + 2(\bar{\tau} - \alpha)\phi_{21} - \bar{\sigma}\phi_{22} & \quad - 2\bar{\tau}\phi_{11} + \bar{\sigma}\phi_{12} - \tau\phi_{20} + \rho\phi_{21} \end{pmatrix} \tag{D-2c}
$$

$$\Delta\begin{pmatrix}\psi_2-\phi_{11}-\Lambda & -\psi_1+\phi_{01}\\ \psi_3-\phi_{21} & -\psi_2+\phi_{11}+\Lambda\end{pmatrix}-\delta\begin{pmatrix}\psi_3 & -\psi_2-2\Lambda\\ \psi_4 & -\psi_3\end{pmatrix}+\bar{\delta}\begin{pmatrix}\phi_{12} & -\phi_{02}\\ \phi_{22} & -\phi_{12}\end{pmatrix}$$

$$=\begin{pmatrix}2\nu\psi_1-3\mu\psi_2+2(\beta-\tau)\psi_3+\sigma\psi_4 & -\nu\psi_0+2(\mu-\gamma)\psi_1+3\tau\psi_2-2\sigma\psi_3\\ 3\nu\psi_2-(4\mu+2\gamma)\psi_3+(4\beta-\tau)\psi_4 & -2\nu\psi_1+3\mu\psi_2+2(\tau-\beta)\psi_3-\sigma\psi_4\end{pmatrix}$$

$$+\begin{pmatrix}\begin{matrix}-\nu\phi_{01}+\lambda\phi_{02}-\bar{\nu}\phi_{10}+2\bar{\mu}\phi_{11}\\ +(\bar{\tau}-2\bar{\beta})\phi_{12}+\tau\phi_{21}-\rho\phi_{22}\end{matrix} & \begin{matrix}\bar{\nu}\phi_{00}+2(\gamma-\bar{\mu})\phi_{01}\\ +(2\bar{\beta}-2\alpha-\bar{\tau})\phi_{02}-2\tau\phi_{11}+2\rho\phi_{12}\end{matrix}\\ \begin{matrix}-2\nu\phi_{11}+2\lambda\phi_{12}-\bar{\nu}\phi_{20}\\ +2(\gamma+\bar{\mu})\phi_{21}+(\bar{\tau}-2\alpha-2\bar{\beta})\phi_{22}\end{matrix} & \begin{matrix}\nu\phi_{01}-\lambda\phi_{02}+\bar{\nu}\phi_{10}-2\bar{\mu}\phi_{11}\\ +(2\bar{\beta}-\bar{\tau})\phi_{12}-\tau\phi_{21}+\rho\phi_{22}\end{matrix}\end{pmatrix} \qquad \text{(D-2d)}$$

As can be seen Eqs. (D-1) and (D-2) are equivalent to the usual equations obtained using the Newman–Penrose formalism.[1]

D-2 COMMUTATION RELATIONS

Finally, we give in this Appendix the explicit equations obtained by applying the commutator $(\nabla_{ab'}\nabla_{cd'}-\nabla_{cd'}\nabla_{ab'})$ on any scalar function ϕ (see Problem 10.4):

$$(D\Delta-\Delta D)\phi=[-(\gamma+\bar{\gamma})D+(\pi+\bar{\tau})\delta+(\tau+\bar{\pi})\bar{\delta}-(\varepsilon+\bar{\varepsilon})\Delta]\phi \qquad \text{(D-3a)}$$

$$(\delta D-D\delta)\phi=[(\bar{\alpha}+\beta-\bar{\pi})D-(\bar{\rho}+\varepsilon-\bar{\varepsilon})\delta-\sigma\bar{\delta}+\kappa\Delta]\phi \qquad \text{(D-3b)}$$

$$(\delta\Delta-\Delta\delta)\phi=[-\bar{\nu}D+(\mu-\gamma+\bar{\gamma})\delta+\bar{\lambda}\bar{\delta}+(\tau-\bar{\alpha}-\beta)\Delta]\phi \qquad \text{(D-3c)}$$

$$(\delta\bar{\delta}-\bar{\delta}\delta)\phi=[(\mu-\bar{\mu})D+(\bar{\beta}-\alpha)\delta+(\bar{\alpha}-\beta)\bar{\delta}+(\rho-\bar{\rho})\Delta]\phi \qquad \text{(D-3d)}$$

If one now substitutes x^μ for the arbitrary function ϕ in these equations, and uses the fact that

$$Dx^\mu=l^\alpha\partial_\alpha x^\mu=l^\mu$$

$$\delta x^\mu=m^\alpha\partial_\alpha x^\mu=m^\mu$$

$$\bar{\delta}x^\mu=\bar{m}^\alpha\partial_\alpha x^\mu=\bar{m}^\mu$$

$$\Delta x^\mu=n^\alpha\partial_\alpha x^\mu=n^\mu$$

[1] F. A. E. Pirani, Introduction to Gravitational Radiation Theory, in *Lectures on General Relativity*, Prentice-Hall, Englewood Cliffs, N.J., 1964, p. 350. (There are two misprints, however, in the Newman–Penrose equations appearing in this reference).

one finally obtains:

$$Dn^{\mu} - \Delta l^{\mu} = -(\gamma + \bar{\gamma})l^{\mu} + (\pi + \bar{\tau})m^{\mu} + (\tau + \bar{\pi})\bar{m}^{\mu} - (\varepsilon + \bar{\varepsilon})n^{\mu} \quad \text{(D-4a)}$$

$$\delta l^{\mu} - Dm^{\mu} = (\bar{\alpha} + \beta - \bar{\pi})l^{\mu} - (\bar{\rho} + \varepsilon - \bar{\varepsilon})m^{\mu} - \sigma\bar{m}^{\mu} + \kappa n^{\mu} \quad \text{(D-4b)}$$

$$\delta n^{\mu} - \Delta m^{\mu} = -\bar{\nu}l^{\mu} + (\mu - \gamma + \bar{\gamma})m^{\mu} + \bar{\lambda}\bar{m}^{\mu} + (\tau - \bar{\alpha} - \beta)n^{\mu} \quad \text{(D-4c)}$$

$$\delta\bar{m}^{\mu} - \bar{\delta}m^{\mu} = (\mu - \bar{\mu})l^{\mu} + (\bar{\beta} - \alpha)m^{\mu} + (\bar{\alpha} - \beta)\bar{m}^{\mu} + (\rho - \bar{\rho})n^{\mu} \quad \text{(D-4d)}$$

Equations (D-4) are identical to the metric equation (9-73) when the latter is written in detail.

APPENDIX

E

TRANSFORMATION PROPERTIES OF THE NEWMAN–PENROSE FIELD VARIABLES

In this Appendix we give the transformation laws of the Newman–Penrose gravitational field variables (see Section 9-3 and Appendix D) under the group $SL(2, C)$. These variables were the tetrad of null vectors, the spin coefficients, and the components of the Riemann tensor.

E-1 GENERAL TRANSFORMATION PROPERTIES

It has been shown in Section 9-3 that under a change of the spin frame $\zeta = S\zeta'$, where S is an element of the group $SL(2, C)$, the matrices B_μ and $F_{\mu\nu}$ transform as follows:

$$B'_\mu = S^{-1}B_\mu S - S^{-1}\,\partial_\mu S \tag{E-1}$$

$$F'_{\mu\nu} = S^{-1}F_{\mu\nu}S \tag{E-2}$$

Under the transformation $\zeta = S\zeta'$, the matrix σ^μ becomes $\sigma^\mu = \zeta\tilde{\sigma}^\mu\zeta^\dagger = S\zeta'\tilde{\sigma}^\mu\zeta'^\dagger S^\dagger = S\sigma'^\mu S^\dagger$, thus $\sigma'^\mu = S^{-1}\sigma^\mu(S^\dagger)^{-1} = S^{-1}\sigma^\mu(S^{-1})^\dagger$ since $(S^\dagger)^{-1} = (S^{-1})^\dagger$. For convenience we introduce the matrix $g = S^{-1}$, whence the transformation law of σ^μ becomes $\sigma'^\mu = g\sigma^\mu g^\dagger$, or

$$\sigma'^\mu{}_{ab'} = g_a{}^c\sigma^\mu{}_{cd'}g^{\dagger d'}{}_{b'} \tag{E-3}$$

in matrix elements notation.

The laws of transformation of $B_{ab'} = \sigma^{\mu}{}_{ab'} B_{\mu}$ and $F_{ab'cd'} = \sigma^{\mu}{}_{ab'} \sigma^{\nu}{}_{cd'} F_{\mu\nu}$ can now be obtained. They are:

$$B'_{cd'} = \sigma'^{\mu}{}_{cd'} B'_{\mu} \tag{E-4}$$

$$F'_{ab'cd'} = \sigma'^{\mu}{}_{ab'} \sigma'^{\nu}{}_{cd'} F'_{\mu\nu} \tag{E-5}$$

where $\sigma'^{\mu}{}_{ab'}$ is given by Eq. (E-3) and B'_{μ} and $F'_{\mu\nu}$ are given by Eqs. (E-1) and (E-2), respectively. In matrix elements notation we obtain:

$$(B'_{cd'})_a{}^b = g_c{}^e g^{\dagger f'}{}_{d'} [g_a{}^p (B_{ef'})_p{}^q (g^{-1})_q{}^b - g_a{}^p (\partial_{ef'} g^{-1})_p{}^b] \tag{E-6}$$

$$(F'_{ab'cd'})_p{}^q = g_a{}^e g^{\dagger f'}{}_{b'} g_c{}^h g^{\dagger n'}{}_{d'} g_p{}^r (F_{ef'hn'})_r{}^s (g^{-1})_s{}^q \tag{E-7}$$

Here the matrices $B_{ab'}$ and $F_{ab'cd'}$ are given explicitly by Eqs. (9-53) and (9-63), respectively.

E-2 TRANSFORMATIONS UNDER ONE-PARAMETER SUBGROUPS

If the general matrix g of the group $SL(2, C)$ is denoted by

$$g_a{}^b = \begin{pmatrix} A & B \\ C & D \end{pmatrix} \tag{E-8}$$

then

$$(g^{-1})_a{}^b = \begin{pmatrix} D & -B \\ -C & A \end{pmatrix} \tag{E-9}$$

and

$$(g^{\dagger})^{a'}{}_{b'} = \begin{pmatrix} \bar{A} & \bar{C} \\ \bar{B} & \bar{D} \end{pmatrix} \tag{E-10}$$

We have seen in Section 8-3 that the matrix g of the group $SL(2, C)$ can be written as a product of three matrices of the form [compare Eq. (8-41)]

$$g_1(z) = \begin{pmatrix} 1 & 0 \\ z & 1 \end{pmatrix}, \quad g_2(z) = \begin{pmatrix} z & 0 \\ 0 & z^{-1} \end{pmatrix}, \quad g_3(z) = \begin{pmatrix} 1 & z \\ 0 & 1 \end{pmatrix} \tag{E-11}$$

where z is a complex variable. The transformations $g_1(z)$ and $g_3(z)$ describe one-(complex) parameter null rotations about l_{μ} and n_{μ}, respectively. The transformation $g_2(z)$ corresponds to an ordinary Lorentz transformation (boost) in the $l_{\mu} - n_{\mu}$ plane along with a spatial rotation in the $m_{\mu} - \bar{m}_{\mu}$ plane. Here we have denoted $\zeta_0{}^A = l^A$ and $\zeta_1{}^A = n^A$, as usual.

We are now in a position to find out the explicit laws of transformations. The transformations will be given for each one of the three one-parameter subgroups described above. Use will be made of the intrinsic derivative operators $D = \partial_{00'}$, $\delta = \partial_{01'}$, $\bar{\delta} = \partial_{10'}$ and $\Delta = \partial_{11'}$, where $\partial_{ab'} = \sigma^{\mu}{}_{ab'} \partial_{\mu}$.

E-3 TRANSFORMATION UNDER NULL ROTATION ABOUT l_μ

Here one has $\zeta' = g_1(z)\zeta$. Hence

$$\begin{aligned} l'_A &= l_A \\ n'_A &= z l_A + n_A \end{aligned} \tag{E-12}$$

and the null tetrad transforms as follows:

$$\begin{aligned} l'_\mu &= l_\mu \\ m'_\mu &= \bar{z} l_\mu + m_\mu \\ n'_\mu &= z\bar{z} l_\mu + z m_\mu + \bar{z}\bar{m}_\mu + n_\mu \end{aligned} \tag{E-13}$$

The spin coefficients transform as follows:[1]

$$\begin{aligned} \rho' &= \rho + z\kappa \\ \alpha' &= \alpha + z(\rho + \varepsilon) + z^2\kappa \\ \lambda' &= \lambda + z(\pi + 2\alpha) + z^2(\rho + 2\varepsilon) + z^3\kappa + \bar{\delta} z + z D z \end{aligned} \tag{E-14a}$$

$$\begin{aligned} \kappa' &= \kappa \\ \varepsilon' &= \varepsilon + z\kappa \\ \pi' &= \pi + 2z\varepsilon + z^2\kappa + Dz \end{aligned} \tag{E-14b}$$

$$\begin{aligned} \sigma' &= \sigma + \bar{z}\kappa \\ \beta' &= \beta + z\sigma + \bar{z}\varepsilon + z\bar{z}\kappa \\ \mu' &= \mu + 2z\beta + \bar{z}\pi + z^2\sigma + 2z\bar{z}\varepsilon + z^2\bar{z}\kappa + \delta z + \bar{z} D z \\ \tau' &= \tau + z\sigma + \bar{z}\rho + z\bar{z}\kappa \end{aligned} \tag{E-14c}$$

$$\begin{aligned} \gamma' &= \gamma + z(\tau + \beta) + \bar{z}\alpha + z^2\sigma + z\bar{z}(\rho + \varepsilon) + z^2\bar{z}\kappa \\ \nu' &= \nu + z(\mu + 2\gamma) + \bar{z}\lambda + z^2(2\beta + \tau) + z\bar{z}(2\alpha + \pi) + z^2\bar{z}(2\varepsilon + \rho) \\ &\quad + z^3\sigma + z^3\bar{z}\kappa + \Delta z + z\,\delta z + \bar{z}\,\bar{\delta} z + z\bar{z} D z \end{aligned} \tag{E-14d}$$

The tetrad components of the Weyl tensor transform as follows:

$$\begin{aligned} \psi'_0 &= \psi_0 \\ \psi'_1 &= z\psi_0 + \psi_1 \\ \psi'_2 &= z^2\psi_0 + 2z\psi_1 + \psi_2 \\ \psi'_3 &= z^3\psi_0 + 3z^2\psi_1 + 3z\psi_2 + \psi_3 \\ \psi'_4 &= z^4\psi_0 + 4z^3\psi_1 + 6z^2\psi_2 + 4z\psi_3 + \psi_4 \end{aligned} \tag{E-15}$$

[1] M. Carmeli and M. Kaye, *Ann. Phys. (N.Y.)* **99**, 188 (1976).

and the components of the trace-free Ricci tensor transform as follows:

$$\begin{aligned}
\phi'_{00} &= \phi_{00} \\
\phi'_{01} &= \bar{z}\phi_{00} + \phi_{01} \\
\phi'_{02} &= \bar{z}^2\phi_{00} + 2\bar{z}\phi_{01} + \phi_{02} \\
\phi'_{11} &= \bar{z}z\phi_{00} + z\phi_{01} + \bar{z}\phi_{10} + \phi_{11} \\
\phi'_{12} &= \bar{z}^2 z\phi_{00} + 2\bar{z}z\phi_{01} + z\phi_{02} + 2\bar{z}\phi_{11} + \bar{z}^2\phi_{10} + \phi_{12} \\
\phi'_{22} &= \bar{z}^2 z^2\phi_{00} + 2\bar{z}z^2\phi_{01} + z^2\phi_{02} + 2\bar{z}^2 z\phi_{10} + 4\bar{z}z\phi_{11} + 2z\phi_{12} \\
&\quad + \bar{z}^2\phi_{20} + 2\bar{z}\phi_{21} + \phi_{22}
\end{aligned} \tag{E-16}$$

Note that these variables satisfy the relation $\phi_{mn} = \bar{\phi}_{nm}$ for $m, n = 0, 1, 2$.

E-4 BOOST IN $l^\mu - n^\mu$ PLANE AND SPATIAL ROTATION IN $m^\mu - \bar{m}^\mu$ PLANE

Here one has $\zeta' = g_2(z)\zeta$. The complex parameter z can be written in the form $z = A^{1/2}\,e^{i\theta/2}$, where A and θ are real. Then A is the boost parameter and θ the spatial rotation parameter. Hence

$$\begin{aligned}
l'_A &= z l_A \\
n'_A &= z^{-1} n_A
\end{aligned} \tag{E.17}$$

and the null tetrad transforms as follows:

$$\begin{aligned}
l'_\mu &= z\bar{z} l_\mu = A l_\mu \\
m'_\mu &= z\bar{z}^{-1} m_\mu = e^{i\theta} m_\mu \\
n'_\mu &= z^{-1}\bar{z}^{-1} n_\mu = A^{-1} n_\mu
\end{aligned} \tag{E-18}$$

The spin coefficients transform as follows:

$$\begin{aligned}
\rho' &= z\bar{z}\rho \\
\alpha' &= z^{-1}\bar{z}(\alpha - z\,\delta z^{-1}) \\
\lambda' &= z^{-3}\bar{z}\lambda
\end{aligned} \tag{E-19a}$$

$$\begin{aligned}
\kappa' &= z^3\bar{z}\kappa \\
\varepsilon' &= z\bar{z}(\varepsilon - zDz^{-1}) \\
\pi' &= z^{-1}\bar{z}\pi
\end{aligned} \tag{E-19b}$$

$$\begin{aligned}
\sigma' &= z^3\bar{z}^{-1}\sigma \\
\beta' &= z\bar{z}^{-1}(\beta - z\,\delta z^{-1}) \\
\mu' &= z^{-1}\bar{z}^{-1}\mu
\end{aligned} \tag{E-19c}$$

$$\tau' = z\bar{z}^{-1}\tau$$
$$\gamma' = z^{-1}\bar{z}^{-1}(\gamma - z\,\Delta z^{-1}) \tag{E-19d}$$
$$\nu' = z^{-3}\bar{z}^{-1}\nu$$

The tetrad components of the Weyl tensor transform as follows:

$$\begin{aligned} \psi'_0 &= z^4\psi_0 \\ \psi'_1 &= z^2\psi_1 \\ \psi'_2 &= \psi_2 \\ \psi'_3 &= z^{-2}\psi_3 \\ \psi'_4 &= z^{-4}\psi_4 \end{aligned} \tag{E-20}$$

and the tetrad components of the trace-free Ricci tensor transform as follows:

$$\begin{aligned} \phi'_{00} &= z^2\bar{z}^2\phi_{00} \\ \phi'_{01} &= z^2\phi_{01} \\ \phi'_{02} &= z^2\bar{z}^{-2}\phi_{02} \\ \phi'_{11} &= \phi_{11} \\ \phi'_{12} &= \bar{z}^{-2}\phi_{12} \\ \phi'_{22} &= \bar{z}^{-2}z^{-2}\phi_{22} \end{aligned} \tag{E-21}$$

Here $\phi_{mn} = \bar{\phi}_{nm}$ for $m, n = 0, 1, 2$.

E-5 TRANSFORMATION UNDER NULL ROTATION ABOUT n_μ

Here one has $\zeta' = g_3(z)\zeta$. Hence

$$\begin{aligned} l'_A &= l_A + zn_A \\ n'_A &= n_A \end{aligned} \tag{E-22}$$

and the null tetrad transforms as follows:

$$\begin{aligned} l'_\mu &= l_\mu + \bar{z}m_\mu + z\bar{m}_\mu + z\bar{z}n_\mu \\ m'_\mu &= m_\mu + zn_\mu \\ n'_\mu &= n_\mu \end{aligned} \tag{E-23}$$

The spin coefficients transform as follows:

$$\rho' = \rho + 2z\alpha + \bar{z}\tau + z^2\lambda + 2z\bar{z}\gamma + z^2\bar{z}\nu - \bar{\delta}z - \bar{z}\,\Delta z$$
$$\alpha' = \alpha + z\lambda + \bar{z}\gamma + z\bar{z}\nu \tag{E-24a}$$
$$\lambda' = \lambda + \bar{z}\nu$$

$$\kappa' = \kappa + z(\rho + 2\varepsilon) + \bar{z}\sigma + z^2(2\alpha + \pi) + z\bar{z}(2\beta + \tau) + z^2\bar{z}(2\gamma + \mu) + z^3\lambda + z^3\bar{z}\nu - Dz - z\,\bar{\delta}z - \bar{z}\,\delta z - z\bar{z}\,\Delta z$$

$$\varepsilon' = \varepsilon + z(\pi + \alpha) + \bar{z}\beta + z^2\lambda + z\bar{z}(\mu + \gamma) + z^2\bar{z}\nu \tag{E-24b}$$

$$\pi' = \pi + z\lambda + \bar{z}\mu + z\bar{z}\nu$$

$$\sigma' = \sigma + z(\tau + 2\beta) + z^2(\mu + 2\gamma) + z^3\nu - \delta z - z\,\Delta z$$

$$\beta' = \beta + z(\mu + \gamma) + z^2\nu \tag{E-24c}$$

$$\mu' = \mu + z\nu$$

$$\tau' = \tau + 2z\gamma + z^2\nu - \Delta z$$

$$\gamma' = \gamma + z\nu \tag{E-24d}$$

$$\nu' = \nu$$

The tetrad components of the Weyl tensor transform as follows:

$$\begin{aligned}\psi'_0 &= \psi_0 + 4z\psi_1 + 6z^2\psi_2 + 4z^3\psi_3 + z^4\psi_4\\ \psi'_1 &= \psi_1 + 3z\psi_2 + 3z^2\psi_3 + z^3\psi_4\\ \psi'_2 &= \psi_2 + 2z\psi_3 + z^2\psi_4\\ \psi'_3 &= \psi_3 + z\psi_4\\ \psi'_4 &= \psi_4\end{aligned} \tag{E-25}$$

and the tetrad components of the trace-free Ricci tensor transform as follows:

$$\begin{aligned}\phi'_{00} &= \phi_{00} + 2\bar{z}\phi_{01} + 2z\phi_{10} + 4\bar{z}z\phi_{11} + \bar{z}^2\phi_{02} + z^2\phi_{20}\\ &\quad + 2\bar{z}z^2\phi_{21} + 2\bar{z}^2z\phi_{12} + \bar{z}^2z^2\phi_{22}\\ \phi'_{10} &= \phi_{10} + 2\bar{z}\phi_{11} + z\phi_{20} + 2\bar{z}z\phi_{21} + \bar{z}^2\phi_{12} + \bar{z}^2z\phi_{22}\\ \phi'_{11} &= \phi_{11} + \bar{z}\phi_{12} + z\phi_{21} + \bar{z}z\phi_{22}\\ \phi'_{20} &= \phi_{20} + 2\bar{z}\phi_{21} + \bar{z}^2\phi_{22}\\ \phi'_{21} &= \phi_{21} + \bar{z}\phi_{22}\\ \phi'_{22} &= \phi_{22}\end{aligned} \tag{E-26}$$

Here $\phi_{mn} = \bar{\phi}_{nm}$ with $m, n = 0, 1, 2$.

In each case the Ricci scalar remains, of course, unchanged, $\Lambda' = \Lambda$.

E-6 TRANSFORMATION UNDER OTHER FACTORIZATION

We conclude this Appendix by finding the transformation laws of the Newman-Penrose gravitational field variables under a different factorization of the group $SL(2, C)$.

In Section 6-2 the group $SL(2, C)$ was factorized in a way different from that given above by Eq. (E-11). The three alternative basic matrices of $SL(2, C)$ were given by Eqs. (6-20):

$$\begin{pmatrix} 1 & 0 \\ z & 1 \end{pmatrix}, \quad \begin{pmatrix} z & 0 \\ 0 & z^{-1} \end{pmatrix}, \quad \begin{pmatrix} 0 & 1 \\ -1 & 0 \end{pmatrix} \tag{E-27}$$

Induced transformation by the three one-parameter subgroups $g_1(z)$, $g_2(z)$ and $g_3(z)$ can then be obtained by transformations under the three matrices (E-27). The first and second matrices of Eq. (E-27) are identical to $g_1(z)$ and $g_2(z)$, respectively, whereas the third one is not identical to $g_3(z)$ (compare Eq. (E-11)). Let us denote the third matrix of Eq. (E-27) by g,

$$g = \begin{pmatrix} 0 & 1 \\ -1 & 0 \end{pmatrix} \tag{E-28}$$

The one-parameter subgroup $g_3(z)$ can then be obtained from $g_1(z)$, $g_2(z)$ and g. In fact one easily finds that

$$g_3(z) = -g\, g_1(-z) g \tag{E-29}$$

This fact can be utilized in order to find out all the transformation laws of the null tetrad, the spin coefficients, and the components of the Riemann tensor obtained in Section E-5 that were induced by a transformation of the spin frame under $g_3(z)$. The desired transformation can be obtained as a product of three transformations of the same field variables under g, $g_1(-z)$ and g. To this end we have to find the transformation laws under the spin frame transformation $\zeta' = g\zeta$, where g is given by Eq. (E-28), using Eqs. (E-6) and (E-7), with

$$g_a{}^b = \begin{pmatrix} 0 & 1 \\ -1 & 0 \end{pmatrix} \tag{E-30a}$$

$$(g^{-1})_a{}^b = \begin{pmatrix} 0 & -1 \\ 1 & 0 \end{pmatrix} \tag{E-30b}$$

$$(g^\dagger)^{a'}{}_{b'} = \begin{pmatrix} 0 & -1 \\ 1 & 0 \end{pmatrix} \tag{E-30c}$$

Since there is no dependence on the parameter z, the calculation is rather simple. The results are:

$$\begin{aligned} l'_A &= n_A \\ n'_A &= -l_A \end{aligned} \tag{E-31}$$

and the null tetrad transforms as

$$\begin{aligned} l'_\mu &= n_\mu \\ m'_\mu &= -\bar{m}_\mu \\ n'_\mu &= l_\mu \end{aligned} \tag{E-32}$$

The spin coefficients transform as follows:

$$\rho' = -\mu \qquad \alpha' = \beta \qquad \lambda' = -\sigma \tag{E-33a}$$

$$\kappa' = \nu \qquad \varepsilon' = -\gamma \qquad \pi' = \tau \tag{E-33b}$$

$$\sigma' = -\lambda \qquad \beta' = \alpha \qquad \mu' = -\rho \tag{E-33c}$$

$$\tau' = \pi \qquad \gamma' = -\varepsilon \qquad \nu' = \kappa \tag{E-33d}$$

The tetrad components of the Weyl tensor transform as follows:

$$\psi'_0 = \psi_4 \qquad \psi'_1 = -\psi_3 \qquad \psi'_2 = \psi_2 \qquad \psi'_3 = -\psi_1 \qquad \psi'_4 = \psi_0 \tag{E-34}$$

and the tetrad components of the trace-free Ricci tensor transform as follows:

$$\phi'_{00} = \phi_{22} \qquad \phi'_{11} = \phi_{11} \qquad \phi'_{22} = \phi_{00} \qquad \phi'_{10} = -\phi_{12} \qquad \phi'_{20} = \phi_{02} \qquad \phi'_{12} = -\phi_{10} \tag{E-35}$$

with $\phi_{mn} = \bar{\phi}_{nm}$ for $m, n = 0, 1, 2$. Finally, one has $\Lambda' = \Lambda$.

As an illustration of the above discussion let us find the transformed value of ψ_0 under $g_3(z)$, or using Eq. (E-29), under successions of three transformations g, $g_1(-z)$ and g since the minus sign does not contribute. One has, using Eqs. (E-34), (E-15) and (E-34), respectively,

$$\begin{aligned}\psi_0''' &= \psi_4'' \\ &= z^4\psi_0' - 4z^3\psi_1' + 6z^2\psi_2' - 4z\psi_3' + \psi_4' \\ &= z^4\psi_4 + 4z^2\psi_3 + 6z^2\psi_2 + 4z\psi_1 + \psi_0\end{aligned} \tag{E-36}$$

a result which is identical to that given by Eq. (E-25).

BIBLIOGRAPHY

This Bibliography includes not merely the papers and books specifically referred to in the text, but many others besides, so that it includes a fairly complete account of the literature relating to the theory of representations of the Lorentz group and its application to the theory of general relativity. The Bibliography does not include all papers published on general relativity. Chapter numbers in this book are given for locations of papers and books referred to in the text, so that the Bibliography can serve as an index of references and names. Cross-references are included to joint authors who are not the first-listed authors, provided they are also authors of other works listed alphabetically in the Bibliography.

Abe, S.: On space–time with a non-symmetric fundamental tensor admitting a six-parameter group of motions, *Tensor*, **18**, 289 (1967).

Abers, E. S. and B. W. Lee: Gauge theories, *Phys. Letters*, **9C**, 1 (1973).

Adler, R. J.: Spinors in a Weyl geometry, *J. Math. Phys.*, **11**, 1185 (1970).

——, M. Bazin and M. Schiffer: *Introduction to General Relativity*, McGraw Hill, New York, 1965.

—— [See also M. M. Schiffer]

—— and C. Sheffield: Classification of space-time in general relativity, *J. Math. Phys.*, **14**, 465 (1973).

Agnese, A. G. and A. Wataghin: On the Schwarzschild exterior space–time, *Lett. al Nuovo Cim.*, **2**, 1041 (1971).

Aichelburg, P. C.: High symmetry fields and the homogeneous field in general relativity, *J. Math. Phys.*, **11**, 1330 (1970).

—— and R. U. Sexl: The gravitational field of a massless particle, *Lett. al Nuovo Cim.*, **4**, 1316 (1970).

Akhiyezer, A. I. and V. B. Berestetskii: *Quantum Electrodynamics*, Moscow, 1953.

—— and I. M. Glazman: *The Theory of Linear Operators in Hilbert Space*, Moscow, 1950.

Alcarás, J. A. and P. L. Ferreira: Algebraic construction of the basis for the irreducible representations of rotation groups and for the homogeneous Lorentz group, *J. Math. Phys.*, **6**, 578 (1965).

Alishauskas, S. I.: On the Clebsch–Gordon coefficients of the $SL(2, C)$ group, *Litov. Fiz. Sb.* (*USSR*), **13**, 829 (1973). [In Russian]

Allcock, G. R.: A geometrized class of Yang–Mills fields, *Nuovo Cim.*, **28,** 858 (1963).
Anderson, J. L.: *Principles of Relativity Physics*, Academic Press, New York, 1967. [Introduction]
—— and R. Gautreau: Possible causal violations at radii greater than the Schwarzschild radius, *Phys. Letters*, **20,** 24 (1966).
—— and R. Gautreau: Operational formulation of the principle of equivalence, *Phys. Rev.*, **185,** 1656 (1969).
Arnowitt, R., S. Deser, and C. W. Misner: Interior Schwarzschild solutions and interpretation of source terms, *Phys. Rev.*, **120,** 321 (1960).
—— and S. I. Fickler: Quantization of the Yang–Mills field, *Phys. Rev.*, **127,** 1821 (1962).
Aronson, B. A.: Twisting null congruences in asymptotically flat space. Thesis, University of Pittsburgh, 1971.
——, R. Lind, J. Messmer, and E. T. Newman: A note on asymptotically flat spaces, *J. Math. Phys.*, **12,** 2462 (1971).
—— and E. T. Newman: Coordinate systems associated with asymptotically shear-free null congruences, *J. Math. Phys.*, **13,** 1847 (1972).
Arzeliès, H.: *General Relativity and Gravitation*, Vols. I and II, (*Trav. Inst. Sci. Cherifien, Ser. Sci. Phys.* (*Morocco*), No. 7, 1961, and No. 8, 1963. [In French])
Ascher, E. and A. Janner: Space–time symmetry of transverse electromagnetic plane waves, *Helv. Phys. Acta*, **43,** 296 (1970).
Audretsch, J. and W. Graf: Neutrino radiation in gravitational fields, *Commun. Math. Phys.*, **19,** 315 (1970).
Avez, A.: The Schwarzschild ds^2 amongst stationary ds^2, *Ann. Inst. Poincaré*, (A)**1,** 291 (1964).
Bade, W. L. and J. Jehle: *Rev. Mod. Phys.*, **25,** 714 (1953). [Chapter 8]
Banerjee, A.: Null electromagnetic fields in general relativity, *J. Phys.*, (A)**3,** 501 (1970).
—— Null electromagnetic fields in general relativity admitting timelike or null killing vectors, *J. Math. Phys.*, **11,** 51 (1970).
—— [See also U.K. De]
Bairerlain, R.: Testing general relativity with laser ranging to the Moon, *Phys. Rev.*, **162,** 1275 (1967).
Bardeen, J. M.: Kerr Metric black holes, *Nature, Lond.*, **226,** 64 (1970).
—— and W. H. Press: Radiation fields in the Schwarzschild background, *J. Math. Phys.*, **14,** 7 (1973).
—— [See also W. H. Press]
Bargmann, V.: Irreducible unitary representations of the Lorentz group, *Ann. Math.*, **48,** 568 (1947). [Chapter 5]
Barker, B. M. and R. F. O'Connell: Derivation of the equations of motion of a gyroscope from the quantum theory of gravitation, *Phys. Rev.*, (D)**2,** 1428 (1970).
Barnes, A.: On Birkhoff's theorem in general relativity, *Commun. Math. Phys.*, **33,** 75 (1973).
Bars, I. and F. Gürsey: Operator treatment of the Gelfand–Naimark basis for $SL(2, C)$, *J. Math. Phys.*, **13,** 131 (1972). [Chapter 5]
—— Projective representations of $SL(3, C)$ in the Z-operator formalism, *J. Math. Phys.*, **14,** 759 (1973).
Barut, A. O. and M. Leiser: Note on gauge transformations in quantum mechanics, *Am. J. Phys.*, **29,** 24 (1961).
Basu, D. and S. D. Majumdar: Complex angular momenta and the Lorentz group, *J. Phys.*, (A)**6,** 1097 (1973).
Bazanski, S. L.: Decomposition of the Lorentz transformation matrix into skew-symmetric tensors, *J. Math. Phys.*, **6,** 1201 (1965).
Beck, G. and H. Weyl: *Math. Z.*, **13,** 142 (1921). [Chapter 7]
Beers, B. L. and R. S. Millman: Analytic vector harmonic expansions on $SU(2)$ and S^2, *J. Math. Phys.*, **16,** 11 (1975).
Beiglböck, W.: Theory of the infinitesimal holonomy group in general relativity theory, *Z. Phys.*, **179,** 148 (1964).
Bel, L.: Sur la radiation gravitationnelle, *C.R. Acad. Sci., Paris*, **247,** 1094 (1958). [Chapter 8]
—— Introduction d'un tenseur du quatrieme order, *C. R. Acad. Sci., Paris*, **248,** 1297 (1959). [Chapter 8]
—— Quelques remarques sur la classification de Petrov, *C.R. Acad. Sci., Paris*, **248,** 2561 (1959).
—— Schwarzschild singularity, *J. Math. Phys.*, **10,** 1501 (1969).

—— Les sources des solutions statiques de Schwarzschild et Curzon, *Gen. Relativ. Gravitn.*, **1,** 337 (1971).

Belinfant, F. J.: Kruskal space without wormholes, *Phys. Letters,* **20,** 251 (1966).

Belinskii, V. A.: General solution of the gravitational equations with a physical singularity, *Zh. Eksp. Teoret. Fiz.*, **57,** 2163 (1969).

—— and I. M. Khalatnikov: A general solution of the gravitational equations with a simultaneous fictious singularity, *Zh. Eksp. Teoret. Fiz.*, **49,** 1000 (1965); *Soviet Physics JETP*, **22,** March 1966.

—— and I. M. Khalatnikov: On the nature of the singularities in the general solutions of the gravitational equations, *Zh. Eksp. Teoret. Fiz.*, **56,** 1700 (1969). [In Russian]

—— and I. M. Khalatnikov: General solution of the gravitational equations with a physical singularity, *Zh. Eksp. Teoret. Fiz.*, **57,** 2163 (1969). [In Russian]

—— and I. M. Khalatnikov: On a generalized solution of the gravitational equations with a physical singularity of an oscillatory nature, *Zh. Eksp. Teoret. Fiz.*, **59,** 3141 (1970).

Bera, K.: Note on an interesting metric of the field equations in general relativity, *J. Phys.*, (A)**2,** 138 (1969).

Berezin, F. A.: Laplace operators on semi-simple Lie groups, *Dokl. Akad. Nauk SSSR*, **107,** 9 (1956); *Trud. Mosk. Mat.*, Ob-va **6,** 371 (1957).

—— and I. M. Gelfand: Some remarks on the theory of functions on symmetric Riemannian manifolds, *Trud Mosk. Mat.*, Ob-va **5,** (1956).

Bergmann, O. and R. Leipnik: Space–time structure of a spherical symmetrical field, *Phys. Rev.*, **107,** 1157 (1957).

Bergmann, P. G.: Gauge-invariant variables in general relativity, *Phys. Rev.*, **124,** 274 (1961).

—— Foundation problems in general relativity, in: *Problems in the Foundations of Physics* (M. Bunge, Ed.), Springer-Verlag, Berlin, 1971.

—— M. Cahen, and A. B. Komar: Spherically symmetric gravitational fields, *J. Math. Phys.*, **6,** 1 (1965).

——, I. Robinson, and E. Schuking: Asymptotic properties of a system with nonzero total mass, *Phys. Rev.*, **126,** 1227 (1962).

Bertotti, B., D. Brill, and R. Krotkov: Experiments on Gravitation, in: *Gravitation: An Introduction to Current Research* (L. Witten, Ed.), John Wiley, New York, 1962. [Chapter 7]

—— and J. Plebanski: Theory of gravitational perturbations in the fast motion approximation, *Ann. Phys. (N.Y.)*, **11,** 169 (1960).

Bhabha, N. J.: Relativistic wave equations for the elementary particles, *Rev. Mod. Phys.*, **17,** 200 (1945).

Bialas, A.: On the continuity of the Petrov classification, *Acta Phys. Polon.*, **23,** 699 (1963).

—— [See also Bialas, E.]

Bialas, E. and A. Bialas: On the relation between the Penrose classification of gravitational fields and the classification according to group of motions, *Acta Phys. Polon.*, **24,** 515 (1963).

Bialynicki-Birula, I.: Gauge invariant variables in the Yang–Mills theory, *Bull. Acad. Polon. Sci., Ser. Math. Astron. Phys.*, **11,** 935 (1963).

Birkhoff, G.: *Relativity and Modern Physics*, Harvard University Press, 1923. [Chapter 7]

Blatt, J. M. and V. F. Weisskopf: *Theoretical Nuclear Physics*, John Wiley, New York, 1952. [Chapter 10]

Boerner, H.: *Representations of Groups*, 2nd edition, North-Holland, Amsterdam, 1970.

Bondi, H.: The space traveller's youth, *Discovery*, p. 505, December 1957.

—— Negative mass in general relativity, *Rev. Mod. Phys.*, **29,** 3 and 423 (1957). [Chapter 7]

——, M. G. J. van der Burg, and A. W. K. Metzner: Gravitational waves in general relativity, VII. Waves from axi-symmetric isolated systems. *Proc. R. Soc. Lond.*, (A)**269,** 21 (1962). [Chapter 12]

Bonnor, W. B.: Certain exact solutions of the equations of general relativity with an electrostatic field, *Proc. Phys. Soc. Lond.*, (A)**66,** 145 (1953).

—— Exact solutions of the Einstein–Maxwell equations, *Z. Phys.*, **161,** 439 (1961).

—— On Birkhoff's theorem, in *Recent Developments in General Relativity*, PWN, Warsaw, 1962.

—— Gravitational waves, *Brit. J. Appl. Phys.*, **14,** 555 (1963).

—— An exact solution of the Einstein–Maxwell equations referring to a magnetic dipole, *Z. Phys.*, **190,** 444 (1966).

—— Null curves in a Minkowski space–time, *Tensor*, **20,** 229 (1969).

—— A new interpretation of the NUT metric in general relativity, *Proc. Camb. Phil. Soc.*, **66,** 145 (1969).
—— On a Robinson–Trautman solution of Einstein's equations, *Phys. Letters*, **31A,** 269 (1970).
—— Charge moving with the speed of light in Einstein–Maxwell theory, *Int. J. Theoret. Phys.*, **3,** 57 (1970).
—— and M. C. Faulkes: An exact solution for oscillating spheres in general relativity, *Mon. Not. R. Astr. Soc.*, **137,** 2391 (1967).
—— and N. S. Swaminarayan: An exact solution for uniformly accelerated particles in general relativity, *Z. Phys.*, **177,** 240 (1964).
—— and N. S. Swaminarayan: An exact solution of Einstein's equations, *Z. Phys.*, **186,** 222 (1965).
—— and P. C. Vaidya: Spherically symmetric radiation of charge in Einstein–Maxwell theory, *Gen. Relativ. Gravitn.*, **1,** 127 (1970).
Bonometto, S.: On gauge invariance for a neutral massive vector field, *Nuovo Cim.*, **28,** 309 (1963).
Borzeskowski, H. H. V. and U. Kasper: Doppler effect and Einstein shift in Schwarzschild space, *Ann. Physik*, **20,** 187 (1967).
Bose, S. K.: Remarks on the Yang–Mills gauge transformation, *Lett. al Nuovo Cim.*, **4,** 985 (1970).
Boulware, D. G. and S. Deser: External sources in gauge theories, *Nuovo Cim.*, **30,** 1009 (1963).
Boyer, R. H. and R. W. Linquist: Maximal analytic extension of the Kerr metric, *J. Math. Phys.*, **8,** 265 (1967).
—— and T. G. Price: An interpretation of the Kerr metric in general relativity, *Proc. Camb. Phil. Soc.*, **61,** 531 (1965).
Braginsky, V. B. and V. I. Panov: The equivalence of inertial and passive gravitational mass, *Gen. Relativ. Gravitn.*, **3,** 403 (1972).
Brans, C. H.: A unique and invariant representation of all local analytic solutions to the vacuum Einstein equations: Petrov type III, *Lett. al Nuovo Cim.*, **1,** 699 (1969).
—— Invariant representation of all analytic Petrov type III solutions of the Einstein equations, *J. Math. Phys.*, **11,** 1210 (1970).
—— Complex 2-form representation of the Einstein equation: the Petrov type III solutions, *J. Math. Phys.*, **12,** 1616 (1971).
—— and R. H. Dicke: Mach's principle and a relativistic theory of gravitation, *Phys. Rev.*, **124,** 925 (1961).
Brehme, R. W.: Geometric representations of the Lorentz transformations, *Am. J. Phys.*, **32,** 233 (1964).
—— Geometrization of the relativistic velocity addition formula, *Am. J. Phys.*, **37,** 360 (1969).
Breit, G., E. Condon, and R. Present: Theory of scattering of protons by protons, *Phys. Rev.*, **50,** 825 (1936). [Chapter 9]
Brill, D. R., P. L. Chrzanowshi, C. M. Pereira, E. D. Fackerell, and J. R. Ipser: Solution of the scalar wave equation in a Kerr background by separation of variables, *Phys. Rev.*, (D)**5,** 1913 (1972).
—— [See also B. Bertotti]
—— [See also J. M. Cohen]
Brumberg, V. A.: On the interpretation of coordinates in Schwarzschild's problem, *Astron. Zh.*, **45,** 1110 (1968).
Buchdahl, H. A.: A special class of solutions of the equations of the gravitational field arising from certain gauge-invariant action principles. *Proc. Natl. Acad. Sci. Wash.*, **34,** 66 (1948).
—— Reciprocal static solutions of the equation of gravitational field, *Austral. J. Phys.*, **9,** (1956); *Quart. J. Math. (Oxford)*, **5,** 116 (1954).
—— Reciprocal static solutions of field equations involving an asymmetric fundamental tensor, *Nuovo Cim.*, **5,** 1083 (1957).
—— Gauge invariant generalization of the field theories with asymmetric fundamental tensor, *Quart. J. Math. (Oxford)*, **8,** 89 (1957), and **9,** 257 (1958).
—— Reciprocal static metrics and scalar fields in the general theory of relativity, *Phys. Rev.*, **115,** 1325 (1959).
—— On the non-existence of a class of static Einstein spaces asymptotic at infinity to a space of constant curvature, *J. Math. Phys.*, **1,** 537 (1960).
Buoncistiani, A. M.: An algebra of the Yang–Mills field, *J. Math. Phys.*, **14,** 849 (1973).
Burcev, P.: On the theory of gravitational radiation, *Czech. J. Phys.*, **11,** 385 (1961). [In Russian]

Burghardt, R.: Yang–Mills fields in gravitation theory, III. Group-theoretical foundations of gravitation theory, *Acta Phys. Aust.*, **39,** 127 (1974). [In German]

Burke, W. L. and K. S. Thorne: Gravitational Radiation Damping, in *Relativity* (M. Carmeli, S. I. Fickler, and L. Witten, Editors), Plenum Press, New York, 1970. [Chapter 7]

Cahen, M.: On a class of homogeneous spaces in general relativity, *Bull. Acad. Roy. Belg., Cl. Sci.*, **50,** 972 (1964).

—— [See also P. G. Bergmann]

—— and R. Debever: On Birkhoff's theorem, *C.R. Acad. Sci., Paris,* **260,** 815 (1965).

—— and J. Leroy: Espace–temps du vide admettant une congruence caracteristique simple integrable, *Bull. Acad. Roy. Belg., Cl. Sci.*, **54,** 232 (1968).

—— and J. Sengier-Diels: Espaces de Classe D, admettant un champ electromagnetiqe, *Bull. Acad. Roy. Belg., Cl. Sci.*, **53,** 801 (1967).

—— and J. Spelkens: Class III spaces, solutions of Maxwell–Einstein equations. *Bull. Acad. Roy. Belg., Cl. Sci.*, **53,** 817 (1967).

Cahill, M. E. and G. C. McVittie: Spherical symmetry and mass–energy in general relativity, I and II, *J. Math. Phys.*, **11,** 1382 and 1392 (1970).

Callaway, J.: Gravitational field of a point charge, *Amer. J. Phys.*, **27,** 469 (1959).

Campbell, W. B.: Tensor and spinor spherical harmonics and the spin-*s* harmonics ${}_sY_{lm}(\theta, \phi)$, *J. Math. Phys.*, **12,** 1763 (1971).

Cantoni, V.: A class of representations of the generalized Bondi–Metzner–group, *J. Math. Phys.*, **7,** 1361 (1966).

—— Reduction of some representations of the generalized Bondi–Metzner–group, *J. Math. Phys.*, **8,** 1700 (1967).

Cap, F., W. Majerotto, W. Raab, and P. Unteregger: Spin calculus in Riemannian manifolds, *Fortschr. Phys.*, **14,** 205, (1966).

Carmeli, M.: The EIH equations of motion up to the 9th order, *Phys. Letters*, **9,** 132 (1964).

—— Has the geodesic postulate any significance for a finite mass?, *Phys. Letters*, **11,** 24 (1964). [Chapter 7]

—— Motion of a charge in gravitational field, *Phys. Rev.*, **138,** B1003 (1965).

—— The equations of motion of slowly moving particles in general theory of relativity, *Nuovo Cim.*, **37,** 842 (1965). [Chapter 7]

—— Equations of motion without infinite self-action terms in general relativity, *Phys. Rev.*, **140,** B1441 (1965). [Chapter 7]

—— Semi-generally covariant equations of motion I and II, *Ann. Phys. (N.Y.)*, **34,** 465 (1965), and **35,** 250 (1965).

—— Representations of the three-dimensional rotation group in terms of direction and angle of rotation, *J. Math. Phys.*, **9,** 1987 (1968). [Chapter 1]

—— Group theoretic approach to the new conserved quantities in general relativity, *J. Math. Phys.*, **10,** 569 (1969). [Chapter 12]

—— Group analysis of Maxwell's equations, *J. Math. Phys.*, **10,** 1699 (1969). [Chapter 12]

—— Hilbert space description of the gravitational field, *Phys. Letters*, **28A,** 683 (1969).

—— Symmetry of the gravitational field, *Lett. al Nuovo Cim.*, **4,** 40 (1970). [Chapter 9]

—— *SL*(2, *C*) symmetry of the gravitational field dynamical variables, *J. Math. Phys.*, **11,** 2728 (1970). [Chapter 9]

—— Constraints in electrodynamics and quantization of the electromagnetic field without charges, *Nuovo Cim.*, **67B,** 103 (1970). [Chapter 12]

—— Infinite-dimensional representations of the Lorentz group, *J. Math. Phys.*, **11,** 1917 (1970). [Chapter 4]

—— Applications of the group *SU*(2) technique in general relativity, in *Relativity and Gravitation* (C. Kuper and A. Peres, Editors), Gordon & Breach, London, 1971.

—— *SL*(2, *C*) invariance of the gravitational field, *Gen. Relativ. Gravitn.*, **3,** 317 (1972).

—— *SL*(2, *C*) symmetry of the gravitational field, *Nuovo Cim.*, **A7,** 9 (1972). [Chapter 9]

—— *SL*(2, *C*) invariance and the gravitational field, *Ann. Phys. (N.Y.)*, **71,** 603, 1972. [Chapter 9]

—— Behavior of fast particles in the Schwarzschild field, *Lett. al Nuovo Cim.*, **3,** 379 (1972). [Chapter 7]

—— Gauge fields and gravitational field equations, *Nucl. Physics*, (B)**38**, 621 (1972). [Chapter 9]
—— *SL*(2, *C*) symmetry of the gravitational field, in *Group Theory in Non-Linear Problems* (A. O. Barut, Ed.), D. Reidel, Dordrecht, Netherlands, 1974, p. 59. [Introduction]
—— General relativity and gauge theories, *Gen. Relativ. Gravitn.*, **5,** 287 (1974).
—— Modified gravitational Lagrangian, *Phys. Rev.*, (D)**14,** 1727 (1976). [Chapter 9]
—— Integral formalism of gauge fields and general relativity, *Phys. Rev. Lett.* **36,** 59 (1976). [Chapter 9]
—— and S. I. Fickler: Gravitational Lagrangian, *Phys. Rev.*, (D)**5,** 290 (1972). [Chapter 9]
—— S. I. Fickler and L. Witten, Ed.: *Relativity*, Plenum Press, New York 1970.
—— and M. Kaye: Metric of a radiating rotating sphere, *Phys. Letters*, **53A,** 14, 1975, (E)**55A,** 495 (1976). [Chapter 11]
—— and M. Kaye: Transformation laws of the Newman–Penrose variables, *Ann. Phys.* (N.Y.) **99,** 188 (1976). [Appendix E]
—— and S. Malin: Infinite-dimensional representations of the Lorentz group: complementary series of representations, *J. Math. Phys.* **12,** 225 (1971). [Chapters 3, 5]
—— and S. Malin: Finite- and infinite-dimensional representations of the Lorentz group, *Fortsrh. Physik*, **21,** 397 (1973). [Chapters 2, 3, 6]
—— and S. Malin: Infinite-dimensional representations of the Lorentz group: the complete series, *Int. J. Theoret. Phys.*, **9,** 145 (1974). [Chapter 6]
—— and S. Malin: *Representations of the Rotation and Lorentz Groups: An Introduction*, Marcel Dekker, Inc., 1976.

Cartan, E.: *The Theory of Spinors*, The M.I.T. Press, Cambridge, MA, 1966.

Carter, B.: Complete analytic extension of the symmetry axis of Kerr's solutions of Einstein's equations, *Phys. Rev.*, **141,** 1242 (1966).
—— The complete analytic extension of the Reissner–Nordstrom metric in the special case $e^2 = m^2$, *Phys. Letters*, **21,** 423 (1966).
—— Global structure of the Kerr family of gravitational fields, *Phys. Rev.*, **174,** 1559 (1968).
—— Hamilton–Jacobi and Schrödinger separable solutions of Einstein's equations, *Commun. Math. Phys.*, **10,** 280 (1968).
—— A new family of Einstein spaces, *Phys. Letters*, **26A,** 399 (1968).
—— Killing horizons and orthogonally transitive groups in space-time, *J. Math. Phys.*, **10,** 70 (1969).
—— The commutation property of a stationary axisymmetric system, *Commun. Math. Phys.*, **17,** 233, (1970).
—— Casual structure in space–time, *Gen. Relativ. Gravitn.*, **1,** 349 (1971).

Case, L. A.: Changes of the Petrov type of a space–time, *Phys. Rev.*, **176,** 1554 (1963).

Cassen, B. and E. Condon: On nuclear forces, *Phys. Rev.*, **50,** 846 (1936). [Chapter 9]

Chang, S.-J. and L. O'Raifeartaigh: Unitary representations of *SL*(2, *C*) in an *E*(2) basis, *J. Math. Phys.*, **10,** 21 (1969).

Chao, K. L. and M. Kohler: Vierbein formulation of the gravitational field, *Z. Naturforsch.*, **20a,** 753 (1965).

Chase, J. E.: Event horizons in static scalar-vacuum space-times, *Commun. Math. Phys.*, **19,** 276 (1970).
—— [See also V. de la Cruz]

Chau, W.: Gravitational radiation from neutron star, *Astrophys. J.*, **147,** 664 (1967).

Chen, S. F.: Yang–Mills field and its applications, thesis, New York State University, 1971.

Chevalier, M.: Goldberg–Sachs theorem and spaces with conformal connection, *C.R. Acad. Sci.*, *Paris*, (A)**261,** 2446 (1965).

Choquet-Bruhat, Y.: Global theorem of uniqueness for the solutions of Einstein's equations, *C.R. Acad. Sci., Paris*, (A)**226,** 182 (1968). [Chapter 7]
—— Construction de solutions radiatives approchees des equations d'Einstein, *Commun. Math. Phys.*, **12,** 16 (1969).
—— The bearings of global hyperbolicity on existence and uniqueness theorems in general relativity, *Gen. Relativ. Gravitn*, **2,** 1 (1971). [Chapter 7]

Christodoulou, D.: Reversible and irreversible transformations in black-hole physics, *Phys. Rev. Letters*, **25,** 1596 (1970).

Chylinski, Z.: The nature of the Lorentz-invariants, *Acta Phys. Polon.*, **30,** 293 (1966).

Coddington, E. and V. Levinson: *Theory of Ordinary Differential Equations*, McGraw Hill, New York, 1955, p. 103. [Chapter 10]

Cohen, J. M.: Note on the Kerr metric and rotating masses, *J. Math. Phys.*, **8,** 1477 (1967).
—— and D. R. Brill: Angular momentum and the Kerr metric, *J. Math. Phys.*, **9,** 905 (1968).
—— and R. M. Wald: Point charge in the vicinity of a Schwarzschild black hole, *J. Math. Phys.*, **12,** 1845 (1971).
Cohn, J.: On a physical interpretation of the Schwarzschild metric, *Int. J. Theoret. Phys.*, **2,** 267 (1969).
—— Additional remarks on the principle of equivalence, electrodynamics and general relativity, *Int. J. Theoret. Phys.*, **2,** 407 (1969).
Collinson, C. D.: Symmetry properties of Harrison space-times, *Proc. Camb. Phil. Soc.*, **60,** 259 (1964).
—— Empty space–times algebraically special on a given world line as a hypersurface, *J. Math. Phys.*, **8,** 1547 (1967).
—— Symmetries of type N empty space-time possessing twisting geodesic rays, *J. Phys.* (A)**2,** 621 (1969).
—— Conservation laws in general relativity based upon the existence of preferred collineations, *Gen. Relativ. Gravitn.*, **1,** 137 (1970).
—— and R. K. Dodd: Symmetries of stationary axisymmetric empty space–times, *Nuovo Cim.*, **B3,** 281 (1971).
—— and P. B. Morris: Space–times admitting nonunique neutrino fields, *Nuovo Cim.*, **16B,** 273 (1973).
Costa de Beauregard, O.: The principle of equivalence, negative mass and gravitational repulsion, *C.R. Acad. Sci., Paris*, **252,** 1737 (1961).
Couch, W. E. and W. H. Hallidy: Radiation scattering in Einstein–Maxwell theory, *J. Math. Phys.*, **12,** 2170 (1971).
——, W. M. Kinnersley, and R. J. Torrence: Nonlinear radiative interactions in general relativity, *Phys. Letters*, **31A,** 576 (1970).
—— and E. T. Newman: Algebraically special perturbations of the Schwarzschild metric, *J. Math. Phys.*, **14,** 285 (1973).
—— [See also E. T. Newman]
—— and R. J. Torrence: Self-interaction of gravitational radiation, *J. Math. Phys.*, **11,** 2096 (1970).
—— and R. J. Torrence: Asymptotic behavior of vacuum space-times, *J. Math. Phys.*, **13,** 69 (1972).
—— R. J. Torrence, A. I. Janis, and E. T. Newman: Tail of a gravitational wave, *J. Math. Phys.*, **9,** 484 (1968).
Courant, R. and D. Hilbert: *Methods of Mathematical Physics*, Interscience, New York, 1953. [Chapter 12]
Cranshaw, T. E.: The Mössbauer effect and experiments in relativity, in: *Evidence for Gravitational Theories*, Academic Press, New York, 1962.
——, S. P. Schiffer, and A. B. Whitehead: Measurement of the gravitational red shift using the Mössbauer effect in Fe^{57}, *Phys. Rev. Letters*, **4,** 163 (1960). [Chapter 7]
—— and S. P. Schiffer: Measurement of the gravitational red shift with the Mössbauer effect, *Proc. Phys. Soc., Lond.*, **84,** 245 (1964).
Crumeyrolle, A.: On the representations of finite degree of the special unitary group and the special linear group, *C.R. Acad. Sci., Paris*, **260,** 1871 (1965).
—— Spinor structure, *Ann. Inst. Poincaré*, (A)**11,** (1969). [In French]
Das, A.: Static gravitational fields, II. Ricci rotation coefficients, *J. Math. Phys.*, **14,** 1099 (1973).
Dass, T.: Gauge groups and Noether's theorem, *Nuovo Cim.*, **A42,** 730 (1966).
Datta, B. K.: Static electromagnetic fields in general relativity, *Ann. Phys.* (*N.Y.*), **12,** 295 (1961), and **15,** 304 (1961).
—— Homogeneous nonstatic electromagnetic fields in general relativity, *Nuovo Cim.*, **36,** 109 (1965).
—— Nonstatic electromagnetic fields in general relativity, *Nuovo Cim.*, **47,** 568 (1967).
—— Stationary electromagnetic fields in general relativity, *J. Math. Phys.*, **9,** 1715 (1968).
—— Nonstatic spherically symmetric clusters of particles in general relativity, I, *Gen. Relativ. Gravitn.*, **1,** 19 (1970).
—— Spinor fields in general relativity, I: Noether theorem and the conservation laws in Riemann–Cartan space, *Nuovo Cim.*, **B6,** 1 (1971).
—— Spinor fields in general relativity, II: Generalized field equations and applications to the Dirac field, *Nuovo Cim.*, **B6,** 16 (1971).

—— and A. K. Raychaudhuri: Stationary electromagnetic fields in general relativity, *J. Math. Phys.*, **9,** 1715 (1968).

Dautcourt, G., A. Papapetrou, and H. Treder: One-dimensional gravitational fields, *Ann. Physik* **9,** 330 (1962). [In German]

Davidson, W. and S. F. Singer: Use of an artificial satellite to test the clock "paradox" and general relativity, *Nature, Lond.*, **188,** 1013 (1960).

Davis, M. and R. Ruffini: Gravitational radiation in the presence of a Schwarzschild black hole: a boundary value search, *Lett. al Nuovo Cim.*, **12,** 1165 (1971).

Davis, T. M. and J. R. Ray: Neutrinos in cylindrically-symmetric space-times, *J. Math. Phys.*, **16,** 80 (1975).

Davis, W. R.: *Classical Fields, Particles and the Theory of Relativity*, Gordon and Breach, London, 1970. [Introduction]

De, U. K. and A. Banerjee: Two classes of solutions of null electromagnetic radiation, *Progr. Theoret. Phys.*, **47,** 1204 (1972).

Debever, R.: Space–times of Petrov type III, *C.R. Acad. Sci., Paris*, **251,** 1352 (1960).

—— [See also M. Cahen]

Debney, G. C.: On vacuum space-times admitting a null Killing bivectors, *J. Math. Phys.*, **12,** 2372 (1971).

—— Null Killing vectors in general relativity, *Lett. al Nuovo Cim.*, **5,** 954 (1972).

—— R. P. Kerr, and A. Schild: Solutions of the Einstein and Einstein–Maxwell equations, *J. Math. Phys.*, **10,** 1842 (1969).

—— [See also R. P. Kerr]

Defrise, L.: Empty space type D matrices containing an isotropy group *G*, acting on space like orbits, *Bull. Acad. Roy. Belg., Cl. Sci.*, **53,** 827 (1967).

de la Cruz, V., J. E. Chase, and W. Israel: Gravitational collapse with asymetries, *Phys. Rev. Letters*, **24,** 423 (1970).

Delbourgo, G., K. Koller, and P. Mahanta: On transformations between basis vectors of unitary *SL*(2, *C*) representations, *Nuovo Cim.*, **52A,** 1254 (1967).

Demianski, M.: New Kerr-like space–time, *Phys. Letters*, **42A,** 157 (1972).

—— and E. T. Newmann: A combined Kerr–NUT solution of the Einstein field equations, *Bull. Acad. Polon Sci., Ser. Math. Astr. Phys.*, **14,** 653 (1966). [Chapter 11]

Demmie, P. N. and A. I. Janis: The characteristic development of trapped surfaces, *J. Math. Phys.*, **14,** 793 (1973).

Derry, L., R. Isaacson, and J. Winicour: Shear-free gravitational radiation, *Phys. Rev.*, **185,** 1647 (1969).

Deser, S.: Self-interaction and gauge invariance, *Gen. Relativ. Gravitn.*, **1,** 9 (1970).

—— [See also R. Arnowitt]

—— [See also D. G. Boulware]

—— and J. Higbie: Essential singularities in general relativity, *Phys. Rev. Letters*, **23,** 1184 (1969).

DeWitt, B. S.: Theory of radiative corrections for non-Abelian gauge fields, *Phys. Rev. Letters*, **12,** 742 (1964).

—— *Phys. Rev.*, **162,** 1195 and 1230 (1967).

DeWitt, C. M. and J. L. Ging: Gravitational radiation damping, *C. R. Acad. Sci., Paris*, **251,** 1868 (1960).

Dicke, R. H.: New research on old gravitation: Are the observed physical constants independent of the position, epoch and velocity of the laboratory?, *Science, N.Y.*, **129,** 621 (1959). [Chapter 7]

—— *Relativity, Groups and Topology* (DeWitt *et al.*, Editors), Gordon and Breach, 1964. [Chapter 7]

—— The Sun's rotation and relativity, *Nature, Lond.*, **202,** 432 (1964).

—— *Theoretical Significance of Experimental Relativity*, Gordon and Breach, New York, 1964.

—— The weak and strong principles of equivalence, *Ann. Phys.* (*N.Y.*), **31,** 235 (1965).

—— [See also C. H. Brans]

Dimmock, J. O.: Representation theory for nonunitary groups, *J. Math. Phys.*, **4,** 1307 (1963).

Dirac, P. A. M.: Unitary representations of the Lorentz group, *Proc. Roy. Soc.* (*Lond.*), (A)**183,** 284 (1945).

Dixon, W. G.: Analysis of the Newman–Unti integration procedure for asymptotically flat space–times, *J. Math. Phys.*, **11,** 1238 (1970).

Dolan, P.: A singularity-free solution of the Maxwell–Einstein equations, *Commun. Math. Phys.*, **9,** 161 (1968).
Domokos, G., S. Kovesi-Domokos, and E. Schonberg: Fully Reggeized scattering amplitudes, *Phys. Rev.*, (D)**2,** 1026 (1970). [Chapter 5]
Doplicher, S. and J. E. Roberts: Fields, statistics and non-abelian gauge groups, *Commun. Math. Phys.*, **28,** 331 (1972).
Dowker, J. S.: Non-local Yang–Mills theory, *J. Phys.*, (A)**3,** 59 (1970).
—— The NUT solution as a gravitational dyon, *Gen. Relativ. Gravitn.*, **5,** 603 (1974).
—— and M. Goldstone: The geometry and algebra of the representations of the Lorentz group, *Proc. R. Soc. Lond.*, (A)**303,** 381 (1968).
Dozmorov, I. M.: Solutions of the Robinson–Trautman type, *Izv. Viss. Ucheb. Zaved. Fiz.*, **9,** 43 (1969). [In Russian]
—— Gravitational waves in general relativity: a wave solution of type III, *Izv. Viss. Ucheb. Zaved. Fiz.*, **11,** 121 (1969). [In Russian]
Duc, D. V. and N. V. Hieu: On the theory of unitary representations of the *SL*(2, *C*) group, *Ann. Inst. Poincaré*, (A)**6,** 17 (1967).
Duff, M. J.: Covariant gauges and point sources in general relativity, *Ann. Phys.* (*N.Y.*), **79,** 261 (1973).
Duffin, R. J.: On the characteristic matrices of covariant systems, *Phys. Rev.*, **54,** 1114 (1938).
Duncombe, R. L.: Relativity effects for the three inner planets, *Astron. J.*, **61,** 174 (1956). [Chapter 7]
Dürr, H. P.: Isospin and parity in nonlinear spinor theory, *Z. Naturforsch.*, **16a,** 327 (1961).
—— Poincaré gauge-invariant spinor theories and the gravitational field, *Nuovo Cim.*, **A4,** 187 (1971).
Eardley, D. and R. K. Sachs: Space–times with a future projective infinity, *J. Math. Phys.*, **14,** 209 (1973).
Earman, J.: Are spatial and temporal congruence conventional?, *Gen. Relativ. Gravitn.*, **1,** 143 (1970).
Edelstein, L. A. and C. V. Vishveshwara: Differential equations for perturbations on the Schwarzschild metric, *Phys. Rev.*, (D)**1,** 3514 (1970).
Efinger, H. J.: Space–time of electrostatic field in general relativity, *Acta Phys. Austr.*, **19,** 361 (1965).
Ehlers, J.: Exakte losing der Einstein–Maxwell'schen feldgleichungen fur statischefelder, *Z. Phys.*, **140,** 394 (1955).
—— Thesis, University of Hamburg, 1957 (unpublished). [Chapter 7]
—— Transformations of static exterior solutions of Einstein's gravitational field equations into different solutions by means of conformal mappings, *Collection Interne de CNRS*, **91,** 275 (1962).
—— [See also P. Jordan]
—— and W. Kundt: Exact solutions of the gravitational field equations, in *Gravitation: An Introduction to Current Research* (L. Witten, Editor), John Wiley, New York, 1962. [Chapter 11]
Einstein, A.: Die Grundlage der allgemeinen Relativitätstheorie, *Ann. Physik*, **49,** 769 (1916). [Chapter 7]
—— *Ann. Physik*, **55,** 241 (1918). [Chapter 7]
—— *S. B. Preuss. Akad. Wiss.*, **414,** (1925). [Chapter 7]
—— *The Meaning of Relativity*, Princeton University Press, Princeton, N.J., 1955.
—— and J. Grommer: Allgemeine Relativitätstheorie und Bewegungsgesetz, *Sitzungsber. Ber. Akad. Wiss.*, **2,** 235 (1928). [Chapter 7]
—— L. Infeld, and B. Hoffmann: Gravitational equations and problems of motion, *Ann. Math.*, **39,** 65 (1938). [Chapter 7]
—— and L. Infeld: On the motion of particles in general relativity theory, *Canad. J. Math.*, **1,** 209 (1949). [Chapter 7]
—— and N. Rosen: The particle problem in the general theory of relativity, *Phys. Rev.*, **48,** 73 (1935). [Chapter 7]
—— and N. Rosen: On gravitational waves, *J. Franklin Inst.*, **223,** 43 (1937). [Chapter 7]
Eisenhart, L. P.: *Riemannian Geometry*, Princeton University Press, New Jersey, 1949. [Chapters 7, 10, 12]
Eötvös, R. V.: *Math. u. Natur. Ber. Ungarn*, **8,** 65 (1890). [Chapter 7]
—— *Beibl. Ann. Physik*, **15,** 688 (1891). [Chapter 7]
—— *Ann. Physik*, **59,** 354 (1896). [Chapter 7]
——, D. Pekar, and E. Fekete: *Ann. Physik*, **68,** 11 (1922). [Chapter 7]
Epstein, S. T.: A note on gauge theories of vector particles, *Nuovo Cim.*, **25,** 222 (1962).

Erdlye, F.: *Asymptotic Expansions*, Dover, New York, 1955. [Chapter 10]
Erez, G. and N. Rosen: The gravitational field of a particle possessing a multipole moment, *Bull. Res. Counc. Isreal*, **8F**, 47 (1959). [Chapter 7]
Erikson, K.-E. and S. Yngström: A simple approach to the Schwarzschild solution, *Phys. Letters*, **5,** 119 and 327 (1963).
Ernst, F. J.: New formulation of the axially symmetric gravitational field I & II, *Phys. Rev.*, **167,** 1175, (1968) and **168,** 1415 (1968).
—— Exterior algebraic derivation of Einstein field equations employing a generalized basis, *J. Math. Phys.*, **12,** 2395 (1971).
—— Charged version of Tomimatsu–Sato spinning-mass field, *Phys. Rev.* (D)**7,** 2520 (1973).
Evans, L., G. Feldman, and P. T. Mathews: Gauge invariance and renormalization constants, *Ann. Phys. (N.Y.)*, **13,** 268 (1961).
Everitt, C. W. F., W. M. Fairbank, and W. O. Hamilton: General relativity experiments using low temperature techniques, in *Relativity* (M. Carmeli, S. I. Fickler, and L. Witten, Editors), Plenum Press, New York, 1970. [Chapter 7]
Evett, A. A.: Static gravitational fields in general relativity, *Nuovo Cim.* **22,** 779 (1961).
Exton, R. A., E. T. Newman, and R. Penrose: Conserved quantities in the Einstein–Maxwell theory, *J. Math. Phys.*, **10,** 1566 (1969).
Faddeev, L. D.: Quantum theory of gauge fields, in *Proceedings of the International Seminar on Vector Mesons and Electromagnetic Interactions, Dubna, 1969*.
Feinberg, G.: Possibility of faster-than-light particles, *Phys. Rev.*, **159,** 1089 (1967).
Feinblum, D. A.: Global singularities and the Taub–NUT metric, *J. Math. Phys.*, **11,** 2713 (1970).
Fermi, E.: Sopra i fenomeni che avvengono in prossimita di una linear oraria, *Atti Accad. Nazl. Lincei*, **21,** 21 and 51 (1922). [Chapter 7]
Feynman, R. P.: The quantum theory of gravitation, in *Warsaw Conference on Relativistic Theories of Gravitation* [L. Infeld, Editor], PWN, Warsaw, 1964; *Acta Phys. Polon.*, **24,** 697 (1963).
Fierz, M. and W. Pauli: On relativistic wave equations for particles of arbitrary spin in an electromagnetic field, *Proc. R. Soc. Lond.*, (A)**173,** 211 (1939).
Finkelstein, D.: Past–future asymmetry of the gravitational field of a point particle, *Phys. Rev.*, **110,** 965 (1958). [Chapter 7]
Finkelstein, F. J.: Electromagnetic interactions of a Yang–Mills field, *Rev. Mod. Phys.*, **36,** 632 (1964).
Fischer, A. E. and J. E. Marsden: The Einstein equations of evolution—a geometric approach, *J. Math. Phys.*, **13,** 546 (1972).
—— and J. E. Marsden: The Einstein evolutions equations as a first order quasilinear symmetric hyperbolic system, I, *Commun. Math. Phys.*, **28,** 1 (1972).
Fitzgerald, R. and S. Schiminovich: Gauge fields with noninvariant interactions, *J. Math. Phys.*, **9,** 73 (1968).
Florides, P. S.: A rotating sphere as a possible source of the Kerr metric, *Nuovo Cim.*, **13B,** 1 (1973).
—— and R. L. Jones: On stationary systems with spherical symmetry consisting of many gravitating masses, *Nuovo Cim.*, **69B,** 41 (1970).
—— and J. L. Synge: Notes on the Schwarzschild line element, *Commun. Dublin Inst. Adv. Stud.*, (A)., No. 14 (1961).
—— and J. L. Synge: Stationary gravitational fields due to single bodies, *Proc. R. Soc. Lond.*, (A)**280,** 459 (1964).
Fock, V.: Einstein statics in conformal space, in *Recent Developments in General Relativity*, PWN, Warsaw, 1962.
—— *Theory of Space, Time and Gravitation*, Pergamon Press, Oxford, 1964.
Foster, J.: S-spaces of Robinson and Trautman: topological identifications and asymptotic symmetry, *Proc. Camb. Phil. Soc.*, **66,** 521 (1969).
—— On Robinson–Trautman solution of Einstein's equations, *Phys. Letters*, (A)**37,** 313 (1971).
—— and E. T. Newman: Note on the Robinson–Trautman solutions, *J. Math. Phys.*, **8,** 189 (1967).
Foyster, J. M. and C. B. G. McIntosh: A class of solutions of Einstein's equations which admit a 3-parameter group of isometries, *Commun. Math. Phys.*, **27,** 241 (1972).
Fradkin, E. S. and I. V. Tyutin: *S* matrix for Yang–Mills and gravitational fields, *Phys. Rev.*, (D)**2,** 2841 (1970).

Frank, B.: Quaternion solution to the Schwarzschild problem and neutrino mass, *Nuovo Cim.*, **61A,** 263 (1969).

Frehland, E.: The general stationary gravitational vacuum field of cylindrical symmetry, *Commun. Math. Phys.*, **23,** 127 (1971).

Fronsdal, C.: Completion and embedding of the Schwarzschild solution, *Phys. Rev.*, **116,** 778 (1959). [Chapter 7]

Gamow, G.: Gravity, *Sci. Amer.*, **204,** 941 (1961).

Gard, P. and N. B. Backhouse: Methods for obtaining symmetrized representations of $SU(2)$ and the rotation group, *J. Phys.*, (A)**7,** 1793 (1974).

Garecki, J.: On the physical interpretation of the Bel–Robinson tensor, *Acta Phys. Polon.*, (B)**4,** 347 (1973).

Gautreau, R.: A Kruskal-like extension of the Kerr metric along the symmetry axis, *Nuovo Cim.*, **50A,** 120 (1967).

—— [See also J. L. Anderson]

—— and R. B. Hoffman: Exact solutions of the Einstein vacuum field equations in Weyl coordinates, *Nuovo Cim.*, **61B,** 411 (1969).

—— and R. B. Hoffman: Class of exact solutions of the Einstein–Maxwell equations, *Phys. Rev.*, (D)**2,** 271 (1970).

—— and R. B. Hoffman: Generating potential for the NUT metric in general relativity, *Phys. Letters*, (A)**39,** 75 (1972).

—— and R. B. Hoffman: The structure of the sources of Weyl-type electrovac fields in general relativity, *Nuovo Cim.*, **B16,** 162 (1973).

Geheniau, J.: Une classification des espaces einteiniens, *Compt-Rend. Acad. Sci., Paris,* **244,** 723 (1957).

Gelfand, I. M.: On one-parameter groups of operators in a normed space, *Dokl. Akad. Nauk SSR*, **25,** 711 (1939).

—— *Lectures on Linear Algebra*, Interscience, New York, 1948.

—— The center of an infinitesimal group ring, *Mat. Sborn.*, **26** (68), 103 (1950).

—— Spherical functions on symmetric Riemann spaces, *Dokl. Akad. Nauk SSSR*, **70,** 5 (1950).

—— [See also F. A. Berezin]

—— and M. I. Graev: On a general method of decomposition of the regular representation of a Lie group into irreducible representations, *Dokl. Akad. Nauk SSSR* **92,** 221 (1953).

—— and M. I. Graev: Analogue to Plancherel's formula for classical groups, *Trudy Mosk. Matem.*, Ob-va **4,** 375 (1955).

——, M. I. Graev, and N. Ya. Vilenkin: *Generalized Functions, Vol. 5: Integral Geometry and Representation Theory*, Academic Press, New York, 1966. [Introduction, Chapters 4, 6, 9; Appendixes B, C]

——, R. A. Minlos, and Z. Ya. Shapiro: *Representations of the Rotation and Lorentz Groups and their Applications*, Pergamon Press, New York, 1963. [Introduction]

—— and M. A. Naimark: Unitary representations of the Lorentz group, *J. Phys.*, **10,** 93 (1946); *Izv. Akad. Nauk SSSR, Ser. Mat.*, **11,** 411 (1947). [Chapters 4, 6]

—— and M. A. Naimark: Normed rings with involutions and their representation, *Izv. Akad. Nauk SSSR, Ser. Mat.*, **12,** 445 (1948).

—— and M. A. Naimark: The connection between unitary representations of a complex unimodular group and its unitary subgroup, *Izv. Akad. Nauk SSSR, Ser. Mat.*, **14,** 239 (1950).

—— and Z. Ya. Shapiro: Representations of the group of rotation in three-dimensional space and their applications, *Usp. Mat. Nauk*, **7,** 3, 1952; English translation: *Amer. Math. Soc. Translations* (2), **2,** 207 (1956).

—— and G. E. Shilov: *Generalized Functions, Vol. 1: Properties and Operations*, Academic Press, New York and London, 1964. [Chapter 4]

—— and N. Ya. Vilenkin: *Generalized Functions, Vol. 4: Applications of Harmonic Analysis*, Academic Press, New York, 1964.

—— and A. M. Yaglom: General relativistic invariant equations and infinite-dimensional representations of the Lorentz group, *Zh. Eksp. Teoret. Fiz.*, **18,** 703 (1948). [Chapter 2]

—— and A. M. Yaglom: Pauli's theorem for general relativistic invariant equations, *Zh. Eksp. Teoret. Fiz.*, **18,** 1096 (1948).

Georgi, H. and S. L. Glashow: Gauge theories without anomalies, *Phys. Rev.*, (D)**6,** 429 (1972).
Geroch, R.: What is a singularity in general relativity? *Ann. Phys.*, **48,** 526 (1968).
—— Spinor structure of space–times in general relativity, I & II, *J. Math. Phys.*, **9,** 1739 (1968), and **11,** 343 (1970).
—— Domain of dependence, *J. Math. Phys.*, **11,** 437 (1970).
—— General Relativity in the Large, *Gen. Relativ. Gravitn.*, **2,** 61 (1971). [Chapter 7]
—— Einstein algebras, *Commun. Math. Phys.*, **26,** 271 (1972).
—— A method for generating solutions of Einstein's equation, *J. Math. Phys.*, **12,** 918 (1971), and **13,** 394 (1972).
——, A Held, and R. Penrose: A space–time calculus based on pairs of null directions, *J. Math. Phys.*, **14,** 874 (1973).
—— and P. S. Jang: Motion of a body in general relativity, *J. Math. Phys.*, **16,** 65 (1975).
—— and E. T. Newman: Application of the semi-direct product of groups, *J. Math. Phys.*, **12,** 314 (1971). [Chapter 12]
Gheorghe, C. and E. Mihul: Casual groups of space–time, *Commun. Math. Phys.*, **14,** 165 (1969).
Ghosh, N. N.: Theory of analytic spinors I, *Proc. Nat. Inst. Sci. India*, (A)**31,** 477 (1965).
—— and R. Sengupta: The gravitation of a cylindrically symmetric magnetic field, *Nuovo Cim.*, **38,** 1579 (1965).
Gille, J. F. and J. Manuceau: Gauge transformation of second type and their implementation, I. Fermions, *J. Math. Phys.*, **13,** 2002 (1972).
Ginzburg, V. L. and I. Ye. Tamm: On the theory of spin, *Zh. Eksp. Teoret. Fiz.*, **17,** 227 (1947).
Girardello, L. and G. Parravicini: Continuous spin in the Bondi–Metzner–Sachs group of asymptotic symmetry in general relativity, *Phys. Rev. Letters*, **32,** 565 (1974).
Glashow, S. L. and M. Gell-Mann: Gauge theories of vector particles, *Ann. Phys. (N.Y.)*, **15,** 437 (1961). [Chapter 9]
—— [See also H. Georgi]
—— and Iliopaulos: Divergencies of massive Yang–Mills theories: higher groups, *Phys. Rev.*, (D)**4,** 1918 (1971).
Glass, E. N. and J. N. Goldberg: Newman–Penrose constants and their invariant transformations, *J. Math. Phys.*, **11,** 3400 (1970).
Godement, R.: A theory of spherical functions, I, *Trans. Am. Math. Soc.*, **73,** 496 (1952).
Godfrey, B. B.: Mach's principle, the Kerr metric and black-hole physics, *Phys. Rev.*, (D)**3,** 2721 (1970).
Goldberg, J. N.: Invariant transformations and Newman–Penrose constants, *J. Math. Phys.*, **8,** 2161 (1967).
—— Conservation of the Newman–Penrose conserved quantities, *Phys. Rev. Letters*, **28,** 1400 (1972).
—— Conservation equations and equations of motion in the null formalism, *Gen. Relativ. Gravitn*, **5,** 183 (1974).
—— [See also E. N. Glass]
—— and R. P. Kerr: Some applications of the infinitesimal holonomy group to the Petrov classification of Einstein spaces, *J. Math. Phys.*, **2,** 327 (1961). [Chapter 9]
—— [See also R. P. Kerr]
——, A. J. MacFarlane, E. T. Newman, F. Rohrlich, and E. A. Sudarshan: Spin-*s* spherical harmonics and $\eth$, *J. Math. Phys.*, **8,** 2155 (1967). [Chapter 12]
—— and R. K. Sachs: A theorem on Petrov types, *Acta Phys. Polon.*, **22,** Suppl. 13 (1962). [Chapter 10]
Goldstein, H.: *Classical Mechanics*, Addison-Wesley, Reading, Mass., 1965. [Chapter 7]
Goodinson, P. A.: An electromagnetic interpretation of the Kerr–Vaidya metric, *Int. J. Theoret. Phys.*, **6,** 47 (1972).
—— and R. A. Newing: Einstein–Maxwell null fields with non-zero twist, *Int. J. Theoret. Phys.*, **7,** 25 (1973).
Graev, M. I.: On a general method of computing traces of infinite-dimensional unitary representations of real simple Lie groups, *Dokl. Akad. Nauk SSSR*, **103,** 357 (1955).
—— [See also I. M. Gelfand]
Griffiths, J. B.: Gravitational radiation and neutrinos, *Commun. Math. Phys.*, **28,** 295 (1972).

—— and R. A. Newing: The two-component neutrino field in general relativity, *J. Phys.*, (A)**3**, 136 (1970).
—— and R. A. Newing: Tetrad equations for the two component neutrino field in general relativity, *J. Phys.*, (A)**3**, 269 (1970).
—— and R. A. Newing: Geometrical aspects of the two-components neutrino field in general relativity, *J. Phys.*, (A)**4**, 208 (1971).
—— and R. A. Newing: Neutrino radiation fields in general relativity, *J. Phys.*, (A)**4**, 306 (1971).
Grigor'eva, L. B.: Chronometrically invariant representation of the classification of Petrov gravitational fields, *Dokl. Akad. Nauk SSSR*, p. 1251; English translation in: *Soviet Phys. Dokl.*, **15**, 579 (1970).
Gürsey, F.: Introduction to group theory, in *Relativity, Groups and Topology*, Gordon and Breach, New York, 1964.
—— [See also I. Bars]
Hafele, J. C. and R. E. Keating: Around the world atomic clocks, *Science, N.Y.*, **177**, 166 and 168 (1972).
Hajicek, P.: Extensions of the Taub and NUT spaces and extensions of their tangent bundles, *Commun. Math. Phys.*, **17**, 109 (1970).
Hall, G. S.: On the Petrov classification of gravitational fields, *J. Phys.* (A)**6**, 619 (1973).
Hallidy, W. H. and A. I. Janis: Gravitational radiation: cutting the tail, *J. Math. Phys.*, **11**, 578 (1970).
—— [See also W. E. Couch]
Hamoui, A.: Deux nouvelles classes de solutions non-static a symetrie spherique des equations d'Einstein–Maxwell, *Ann. Inst. Henri Poincaré*, **10**, 195 (1969).
Hara, O.: Possibility of giving a finite mass to the gauge field, *Nuovo Cim.*, **29**, 565 (1963).
Harish-Chandra: Infinite irreducible representations of the Lorentz group, *Proc. R. Soc. (Lond.)*, (A)**189**, 372 (1947). [Chapter 2]
—— On relativistic wave equations, *Phys. Rev.*, **71**, 793 (1947). [Chapter 2]
—— Plancherel formula for complex semi-simple Lie groups, *Proc. Nat. Acad. Sci., Wash.*, **37**, 813 (1951).
—— The Plancherel formula for complex semi-simple Lie groups, *Trans. Am. Math. Soc.*, **76**, 485 (1954).
Harrison, B. K.: Electromagnetic solutions of the field equations of general relativity, *Phys. Rev.*, **138**, B488 (1965).
—— New solutions of the Einstein–Maxwell equations from old, *J. Math. Phys.*, **9**, 1744 (1968).
Hartle, J. B. and S. W. Hawking: Solutions of the Einstein–Maxwell equations with many black holes, *Commun. Math. Phys.*, **26**, 87 (1972).
Hatsukade, S.: New representation of rotation and Lorentz groups based on a model of non-Cantorian set theory, *Rep. Math. Phys.*, **3**, 173 (1972).
Hauser, I.: Type N gravitational field with twist, *Phys. Rev. Letters*, **33**, 1112 (1974).
Hawking, S. W.: Stable and generic properties in general relativity, *Gen. Relativ. Gravitn.*, **1**, 393 (1971).
—— Black holes in general relativity, *Commun. Math. Phys.*, **25**, 152 (1972).
—— and G. F. R. Ellis, *The Large Scale Structure of Space–Time*, Cambridge University Press, Cambridge, England, 1973. [Introduction]
—— [See also J. B. Hartle]
Hayashi, K. and A. Berman: Poincaré gauge invariance and the dynamical role of spin in gravitational theory, *Ann. Phys. (N.Y.)*, **75**, 562 (1973).
Heine, W.: *Group Theory in Quantum Mechanics*, Pergamon Press, 1970.
Heintzmann, H.: New exact static solutions of Einstein's field equations, *Z. Physik*, **228**, 489 (1969).
Heisenberg, W.: *Z. Physik*, **77**, 1 (1932). [Chapter 9]
Held, A.: [See also R. Geroch]
——, E. T. Newman, and R. Posadas: The Lorentz group and the sphere, *J. Math. Phys.*, **11**, 3145 (1970).
Hely, J.: A new set of states of pure radiation, *Compt. Rend. Acad. Sci., Paris*, **257**, 2083 (1963).
Hernandez, W. C.: Static axially symmetric, interior solution in general relativity, *Phys. Rev.*, **153**, 1359 (1967).
—— Material sources for the Kerr metric, *Phys. Rev.*, **159**, 1070 (1967).
—— Kerr metric, rotating sources and Machian effects, *Phys. Rev.*, **167**, 1180 (1968).

Herrera, L. A.: A gravitational Lagrangian and some identities in the Newman–Penrose formalism *Nuovo Cim.*, **17A,** 48 (1973).

Hildebrand, R. H.: Neutral meson production in *n-p* collisions, *Phys. Rev.*, **89,** 1090 (1953). [Chapter 9]

Hillion, P. and J. P. Vigier: On the Yang–Mills field, *Compt. Rend. Acad. Sci., Paris,* **252,** 1113 (1961).

—— and J. P. Vigier: An extension of Utiyama's formalism, *Cahiers de Phys.*, **15,** 435 (1961); **16,** 2 (1962).

—— and J. P. Vigier: Gauge group and isospace symmetries associated with the relativistic rotator theory of weak interaction, *Cahiers de Phys.*, **16,** 381 (1962). [In French]

Hilton, E.: The singularity in the Schwarzschild space–time, *Proc. R. Soc. Lond.*, (A)**283,** 491 (1965).

Hoffman, R. B.: Stationary axially symmetric generalizations of the Weyl solutions in general relativity, *Phys. Rev.*, **182,** 1361 (1969).

—— Stationary "noncanonical" solutions of the Einstein vacuum field equations, *J. Math. Phys.*, **10,** 953 (1969).

—— [See also R. Gautreau]

Hoffmann, B.: On the spherically symmetric field in relativity, *Quart. J. Math., Oxford,* **3,** 226 (1932), and **4,** 179 (1933).

—— [See also Einstein, A.]

—— Aspects of negative mass and the quasers, *Conference on Relativistic Theories of Gravitation,* Vol. II, London, 1965.

—— Negative mass, *Science Jl,* **1,** 74 (1965).

Hooft, G.: Renormalization of massless Yang–Mills fields, *Nucl. Phys.*, (B)**33,** 173 (1971).

Horak, R.: Analysis of Schwarzschild's solution, *Phys. Letters,* **13,** 318 (1964).

Horsky, J. and M. Lenc: A comment on the plane symmetric solution of Einstein equations, *Czech. J. Phys.*, (B)**20,** 1053 (1970).

—— and J. Novotny: The exact static exterior and interior metric of a thick plane plate, *J. Phys.*, (A)**2,** 251 (1969).

Hsu, J. P. and E. C. G. Sudarshan: Theory of massive and massless Yang–Mills fields, *Phys. Rev.*, (D)**9,** 1678 (1974).

Hubaut, X.: A class of spaces with metric connection connected to vacuum spaces in relativity, *Bull. Acad. Roy. Belg., Cl. Sci.,* **5,** 1094 (1962).

Hughston, L. P.: Generalized Vaidya metrics, *Int. J. Theoret. Phys.*, **4,** 267 (1971).

——, R. Penrose, P. Sommers, and M. Walker: On a quadratic first integral for the charged particle orbits in the charged Kerr solution, *Commun. Math. Phys.*, **27,** 303 (1972).

Huszar, M.: Angular momentum and unitary spinor bases of the Lorentz group, *Acta Phys. Acad. Sci. Hung.*, **30,** 241 (1971).

—— Unitary representation of the *SL*(2, *C*) group in horospheric basis, *J. Math. Phys.* **14,** 1620 (1973).

—— Unitary spinors in the Lorentz group, *Acta Phys. Slovaca,* **24,** 3 (1974).

—— Deformation of the *SO*(2, *C*) subgroup of the Lorentz group, *J. Math. Phys.*, **15,** 654 (1974).

Ihrig, E.: An exact determination of the gravitational potentials g_{ij} in terms of the gravitational fields R_{lmns}, *J. Math. Phys.*, **16,** 54 (1975).

Ikeda, M. and Y. Miyachi: On the static and spherically symmetric solution of the Yang–Mills field, *Progr. Theoret. Phys. (Japan),* **27,** 474 (1962).

Infeld, L.: Equations of motion in general relativity theory and the action principle, *Rev. Mod. Phys.*, **29,** 398 (1957). [Chapter 7]

—— Equations of motion and gravitational radiation, *Ann. Phys. (N.Y.),* **6,** 341 (1959). [Chapter 7]

—— The EIH and the *k*-approximation methods, *Bull. Acad. Polon. Sci., Ser. Math. Astron. Phys.*, **9,** 93 (1961).

—— Is Planck's constant a constant in a gravitational field?, *Bull. Akad. Polon. Sci., Ser. Math. Astron. Phys.*, **9,** 617 (1961).

—— [See also A. Einstein]

—— and J. Plebanski: *Motion and Relativity,* Pergamon Press, London, and PWN, Warsaw, 1960.

—— and A. Schild: On the motion of test particles in general relativity, *Rev. Mod. Phys.*, **21,** 408 (1949). [Chapter 7]

—— and B. L. van der Waarden: The wave equation of the electron in the general relativity theory, *Sb. preuss. Akad. Wiss., Phys.-mat. Kl.*, **380** (1933). [Chapter 8]

Inglis, D. R.: The energy levels and the structure of light nuclei, *Rev. Mod. Phys.*, **25,** 390 (1953). [Chapter 9]
Ipser, J. R.: Electromagnetic test fields around a Kerr-metric black hole, *Phys. Rev. Letters*, **27,** 5291 (1971).
—— [See also D. R. Brill]
Isham, C. J., A. Salam, and J. Strathdee: *SL*(6, *C*) gauge invariance of Einstein-like Lagrangians, *Lett. al Nuovo Cim.*, **5,** 969 (1972).
Israel, W.: New interpretation of the extended Schwarzschild manifold, *Phys. Rev.*, **143,** 1016 (1966).
—— Causal anomalies in the extended Schwarzschild manifold, *Phys. Letters*, **21,** 47 (1966).
—— Event horizons in static vacuum space–times, *Phys. Rev.*, **164,** 1776 (1967).
—— Event horizons in static electrovac space–times, *Commun. Math. Phys.*, **8,** 245 (1968).
—— Source of the Kerr metric, *Phys. Rev.*, (D)**2,** 641 (1970).
—— [See also V. de la Cruz]
—— and G. A. Wilson: A class of stationary electromagnetic vacuum fields, *J. Math. Phys.*, **13,** 865 (1972).
Iverson, G. J. and G. Mack: *E*2 parametrization of *SL*(2, *C*), *J. Math. Phys.*, **11,** 1581 (1970).
Izmailov, S.: On the quantum theory of particles possessing internal rotational degrees of freedom, *Zh. Eksp. Teoret. Fiz.*, **17,** 629 (1947).
Jackson, J. C.: Infilling for NUT space, *J. Math. Phys.*, **11,** 924 (1970).
Jaffe, J. and I. I. Shapiro: Lightlike behavior of particles in a Schwarzschild field, *Phys. Rev.* (D), **6,** 405 (1972). [Chapter 7]
Janis, A. I.: [See also W. E. Couch]
—— [See also P. N. Demmie]
—— [See also W. H. Halliday]
—— and E. T. Newman: Structure of gravitational sources, *J. Math. Phys.*, **6,** 902 (1965). [Chapter 10]
——, E. T. Newman, and J. Winicour: Reality of the Schwarzschild singularity, *Phys. Rev. Letters*, **166,** 878 (1968).
—— [See also E. T. Newman]
——, D. C. Robinson, and J. Winicour: Comments on Einstein scalar solutions, *Phys. Rev.*, **186,** 1729 (1969).
—— [See also J. Winicour]
Janossy, L.: Two subgroups of the Lorentz group and their physical significance, *Acta Phys. Hung.*, **20,** 115 (1966).
Jauch, J. M.: Gauge invariance as a consequence of Galilei invariance for elementary particles, *Helvet. Phys. Acta*, **37,** 284 (1964).
Jordan, P., J. Ehlers, and W. Kundt: Exact solutions of the field equations of general relativity, *Akad. Wiss. Lit. Mainz Abhandl., Mat.-Nat. Kl.*, (No. 2), 3 (1960). [In German]
——, J. Ehlers, and R. K. Sachs: Exact solutions of the field equations of general relativity, II. Contributions to the theory of pure gravitational radiation, *Akad. Wiss. Lit. Mainz Abhandl., Mat.-Nat. Kl.*, (No. 1), 3 (1961).
—— and W. Kundt: Exact solutions of the field equations of general relativity, III. Geometrodynamics in the null case, *Akad. Wiss. Lit. Mainz Abhandl., Mat.-Nat. Kl.*, (No. 3), 3 (1961).
Joseph, V., A spatially homogeneous gravitational field, *Proc. Camb. Phil. Soc.*, **62,** 87 (1966).
Jost, R.: The Normal form of a complex Lorentz transformation, *Helvet. Phys. Acta*, **33,** 773 (1960).
Jouvet, B.: Gauge invariance, the mass of the photon, and the asymptotic form of the photom, *Compt. Rend. Acad. Sci., Paris*, **251,** 1119 (1960).
—— Classical gauge invariance of quantum-electrodynamics, *Nuovo Cim.*, **20,** 28 (1961).
Just, K.: Gauge invariance and quantum measurements, *Nuovo Cim.*, **36,** 294 (1965).
Karlov, L.: How to get the quasi-uniform gravitational field from the Schwarzschild metric, *Am. J. Phys.*, **37,** 667 (1969).
—— and W. Rindler: Torsion-free world lines in curved space–time, *Phys. Rev.*, (D)**13,** 1035 (1971).
Kasner, E.: Finite representation of the solar gravitational field in flat space of six dimensions, *Am. J. Math.*, **43,** 130 (1921). [Chapter 7]

Kastrup, H. A.: Position operators, gauge transformations and the conformal group, *Phys. Rev.*, **143,** 1021 (1966).

Katz, A.: Derivation of Newton's law of gravitation from general relativity, *J. Math. Phys.*, **9,** 983 (1968).

Katz, R.: *An Introduction to the Special Theory of Relativity*, Van Nostrand, Princeton, New York and London, 1964.

Kaye, M.: [See M. Carmeli]

Kerbrat-Lunc, H.: Mathematical introduction to the study of the Yang–Mills field on a space–time curve, *Compt. Rend. Acad. Sci., Paris,* **259,** 3449 (1964).

—— The Yang and Mills field on a curved space–time, *Compt. Rend. Acad. Sci., Paris,* **260,** 6809 (1965).

—— Conformal group on a space–time curve and associated Yang–Mills field, *Compt. Rend. Acad. Sci., Paris,* **266,** 29 (1968).

—— Varied Yang–Mills field, *Compt. Rend. Acad. Sci., Paris,* **276,** 711 (1973). [In French]

Kerner, R.: On the Cauchy problem for the Yang–Mills field equations, *Ann. Inst. Henri Poincaré,* (A)**20,** 279 (1974).

Kerr, R. P.: Gravitational field of a spinning mass as an example of algebraically special metric, *Phys. Rev. Letters,* **11,** 237 (1963). [Chapter 11]

—— and G. C. Debney, Jr.: Einstein spaces with symmetry groups, *J. Math. Phys.*, **11,** 2807 (1970).

—— [See also G. C. Debney]

—— and J. N. Goldberg: Einstein spaces with four-parameter holonomy groups, *J. Math. Phys.*, **2,** 332 (1961). [Chapter 9]

—— [See also J. N. Goldberg]

—— and A. Schild: A new class of vacuum solutions of the Einstein field equations, *Centenario della nascita di Galileo, Firenze,* Vol. II, Tome 1, p. 22 (1965).

Keswani, G. H.: Pulsars and possible new test of general relativity, *Nature, Lond.*, **220,** 148 (1968).

Khrapko, R. I.: On the uniqueness of the central-symmetric gravitational field in vacuum, *Zh. Eksp. Teoret. Fiz.*, (USSR) **49,** 1884 (1965).

—— Propagation of light through a singular Schwarzschild sphere, *Soviet Phys. JETP,* **23,** 4 (1966).

Kibble, T. W. B.: Lorentz invariance and the gravitational field, *J. Math. Phys.*, **2,** 212 (1961). [Chapter 9]

Kihlberg, A.: On a class of explicit representations of the homogeneous Lorentz group, *Ark. Fys.*, **27,** 373 (1964).

Kimura, T.: On a fixation of gauge and true observables of Yang–Mills field, *Progr. Theoret. Phys.* (*Japan*), **26,** 790 (1961).

Kinnersley, W.: Type D vacuum metrics, *J. Math. Phys.*, **10,** 1195 (1969). [Chapter 11]

—— Field of an arbitrary accelerating point mass," *Phys. Rev.*, **186,** 1335 (1969).

—— Generation of stationary Einstein–Maxwell fields, *J. Math. Phys.*, **14,** 651 (1973).

—— Recent progress in exact solutions, *Proceedings of Tel Aviv Conference, 1974.*

—— [See also W. E. Couch]

—— and M. Walker: Uniformly accelerating charged mass in general relativity, *Phys. Rev.*, (D)**2,** 1359 (1970).

Klein, O.: Some remarks on general relativity and the divergence problem of quantum field theory, *Nuovo Cim. Suppl.*, **6,** 344 (1957).

Klink, W. H.: Mixed D-functions and Clebsch–Gordon coefficients of non-compact group, *J. Math. Phys.*, **11,** 3210 (1970).

Komar, A.: On the necessity of singularities in the solutions of the field equations of general relativity, *Phys. Rev.*, **104,** 544 (1956).

—— [See also P. G. Bergmann]

Konuma, M., H. Umezawa, and M. Wada: General theory of the gauge transformation for vector meson fields, *Nucl. Phys.*, **31,** 507 (1962).

Koppel, A.: Some exact axially symmetric solutions of Einstein's field equations, *Füüsika ja Astron. Inst. Uurimused* (*Tartu*), **20,** 58 (1963).

—— On the exact axial-symmetric static one-coordinate solutions of the Einstein equations, *Füüsika ja Astron. Inst. Uurimused* (*Tartu*), **22,** 53 (1963).

Korkina, M. P.: The solution of the Einstein equations in the Gaussian coordinates, *Ukrayin Fiz. Zh.* (*USSR*), **13,** 647 (1968). [In Russian]

Kota, J. and Z. Perjes: All stationary vacuum metrics with shearing geodesic eigenrays, *J. Math. Phys.*, **13,** 1695 (1972).
Kozarzewski, B.: A theorem concerning the gravitational field with shear-free, rotation-free and expansion-free geodesic rays, *Acta Phys. Polon.*, **25,** 437 (1964).
Kramer, D.: A class of non-stationary solutions of the Einstein–Maxwell equations, *Wiss. Z. Friedrich-Schiller Univ., Jena, Math. Naturwiss. Reihe*, **21,** 45 (1972). [In German]
—— Group structure and field equations in Einstein universe, *Acta Phys. Polon.*, (B)**4,** 11 (1973). [In German]
—— and G. Neugebauer: Zu axialsymmetrischen stationärer lösungen der Einstein'schen feldgleichungen für das vakuum, *Commun. Math. Phys.*, **10,** 132 (1968).
—— and G. Neugebauer: Eine exakte stationare lösung der Einstein–Maxwell gleichungen, *Ann. Physik*, **24,** 59 (1969).
—— and G. Neugebauer: Internal Reissner–Weyl solution, *Ann. Physik*, **27,** 129 (1971). [In German]
Kraus, K.: Remarks on a class of almost solutions of the gravitational field equations, *Ann. Phys. (N.Y.)*, **50,** 102 (1968).
Kretschmann, E.: *Ann. Physik*, **53,** 575 (1917). [Chapter 7]
Krokina, M. P.: The solution of the Einstein equations in the Gaussian coordinates, *Ukrayin. Fiz. Zh.*, **13,** 647 (1968). [In Russian]
Krori, K. D.: On motions of test particles in Kerr metric, *Indian J. Phys.*, **44,** 227 (1970).
Kruskal, M. D.: Maximal extension of Schwarzschild metric, *Phys. Rev.*, **119,** 1743 (1960). [Chapter 7]
—— [See also J. G. Miller]
Kuchowicz, B.: A simplified method of obtaining the infinitesimal operators of the homogeneous Lorentz group I, *Bull. Acad. Polon. Sci., Ser. Math. Astron. Phys.*, **10,** 221 (1962).
—— Neutrinos in general relativity: four(?) levels of approach, *Gen. Relativ. Gravitn.*, **5,** 201 (1974).
Kuenzle, H. P.: Construction of singularity-free spherically symmetric space–time manifolds, *Proc. R. Soc. Lond.*, (A)**297,** 244 (1967).
Kummer, W.: Gauge-transformation and quantization of free electromagnetic field, *Acta Phys. Austr.*, **14,** 149 (1961).
Kundt, W.: The plane-fronted gravitational waves, *Z. Phys.*, **163,** 77 (1961), [Chapter 11]
—— [See also J. Ehlers]
—— [See also P. Jordan]
—— and E. T. Newman: Hyperbolic differential equations in two dimensions, *J. Math. Phys.*, **9,** 2193 (1969).
—— and A. Thompson: Weyl tensor and an associated shear-free geodesic congruence, *Compt. Rend. Acad. Sci., Paris*, **254,** 4257 (1962).
—— and M. Trümper: Contributions to the theory of gravitational waves, *Akad. Wiss. Lit. Mainz Abhandl., Mat.-Nat. Kl.*, No. **12,** 36 (1962).
—— and M. Trümper: Orthogonal decomposition of axisymmetric stationary spacetimes, *Z. Phys.*, **192,** 419 (1966).
Kurosh, A. G.: *Theory of Groups*, Chelsea, London, 1955.
Kursunoglu, B.: Complex orthogonal and antiorthogonal representation of Lorentz group, *J. Math. Phys.*, **2,** 22 (1961).
—— *Modern Quantum Theory*, W. H. Freeman, San Francisco, 1962. [Chapter 8]
Kustaanheimo, P.: A note on the transformability of spherically symmetric metrics, *Proc. Edinb. Math. Soc.*, **9,** 13 (1953).
Lal, K. B. and H. Prasad: Cylindrical wave solutions of field equations of general relativity containing electromagnetic fields, *Tensor*, **20,** 45 (1969).
Landau, L. D.: in *Niels Bohr and the Development of Physics* (W. Pauli, Editor), McGraw Hill, New York, 1955. [Chapter 7]
—— and E. M. Lifshitz: *The Classical Theory of Fields*, Pergamon Press, London, 1959.
—— and E. M. Lifshitz: *Quantum Mechanics*, Pergamon Press, London, 1960.
Lauritsen, T.: Energy levels of light nuclei, *Ann. Rev. Nucl. Sci.*, **1,** 67 (1952). [Chapter 8]
Léauté, B.: A study of the Kerr metric, *Ann. Inst. Henri Poincaré*, (A)**8,** 93 (1968).
Lee, K. K.: Global spinor fields in space–time, *Gen. Relativ. Gravitn.*, **4,** 421 (1973).

Lehman, E.: The Bel–Petrov classification of a pure electromagnetic field in general relativity, *Compt. Rend. Acad. Sci., Paris*, **262,** 806 (1966).

Leibovitz, C.: Spherically symmetric static solutions of Einstein's equations, *Phys. Rev.*, **185,** 1664 (1969).

Lemaitre, G.: *Ann. Soc. Sci. Bruxelles*, (A)**53,** 51 (1933). [Chapter 7]

Lense, J. and H. Thirring: Über den einflüss der eigenrotation der zentralkörper auf die bewegung der planeten und monde nach der Einsteinschen gravitationstheorie, *Physik. Z.*, **19,** 156 (1918). [Chapter 11]

Levi-Civita, T.: *Rend. Acc. Lincei*, Several Notes (1918–19). [Chapter 7]

Levinson, N.: The asymptotic behavior of a system of linear differential equations, *Am. J. Math.*, **68,** 1 (1946). [Chapter 10]

Levy, H.: Classification of stationary axisymmetric gravitational fields, *Nuovo Cim.*, **56B,** 253 (1968).

—— Stationary gravitational fields from static fields, *Phys. Letters*, **42A,** 176 (1972).

Lichnerowicz, A.: *Theories Relativistes de la Gravitation et de l'Electromagnetisme*, Masson, Paris, 1955.

—— Espaces fibres et espace–temps, *Gen. Relativ. Gravitn.*, **1,** 235 (1971).

Lind, R. W.: Algebraically special Einstein–Maxwell fields in the spin coefficient formalism, Thesis, University of Pittsburgh, 1970.

—— Shear-free, twisting Einstein–Maxwell metrics in the Newman–Penrose formalism, *Gen. Relativ. Gravitn.*, **5,** 251 (1974).

—— Gravitational and electromagnetic radiation in Kerr–Maxwell spaces, *J. Math. Phys.*, **16,** 34 (1975).

—— Stationary Kerr–Maxwell spaces, *J. Math. Phys.*, **16,** 39 (1975).

—— and E. T. Newman: Complexification of the algebraically special gravitational fields, *J. Math. Phys.*, **15,** 1103 (1974).

—— [See also B. A. Aronson]

——, J. Messmer, and E. T. Newman: Applications of the intertwining operators for representations of the restricted Lorentz group, *J. Math. Phys.*, **13,** 1879 (1972).

——, J. Messmer, and E. T. Newman: Equations of motion for the sources of asymptotically flat spaces, *J. Math. Phys.*, **13,** 1884 (1972).

Lindquist, R. W., R. A. Schwarz, and C. W. Misner: Vaidya's radiating Schwarzschild metric, *Phys. Rev.*, **137,** B1364 (1965).

—— [See also R. H. Boyer]

Lomont, J. S. and H. E. Moses: Exponential representation of complex Lorentz matrices, *Nuovo Cim.*, **29,** 1059 (1963).

Look, K. H.: The Yang–Mills fields and the connections of principal fibre bundles, *Acta Phys. Sin.*, **23,** 249 (1974).

Loomis, L.: *An Introduction to Abstract Harmonic Analysis*, Van Nostrand, New York, 1953.

Loos, H. G.: The range of gauge fields, *Nucl. Phys.*, **72,** 677 (1965).

—— Rudimentary geometric particle theory, *Ann. Phys.*, **36,** 486 (1966).

—— Free ghost gauge fields, *Nuovo Cim.*, **52A,** 1085 (1967).

—— Gauge field of a point charge, *J. Math. Phys.*, **8,** 1870 (1967).

—— Internal holomy groups of Yang–Mills fields, *J. Math. Phys.*, **8,** 2114 (1967). [Chapter 9]

—— Nonexistence of classes of Yang–Mills fields, *Nuovo Cim.*, **53A,** 201 (1968).

—— Canonical gauge- and Lorentz-invariant quantization of the Yang–Mills field, *Phys. Rev.*, **188,** 2342 (1969).

—— Existence of charged states of the Yang–Mills field, *J. Math. Phys.*, **11,** 3258 (1970).

Lord, E. A.: General relativity from gauge invariance, *Proc. Camb. Phil. Soc.*, **69,** 423 (1971).

Lorente, M., P. L. Huddleston, and P. Roman: Some properties of matrices occuring in $SL(2, C)$ invariant wave equations, *J. Math. Phys.*, **14,** 1495 (1973).

Lovelock, D.: A spherically symmetric solution of the Maxwell–Einstein equations, *Commun. Math. Phys.*, **5,** 257 (1967).

Lubkin, E.: Geometric definition of gauge invariance, *Ann. Phys.*, **23,** 233 (1963).

Ludwig, G.: Classification of electromagnetic and gravitational fields, *Am. J. Phys.*, **37,** 1225 (1969). [Chapter 8]

—— Geometrodynamics of electromagnetic fields in the Newman–Penrose formalism, *Commun. Math. Phys.*, **17,** 98 (1970).
—— On event horizons in static space–times, *Commun. Math. Phys.*, **23,** 255 (1971).
Lugiato, L. A. and V. Gorini: On the structure of relativity groups, *J. Math. Phys.*, **13,** 665 (1972).
Lukierski, J.: Gauge transformation in quantum field theory, *Nuovo Cim.*, **29,** 561 (1963).
Lyusternik, L. A. and V. I. Sobolev: *Elements of Functional Analysis*, Moscow, 1951.
McCarthy, P. J.: Asymptotically flat space–times and elementary particles, *Phys. Rev. Letters*, **29,** 817 (1972). [Chapter 12]
—— Representations of the Bondi–Metzner–Sachs group I. Determination of the representations, *Proc. R. Soc. Lond.*, (A)**330,** 517 (1972).
—— Structure of the Bondi–Metzner–Sachs group, *J. Math. Phys.*, **13,** 1837 (1972). [Chapter 12]
—— Representations of the Bondi–Metzner–Sachs group II. Properties and classification of the representations, *Proc. R. Soc. Lond.*, (A)**333,** 317 (1973).
McCrea, W. H.: The interpretation of the Schwarzschild metric and the release of gravitational energy, *Astrophys. Norvegica*, **9,** 89 (1964).
MacFarlane, A. J.: On the restricted Lorentz group and groups homomorphically related to it, *J. Math. Phys.*, **3,** 1116 (1962).
—— [See also J. N. Goldberg]
McVittie, G. C. and R. Stabell: Spherically symmetric nonstatical solutions of Einstein's equations, *Ann. Inst. Henri Poincaré*, **7,** 103 (1967).
—— [See also M. E. Cahill]
Madore, J.: Yang–Mills fields with non-zero mass, *Nuovo Cim.*, **4A,** 599 (1971).
Maheshwari, A.: Schwinger equations for the Yang–Mills and the gravitational fields, *Progr. Theoret. Phys.*, **45,** 1662 (1971).
Maia, M. D.: Isospinors, *J. Math. Phys.*, **14,** 882 (1973).
Malin, S.: Gauge-free quantization of the linearized equations of general relativity, *Phys. Rev.*, (D)**10,** 2338 (1974). [Chapter 12]
—— The Weyl and Dirac equations in terms of functions over the group SU_2, *J. Math. Phys.*, **16,** 679 (1975).
—— [See also M. Carmeli]
Marathe, K. B.: Spaces admitting gravitational fields, *J. Math. Phys.*, **14,** 228 (1973).
Marder, L.: On uniform acceleration in special and general relativity, *Proc. Camb. Phil. Soc., Math. Phys. Sci.*, **53,** 194 (1957).
—— Gravitational waves in general relativity, V. An exact spherical wave, *Proc. R. Soc. Lond.*, (A)**261,** 91 (1961).
—— On space–times with bounded empty regions, *Proc. Camb. Phil. Soc.*, **60,** 97 (1974).
—— Gravitational waves in general relativity, XI. Cylindrical-spherical waves, *Proc. R. Soc. Lond.*, (A)**313,** 83 (1969).
—— Sources for Kerr type space–times in general relativity, *J. Phys.*, (A)**5,** 605 (1972).
Marek, J. J.: Static axisymmetric interior solution in general relativity, *Phys. Rev.*, **163,** 1373 (1967).
—— Some solutions of Einstein's equations in general relativity, *Proc. Camb. Phil. Soc.*, **64,** 167 (1968).
Marinov, M. S. and V. I. Roginsky: On invariants of the rotation group, *Fortschr. Phys.*, **19,** 509 (1971).
Martucci, G. and M. Modygno: Asymptotic deflection of a light ray in a Kerr field, *Lett. al Nuovo Cim.*, **3,** 553 (1970).
Massa, E.: A simple relation between the restricted Lorentz group L^{+} and the unimodular group of order two, *Lett. al Nuovo Cim.*, **1,** 806 (1969).
Matzner, R. A.: On stationary axisymmetric solutions of the Einstein field equations, *Astrophys. J.*, **167,** 149 (1971).
—— and C. W. Misner: Gravitational field equations for sources with axial symmetry and angular momentum, *Phys. Rev.*, **154,** 1229 (1967).
Mehra, A. L., P. C. Vaidya, and R. S. Kushwaha: Gravitational field of a sphere composed of concentric shells, *Phys. Rev.*, **186,** 1333 (1969).
Meister, H. J. and A. Papapetrou: Die Schwarzschild'sche lösung in De Donder'schen Koordinated, *Z. Phys.*, **145,** 403 (1956).

Melvin, M. A.: Pure magnetic and electric geons, *Phys. Letters*, **8,** 65 (1964).
—— Dynamics of cylindrical electromagnetic universes, *Phys. Rev.*, **139,** B225 (1965).
—— and J. Wallingford: Orbits in magnetic universe, *J. Math. Phys.*, **6,** 333 (1965).
Miller, J. G.: Global analysis of the Kerr–Taub–NUT metric, *J. Math. Phys.*, **14,** 486 (1973).
—— and M. D. Kruskal: Taub–NUT metric and incompatible extensions, *Phys. Rev.*, (D)**4,** 2945 (1971).
—— and M. D. Kruskal: Extension of a compact Lorentz manifold, *J. Math. Phys.*, **14,** 484 (1973).
Miller, W.: Clebsch–Gordan coefficients and special function indentities, II: the rotation and Lorentz groups in 3-space, *J. Math. Phys.*, **13,** 827 (1972).
Misner, C. W.: The flatter regions of Newman, Unti and Tamburino's generalized Schwarschild space, *J. Math. Phys.*, **4,** 924 (1963). [Chapter 11]
—— [See also R. Arnowitt]
—— [See also R. W. Lindquist]
—— [See also R. A. Matzner]
——, K. Thorne, and J. A. Wheeler: *Gravitation*, Freeman, San Francisco, 1973. [Introduction]
—— and J. A. Wheeler: Classical physics as geometry: Gravitation, electromagnetism, unquantized charge, and mass as properties of curved empty space, *Ann. Phys. (N.Y.)*, **2,** 525 (1957). [Chapter 7]
Misra, M.: Some axially symmetric empty gravitational fields, *Proc. Nat. Inst. Sci., India*, (A)**26,** 673 (1960).
—— A class of solutions representing plane electromagnetic waves in general relativity, *Proc. Nat. Inst., India*, (A)**29,** 104 (1963).
—— Some exact solutions of the field equations of general relativity, *J. Math. Phys.*, **7,** 155 (1966).
—— Type-null vacuum solutions in general relativity, *J. Math. Phys.*, **9,** 1052 (1968).
—— On the generalized Goldberg–Sachs theorem, *Lett. al Nuovo Cim.*, **1,** 615 (1969).
—— A unified treatment of the Kerr and Vaidya solutions in general relativity, *Proc. R. Irish Acad.*, (A)**69,** 39 (1970).
—— A new interpretation of the Kerr–Schild metric, *Lett. al Nuovo Cim.*, **6,** 715 (1973).
—— A property of the Schild–Kerr metric, *Curr. Sci. (India)*, **42,** 349 (1973).
Misra, R. M.: On stationary systems with spherical symmetry, *Nuovo Cim.*, **32,** 939 (1964).
—— Axially symmetric fields in general relativity, *Phys. Rev.*, (D)**2,** 410 (1970).
—— and D. B. Pandey: Stationary axially symmetric fields and the Kerr metric, *J. Math. Phys.*, **13,** 1538 (1972).
—— and D. B. Pandey: Class of electromagnetic fields, *J. Phys.*, (A) **6,** 924 (1973).
——, D. B. Pandey, D. C. Srivastava, and S. N. Tripathi: New class of solutions of the Einstein–Maxwell fields, *Phys. Rev.*, (D)**7,** 1587 (1973).
—— and R. A. Singh: Three-dimensional formulation of gravitational null fields I and II, *J. Math. Phys.*, **7,** 1836, 1966, and **8,** 1065 (1967).
——, U. Narain, and R. S. Mishra: A solution of the field equations representing a gravitational field of magnetic type, *Tensor*, **19,** 303 (1968).
Missana, M.: Solutions of Einstein's equations in spherical symmetry, *Mem. Soc. Astron. Ital.*, **39,** 275 (1968).
Mohapatra, R. N.: Feynman rules for the Yang–Mills fields: a canonical quantization approach, *Phys. Rev.*, (D)**4,** 378, 1007 and 2215 (1971).
Moller, C.: On the possibility of terrestrial test of the general theory of relativity, *Nuovo Cim. Suppl.* **6,** 381 (1957). [Chapter 7]
—— Further remarks on the localization of the energy in the general theory of relativity, *Ann. Phys. (N.Y.)*, **12,** 118 (1961).
—— Gravitational energy radiation, *Phys. Lett.*, **3,** 329 (1963).
Moret-Bailly, I.: Structure du champ gravitationnel non-radiatif sans hypothese de symetrie en relativite générale, *Ann. Inst. Poincaré*, **9,** 395 (1969).
—— The structure of non-radiative gravitational fields in general relativity using Newman–Penrose formalism, *Ann. Inst. Poincaré*, (A)**11,** 415 (1969). [In French]
Moses, H. E.: Irreducible representations of the rotation group in terms of Euler's theorem, *Nuovo Cim.*, **40A,** 1120 (1965). [Chapter 1]

—— Irreducible representations of the rotation group in terms of the axis and angle of rotation, *Ann. Phys.*, **37,** 224 (1966), and **42,** 343 (1967). [Chapter 1]
—— [See also J. S. Lomont]
Motz, L.: Gauge invariance and the Lorentz ponderomotive force, *Phys. Rev.*, **119,** 1102 (1960).
—— Gauge invariance and the structure of charged particles, *Nuovo Cim.*, **26,** 672 (1962).
—— Gauge invariance and the Hamilton–Jacobi equation, *Nuovo Cim.*, **B69,** 95 (1970).
Mukunda, N.: Unitary representations of the homogeneous Lorentz group in an 0(2, 1) basis, *J. Math. Phys.*, **9,** 50 (1968).
—— and B. Radhakrishnan: New forms for the representations of the three-dimensional Lorentz group, *J. Math. Phys.*, **14,** 254 (2973).
Müller zum Hagen, H.: On the analyticity of static vacuum solutions of Einstein's equations, *Proc. Camb. Phil. Soc.*, **67,** 415 (1970).
—— and D. C. Robinson: Black holes in static vacuum space–times, *Gen. Relativ. Gravitn.*, **4,** 53 (1973).
Murenbeeld, M. and J. R. Trollope: Slowly rotating radiation sphere and a Kerr–Vaidya metric, *Phys. Rev.*, (D)**1,** 3220 (1970).
Mysak, L. and G. Szekeres: Behavior of the Schwarzschild singularity in superimposed gravitational fields, *Canad. J. Phys.*, **44,** 617 (1966).
Naimark, M. A.: Rings with involutions, *Usp. Mat. Nauk.*, **3,** 52 (1948); English translation: *Amer. Math. Soc. Translation*, No. 25 (1950).
—— On linear representations of the proper Lorentz group, *Dokl. Akad. Nauk SSSR*, **97,** 969 (1954). [Chapter 6]
—— On the description of all unitary representations of the complex classical groups I and II, *Mat. Sborn.*, **35,** (77), 317 (1954); **37,** (79), 121 (1955).
—— Linear representation of the Lorentz group, *Usp. Mat. Nauk*, **9,** 19 (1954); English translation: *Amer. Math. Soc. Translation*, Series 2, 6 (1957).
—— *Normed Rings*, Noordhoff, Groningen, Netherlands, 1959. [Chapters 4, 5]
—— On irreducible linear representations of the complete Lorentz group, *Dokl. Akad. Nauk SSSR*, **112,** 583 (1957).
—— *Linear Representations of the Lorentz Group*, Pergamon Press, New York, 1964. [Introduction, Chapters 2, 4, 5, 6; Appendixes B, C]
—— [See also I. M. Gelfand]
Nambu, Y.: Infinite-component wave equations with hydrogenlike spectra, *Phys. Rev.*, **160,** 1171 (1967). [Chapter 5]
Namyslowski, J.: The Dirac equation in general relativity in the vierbein formalism, *Acta Phys. Polon.*, **20,** 927 (1961).
Nariai, H.: On some static solutions of Einstein's gravitational field equations in a spherically symmetric case, *Sci. Rep. Tohoku Univ.*, **34,** 160 (1950).
Ne'eman, Y.: Derivation of strong interactions from a gauge invariance, *Nuclear Phys.*, **26,** 222 (1961).
—— Yukawa terms in the unitary gauge theory, *Nuclear Phys.*, **26,** 230 (1961).
Nemytskii, V., M. Sludskaya and A. Cherkasov: *A Course in Mathematical Analysis*, Moscow, 1944
Newman, E. T.: Some properties of empty space–time, *J. Math. Phys.*, **2,** 324 (1961).
—— New approach to the Einstein and Maxwell–Einstein field equations, *J. Math. Phys.*, **2,** 674 (1961).
—— A generalization of the Schwarzschild metric, *Conference on Relativistic Theories of Gravitation*, PWN, Warsaw, 1962.
—— The nature of gravitational sources, Gravity Research Foundations, 1964.
—— A possible connection between the gravitational field and elementary particle physics, *Nature, Lond.*, **206,** 811 (1965). [Chapter 12]
—— Singularities: their structure and motion, *Gen. Relativ. Gravitn.*, **1,** 401 (1971).
—— Complex coordinate transformation and the Schwarzschild–Kerr metrics, *J. Math. Phys.*, **14,** 774 (1973). [Chapter 11]
—— Lienard–Wiechert fields and general relativity, *J. Math. Phys.*, **15,** 44 (1974).
—— Complex Maxwell and Einstein fields, IAU Symposium No. 64: *Gravitational Radiation and Gravitational Collapse*, Warsaw, Poland, 5–8 September 1974. Reidel, Dordrecht, Netherlands, 1974, p. 105.

—— [See also B. A. Aronson]
—— W. E. Couch, K. Chinnapared, A. Exton, A. Prakash, and R. Torrence: Metric of a rotating charged mass, *J. Math. Phys.*, **6**, 918 (1965). [Chapter 11]
—— [See also W. E. Couch]
—— [See also M. Demianski]
—— [See also R. A. Exton]
—— [See also J. Foster]
—— [See also R. Geroch]
—— [See also J. N. Goldberg]
—— [See also A. Held]
—— and A. Janis: Note on the Kerr spinning particle metric, *J. Math. Phys.*, **6**, 915 (1965). [Chapter 11]
—— [See also A. I. Janis]
—— [See also R. W. Lind]
—— and R. Penrose: An approach to gravitational radiation by a method of spin coefficients, *J. Math. Phys.*, **3**, 566 (1962), **4**, 998 (1963). [Chapters 9, 10]
—— and R. Penrose: 10 exact gravitationally conserved quantities, *Phys. Rev. Letters*, **15**, 231 (1965).
—— and R. Penrose: Note on the Bondi–Metzner–Sachs group, *J. Math. Phys.*, **7**, 863 (1966). [Chapter 12]
—— and R. Penrose: *Some new gravitationally conserved quantities*, Gravity Research Foundation, (1966).
—— and R. Posadas: Equations of motion and the structure of singularities, *Phys. Rev. Letters*, **22**, 1196 (1969).
—— and R. Posadas: Motion and structure of singularities in general relativity, *Phys. Rev.*, **187**, 1784 (1969).
—— and R. Posadas: Motion and structure of singularities in general relativity, II, *J. Math. Phys.*, **12**, 2319 (1971).
—— and L. A. Tamburino: New approach to Einstein's empty space field equations, *J. Math. Phys.*, **2**, 667 (1961).
—— and L. A. Tamburino: Empty space metrics containing hypersurface orthogonal geodesic rays, *J. Math. Phys.*, **3**, 902 (1962). [Chapter 11]
——, L. Tamburino, and T. Unti: Empty-space generalization of the Schwarzschild metric, *J. Math. Phys.*, **4**, 915 (1963). [Chapter 11]
—— and T. W. J. Unti: Behavior of asymptotically flat empty spaces, *J. Math. Phys.*, **3**, 891 (1962). [Chapter 10]
—— and T. W. J. Unti: A class of null flat-space coordinate systems, *J. Math. Phys.*, **4**, 1467 (1963).
—— and T. W. J. Unti: Note on the dynamics of gravitational sources, *J. Math. Phys.*, **6**, 1806 (1965).
—— and J. Winicour: The Kerr congruence, *J. Math. Phys.*, **15**, 426 (1975).
—— and R. Young: A new approach to the motion of a pole–dipole particle, *J. Math. Phys.*, **11**, 3154 (1970).
Nordtvedt, K., Jr.: Solution of the Einstein–Maxwell equations for a static distribution of massive charged particles, *Gen. Relativ. Gravitn.*, **3**, 95 (1972).
—— and H. Pagles: Electromagnetic plane wave solutions in general relativity, *Ann. Phys. (N.Y.)*, **17**, 426 (1962).
Novikov, I. D.: Note on the space–times metric inside the Schwarzschild singular sphere, *Astron. Zh. USSR*, **38**, 564 (1961).
—— Light propagation outside and inside the singular Schwarzschild sphere, *Dokl. Akad. Nauk SSSR*, **150**, 1019 (1963).
O'Connel, R. F.: Precession of Schiff's proposed gyroscope in an arbitrary force field, *Lett. al Nuovo Cim.*, **18**, 933 (1969).
—— Present status of the theory of the relativity–gyroscope experiment, *Gen. Relativ. Gravitn.*, **3**, 123, (1972).
—— [See also B. M. Barker]
Oektem, F.: On an extension of electromagnetic and gravitational tensors and the Bel–Robinson tensor, *Nuovo Cim.*, **58B**, 167 (1968).

Ogievetskii, V. I. and I. V. Polubarinov: Spinors in gravitation theory, *Zh. Eksp. Teoret. Fiz.*, **48,** 1625 (1965).

—— and I. V. Polubarinov: Group theoretical approach to spinors in the Einstein theory of gravitation, in *International Conference on Relativistic Theories of Gravitation*, King's College, London, 1965.

Ohanian, H. C.: Gravitons as Goldstone Bosons, *Phys. Rev.*, **184,** 1305 (1969).

Ohta, T., H. Okamura, T. Kimura, and K. Hiida, Physically acceptable solution of Einstein's equation for many body system, *Progr. Theoret. Phys.*, **50,** 492 (1973).

Oksak, A. I.: Trilinear Lorentz invariant forms, *Commun. Math. Phys.*, **29,** 189 (1973).

Oktem, F.: Are the neutrinos an aspect of Riemannian geometry?, *Nuovo Cim.*, **A1,** 38 (1971).

Olijnychenko, P.: On static solution in general relativity, *Nuovo Cim.*, **21,** 389 (1961).

Oliveira, C. G. and C. M. do Amaral: Spinor curvatures in the theory of gravitation, *Phys. Letters*, **22,** 64 (1966).

—— and C. M. do Amaral: Spinor formalism in gravitation, *Nuovo Cim.*, **A47,** 9 (1967).

O'Raifeartaigh, L. and J. L. Synge: A property of empty space–time, *Proc. R. Soc. Lond.*, (A)**246,** 299 (1958).

—— [See also S.-J. Chang]

Ozsvath, I.: Solutions of the Einstein field equations with simply transitive groups of motions, *Akad. Wiss. Lit. Mainz Abhandl., Mat.-Nat. Kl.*, No. 13, 22 (1962). [In German]

—— Homogeneous solutions of the Einstein–Maxwell equations, *J. Math. Phys.*, **6,** 1255 (1965).

Pais, A.: On spinors in *n*-dimensions, *J. Math. Phys.*, **3,** 1135 (1962).

Pajas, P. and P. Winternitz: Representations of the Lorentz group: new integral relations between Legendre functions, *J. Math. Phys.*, **11,** 1505 (1970).

Pajerski, D. W. and E. T. Newman: Trapped surfaces and the development of singularities, *J. Math. Phys.*, **12,** 1929 (1971).

Palatini, A.: *Rend. circ. mat. Palermo*, **43,** 203 (1919). [Chapter 7]

Pandya, I. M. and P. C. Vaidya: Wave solutions in general relativity, I, *Proc. Nat. Inst. Sci. India*, (A)**27,** 620 (1961).

Papapetrou, A.: Spinning test-particles in general relativity, I, *Proc. R. Soc. Lond.*, (A)**209,** 248 (1951). [Chapter 7]

—— Some remarks on stationary gravitational fields, *Compt. Rend. Acad. Sci., Paris*, **257,** 2797 (1963).

—— Stationary gravitational fields of axial symmetry, *Compt. Rend. Acad. Sci., Paris*, **258,** 90 (1964).

—— Theorem on radiative electromagnetic and gravitational fields, *J. Math. Phys.*, **6,** 1403 (1965).

—— Spherically symmetric gravitational field with electromagnetic radiation, *Compt. Rend. Acad. Sci., Paris*, **258,** 6081 (1964); **262,** 162 (1966).

—— Stationary gravitational fields with axial symmetry, *Ann. Inst. Henri Poincaré*, (A)**4,** 83 (1966). [In French]

—— Shock waves in the Newman–Penrose formalism, *Commun. Math. Phys.*, **34,** 229 (1973).

—— *Lectures on General Relativity*, D. Reidel, Dordrecht, Netherlands, 1974. [Introduction]

—— [See also G. Dautcourt]

—— [See also H. J. Meister]

—— and H. Treder: Non-existence of static, everywhere regular solutions in general relativity, *Compt. Rend. Acad. Sci., Paris*, **254,** 4254 (1962).

—— and H. Treder: Problem of everywhere regular stationary solutions in general relativity, *Compt. Rend. Acad. Sci., Paris*, **254,** 4431 (1962).

Parker, L., R. Ruffini, and D. Wilkins: Metric of two spinning charged sources in equilibrium, *Phys. Rev.*, (D)**7,** 2874 (1973).

Parvizi, D.: Representations of the *SL*(2, *C*) Lie algebra, *Compt. Rend. Acad. Sci., Paris*, (A)**276,** 905 (1973).

Patnaik, S.: Einstein–Maxwell fields with plane symmetry, I, *Proc. Camb. Phil. Soc.*, **67,** 127 (1970).

Pauri, M. and G. M. Prosperi: Canonical realization of the rotation group, *J. Math. Phys.*, **8,** 2256 (1967).

Pavelle, R.: Yang's gravitational field equations, *Phys. Rev. Letters*, **33,** 1461 (1974).

—— Unphysical solutions of Yang's gravitational field equations, *Phys. Rev. Letters*, **34,** 1114 (1975).

Pavlopoulos, T. G.: Breakdown of Lorentz invariance, *Phys. Rev.*, **159,** 1106 (1967).
Peak, D.: Foundation of gauge theory, Thesis, State University of New York, Albany, 1970.
Pechlaner, E. and J. L. Synge: Model of a spinning body without gravitational radiation, *Proc. R. Irish Acad.*, **66,** 93 (1968).
Peebles, J.: The Eötvös experiment, spatial isotropy, and generally covariant field theories of gravity, *Ann. Phys. (N.Y.)*, **20,** 240 (1962).
Penney, R.: Generalization of the Reisner–Nordstrom solution to the Einstein-field equations, *Phys. Rev.*, **182,** 1383 (1969).
Penrose, R.: A spinor approach to general relativity, *Ann. Phys. (N.Y.)*, **10,** 171 (1960). [Chapter 8]
—— Zero rest-mass fields including gravitation: asymptotic behaviour, *Proc. R. Soc. Lond.*, (A)**284,** 1397 (1965).
—— Relativistic symmetry groups, in *Group Theory in Nonlinear Problems*, (A. O. Barut, Editor), Riedel, Dordrecht, Netherlands; Boston, USA, 1974, p. 1. [Introduction]
—— [See also R. A. Exton]
—— [See also R. Geroch]
—— [See also L. P. Hughston]
—— and M. A. H. MacCallum: Twistor theory: an approach to the quantization of fields and spacetime, *Phys. Letters*, **6C,** 241 (1973).
—— [See also E. T. Newman]
Peres, A.: Spinor fields in generally covariant theories, *Nuovo Cim. Suppl.* **24,** 389 (1962).
—— Non-geodesic motion in general relativity, *Phys. Rev.*, **137,** B1126 (1965).
Perjes, Z.: A model method of constructing certain axially-symmetric Einstein–Maxwell fields, *Nuovo Cim.*, **55B,** 600 (1968), Erratum: *Nuovo Cim.*, **56B,** 203 (1968).
—— Three-dimensional relativity for axisymmetric stationary space–time, *Commun. Math. Phys.*, **12,** 275 (1969).
—— Spinor treatment of stationary space–times, *J. Math. Phys.*, **11,** 3383 (1970).
—— Solutions of the coupled Einstein–Maxwell equations representing the fields of spinning sources, *Phys. Rev. Letters*, **27,** 1668 (1971).
—— [See also J. Kota]
Persides, S.: Asymptotically flat spaces in asymptotically null-spherical coordinates, *Proc. R. Soc. Lond.*, (A)**320,** 1542 (1970).
Perveen, S.: Renormalization problem of the Yang–Mills theory with non-zero mass: the spectral approach, *J. Phys.*, (A)**3,** 625 (1970).
Petrov, A. Z.: About spaces which determine gravitational fields, *Dokl. Akad. Nauk SSSR*, **81,** 149 (1951).
—— *Sci. Not.* Kazan State University, **114,** No. 8, 55 (1954), (Trans. No. 29, Jet Propulsion Laboratory, Pasadena, Cal., 1963).
—— Classification of gravitational fields of general form by means of group motions I, II and III, *Izv. Viss. Ucheb. Zaned. Fizika, Ser. Mat.*, No. 6 (1959), No. 1 (1960), and No. 6 (1960).
—— Invariant classifications of gravitational fields, gravitational field geometry as the geometry of automorphisms, in *Recent Developments in General Relativity*, PWN, Warsaw, 1962, pp. 371 and 379.
—— Classification invariante des champs de gravitation, *Collection Interne de CNRS*, **91,** 107 (1962).
—— Centrally-symmetric gravitational fields, *Soviet Phys. JETP*, **17,** 1026 (1963).
—— Central-symmetric gravitational fields, *Zh. Eksp. Teoret. Fiz.*, **44,** 1525 (1963).
—— On the gravitational fields with spherical symmetry, *Cahiers Phys.*, **18,** 1 (1964).
—— *Einstein Spaces*, Pergamon Press, Oxford, 1969.
—— Gravitational fields of the third type, *JETP Letters*, **11,** 141 (1970).
Petrovski, I. G.: *Lectures on the Theory of Integral Equations*, Graylock, London, 1957.
Phat, T. H.: A universal regulator?, *Ann. Physik*, **27,** 321 (1972).
Phillops, P. R.: Is the graviton a Goldstone boson?, *Phys. Rev.*, **146,** 966 (1966).
Pirani, F. A. E.: *Phys. Rev.*, **105,** 1089 (1957).
—— *Lectures on General Relativity*, Prentice Hall, Englewood Cliffs, N.J., 1965. [Chapter 9, Appendix D]
—— A note on Fermi coordinates, *Proc. R. Irish Acad.*, (A)**61,** 9 (1960).
—— and A. Schild: Geometrical and physical interpretation of the Weyl conformal curvature-tensor, *Bull. Acad. Polon. Sci., Ser. Math. Astron. Phys.*, **9,** 543 (1961).
Pochoda, P. and M. Schwarzschild: Variation of the gravitational constant and the evolution of the Sun, *Astrophys. J.*, **139,** 587 (1964).

Pons, R.: Integration of the Newman–Penrose equations in the Minkowski space case, *Ann. Inst. Henri Poincaré*, (A)**16**, 51 (1972). [In French]
Pontrjagin, L. S.: *Topological Groups*, Princeton University Press, New Jersey, 1946. [Introduction, Chapters 1, 4, 12, Appendix A]
Potier, R.: Sur les systèmes d'equation aux dérivées partielles linéaires et du premier ordre, à quatre variables invariantes dans toute transformation de Lorentz, *Compt. Rend. Acad. Sci., Paris*, **222,** 638 (1946).
—— Sur la définition du vecteur-courant en théorie des corpuscules. Cas du spin demi entier, *Compt. Rend. Acad. Sci., Paris*, **222,** 855 (1946).
—— Sur la définition et les propriétés du vecteur-courant associe a un corpuscule de spin quelconque, *Compt. Rend. Acad. Sci., Paris*, **222,** 1076 (1946).
Pound, R. V. and G. A. Rebka, Jr.: Apparent weight of photons, *Phys. Rev. Letters*, **4,** 337 (1960). [Chapter 7]
Press, W. H. and J. M. Bardeen: Non-conservation of the Newman–Penrose conserved quantities, *Phys. Rev. Letters*, **27,** 1303 (1971).
—— [See also J. M. Bardeen]
Raab, W.: On the concept of spinors in general relativity, *Acta Phys. Austr.*, **19,** 222 (1965). [In German]
—— [See also F. W. Cap]
Rao, J. K.: Type-N gravitational waves in non-empty space–time, *Current Sci., India*, **32,** 350 (1963).
—— Cylindrical waves in general relativity, *Proc. Nat. Inst. Sci., India*, (A)**30,** 439 (1964).
—— Cylindrically symmetric null fields in general relativity, *Indian J. Pure Appl. Math.*, **1,** 367 (1970).
—— Radiating Levi–Cevita metric, *J. Phys.*, (A)**4,** 17 (1971).
Rao, J. R., A. R. Roy, and R. N. Tiwari: A class of exact solutions for coupled electromagnetic and scalar fields for Einstein–Rosen metric, I, *Ann. Phys. (N.Y.)*, **69,** 473 (1972).
——, R. N. Tiwari, and R. N. Roy: A class of exact solutions for coupled electromagnetic and scalar fields for Einstein–Rosen metric, Part I A, *Ann. Phys. (N.Y.)*, **78,** 553 (1973).
Rasband, S. N.: Black holes and spinning test bodies, *Phys. Rev. Letters*, **30,** 111 (1973).
Rastall, P.: A solution of the Einstein Field equations, *Canad. J. Phys.*, **38,** 1661 (1960).
—— Rotations and Lorentz transformations, *Nucl. Phys.*, **57,** 191 (1964).
—— A new spinor calculus, *Gen. Relativ. Gravitn.*, **3,** 281 (1972).
Raychaudhuri, A. K.: Arbitrary concentrations of matter and the Schwarzschild singularity, *Phys. Rev.*, **89,** 417 (1953).
—— [See also B. K. Datta]
Razgovorov, N. N.: Electromagnetic radiation in the general theory of relativity, II, *Izv. Viss. Ucheb. Zaved. Fiz. USSR*, **5,** 32 (1973).
Regge, T.: General relativity without coordinates, *Nuovo Cim.* , **19,** 558 (1961).
Reinhart, B. L.: Maximal foliations of extended Schwarzschild space, *J. Math. Phys.*, **14,** 719 (1973).
Remler, E. A.: Scattering of Yang–Mills quanta, *Phys. Rev.*, **133,** B1267 (1964).
Reuss, J. D.: A nonstationary, cylindrical gravitational field, *Compt. Rend. Acad. Sci., Paris*, **266,** 794 (1968).
Reissner, H.: *Ann. Physik*, **50,** 106 (1916). [Chapter 7]
Rindler, R.: *Special Relativity*, Oliver and Boyd, Edinburgh, 1960.
—— Elliptic Kruskal–Schwarzschild space, *Phys. Rev. Letters*, **15,** 1001 (1965).
—— What are spinors?, *Am. J. Phys.*, **34,** 937 (1966).
—— Kruskal space and the uniformly accelerated frame, *Am. J. Phys.*, **34,** 1174 (1966).
—— [See also L. Karlov]
Robertson, H. P. Lecture in Toronto, 1939, (unpublished), cited by J. L. Synge. [Chapter 7]
Robinson, D. C.: Conserved quantities of Newman and Penrose, *J. Math. Phys.*, **10,** 1745 (1969).
—— Uniqueness of the Kerr black hole, *Phys. Rev. Letters*, **34,** 905 (1975).
—— [See also A. I. Janis]
—— [See also H. Müller zum Hagen]
Robinson, I. [See also P. G. Bergmann]
—— and J. R. Robinson: Vacuum metrics without symmetry, *Int. J. Theoret. Phys.*, **2,** 231 (1969).
—— and A. Schild: Generalization of a theorem by Goldberg and Sachs, *J. Math. Phys.*, **4,** 489 (1963).

——, A. Schild, and H. Strauss: A generalized Reissner–Nordtröm solution, *Int. J. Theoret. Phys.*, **2,** 243 (1969).
—— and A. Trautman: Spherical gravitational waves, *Phys. Rev. Letters.*, **4,** 431 (1960). [Chapters 10, 11]
—— and A. Trautman: Some spherical gravitational waves in general relativity, *Proc. R. Soc. Lond.*, (A)**265,** 463 (1962). [Chapter 11]
—— and J. D. Zund: Degenerate gravitational fields with twisting rays, *J. Math. Mech.*, **18,** 881 (1969).
Roche, J. A. and J. S. Dowker: The algebraic structure of the vacuum Riemann tensor, *J. Phys.*, (A)**1,** 527 (1968).
Rodichev, V. I. and I. M. Dozmorov: Plane waves in the general theory of relativity, *Izv. Viss. Ucheb. Zaved. Fiz.*, **4,** 20 (1969). [In Russian]
Rohrlich, F.: The principle of equivalence, *Ann. Phys. (N.Y.)*, **22,** 169 (1963).
—— On relativistic theories, *Am. J. Phys.*, **34,** 897 (1966).
—— [See also J. N. Goldberg]
Roll, P. G., R. Krotkov, and R. H. Dicke: The equivalence of inertial and passive gravitational mass, *Ann. Phys. (N.Y.)*, **26,** 442 (1964). [Chapter 7]
Rollnik, H.: Operator gauge transformations in quantum electrodynamics, *Z. Phys.*, **161,** 370 (1961).
Roman, P.: *Theory of Elementary Particles*, North-Holland, Amsterdam, 1960. [Chapter 12]
Rosen, G.: Spatially homogeneous solutions to the Einstein–Maxwell equations, *Phys. Rev.*, (B)**136,** 297 (1964).
—— Exact solutions to the Yang–Mills field equations, *J. Math. Phys.*, **13,** 595 (1972).
Rosen, N.: Some cylindrical gravitational waves, *Bull. Res. Counc. Israel*, **3,** 328 (1954). [Chapter 7]
—— The complete Schwarzschild solution, *Ann. Phys. (N.Y.)*, **63,** 127 (1971).
—— Theory of Gravitation, *Phys. Rev.*, (D)**3,** 2317 (1971).
—— [See also A. Einstein]
—— [See also G. Erez]
—— and D. Yaakobi: A particle at rest in a static axially symmetric gravitational field, *Bull. Res. Counc. Israel*, **10F,** 49 (1961).
Ross, D. K. and L. I. Schiff: Analysis of the proposed planetary radar reflection experiment, *Phys. Rev.*, **141,** 1215 (1966).
Roy, A. R., J. R. Rao, and R. N. Tiwari: A class of exact solutions for coupled electromagnetic and scalar fields for Einstein–Rosen metric, II, *Ann. Phys. (N.Y.)*, **79,** 276 (1973).
—— [See also J. R. Rao]
Roy, S. R. and V. N. Tripathi: On non-static axially symmetric space–time, *Gen. Relativ. Gravitn.*, **2,** 121 (1971).
Ruffini, R.: [See also M. Davis]
—— and J. Tiommo: Electromagnetic field of a particle moving in a spherically symmetric black hole background, *Lett. al Nuovo Cim.*, **3,** 211 (1972).
—— [See also L. Parker]
Rühl, W.: Multilinear invariant forms built of an arbitrary number of unitary representations of *SL*(*n*, *C*), *Nuovo Cim.*, **A44,** 659 (1966).
—— Tensor operators of rank two for the homogeneous Lorentz group, *Nuovo Cim.*, **A68,** 2131 (1969).
—— *The Lorentz Group and Harmonic Analysis*, Benjamin, New York, 1970. [Introduction, Chapters 3, 6]
Ryabushko, A. P.: Centrally symmetric Einstein spaces and Birkhoff's theorem, *Soviet Phys. JETP*, **19,** 1379 (1964).
Rylov, Yu. A.: The singularity of the Schwarzschild solution of the gravitational equations, *Zh. Eksp. Teoret. Fiz.*, **40,** 1755 (1961).
Sachs, R. K.: Gravitational waves in general relativity, VI. The outgoing radiation condition, *Proc. R. Soc. Lond.*, (A)**264,** 309 (1961). [Chapters 10, 11]
—— Asymptotic symmetries in gravitational theory, *Phys. Rev.*, **128,** 2851 (1962). [Chapter 12]
—— On the characteristic intial value problem in gravitational theory, *J. Math. Phys.*, **3,** 908 (1962).
—— Gravitational waves in general relativity, VIII. Waves in asymptotically flat space–time, *Proc. R. Soc. Lond.*, (A)**270,** 103 (1962). [Chapters 10, 12]

—— Gravitational radiation, in *Relativity, Groups and Topology*, (C. DeWitt *et al.*, Editors), Gordon and Breach, New York, 1964.
—— [See also D. Eardly]
—— [See also J. N. Goldberg]
—— [See also P. Jordan]
Sackfield, A.: Physical interpretation of NUT metric, *Proc. Camb. Phil. Soc.*, **70,** 89 (1971).
Safko, J. L. and L. Witten: Some properties of cylindrically symmetric Einstein–Maxwell fields, *J. Math. Phys.*, **12,** 257 (1971).
—— and L. Witten: Models of static, cylindrically symmetric solutions of the Einstein–Maxwell field, *Phys. Rev.*, (D)**5,** 293 (1972).
Sahakian, G. S. and Y. L. Vartanian: On Weyl's solution for axial-symmetric gravitational fields, *Nuovo Cim.*, **27,** 1497 (1963).
Salam, A.: Renormalizibility of gauge theories, *Phys. Rev.*, **127,** 331 (1962).
—— The formalism of Lie groups, in *Theoretical Physics*, International Atomic Agency, Vienna, 1963, p. 173. [Introduction]
—— On *SL*(6, *C*) gauge invariance, in *Proceedings of the Coral Cables Conference on Fundamental Interactions*, Center for Theoretical Studies, Miami, Florida, January 1973. [Chapter 9]
—— [See also C. J. Isham]
—— and J. C. Ward: On a gauge theory of elementary interactions, *Nuovo Cim.*, **19,** 165 (1961).
Sannikov, S. S.: Representations of the Lorentz group with complex spin, *Yadernaya Fiz. USSR*, **2,** 570 (1965). [In Russian]
—— A realization of infinite-dimensional representations of the rotation group with complex spin by entire functions, *Dokl. Akad. Nauk SSSR*, **176,** 801 (1967).
Schell, J. F.: Classification of four-dimensional Riemannian spaces, *J. Math. Phys.*, **2,** 202 (1961). [Chapter 9]
Schiff, L. I.: Sign of the gravitational mass of a positron, *Phys. Rev. Letters*, **1,** 254 (1958). [Chapter 7]
—— Gravitational properties of antimatter, *Proc. Nat. Acad. Sci. Wash.*, **45,** 69 (1959). [Chapter 7]
—— Motion of a gyroscope according to Einstein's theory of gravitation, *Proc. Nat. Acad. Sci. Wash.*, **46,** 871 (1960). [Chapter 7]
—— On experimental tests of the general theory of relativity, *Am. J. Phys.*, **28,** 340 (1960). [Chapter 7]
Schiffer, M. M., R. J. Adler, J. Mark, and C. Sheffield: Kerr geometry as complexified Schwarzschild geometry, *J. Math. Phys.*, **14,** 52 (1973). [Chapter 11]
—— [See also R. J. Adler]
Schild, A.: Discrete space–time and integral Lorentz transformations, *Phys. Rev.*, **73,** 414 (1948).
—— Equivalence principle and red-shift measurements, *Am. J. Phys.*, **28,** 778 (1960).
—— On gravitational radiation reaction forces, *Bull. Acad. Polon. Sci.*, *Ser. Math. Astron. Phys.*, **9,** 103 (1961).
—— [See also G. C. Debney]
—— [See also R. P. Kerr]
—— [See also F. A. E. Pirani]
Schmidt, B. G.: Homogeneous Riemannian spaces and Lie algebras of Killing fields, *Gen. Relativ. Gravitn.*, **2,** 105 (1971).
Schmutzer, E.: Some remarks on the theory of spinors and bispinors in curved space–time, *Acta Phys. Hung.*, **17,** 57 (1964).
Schrodinger, E. M.: *Expanding Universe*, Cambridge University Press, New York, 1956. [Chapter 10]
Schwarzschild, K.: *S.B. Preuss. Akad. Wiss.*, 189 (1916). [Chapter 12]
Schweber, S.: *Relativistic Field Theory*, Row-Peterson, Evanston, Illinois, 1961. [Chapter 12]
Schwinger, J.: On the charge independence of nuclear forces, *Phys. Rev.*, **78,** 135 (1950). [Chapter 9]
—— Gauge invariance and mass, *Phys. Rev.*, **126,** 397 (1962).
—— Non-abelian gauge fields: commutation relations, *Phys. Rev.*, **125,** 1043 (1962).
—— Non-abelian gauge fields: relativistic invariance, *Phys. Rev.*, **126,** 2261 (1962).
—— Gauge theories of vector particles, in *Theoretical Physics*, International Atomic Energy Agency, Vienna, 1963.
Sciama, D. W.: Recurrent radiation in general relativity, *Proc. Camb. Phil. Soc.*, **57,** 436 (1961).

—— Recent developments in the theory of gravitational radiation, *Gen. Relativ. Gravitn.*, **3**, 167 (1972).

——, P. C. Wayler, and R. C. Gilman: Generally covariant integral formulation of Einstein's field equations, *Phys. Rev.*, **187**, 1762 (1969).

Sciarrino, A. and M. Toller: Decomposition of the unitary irreducible representations of the group $SL(2, C)$ restricted to the subgroup $SU(1, 1)$, *J. Math. Phys.*, **8**, 1252 (1967).

Sedrakyan, D. M. and E. V. Chubaryan: On the internal solution of stationary axial-symmetric gravitational fields, *Astrofizika*, **4**, 551 (1968).

—— Stationary axial-symmetric gravitational fields, *Astrofizika*, **4**, 239 (1968).

Segercrantz, J.: On null tetrads and eigenvectors of Lorentz transformations in spinor treatment, *Commun. Phys.-Math.* (*Finland*), **32**, 8 (1966).

Sexl, R. U.: Universal conventionalism and space–time, *Gen. Relativ. Gravitn.*, **1**, 159 (1970).

—— [See also P. C. Aichelburg]

Shapiro, I. I.: Fourth test of general relativity, *Phys. Rev. Letters*, **13**, 789 (1964). [Chapter 7]

—— Testing general relativity with radar, *Phys. Rev.*, **141**, 1219 (1966).

—— Testing general relativity: progress, problems and prospects, *Gen. Relativ. Gravitn.*, **3**, 135 (1972). [Chapter 7]

——, W. B. Smith, M. B. Ash, R. P. Ingalls, and G. H. Pettengill: Gravitational constant: experimental bound on its time variation, *Phys. Rev. Letters*, **26**, 27 (1971).

Shekhar, K. C.: Solutions of Einstein's field equations (singular case), *Nuovo Cim.*, **25**, 1205 (1962).

—— Particular solution of Einstein's field equations, *Nuovo Cim.*, **28**, 902 (1963).

Sheldon, L., S. L. Glashow, and M. Gell-Mann: Gauge theories of vector particles, *Ann. Phys.* (*N.Y.*), **15**, 437 (1961). [Chapter 9]

Shima, K.: Local isotopic gauge transformation, *Progr. Theoret. Phys.*, **31**, 139 (1964).

Shirokov, M. F.: Solutions of the Schwarzschild–Nordstrom type for a point charge without singularities, *Soviet Phys. JETP*, **18**, 236 (1948).

Singer, S. F.: Application of an artificial satellite to the measurement of the general relativistic red shift, *Phys. Rev.*, **104**, 11 (1956). [Chapter 7]

Singh, K. P., L. Radhakrishna, and R. Sharan: Electromagnetic fields and cylindrical symmetry, *Ann. Phys.* (*N.Y.*), **32**, 46 (1965).

—— [See also R. M. Misra]

Sinsky, J. and J. Weber: New source for dynamical gravitational fields, *Phys. Rev. Letters*, **18**, 795 (1967). [Chapter 7]

Slavnov, A. A.: Massive gauge fields, *Theoret. Math. Phys.*, **10**, 201 (1972).

Sneddon, I. N.: in *Handbook der Physik* (Edited by S. Flügge), Springer-Verlag, Berlin, 1955. Vol. 2, p. 299. [Chapter 12]

Sokolik, G. A.: A spinor formalism of gravitational field, *Dokl. Akad. Nauk SSSR*, **154**, 565 (1964).

Sommerfeld, A.: *Partial Differential Equations in Physics*, Academic Press, New York, 1949. [Chapter 12]

Sommers, P.: On Killing tensors and constants of motion, *J. Math. Phys.*, **14**, 787 (1973).

—— [See also L. P. Hugston]

Stachel, J.: Behavior of Weyl–Levi–Civita coordinates for a class of solutions approximating the Schwarzschild metric, *Nature, Lond.*, **219**, 1346 (1968).

Stein, E. M.: Analysis in matrix spaces and some new representations of $SL(N, C)$, *Ann. Math.*, **86**, 461 (1967). [Chapter 6]

Stepanyouk, N. M.: An exact static axial solution of the Einstein equations, *Zh. Eksp. Teoret. Fiz.*, **53**, 570 (1967), [in Russian]; *JETP*, **26**, 369 (1968). [in English]

Stephenson, G.: Variational principles for the gravitational field, *Lett. al Nuovo Cim.*, **1**, 97 (1969).

Strauss, M. D. H.: A simple proof of the Lorentz transformation, *Nature, Lond.*, **157**, 516 (1946).

Streater, R. F. and A. S. Wightman: *PCT, Spin and Statistics, and All That*, Benjamin, New York, 1964. [Chapter 2]

Strocchi, F.: On the physical implications of gauge invariance, *Nuovo Cim.*, **A42**, 9 (1966).

Sigal, R.: A symmetry of Petrov type N spaces, *J. Math. Phys.*, **14**, 1434 (1973).

Ström, S.: On the contraction of representations of the Lorentz group to representations of the Euclidean group, *Ark. Fys.*, **30,** 267 (1965).

Swaminarayan, N. S.: An exact solution of Einstein's equations, *Z. Phys.*, **188,** 222 (1965).

—— An exact solution for uniformly accelerated particles in general relativity, III. Singular expending (contracting) null surface, *Commun. Math. Phys.*, **2,** 59 (1966).

—— [See also W. B. Bonnor]

Synge, J. L.: The gravitational field of a particle, *Proc. R. Irish Acad.*, (A)**53,** 83 (1950). [Chapter 7]

—— The Petrov classification of gravitational fields, *Commun. Dublin Inst. Adv. Stud.*, **15,** 51 (1964).

—— [See also P. S. Florides]

—— [See also L. O'Raifeartaigh]

—— [See also E. Pechlaner]

—— and A. Schild: *Tensor Calculus*, University of Toronto Press, 1949. [Chapter 7]

Szego, K. and K. Toth: *SL*(2, *C*) representations in explicitly energy-dependent basis, *J. Math. Phys.*, **12,** 846 and 853 (1971).

Szekeres, P.: Spaces conformal to a class of spaces in general relativity, *Proc. R. Soc. Lond.*, (A)**274,** 206 (1963).

—— [See also L. Mysak]

Takeno, H.: On plane wave solutions of field equations in general relativity, *Tensor*, **7,** 97 (1957).

—— The theory of spherically symmetric space–times, *Sci. Rep. Res. Inst. Theoret. Phys., Hiroshima Univ.*, No. 5 (1965).

—— On the theory of spherically symmetric space–times, I, II, III and IV, *Tensor*, **16,** 198 (1965); **16,** 247 (1965); **17,** 1 (1966); and **17,** 56 (1966).

—— and S. Kitamura: On the theory of spherically symmetric space–times, *Tensor*, **17,** 190 and 266 (1966).

Talbot, C. J.: Newman–Penrose approach to twisting degenerate metrics, *Commun. Math. Phys.*, **13,** 45 (1969).

Talman, J. D.: *Special Functions, A Group Theoretic Approach*, Benjamin, New York, 1968.

Tamburino, L. A.: Exact empty space metrics containing geodesic rays, Ph.D. Dissertation, University of Pittsburgh, 1962. [Chapter 11]

—— and J. H. Winicour: Gravitational fields in finite and conformal Bondi frames, *Phys. Rev.*, **150,** 1039 (1966).

—— [See also E. T. Newman]

Tangherlini, F. R.: Postulational approach to Schwarzschild's exterior solution with application to a class of interior solutions, *Nuovo Cim.*, **25,** 1081 (1962).

—— Source of the Schwarzschild field, *Nuovo Cim.*, **38,** 153 (1965).

Taub, A. H.: Empty space–time admitting a three parameter group of motions, *Ann. Math.*, **53,** 472 (1951). [Chapter 11]

Tauber, G. E.: The gravitational fields of electric and magnetic dipoles, *Canad. J. Phys.*, **35,** 477 (1957).

Thanh-Phong, L.: The permanence of the Petrov classification, *Compt. Rend. Acad. Sci., Paris*, **254,** 1926 (1962).

Thompson, A. H.: Yang's gravitational field equations, *Phys. Rev. Letters*, **34,** 507 (1975).

Thompson, J. and G. Schrank: Algebraic classification of four-dimensional Riemannian spaces, *J. Math. Phys.*, **10,** 166 (1969).

Thompson, I. H. and G. J. Whitrow: Time-dependent internal solutions for spherically symmetrical bodies in general relativity, *Mon. Not. R. Astron. Soc.*, **136,** 207 (1967), and **139,** 499 (1968).

Thorne, K. S.: [See also C. W. Misner]

—— and C. M. Will: High-precision tests of general relativity, *Comm. Astrophys. Space Phys.*, **11,** 35 (1970).

Tiomno, J.: Electromagnetic field of rotating charged bodies, *Phys. Rev.*, (D)**7,** 992 (1973).

—— [See also R. Ruffini]

Tiwari, R. and D. N. Pant: Non existence of spherically symmetric solution representing vacuum electrostatic field, *Phys. Letters*, (A)**33,** 505 (1970).

—— [See also J. R. Rao]
—— [See also A. R. Roy]

Toller, M.: An expansion of the scattering amplitude at vanishing four-momentum transfer using the representations of the Lorentz group, *Nuovo Cim.*, **53A,** 671 (1968). [Chapter 5]

—— [See also A. Sciarrino]

Tomimatsu, A. and H. Sato: New exact solution for the gravitational field of a spinning mass, *Phys. Rev. Letters*, **29,** 1344 (1972).

—— New series of exact solutions for gravitational fields of spinning masses, *Progr. Theoret. Phys.* (*Japan*), **50,** 95 (1973).

Tonnelat, M. A.: The recent experiments on the red-shift as a proof of general relativity, *Ann. Inst. Poincare*, **17,** 59 (1961).

—— *Experimental Tests in General Relativity*, Masson, Paris, 1964.

Toro, I. I.: A nonlocal extension of the formalism of Utiyama, *Compt. Rend. Acad. Sci., Paris*, **261,** 2593 (1965).

—— A generalization of the gauge transformation, *Compt. Rend. Acad. Sci., Paris*, **263,** 1 (1965).

—— The Yang–Mills field, *Compt. Rend. Acad. Sci., Paris*, **262,** 1413 (1966).

Torrence, R. and A. Janis: An approach to gravitational radiation scattering, *J. Math. Phys.*, **8,** 1355 (1967).

—— [See also W. E. Couch]
—— [See also E. T. Newman]
—— [See also T. W. J. Unti]

Trautman, A.: *Lectures on General Relativity*, Prentice-Hall, Englewood Cliffs, New Jersey, 1965. [Introduction, Chapter 7]

—— Fibre bundles associated with space–time, *Rep. Math. Phys.*, **1,** 29 (1970).

—— On the Einstein–Cartan equations, *Bull. Acad. Polon. Sci., Ser. Math. Astron. Phys.*, **20,** 503 and 895 (1972).

—— [See also I. Robinson]

Treat, R. P.: Possible link between isospace and the geometry of space–time, *Phys. Rev. Letters*, **12,** 407 (1964).

—— Local gauge field in general relativity, *J. Math. Phys.*, **11,** 2176 (1970).

—— Plane non-abelian Yang–Mills waves, *Nuovo Cim.* **A6,** 121 (1971).

—— Gauge invariant decomposition of Yang–Mills potentials, *J. Math. Phys.*, **13,** 1704 (1972).

Treder, H. J.: Einstein and symmetry groups of gravitational theory, *Ann. Phys.*, **28,** 333 (1973). [In German]

Trollope, J. R.: Null field in Einstein–Maxwell field theory, II, *J. Math. Phys.*, **8.** 938 (1967).

—— [See also M. Murenbeeld]

Trümper, M.: On a special class of type-I gravitational fields, *J. Math. Phys.*, **6,** 584 (1965).

Trim, P. W. and J. Wainwright: Nonradiative algebraically special spacetimes, *J. Math. Phys.*, **15,** 535 (1974).

—— [See also W. Kundt]

Tyutin, I. V. and E. S. Fradkin: On the theory of Yang–Mills fields, *Yadernaya Fiz.* (*USSR*), **13,** 433 (1971).

—— [See also E. S. Fradkin]

Tzou, K.-H.: A new gauge invariance, *Compt. Rend. Acad. Sci., Paris*, **253,** 2868 (1961).

—— Isotopic gauge-invariance of a vector field of zero proper mass, *Compt. Rend. Acad. Sci., Paris*, **254,** 1210 (1962).

—— Formulations of gauge invariance, *Compt. Rend. Acad. Sci., Paris*, **254,** 3819 (1962).

Unt, V.: On the connection of news function with source in axi-symmetric case, *Füüsika Mat.*, **18,** 421 (1969).

Unti, T. W. J.: Note on asymptotically flat empty spaces, *Gen. Relativ. Gravitn.*, **3,** 43 (1972).

—— [See also E. T. Newman]

—— and R. J. Torrence: Theorem on gravitational fields with geodesic rays, *J. Math. Phys.*, **7,** 535 (1966).

Urbantke, H. K.: Note on Kerr–Schild type vacuum gravitational fields, *Acta Phys. Austr.*, **35,** 396 (1972).

Utiyama, R.: Invariant theoretical interpretation of interaction, *Phys. Rev.*, **101,** 1597 (1956). [Chapter 9]

—— Quantum theory and general relativity, *Progr. Theoret. Phys.*, **33,** 524 (1965).

—— Generalized isospinspace, generalized gauge transformation and derivation of meson, *Progr. Theoret. Phys. Suppl.* Extra No. 332 (1965).

Uzes, C. A.: Internal charge of the gauge field, *Ann. Phys. (N.Y.)*, **50,** 534 (1968).

—— The long range self-interaction of the gauge fields, *Nuovo Cim.*, **62,** 157 (1969).

—— Gauge fields with positive–definite energy density, *J. Math. Phys.*, **12,** 716 (1971).

Vaidya, P. C.: The external field of a radiating star in general relativity, *Curr. Sci.*, **12,** 183 (1943).

—— Non-static solutions of Einstein's field equations for sphere of fluids radiating energy, *Phys. Rev.*, **83,** 10 (1951).

—— The general relativity field of a radiating star, *Bull. Calcutta Math. Soc.*, **47,** 77 (1955).

—— The gravitational field of radiating star, *Proc. Indian Acad. Sci.*, **33,** 264 (1957).

—— Non-static analogs of Schwarzschild's interior solution on general relativity, *Phys. Rev.*, **174,** 1615 (1968).

—— [See also W. B. Bonnor]

—— [See also A. L. Mehra]

—— and I. M. Pandya: Axially symmetric fields of pure radiation in the general theory of relativity, *Proc. Nat. Inst. Sci. India*, (A)**26,** 459 (1960).

—— and I. M. Pandya: Gravitational field of a radiating spheroid, *Progr. Theoret. Phys.*, **35,** 129 (1966).

—— and L. K. Patel: Radiating Kerr metric, *Phys. Rev.*, (D)**7,** 3590 (1973).

—— [See also I. M. Pandya]

van Dam, H.: On the mass of the graviton, *Gen. Relativ. Gravitn.*, **3,** 215 (1972).

—— and M. Veltman: Massive and massless Yang–Mills and gravitational fields, *Nucl. Phys.*, (B)**22,** 397 (1970).

van der Burg, M. G. J.: Multipole expansions for stationary fields in general relativity, *Proc. R. Soc. Lond.*, (A)**303,** 37 (1968).

—— [See also H. Bondi]

van der Waerden, B. L.: *Die Gruppentheoretische Methode in der Quantenmechanik*, Springer, Berlin, 1932. [Chapter 6]

—— *Modern Algebra*, Ungar, New York, 1953. [Introduction, Chapter 6, Appendix A]

Vashchuk, V. I.: Conformal plane solution of the Einstein–Maxwell equations, *Izv. Viss. Ucheb. Zaved., Ser. Fiz.*, **8,** 76 (1968). [In Russian]

Verde, M. and I. Ferretti: On the unitary representations of the three-dimensional rotation group, *Atti Accad. Sci. Torino*, (I)**103,** 905 (1969).

Vigier, J. P.: Mass of the Yang–Mills vector field, *Nuovo Cim.*, **23,** 1171 (1962).

—— [See also P. Hillion]

Vilenkin, N. Ya.: Bessel functions and representations of the group of Euclidean motions, *Usp. Mat. Nauk*, **9,** 69 (1956).

—— [See also I. M. Gelfand]

Vishveshwara, C. V.: Generalization of the Schwarzschild surface to arbitrary static and stationary metrics, *J. Math. Phys.*, **9,** 1319 (1968).

—— Stability of the Schwarzschild metric, *Phys. Rev.*, (D)**1,** 2870 (1970).

Voorhees, B. H.: Static axially symmetric gravitational fields, *Phys. Rev.*, (D)**2,** 2119 (1970).

Wagh, R. V.: Gravitational field of distant rotating masses, *Indian J. Phys.*, **34,** 211 (1960).

Wainwright, J.: Geometric properties of neutrino fields in curved space–times, *J. Math. Phys.*, **12,** 828 (1971).

Wald, R.: On uniqueness of the Kerr–Newman black holes, *J. Math. Phys.*, **13,** 490 (1972).

—— On perturbation of a Kerr black hole, *J. Math. Phys.*, **14,** 1453 (1973).

—— [See also J. M. Cohen]

Watson, G. N.: *A Treatise on the Theory of Bessel Functions*, Cambridge, 1966. [Chapter 7]

Weber, J.: *General Relativity and Gravitational Waves*, Interscience, New York, 1961. [Chapter 7]

—— Gravitational radiation experiments, in: *Relativity*, (M. Carmeli, S. I. Fickler, and L. Witten, Editors), Plenum Press, N.Y., 1970. [Chapter 7]

—— Advances in gravitational radiation detection, *Gen. Relativ. Gravitn.*, **3,** 59 (1972).

—— [See also J. Sinsky]

—— and D. Zipoy: Classification of gravitational radiation, *Nuovo Cim.*, **18,** 191 (1960).

Weil, A.: L'integration dans les groupes topologiques et ces applications, *Actual. Sci. Industr.*, **869,** Paris (1938). [Chapter 1]

Weinberg, S.: Photons and gravitons in *S*-matrix theory: derivation of charge conservation and equality of gravitational and inertial mass, *Phys. Rev.*, **135,** B1049 (1964).

—— Photons and gravitons in perturbation theory: derivation of Maxwell's and Einstein's equations, *Phys. Rev.*, **138,** B988 (1965). [Chapter 9]

—— Infrared photons and gravitons, *Phys. Rev.*, **140,** B516 (1965).

—— A model of leptons, *Phys. Rev. Letters*, **19,** 1264 (1967). [Introduction]

—— *Gravitation and Cosmology*, Wiley, New York, 1973. [Introduction, Chapter 7]

—— Problems in gauge field theories, *Proceedings of the 17th International Conference on High Energy Physics, London, England, 1–10 July 1974.* Science Research Council, Didcot, Berkshire, 1974.

—— Recent progress in gauge theories of the weak, electromagnetic and strong interactions, *Rev. Mod. Phys.*, **46,** 255 (1974).

Weyl, H.: *Ann. Physik*, **54,** 117 (1917). [Chapter 7]

—— *Ann. Physik*, **59,** 185 (1919). [Chapter 7]

—— *The Theory of Groups and Quantum Mechanics*, Dover, New York, 1931. [Chapter 10]

—— *The Classical Groups*, Princeton University Press, New Jersey, 1946.

—— [See also G. Beck]

Wheeler, J. A.: Geons, *Phys. Rev.*, **97,** 511 (1955). [Chapter 7]

—— [See also C. W. Misner]

—— and T. Regge: Stability of Schwarzschild singularity, *Phys. Rev.*, **108,** 1063 (1957).

Whittaker, J. M.: An interior solution in general relativity, *Proc. R. Soc. Lond.*, (A)**306,** 1 (1968).

Wigner, E. P.: On the consequences of the symmetry of the nuclear Hamiltonian on the spectroscopy of nuclei, *Phys. Rev.*, **51,** 106 (1937). [Chapter 9]

—— On unitary representations of the inhomogeneous Lorentz group, *Ann. Math.*, **40,** 149 (1939). [Introduction, Chapter 12]

—— *Group Theory and its Application to the Quantum Mechanics of Atomic Spectra*, Academic Press, New York, 1959. [Introduction, Chapter 1, Appendix A, Appendix B]

—— Geometry of light paths between two material bodies, *J. Math. Phys.*, **2,** 207 (1961).

—— Invariant quantum mechanical equations of motion, in *Theoretical Physics*, International Atomic Agency, Vienna, 1963, p. 59. [Introduction]

Wild, E.: On first order wave equations for elementary particles without subsidiary conditions, *Proc. R. Soc. Lond.*, (A)**191,** 253 (1947).

Winicour, J. H.: [See also L. Derry]

—— A. I. Janis, and E. T. Newman: Static, axially symmetric point horizons, *Phys. Rev.*, **176,** 1507 (1968).

—— [See also A. I. Janis]

Witten, L.: Invariants of general relativity and the classification of spaces, *Phys. Rev.*, **113,** 357 (1959). [Chapter 8]

—— Static cylindrically symmetric solutions of the Einstein–Maxwell field equations, *Colloque sur la Theorie de la Relativite*, CBRM, University Press, Louvain, 1960

—— [See also J. L. Safko]

Wong, S. K.: Gauge invariance and Galilean invariance, *Nuovo Cim.*, **4B,** 300 (1971).

Woolley, M. L.: The structure of groups of motion admitted by Einstein–Maxwell space–times, *Commun. Math. Phys.*, **31,** 75 (1973).

Wu, A. C. T.: Studies in the Yang–Mills field, *Phys. Rev.*, (D)**2,** 3081 (1970).

Wu, T. T. and C. N. Yang: Some solutions of the classical gauge field equation, in *Matter, Properties under Unusual Conditions*, Interscience, London, 1969.

Wyman, M.: Static isotropic solutions of Einstein's field equations, *Phys. Rev.*, **66,** 267 (1944).

—— Isotropic solutions of Einstein's field equations, *Canad. Math. Congress*, pp. 90 (1946).

—— Schwarzschild interior solution in an isotropic coordinate system, *Phys. Rev.*, **70,** 74 (1946).

—— and R. Trollope: Null fields in Einstein–Maxwell field theory, *J. Math. Phys.*, **6,** 1995 (1965).

Yadav, B. P.: Distorting metric with vanishing twist and divergence, *Phys. Letters*, **46A,** 253 (1973).

Yang, C. N.: Integral formalism for gauge fields, *Phys. Rev. Letters*, **33,** 445 (1974). [Chapter 9]

—— and R. L. Mills: Conservation of isotopic spin and isotopic gauge invariance, *Phys. Rev.*, **96,** 191 (1954). [Chapter 9]

Yefimov, R. J.: *A Short Course in Analytical Geometry*, Moscow, 1954.

Yilmaz, H.: New theory of gravitation, *Phys. Rev. Letters*, **27,** 1399 (1971), and *Nuovo Cim.*, **B10,** 79 (1972).

Zakharov, V. D.: A physical characteristic of Einsteinian spaces of degenerate type II in the classification of Petrov, *Dokl. Akad. Nauk SSSR*, **161,** 563 (1965). English translation in: *Soviet Phys. Dokl.*, **10,** 242 (1965).

Zaykov, R.: Static and spheric solutions of Einstein equations, *Compt. Rend. Acad. Bulg. Sci.*, **18,** 725 (1965).

Zeeman, E. C.: Causality implies the Lorentz group, *J. Math. Phys.*, **5,** 490 (1964).

Zerilli, F. J.: Effective potential for even-parity Regge–Wheeler gravitational perturbation equations, *Phys. Rev. Letters*, **24,** 737 (1970).

Zipoy, D. M.: Topology of some spheroical metrics, *J. Math. Phys.*, **7,** 1137 (1966).

—— [See also J. Weber]

Zumino, B.: Theories with gauge groups, *Acta Phys. Austr., Suppl. II*, **212** (1965).

Zund, J. D.: Degenerate gravitational fields with twisting rays, II, *Lett. al Nuovo Cim.*, **4,** 879 (1972).

—— and W. Maher: A spinor approach to some problems in Lorentzian geometry, *Tensor*, **21,** 70 (1969).

—— and W. Maher: A spinor approach to the generalized singular electromagnetic field theory of Bel, Lapiedra and Montserrat, *J. Math. Phys.*, **14,** 168 (1973).

—— [See also I. Robinson]

INDEX

Since the bibliography immediately preceeding this index contains references to chapters in which each paper is mentioned, some names referred to in the text are omitted here. In particular, names of authors referred to only in the footnotes of the text are usually not given.